Modern Formulas for Statics and Dynamics

Modern Formulas for Statics and Dynamics

A Stress-and-Strain Approach

WALTER D. PILKEY
Professor and Chairman
Division of Applied Mechanics
University of Virginia

PIN YU CHANG
Principal Scientist
Hydronautics, Inc.

McGRAW-HILL BOOK COMPANY

New York St. Louis San Francisco Auckland Bogotá
Düsseldorf Johannesburg London Madrid
Mexico Montreal New Delhi Panama
Paris São Paulo Singapore
Sydney Tokyo Toronto

To our wives, Barbara Pilkey and Francis Chang

Library of Congress Cataloging in Publication Data

Pilkey, Walter D
Modern formulas for statics and dynamics.

Includes index.
1. Structures, Theory of—Tables, calculations, etc.
2. Structures, Theory of—Computer programs. I. Chang, Pin Yu, 1928– joint author. II. Title.
TA646.P54 624'.171 77-15093
ISBN 0-07-049998-5

34567890 KPKP 7654321098

The editors for this book were Tyler G. Hicks and Joseph Williams, the designer was Naomi Auerbach, and the production supervisor was Frank Bellantoni. It was set in Times Roman by Computype, Inc.

Printed and bound by The Kingsport Press

Contents

Preface

This book is intended for engineers, managers, and educators concerned with the analysis or design of shafts, beams, plates, and shells. This book begins where the simple stress and strain formula books leave off. It attempts to fill the gap between handbook formulas and general-purpose computer programs. Here the information essential for a rapid static, stability, or dynamic analysis of a beam, plate, or shell of variable or composite cross section, with arbitrary mechanical or thermal loading and with any type of support or foundation, can be found. In the case of the dynamic analysis, not only can the response to steady-state and transient loading be found, but the formulas provide natural frequencies and mode shapes as well. The material can be elastic or viscoelastic. Damping mechanisms include proportional and nonproportional viscous damping.

The book is designed to be useful to engineers who decide that they cannot afford the time or money necessary to use a general-purpose structural analysis computer program, and that it is unrealistic to treat their problems in as simple a fashion as required by handbooks. Computer programs based on the information in this book are available on networks in many parts of the world. The programs, which are referenced occasionally in the book, are also available at nominal cost from

> The Structural Members Users Group
> P. O. Box 3958
> The University of Virginia Station
> Charlottesville, Virginia 22903

All of the programs can be used in batch or time-sharing form with fixed, free, or prompted input formats.

Many example problems have been integrated into the book to illustrate the use of the formulas and solution procedures. Both U.S. customary and SI units are used. Most chapters contain several benchmark example problems against which a reader's own computer program can be checked. Advanced strength of materials and introduction to elastic structures courses can and have been based on this book. A booklet providing the underlying theory and the derivations of the formulas is available from The Structural Members Users Group.

In brief, the contents are:

Beams The formulas are for the stresses, deflection, slope, bending moment, and shear force of beams for static, steady-state, and transient conditions. The critical axial load and mode shape are found for stability. The natural frequencies and mode shapes are computed for free transverse vibrations. The beam can be formed of variable cross-section segments with any loading, in-span supports, foundations, and boundary conditions. The reader can include any or all of bending, shear deformation, and rotary inertia effects.

Torsional systems For static, steady-state, and transient torsional loads, the torsional stresses, the angle of twist, and the twisting moment of a shaft can be calculated. The natural frequencies and mode shapes are found for torsional vibration. The torsion system can be a bar formed of variable-cross-section segments with any loading, gears, branches, foundations, and boundary conditions.

Extension systems For static, steady-state, and transient axial loads, the formulas find the axial displacement and force. The natural frequencies and mode shapes are found for longitudinal vibrations. The system can be a sequence of springs, dashpots, and masses, or a bar of variable cross-section segments with arbitrary loading, foundations, and boundary conditions.

Torsion of thin-walled beams The formulas are for stresses, angles of twist, bimoments, warping torques, and internal twisting moments, as well as unstable loads and natural frequencies.

Rotating shafts For a shaft with unbalance forces, the displacements and forces can be calculated. The bearings can be modeled with a spring, dashpot, and mass system. The critical speeds, whirl instability, and transient responses are also found. For all analyses the shaft system can be formed of lumped or continuous segments with supports, foundations, and any boundary conditions.

Grillages Most of the formulas deal with uniform grillages subjected to uniform, hydrostatic, and concentrated forces. Stability and free vibration analyses can also be handled.

Disks The radial displacement, radial force, and tangential force can be found for static and steady-state conditions. The disk can be rotating. The natural frequencies and mode shapes are calculated for radial vibrations. The theory is based on a plane stress assumption. The loading and responses are axially symmetric.

Thick cylinders Formulas for the radial displacement, radial stress, tangential stress, and axial stress are given for static, steady-state, and transient conditions. The natural frequencies and mode shapes are given for radial vibrations. The theory is based on a plane strain assumption. The loadings and responses are axially symmetric.

Thick spherical shells The radial displacement, radial stress, and tangential stress are provided for static, steady-state, and transient loading. The natural frequencies and mode shapes are given for radial vibrations. The loading and responses are spherically symmetric.

Circular plates The formulas are for stresses, deflection, slope, radial moment, shear, transverse moment, and twisting moments for static, steady-state, and transient conditions. The critical load and mode shape for stability, and the natural frequencies and mode shapes (symmetric and unsymmetric) for transverse vibration are included. The plate can be formed of variable thickness rings with any loading, in-span supports, foundations, and boundary conditions. The reader can include unsymmetric loading for applied forces and moments.

Rectangular plates Two opposing edges of the plate must be simply supported. Other boundary conditions, loading, etc., are arbitrary. Static, steady-state, stability, free vibration, and transient formulas are included. The plate may be isotropic or orthotropic.

Thin-walled cylinders The response variables are given for axisymmetric motion of axisymmetric thin-walled circular cylinders. The displacement, slope, bending moment, and shear force are radially oriented. Static, stability, and dynamic analyses can be performed.

Cross-sectional properties and combined stresses in bars Cross-sectional properties of a bar of any cross-sectional shape are included. Properties include area, centroid, moments of inertia about any axes, radii of gyration, shear center, shear deformation coefficients, torsional constant, and warping constant. Modulus weighted properties can be calculated for composite sections. The stresses include normal stresses due to bending and warping, shear stress due to torsion, warping, and transverse loads.

A background in elementary strength of materials or the introduction to the mechanics of solids is necessary to use this book. This background could be obtained from a book such as *Mechanics of Solids*, W. D. Pilkey and O. H. Pilkey, Quantum Publishers, 1974, which is also useful as a reference, providing an introduction to the solution techniques of the present book.

We have been actively preparing this book for many years. Some of this material appeared in 1969 in the Office of Naval Research reports *Manual for the Response of Structural Members*, AD 693 141, AD 693 142. In addition to acknowledging the long-time support from this agency, under the technical direction of John Crowley, Kenneth Saczalski, and Nicholas Perrone, the authors acknowledge the support of the Army Research Office–Durham for the work which permitted the development of some of the formulas given in the rotor dynamics chapter. Considerable assistance in our early efforts was provided by Dr. Richard Nielsen and John Tylke.

We are particularly indebted to Drs. Y. H. Chen, Garnett Horner, John Strenkowski, and Chirasak Thasanatorn who actually helped write preliminary versions of portions of this material. Some of the formulas and tables were derived by Drs. David Hsu, Abdur Rahim, and Bo Ping Wang. Thorough verification, both analytical and computational, of formulas, text, and example problems was performed by Drs. Fei Hon Chu and Y. H. Chen.

The figures and tables were ably prepared in photo-ready form by Peggy Hamm and Jim Hamm. The whole manuscript was patiently typed by Jere Hawkins. Portions were later supplemented by Sue DeMasters, Cathi Miller, and Shoshana Turk.

The authors wish to express especial gratitude to Barbara Pilkey, who has for seemingly unending years patiently and skillfully edited this manuscript.

WALTER D. PILKEY
PIN YU CHANG

ONE

Introduction

Designers and engineers are often required to solve stress analysis problems within severe constraints. Depending on the time available, budget, and degree of accuracy required, they can choose to solve their problems by using simple formulas from handbooks or textbooks, or by using one of the many available general-purpose computer programs. It is not unusual for engineers to conclude that they cannot afford to use a general-purpose program, usually a finite element program, and that it is unrealistic to treat the structure in as simple a fashion as required by the handbooks.

The purpose of this book is to meet the need of engineers for simple but accurate extensions of the handbook formulas to encompass more complex situations. This book contains tables for use in finding stresses and deformations of the common structural members and mechanical elements. All computations required by the methods of this book can be completed within two minutes by a computer and within two hours by hand calculation. These tables permit the engineer to model problems more realistically and thus to obtain greater accuracy. In addition to the tables for use with complex models, the usual simple stress-analysis formulas are provided for completeness. Moreover, general computer programs based on the information in this book are available on several national computer networks or can be obtained from the Structural Members Users Group. All of the programs can be used in batch or time-sharing form with fixed, free, or prompted input formats.

Although this book is based on familiar strength-of-materials theories for linearly elastic materials, the transfer matrix computation procedure permits realistic models to be handled with ease. The fundamentals of the procedure are provided in each chapter. It is not essential for users to fully understand either the theory or the computational procedure to solve their problems. Many step-by-step example problems are provided.

SIMPLE MEMBERS

The first section of most of the chapters contains formulas for a uniform member with arbitrary static loading and any end conditions. These formulas provide the stresses and deformation at any point along the member. Examples are included to illustrate the application of the formulas. Formulas for critical loads and natural frequencies are also given in this section.

COMPLEX MEMBERS

Members with variable cross sections or in-span supports are treated in the second section of most chapters. Transfer matrices are listed for virtually all cases that can physically occur. These can be used to calculate the static or steady-state stresses or deformations, the critical loads, the natural frequencies and mode shapes, and the transient dynamic response.

COMPUTER PROGRAM BENCHMARK PROBLEMS

The final section of each chapter contains example problems solved by the computer programs that are based on this book. These provide benchmark solutions against which readers can check their own computer programs.

NOTATION AND CONVENTIONS

The notation for each chapter is defined within the chapter. The sign conventions are also carefully spelled out. In general, both notation and sign conventions conform to standard practice.

Mathematical Symbols

Certain symbols are common to most chapters. The following singularity functions are frequently used:

$$\langle x-a\rangle^n = \begin{cases} 0 & \text{if } x < a \\ (x-a)^n & \text{if } x \geqslant a \end{cases} \qquad \text{for } n \geqslant 0 \tag{1.1}$$

In particular,

$$\langle x-a\rangle^0 = \begin{cases} 0 & \text{if } x < a \\ 1 & \text{if } x \geqslant a \end{cases} \tag{1.2}$$

is the unit *step function*. The brackets $\langle\ \rangle$ are the same as the customary brackets ( ), except that they are set equal to zero for any value of coordinate x less than a. This notation leads to substantial savings in effort and space by permitting equations to be combined.

There are some other definitions and functions of singularity functions that are used very sparingly in this work. For example, the integral of $\langle x - a\rangle^n$ is defined as

$$\int_0^x \langle x - a\rangle^n\, dx = \frac{\langle x - a\rangle^{n+1}}{n+1} \qquad \text{for} \quad n \geqslant 0 \tag{1.3}$$

This notation can be extended to encompass point occurrences. Since most elementary strength-of-materials textbooks can be consulted for details on singularity functions, only a few essential properties will be presented here. A concentrated occurrence, for example, force or mass, can be expressed as a distributed occurrence with the assistance of a *delta function* $\langle x - a\rangle^{-1}$, which is defined to be zero everywhere except at $x = a$, where it is undefined or infinite. In addition, it is essential to define the integral

$$\int_0^x \langle x - a\rangle^{-1}\, dx = \langle x - a\rangle^0 \tag{1.4}$$

With these definitions, a concentrated force W and concentrated mass M at $x = a$ can be written as $w = W\langle x - a\rangle^{-1}$ and $\rho = M\langle x - a\rangle^{-1}$, respectively, where w and ρ are the distributed loading and mass per unit length. The delta function also possesses the important sifting property of

$$\int_0^L f(x)\langle x - a\rangle^{-1}\, dx = f(a) \tag{1.5}$$

where f is some function of x.

Units

There are no inherent units assigned to any variables in the tables or computer routines. It is important, however, that users employ a consistent set of units while making calculations or preparing input for the computer programs. Many of the most frequent errors result from the use of inconsistent units.

EFFECTIVE USE OF THIS HANDBOOK

The effective use of the information in this handbook depends on an understanding of the class of problems that can be solved and on the selection of the correct formulas or calculation procedure.

Model the Problem as a Structural Member

The information in this book applies to structural members or mechanical elements. These members or elements are characterized by a major axis or direction along which geometric or material properties can vary.

For problems involving beams, bars, rotating shafts, and thin-walled beams, the axis along the length is the major direction.

For circular plates (Fig. 1-1), circular membranes, thick cylinders, disks, and thick spheres, the radial axis should be chosen as the major direction. The geometric and material properties may vary in the radial direction but remain constant in the circumferential direction. In most cases the applied loadings can vary in both directions.

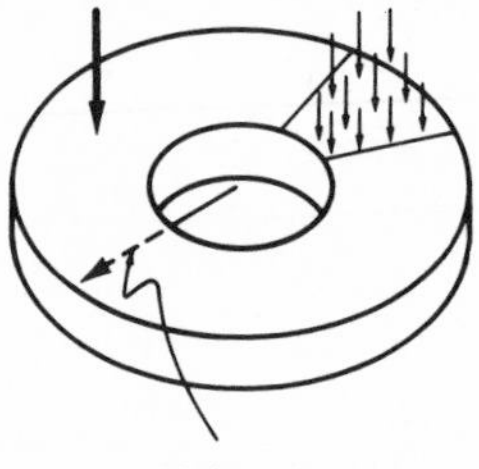

Fig. 1-1 Axially symmetric circular plate.

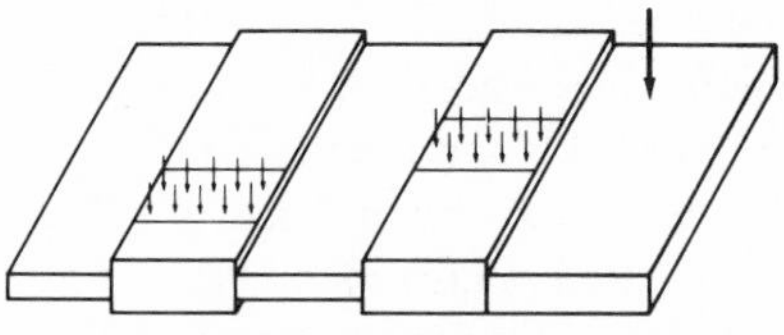

Fig. 1-2 Rectangular plate.

Rectangular plates (Fig. 1-2) and grillages must have uniform material and geometry along a direction parallel to one of the edges, although these properties may vary in the direction parallel to an adjacent edge.

Choose a Method of Calculation

Members with constant geometric and material properties along a major direction can usually be treated with the formulas of the first section of each chapter. Members with variable properties along the major direction should be handled with the transfer matrix tables of the second section of the chapters.

Complex problems with many variations in loads, material, or cross section should be solved with the aid of computer programs.

MODELING OF COMPLEX STRUCTURES

A structure with more than one major direction can frequently be modeled as a member with a single major direction. This is often the case if (1) portions of the

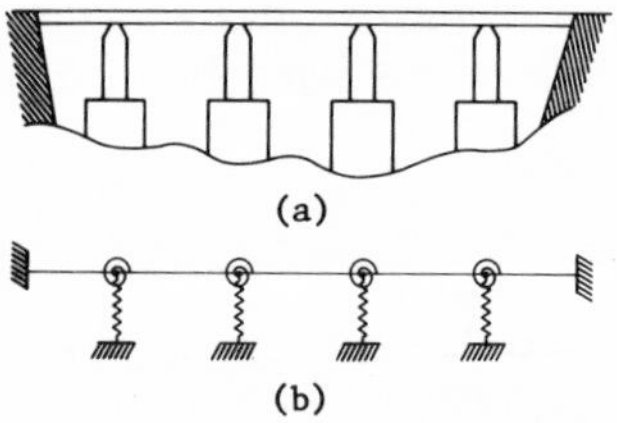

Fig. 1-3 Bridge modeled as beam or plate on spring supports.

structure can be treated as flexible supports or (2) the structure exhibits certain kinds of symmetry.

EXAMPLE 1.1 Multispan Structure A multispan structure such as the bridge of Fig. 1-3*a* can often be represented as a beam, plate, or grillage on supports, for example, Fig. 1-3*b*.

EXAMPLE 1.2 Offshore Drilling Platform The offshore drilling platform of Fig. 1-4 can be modeled as a beam, as shown in Fig. 1-5.

The platform is a four-legged symmetrical structure intended for oil drilling in water depths of several hundred feet. On the top is a large, heavy working area. At the middle and bottom of the legs are truss assemblies that hold the legs together. The structure, especially the legs, must be able to withstand storm and wave loads of the ocean, including fatigue effects.

The structure has four major directions along the four legs. If waves are perpendicular to one side of the structure the deformation is symmetrical. Then only one-half of the structure (Fig. 1-5*b*), including two legs and the trusses connecting the legs, need be considered. Since the difference between wave loads on the windward and leeward sides is small, and the effect of the load difference on the bending moment and displacements is still smaller, we assume the deformations of these two legs are identical. This permits us to treat just one leg with truss supports and the load from the working area (Figs. 1-5*c*, *d*, *e*).

The resulting model for this complex drilling platform is a complex beam column with variable cross section and various elastic supports representing the truss connections. The rotary spring constants of Fig. 1-5*e* can be calculated

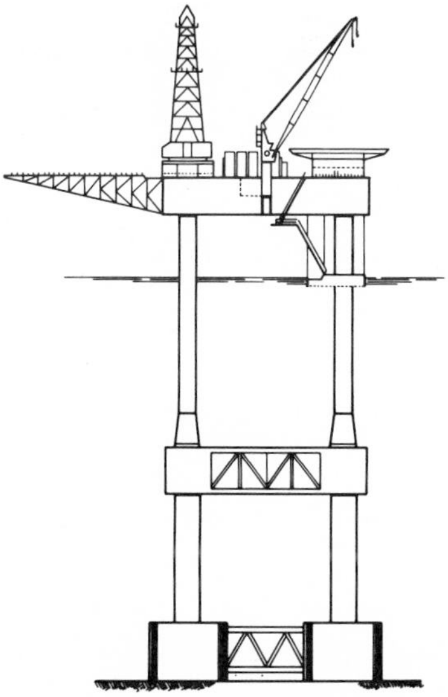

Fig. 1-4 Drilling platform.

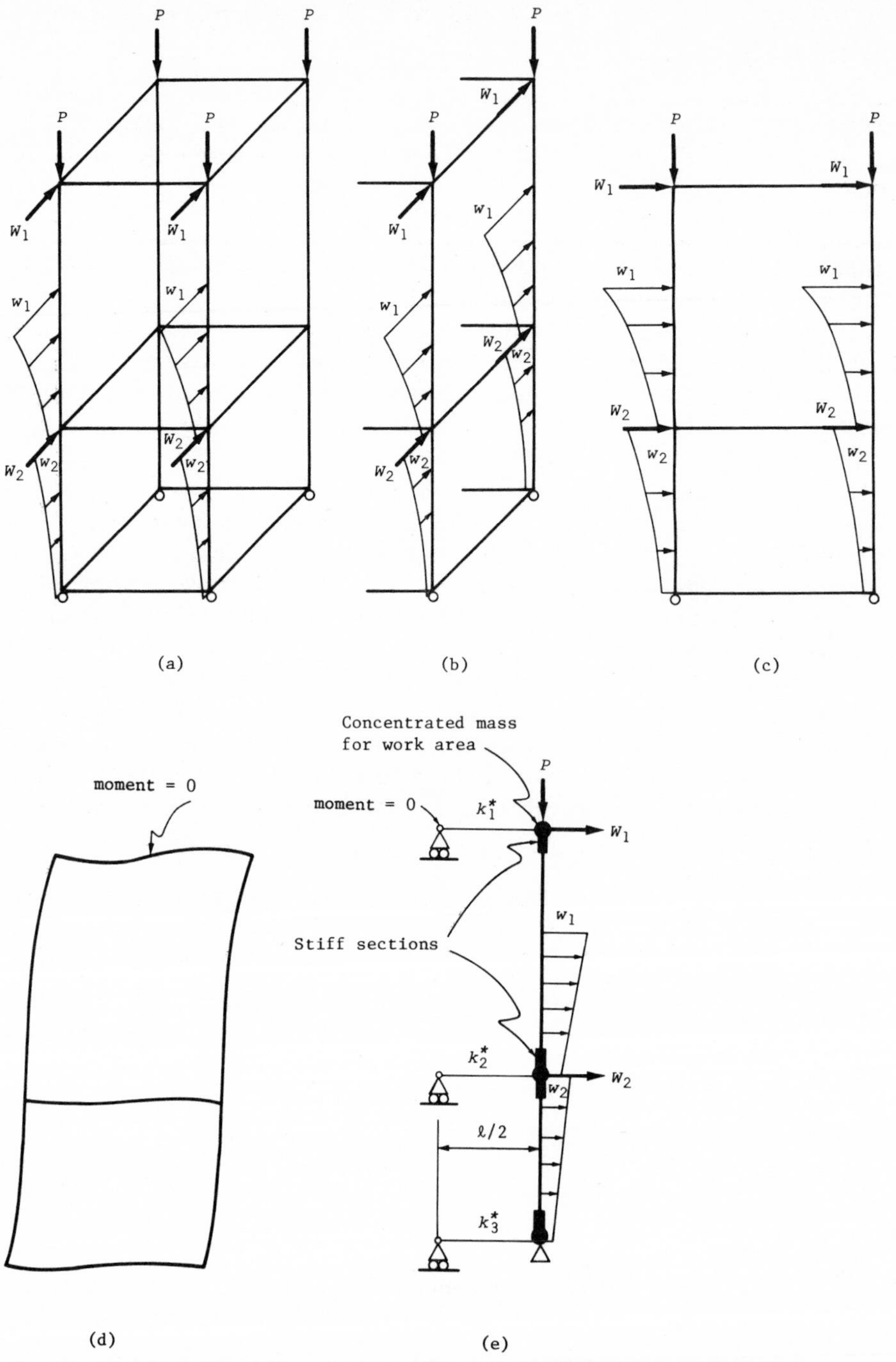

Fig. 1-5 Modeling of the drilling platform of Fig. 1-4. (*a*) Whole structure. (*b*) Half structure. (*c*) Front view of half structure. (*d*) Deformation pattern. (*e*) Final beam model.

using the simple formula $k^* = 6EI/\ell^2$, where ℓ is the distance between two legs and I is the moment of inertia of the truss.

The computations based on this simple model will provide results that are quite close to those found by a two-dimensional frame analysis. For example, the first natural frequency and mode shape for the beam model will correspond closely to the deformation of Fig. 1-5*d*.

TWO

Beams

This chapter treats the static, stability, and dynamic analyses of beams. The loading, which can be mechanical or thermal, is transverse to the beam. Deflection, stability, natural frequency, and stress formulas for simple, single-span beams are handled in Part A. The formulas for more complex beams are provided in Part B. Included are beams of variable cross section and composite materials, incorporating shear deformation and rotary inertia effects. Part C contains some examples of problems solved by a computer program.

The formulas of this chapter apply to beams represented by the classical (Euler-Bernoulli) theory of beam bending. This theory is based on the assumptions that plane cross sections remain plane, stress is proportional to strain, bending takes place in a principal plane, and slopes due to bending are small compared with unity.

Applications and Modeling

In the preliminary design stage, engineers are often required to determine the basic dimensions and major framework of a structure without embarking on detailed calculations. Managers of technical projects and officials of regulatory agencies must be able to check voluminous structural analyses without redoing the complete analyses. Accomplishing this requires much mature engineering experience and judgment. The real structure must be idealized as a much simpler model and analyzed by simple methods. The beam model is probably the most frequently employed idealization in structural mechanics.

Typical complicated structural problems that can be modeled and analyzed as complex beams are

- Deflections, bending stresses, and natural frequencies of ships
- Multiple-stage missiles
- Bridges with rigid or elastic support systems
- Response and critical speeds of rotating shafts
- Piles placed in soil with moduli varying with depth
- Columns with variable cross sections and variable axial loads.

In the case of a ship, it has been shown, for example, that the model of Fig. 2-1*b* leads to an accurate computation of deflections, bending stresses, and natural frequencies (for the first few modes) for the ship of Fig. 2-1*a*.

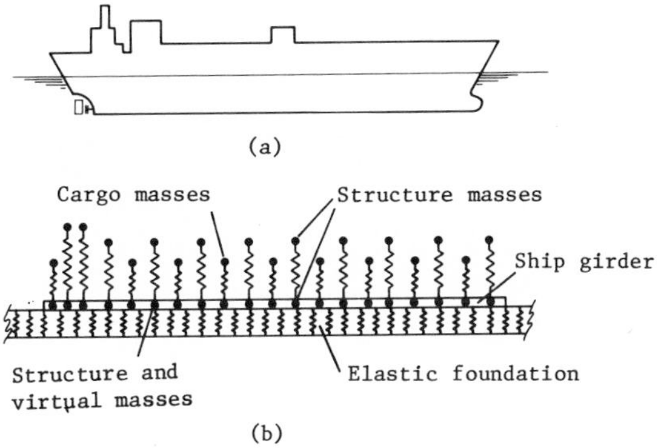

Fig. 2-1 (*a*) Ship hull. (*b*) Ship modeled as beam on elastic foundation.

A. STRESS FORMULAS AND SIMPLE BEAMS

2.1 Notation and Conventions

v	Lateral deflection (length)
θ	Angle or slope of the deflection curve (radians)
M	Bending moment at any section (force · length)
V	Shear force at any section (force)
E	Modulus of elasticity of the material (force/length^2)
I	Moment of inertia taken about the neutral axis (centroidal axis) (length^4)
L	Length of beam
w	Transverse loading intensity (force/length)
P	Axial force
W	Concentrated force
C	Concentrated applied moment (force · length)
w_1	Magnitude of distributed force, uniform in x direction (force/length)
$\dfrac{\Delta w}{\Delta \ell}$	Gradient of distributed force, linearly varying in x direction (force/length^2)
σ	Normal bending stress (force/length^2)

τ	Transverse shear stress (force/length^2)
A	Area (length^2)
I, I_y, I_z, I_x, I_{xy}	Moments of inertia (length^4)
x, y, z	Right-handed coordinate system
$\bar{y}, \bar{z}$	Centroidal coordinates
r_y, r_z, r_p	Radii of gyration
ω	Natural frequency (radians/second)
ρ	Mass per unit length (mass/length), (force-time^2/length^2)

Positive deflection v and slope θ are shown in Fig. 2-2. Positive internal bending moments M and shear forces V are illustrated in Fig. 2-3.

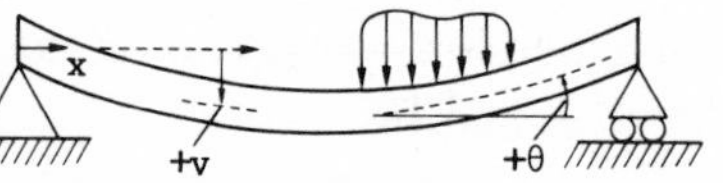

Fig. 2-2 Positive displacement v and slope θ.

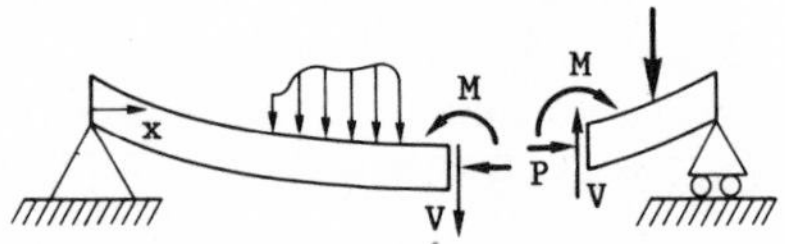

Fig. 2-3 Positive bending moment M and shear force V.

2.2 Stresses on a Cross Section

The tables of this chapter provide the deflection, slope, bending moment, and shear force at any point along a beam. Once the bending moment and shear force are known, the stresses can be derived from the stress formulas given in this section.

Normal Stress The flexural or normal stress σ due to bending is

$$\sigma = \frac{Mz}{I} \tag{2.1}$$

where z is the vertical coordinate measured from the neutral axis. Positive z is shown in Fig. 2-4. The moment of inertia I is given by

$$I = \int_A z^2 \, dA \tag{2.2}$$

where A is the cross-sectional area.

Commonly occurring moments of inertia are listed in Table 2-1.

If a compressive axial force P is considered, then

$$\sigma = -\frac{P}{A} + \frac{Mz}{I} \tag{2.3}$$

Replace P by $-P$ if the axial force is tensile.

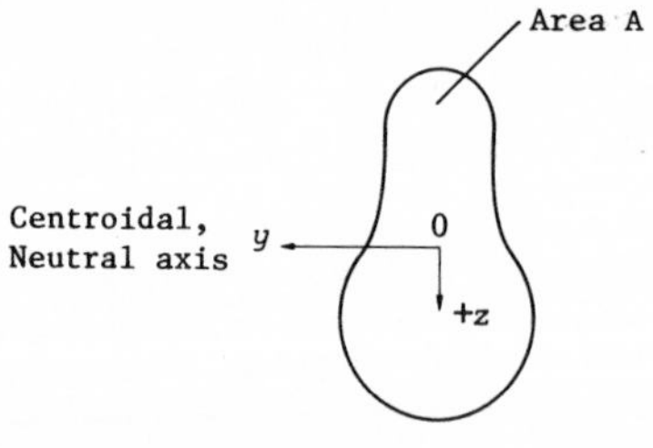

Fig. 2-4 Cross section.

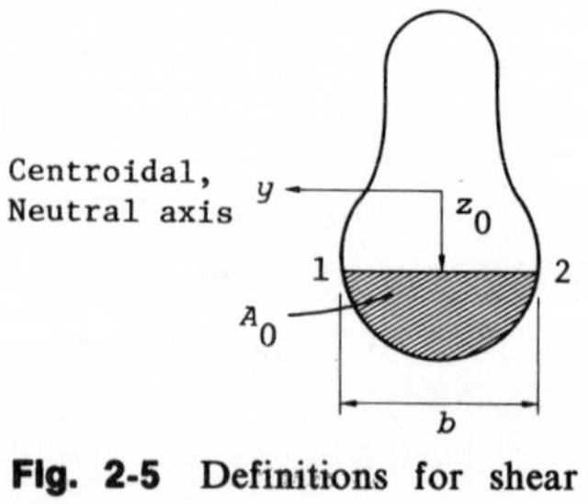

Fig. 2-5 Definitions for shear stress.

The other entries in Table 2-1 are the area A of the cross section, the coordinates of the centroid, the moments of inertia

$$I = I_y = \int_A z^2 \, dA \qquad I_z = \int_A y^2 \, dA \qquad I_{yz} = \int_A yz \, dA \qquad I_x = I_y + I_z \quad (2.4a)$$

with respect to the centroidal axes, and the corresponding radii of gyration

$$r_y = \sqrt{I_y / A} \qquad r_z = \sqrt{I_z / A} \qquad r_p = r_x = \sqrt{I_x / A} \qquad (2.4b)$$

Shear Stress The average shear stress τ along a width of a cross section, for example, anywhere on line 1-2 of Fig. 2-5, is

$$\tau = \frac{VQ}{Ib} \qquad (2.5)$$

where b is the width along which the stress is being computed and

$$Q = \int_{A_0} z \, dA \qquad (2.6)$$

The integration is taken over the area A_0 that lies between the position at which the stress is desired (z_0) and the outer fiber of the cross section. Thus Q represents the first moment of the area between z_0 and the outer fiber.

2.3 Ordinary Beam

The fundamental equations of motion for the bending of a uniform ordinary (Euler-Bernoulli or engineering theory) beam are

$$\begin{aligned} EI \frac{d^4 v}{dx^4} &= w \\ EI \frac{d^3 v}{dx^3} &= -V \\ EI \frac{d^2 v}{dx^2} &= -M \\ \frac{dv}{dx} &= -\theta \end{aligned} \qquad (2.7)$$

These relations are integrated to provide the deflection, slope, bending moment, and shear force.

Deflection:	$v = v_0 - \theta_0 x - M_0 \dfrac{x^2}{2EI} - V_0 \dfrac{x^3}{3!EI} + F_v$	(2.8*a*)
Slope:	$\theta = \theta_0 + M_0 \dfrac{x}{EI} + V_0 \dfrac{x^2}{2EI} + F_\theta$	(2.8*b*)
Bending moment:	$M = M_0 + V_0 x + F_M$	(2.8*c*)
Shear force:	$V = V_0 + F_V$	(2.8*d*)

The F_v, F_θ, F_M, F_V are loading functions given in Table 2-2. If there is more

TABLE 2-1 Geometric Properties of Common Cross Sections

Shape of Section	Area, Location of Centroid $(\bar{y},\bar{z})$	Moments of Inertia ($I_x = I_{\bar{y}} + I_{\bar{z}}$, the polar moment of inertia, is with respect to centroidal axial axis)	Radii of Gyration (r_p is the polar radius of gyration, i.e. $r_p = r_x$)	Shear Form Factor (Refs. 2.3, 2.4) k_s
1. Rectangle 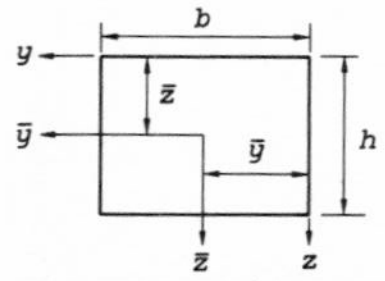	$A = bh$ $\bar{y} = b/2$ $\bar{z} = h/2$	$I = I_{\bar{y}} = bh^3/12$ $I_{\bar{z}} = hb^3/12$ $I_{\overline{yz}} = 0$ $I_x = bh(b^2 + h^2)/12$	$r_{\bar{y}} = h/\sqrt{12}$ $r_{\bar{z}} = b/\sqrt{12}$ $r_p = \sqrt{(b^2 + h^2)/12}$	$\dfrac{10(1+\nu)}{12 + 11\nu}$
2. Triangle 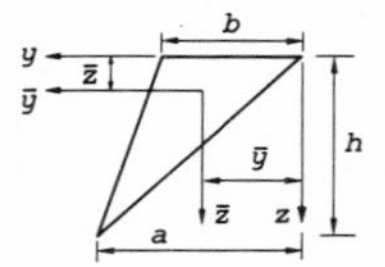	$A = bh/2$ $\bar{y} = \frac{1}{3}(a + b)$ $\bar{z} = h/3$	$I = I_{\bar{y}} = bh^3/36$ $I_{\bar{z}} = bh(b^2 - ab + a^2)/36$ $I_{\overline{yz}} = bh^2(2a - b)/72$ $I_x = bh(h^2 + b^2 - ab + a^2)/36$	$r_{\bar{y}} = h/\sqrt{18}$ $r_{\bar{z}} = \left(\dfrac{b^2 - ab + a^2}{18}\right)^{1/2}$ $r_p = \sqrt{(b^2 + h^2 - ab + a^2)/18}$	Use computer program
3. Trapezoid 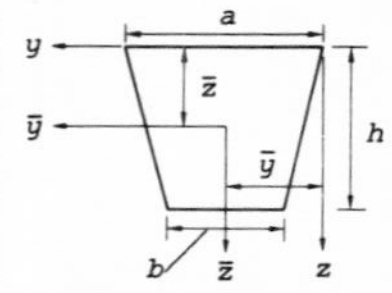	$A = h(a + b)/2$ $\bar{y} = a/2$ $\bar{z} = \dfrac{h}{3}\left(\dfrac{a + 2b}{a + b}\right)$	$I = I_{\bar{y}} = \dfrac{h^3}{36}\left(\dfrac{a^2 + 4ab + b^2}{a + b}\right)$ $I_{\bar{z}} = h(a + b)(a^2 + b^2)/48$ $I_{\overline{yz}} = 0$ $I_x = I_{\bar{y}} + I_{\bar{z}}$	$r_{\bar{y}} = \dfrac{h(a^2 + 4ab + b^2)^{1/2}}{\sqrt{18}(a + b)}$ $r_{\bar{z}} = \left(\dfrac{a^2 + b^2}{24}\right)^{1/2}$ $r_p = \sqrt{I_x/A}$	Use computer program

4. Circle 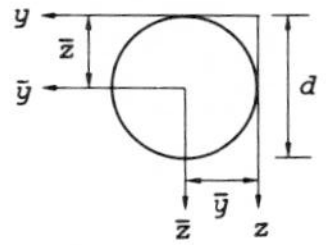	$A = \pi d^2/4$ $\bar{y} = d/2$ $\bar{z} = d/2$	$I = I_{\bar{y}} = I_{\bar{z}} = \pi d^4/64$ $I_{\overline{yz}} = 0$ $I_x = \pi d^4/32$	$r_{\bar{y}} = r_{\bar{z}} = d/4$ $r_p = d/\sqrt{8}$	$\dfrac{6(1+\nu)}{7+6\nu}$
5. Annulus 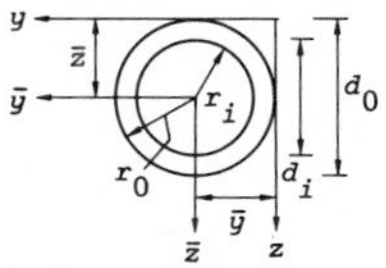	$A = \pi(d_0^2 - d_i^2)/4$ $\bar{y} = d_0/2$ $\bar{z} = d_0/2$	$I = I_{\bar{y}} = I_{\bar{z}} = \pi(r_0^4 - r_i^4)/4$ $I_{\overline{yz}} = 0$ $I_x = \pi(d_0^4 - d_i^4)/32$	$r_{\bar{y}} = r_{\bar{z}} = (d_0^2 + d_i^2)^{1/2}/4$ $r_p = \sqrt{(d_0^2 + d_i^2)/8}$	$m = d_i/d_0, \quad C = (20 + 12\nu)m^2$ $\dfrac{6(1+\nu)(1+m^2)^2}{(7+6\nu)(1+m^2)^2 + C}$
6. Ellipse 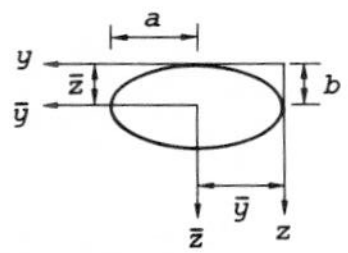	$A = \pi ab$ $\bar{y} = a$ $\bar{z} = b$	$I = I_{\bar{y}} = \pi ab^3/4$ $I_{\bar{z}} = \pi ba^3/4$ $I_{\overline{yz}} = 0$ $I_x = \pi ab(b^2 + a^2)/4$	$r_{\bar{y}} = b/2$ $r_{\bar{z}} = a/2$ $r_p = \sqrt{(a^2 + b^2)/4}$	$C = a^2b^2(16 + 10\nu) + \nu b^4$ $\dfrac{12a^2(1+\nu)(3a^2 + b^2)}{a^4(40 + 37\nu) + C}$
7. Semicircle 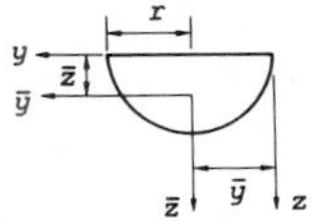	$A = \pi r^2/2$ $\bar{y} = r$ $\bar{z} = 4r/3\pi$	$I = I_{\bar{y}} = 0.11r^4$ $I_{\bar{z}} = 0.393r^4$ $I_{\overline{yz}} = 0$ $I_x = 0.503r^4$	$r_{\bar{y}} = 0.264r$ $r_{\bar{z}} = r/2$ $r_p = 0.581r$	$\dfrac{1+\nu}{1.305 + 1.273\nu}$

TABLE 2-1 Geometric Properties of Common Cross Sections (*Continued*)

8. Angle 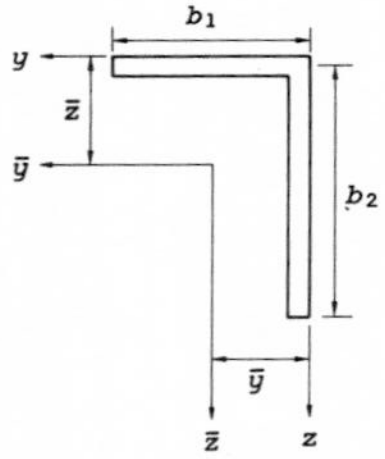	$B_1 = b_1 + t/2,\ B_2 = b_2 + t/2$ $C_1 = b_1 - t/2,\quad C_2 = b_2 - t/2$ $A = t(b_1 + b_2)$ $\bar{y} = \dfrac{B_1^2 + C_2 t}{2(b_1 + b_2)}$ $\bar{z} = \dfrac{B_2^2 + C_1 t}{2(b_1 + b_2)}$	$I_{\bar{y}} = \frac{1}{3}[t(B_2 - \bar{z})^3 + B_1\bar{z}^3 - C_1(\bar{z} - t)^3]$ $I_{\bar{z}} = \frac{1}{3}[t(B_1 - \bar{y})^3 + B_2\bar{y}^3 - C_2(\bar{y} - t)^3]$ $I_{\overline{yz}} = -\frac{t}{2}[b_1\bar{z}(b_1 - 2\bar{y}) + b_2\bar{y}(b_2 - 2\bar{z})]$	$r_{\bar{y}} = \sqrt{I_{\bar{y}}/A}$ $r_{\bar{z}} = \sqrt{I_{\bar{z}}/A}$ $r_p = \sqrt{I_x/A}$	Use computer program
9. **I** Section 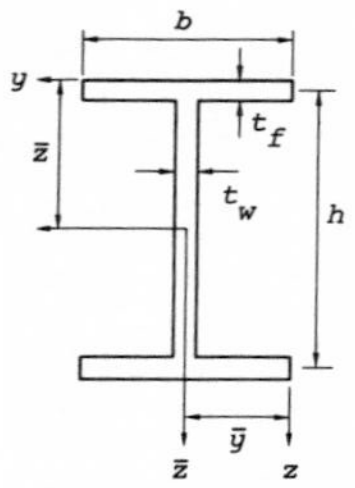	$H_1 = h + t_f,\quad H_2 = h - t_f$ $A = 2bt_f + H_2 t_w$ $\bar{y} = b/2$ $\bar{z} = \frac{1}{2}H_1$	$I_{\bar{y}} = \dfrac{bH_1^3 - (b - t_w)H_2^3}{12}$ $I_{\bar{z}} = \dfrac{H_2 t_w^3 + 2t_f b^3}{12}$ $I_{\overline{yz}} = 0,\ I_x = I_{\bar{y}} + I_{\bar{z}}$	$r_{\bar{y}} = \sqrt{I_{\bar{y}}/A}$ $r_{\bar{z}} = \sqrt{I_{\bar{z}}/A}$ $r_p = \sqrt{I_x/A}$	$m = \dfrac{2bt_f}{ht_w},\quad n = \dfrac{b}{h}$ $C_1 = 6(2 + 12m + 25m^2 + 15m^3)$ $C_2 = \nu(11 + 66m + 135m^2 + 90m^3)$ $C_3 = 5\nu mn^2(8 + 9m)$ $k_s = \dfrac{10(1 + \nu)(1 + 3m)^2}{30mn^2(1 + m) + C_1 + C_2 + C_3}$
10. **Z** Section 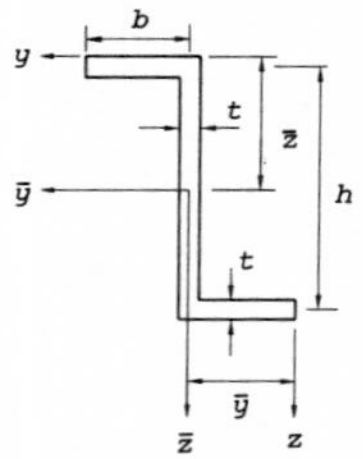	$H = h + t,\ B = b + t/2$ $C = b - t/2$ $A = t(h + 2b)$ $\bar{y} = b,\ \bar{z} = \frac{1}{2}(h + t)$	$I_{\bar{y}} = \dfrac{BH^3 - C(H - 2t)^3}{12}$ $I_{\bar{z}} = \frac{1}{12}[H(B + C)^3 - 2hC^3 - 6B^2hC]$ $I_{\overline{yz}} = -\dfrac{htb^2}{2}$	$r_{\bar{y}} = \sqrt{I_{\bar{y}}/A}$ $r_{\bar{z}} = \sqrt{I_{\bar{z}}/A}$ $r_p = \sqrt{I_x/A}$	Use computer program

11. Cross 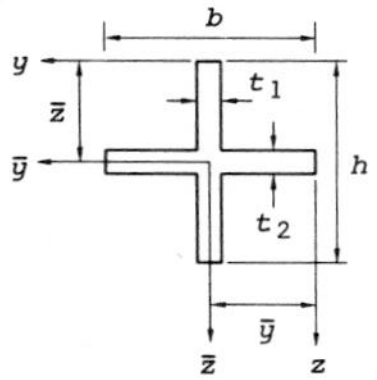	$A = ht_1 + (b - t_1)t_2$ $\bar{y} = b/2$ $\bar{z} = h/2$	$I_{\bar{y}} = \dfrac{t_1h^3 + (b - t_1)h^3}{12}$ $I_{\bar{z}} = \dfrac{t_2b^3 + (h - t_2)t_1^3}{12}$ $I_{\overline{yz}} = 0, \quad I_x = I_{\bar{y}} + I_{\bar{z}}$	$r_{\bar{y}} = \sqrt{I_{\bar{y}}/A}$ $r_{\bar{z}} = \sqrt{I_{\bar{z}}/A}$ $r_p = \sqrt{I_x/A}$	Use computer program
12. Channel 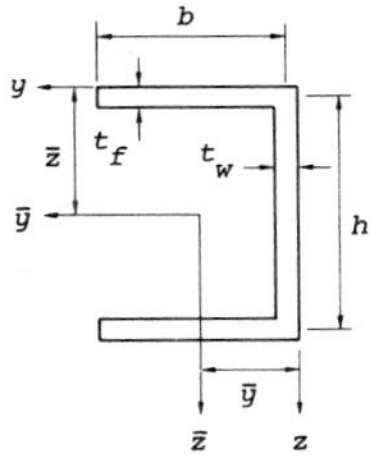	$B = b + t_w/2, \quad C = B - t_w$ $H = h + t_f, \quad D = h - t_f$ $A = ht_w + 2bt_f$ $\bar{y} = B - \dfrac{2B^2t_f + Dt_w^2}{2BH - 2DC}$ $\bar{z} = \dfrac{1}{2}(h + t_f)$	$I_{\bar{y}} = \dfrac{BH^3 - CD^3}{12}$ $I_{\bar{z}} = \dfrac{2t_fB^3 + Dt_w^3}{3}$ $- A(B - \bar{y})^2$ $I_{\overline{yz}} = 0, \; I_x = I_{\bar{y}} + I_{\bar{z}}$	$r_{\bar{y}} = \sqrt{I_{\bar{y}}/A}$ $r_{\bar{z}} = \sqrt{I_{\bar{z}}/A}$ $r_p = \sqrt{I_x/A}$	Use computer program
13. T Section 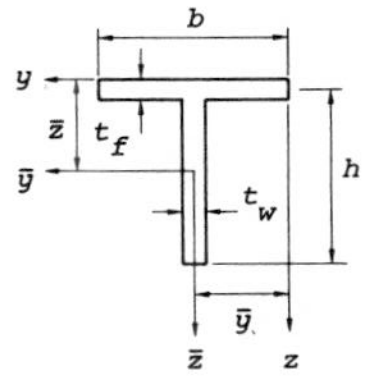	$H = h + t_f/2, \quad C = B - t_w$ $A = bt_f + t_wD$ $\bar{y} = b/2, \quad D = h - t_f/2$ $\bar{z} = H - \dfrac{H^2t_w + Ct_f^2}{2[Bt_f + Dt_w]}$	$I_{\bar{y}} = \dfrac{1}{3}[t_w(H - \bar{z})^3 + B\bar{z}^3$ $- C(\bar{z} - t_f)^3]$ $I_{\bar{z}} = \dfrac{B^3t_f + Dt_w^3}{12}$ $I_{\overline{yz}} = 0, \; I_x = I_{\bar{y}} + I_{\bar{z}}$	$r_{\bar{y}} = \sqrt{I_{\bar{y}}/A}$ $r_{\bar{z}} = \sqrt{I_{\bar{z}}/A}$ $r_p = \sqrt{I_x/A}$	$m = \dfrac{bt_f}{ht_w}, \quad n = \dfrac{b}{h}$ $C_1 = 12(1 + 8m + 23m^2 + 16m^3)$ $C_2 = \nu(11 + 88m + 248m^2 + 216m^3)$ $C_3 = 10\nu mn^2(4 + 5m + m^2)$ $k_s = \dfrac{10(1 + \nu)(1 + 4m)^2}{30mn^2(1 + m) + C_1 + C_2 + C_3}$

than one load on a beam, the F_v, F_θ, F_M, F_V functions are formed by adding the terms given in Table 2-2 for each load. The initial parameters v_0, θ_0, M_0, V_0 (values of v, θ, M, V at the left end of the beam, $x = 0$) are provided in Table 2-3. These formulas are applicable to both statistically determinate and indeterminate beams. Note that the beam ends in Table 2-3 do not display loadings. This is because all loadings should be accounted for by using the loading functions of Table 2-2, and hence such loadings do not affect the selection of the proper boundary conditions.

EXAMPLE 2.1 Cantilevered Beam Find the deflection curve for the beam of Fig. 2-6. From Eq. (2.8*a*), the beam deflection is

$$v = v_0 - \theta_0 x - M_0 \frac{x^2}{2EI} - V_0 \frac{x^3}{3!EI} + F_v \tag{1}$$

where F_v is found from Table 2-2 as $Wx^3/(3!EI)$ for the concentrated load W at $x = 0$ plus $w_1\langle x - a_1\rangle^4/4!EI$ for the distributed load w_1 beginning at $x = a_1$. Thus

$$F_v = W\frac{x^3}{3!EI} + w_1 \frac{\langle x - a_1\rangle^4}{4!EI} \tag{2}$$

The constants v_0, θ_0, M_0, V_0 are given in Table 2-3 for this free-fixed beam as

$$\begin{aligned} v_0 &= -F_v|_{x=L} - LF_\theta|_{x=L} \\ \theta_0 &= -F_\theta|_{x=L} \\ M_0 &= 0 \\ V_0 &= 0 \end{aligned} \tag{3}$$

Note that the beam is considered to be free on the left end even though a load is located there. This is the case because the load has already been taken into account by the loading functions. Physically the load W can be considered as being just to the right of the end ($x = 0^+$), rather than at $x = 0$. Then, $V_0 = 0$, but

$$V_{0^+} = V_0 + F_V|_{x=0^+} = -W$$

In addition to F_v at $x = L$, relations (3) require F_θ, which is found in Table 2-2 to be

$$F_\theta = -W\frac{x^2}{2EI} - w_1 \frac{\langle x - a_1\rangle^3}{3!EI} \tag{4}$$

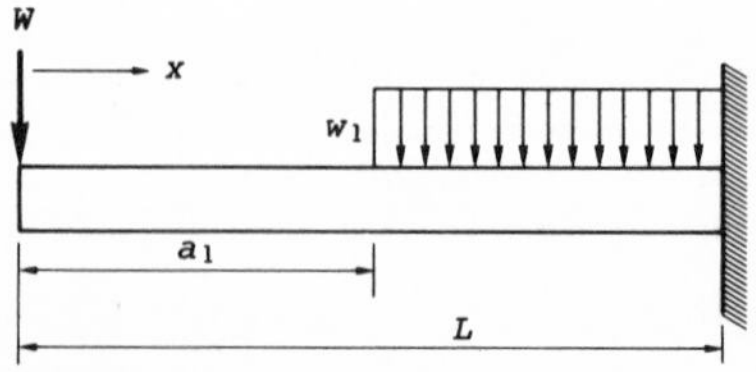

Fig. 2-6

TABLE 2-2 Loading Functions $F_v(x)$, $F_\theta(x)$, $F_M(x)$, $F_V(x)$ for Ordinary Beams (Eqs. 2.8)

	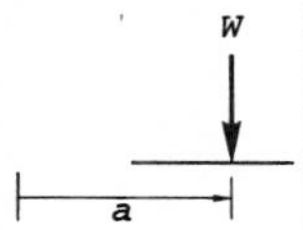	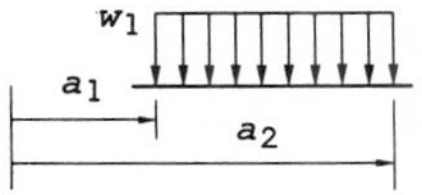	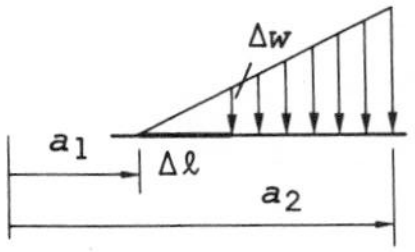	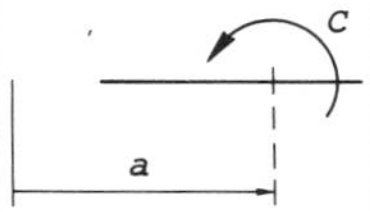	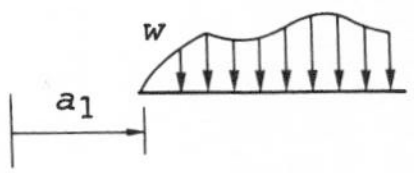
$F_v(x)$	$\frac{W\langle x - a\rangle^3}{3!EI}$	$\frac{w_1}{4!EI}\left(\langle x - a_1\rangle^4 - \langle x - a_2\rangle^4\right)$	$\frac{1}{5!EI}\frac{\Delta w}{\Delta \ell}\left(\langle x - a_1\rangle^5 - \langle x - a_2\rangle^5\right)$ $- \frac{\Delta w}{\Delta \ell}(a_2 - a_1)\frac{\langle x - a_2\rangle^4}{4!EI}$	$\frac{C\langle x - a\rangle^2}{2EI}$	$\frac{1}{EI}\int_{a_1}^{x} dx\int^{x} dx\int^{x} dx\int^{x} w\, dx$
$F_\theta(x)$	$-\frac{W\langle x - a\rangle^2}{2EI}$	$-\frac{w_1}{3!EI}\left(\langle x - a_1\rangle^3 - \langle x - a_2\rangle^3\right)$	$-\frac{1}{4!EI}\frac{\Delta w}{\Delta \ell}\left(\langle x - a_1\rangle^4 - \langle x - a_2\rangle^4\right)$ $+ \frac{\Delta w}{\Delta \ell}(a_2 - a_1)\frac{\langle x - a_2\rangle^3}{3!EI}$	$-\frac{C\langle x - a\rangle}{EI}$	$-\frac{1}{EI}\int_{a_1}^{x} dx\int^{x} dx\int^{x} w\, dx$
$F_M(x)$	$-W\langle x - a\rangle$	$-\frac{w_1}{2}\left(\langle x - a_1\rangle^2 - \langle x - a_2\rangle^2\right)$	$-\frac{1}{3!}\frac{\Delta w}{\Delta \ell}\left(\langle x - a_1\rangle^3 - \langle x - a_2\rangle^3\right)$ $+ \frac{\Delta w}{\Delta \ell}(a_2 - a_1)\frac{\langle x - a_2\rangle^2}{2}$	$-C\langle x - a\rangle^0$	$-\int_{a_1}^{x} dx\int^{x} w\, dx$
$F_V(x)$	$-W\langle x - a\rangle^0$	$-w_1\left(\langle x - a_1\rangle - \langle x - a_2\rangle\right)$	$-\frac{1}{2}\frac{\Delta w}{\Delta \ell}\left(\langle x - a_1\rangle^2 - \langle x - a_2\rangle^2\right)$ $+ \frac{\Delta w}{\Delta \ell}(a_2 - a_1)\langle x - a_2\rangle$	0	$-\int_{a_1}^{x} w\, dx$

TABLE 2-3 Initial Parameters for Ordinary Beams (Eqs. 2.8)

Right End → Left End ↓	1. Pinned 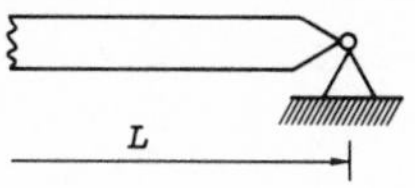	2. Fixed 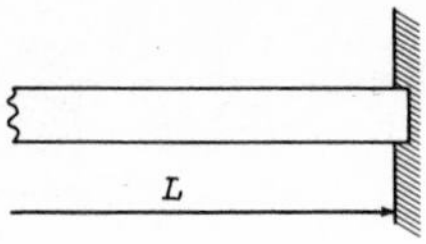
1. Pinned 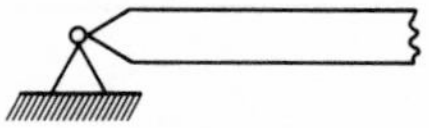 $v_0 = 0,\ M_0 = 0$	$\theta_0 = \frac{1}{L}F_V + \frac{L}{6EI}F_M$ $V_0 = -\frac{1}{L}F_M$	$\theta_0 = \frac{3}{2L}F_V + \frac{1}{2}F_\theta$ $V_0 = -\frac{3EI}{L^3}F_V - \frac{3EI}{L^2}F_\theta$
2. Fixed 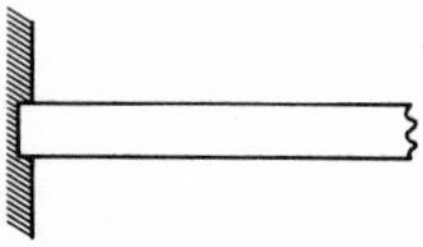 $v_0 = 0,\ \theta_0 = 0$	$M_0 = \frac{3EI}{L^2}F_V + \frac{1}{2}F_M$ $V_0 = -\frac{3EI}{L^3}F_V - \frac{3}{2L}F_M$	$M_0 = \frac{6EI}{L^2}F_V + \frac{2EI}{L}F_\theta$ $V_0 = -\frac{12EI}{L^3}F_V - \frac{6EI}{L^2}F_\theta$
3. Free 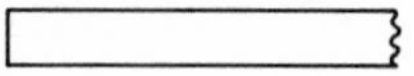 $M_0 = 0,\ V_0 = 0$	Kinematically unstable	$v_0 = -F_V - LF_\theta$ $\theta_0 = -F_\theta$
4. Guided 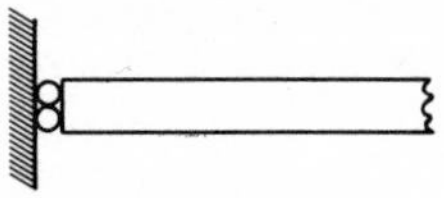 $\theta_0 = 0,\ V_0 = 0$	$v_0 = -F_V - \frac{L^2}{2EI}F_M$ $M_0 = -F_M$	$v_0 = -F_V - \frac{L}{2}F_\theta$ $M_0 = -\frac{EI}{L}F_\theta$
5. Partially Fixed  $v_0 = 0,\ M_0 = k_1^*\theta_0$	$\theta_0 = \left(\frac{1}{L}F_V + \frac{L}{6EI}F_M\right)/A_7$ $V_0 = \left(\frac{k_1^*}{L^2}F_V + A_8F_M\right)/A_7$	$\theta_0 = \left(\frac{3}{2L}F_V + \frac{1}{2}F_\theta\right)/A_9$ $V_0 = -(A_{11}F_V + \frac{3EI}{L}A_8F_\theta)/A_9$

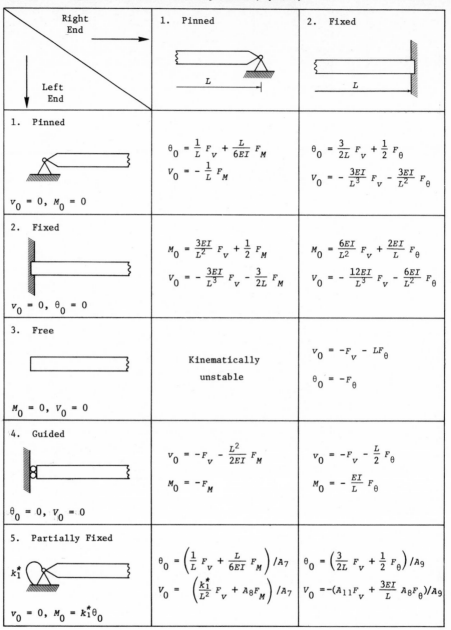

A kinematically unstable beam can only be analyzed for its natural frequencies and buckling loads.

3. Free	4. Guided	5. Partially Fixed
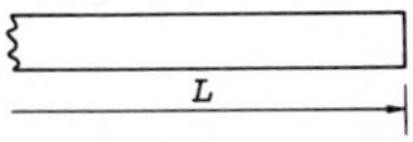	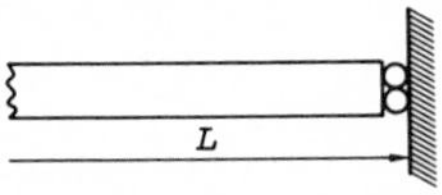	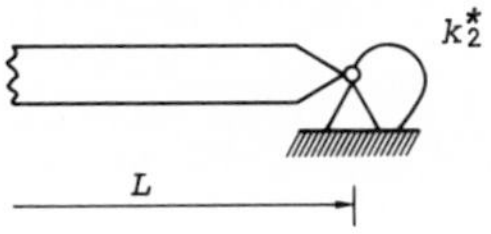
Kinematically unstable	$\theta_0 = \frac{L^2}{2EI} F_V - F_\theta$ $V_0 = -F_V$	$\theta_0 = \left(A_1 F_V - \frac{A_2 L}{6EI}\right) / A_3$ $V_0 = \left(\frac{k_2^*}{L^2} F_V + \frac{A_2}{L}\right) / A_3$
$M_0 = LF_V - F_M$ $V_0 = -F_V$	$M_0 = -\frac{EI}{L} F_\theta + \frac{L}{2} F_V$ $V_0 = -F_V$	$M_0 = \left(\frac{3EI}{L^2} A_6 F_V - \frac{A_2}{2}\right) / A_4$ $V_0 = \left(-\frac{3EIA_5}{L^3} F_V + \frac{3A_2}{2L}\right) / A_4$
Kinematically unstable	Kinematically unstable	$v_0 = -F_V - A_2 \frac{L}{k_2^*}$ $\theta_0 = -A_2 \frac{1}{k_2^*}$
Kinematically unstable	Kinematically unstable	$v_0 = \left(-A_6 F_V + \frac{L^2}{2EI} A_2\right) / A_6$ $M_0 = A_2/A_6$
$\theta_0 = -\frac{1}{k_1^*}(F_M - LF_V)$ $V_0 = -F_V$	$\theta_0 = (-F_\theta + \frac{L^2}{2EI} F_V)/A_{10}$ $V_0 = -F_V$	$\theta_0 = \left(\frac{A_5}{L} F_V - \frac{LA_2}{6EI}\right) / A_{13}$ $V_0 = (A_{12} F_V + A_2 A_8)/A_{13}$

Definitions: $F_V = F_V\big|_{x=L}$, $F_\theta = F_\theta\big|_{x=L}$, $F_M = F_M\big|_{x=L}$, $F_V = F_V\big|_{x=L}$

$A_1 = \frac{1}{L} - \frac{k_2^*}{2EI}$ $\quad A_4 = 1 - \frac{k_2^* L}{4EI}$ $\quad A_7 = 1 + \frac{k_1^* L}{3EI}$ $\quad A_{10} = 1 + \frac{k_1^* L}{EI}$ $\quad A_{12} = \frac{k_2^* - k_1^*}{L^2} + \frac{k_1^* k_2^*}{EIL}$

$A_2 = k_2^* F_\theta - F_M$ $\quad A_5 = 1 - \frac{k_2^* L}{2EI}$ $\quad A_8 = \frac{1}{L} + \frac{k_1^*}{2EI}$ $\quad A_{11} = \frac{3EI}{L^3} + \frac{3k_1^*}{L^2}$ $\quad A_{13} = 1 + \frac{1}{3}\frac{k_1^* L}{EI} - \frac{1}{3}\frac{k_2^* L}{EI} - \frac{k_1^* k_2^* L^2}{12(EI)^2}$

$A_3 = 1 - \frac{k_2^* L}{3EI}$ $\quad A_6 = 1 - \frac{k_2^* L}{EI}$ $\quad A_9 = 1 + \frac{k_1^* L}{4EI}$

If (2) and (4) are substituted in (3)

$$v_0 = -F_v|_{x=L} - LF_\theta|_{x=L}$$

$$= -\left[W\frac{L^3}{3!EI} + w_1\frac{(L-a_1)^4}{4!EI}\right] - L\left[-W\frac{L^2}{2EI} - w_1\frac{(L-a_1)^3}{3!EI}\right]$$

$$= W\frac{L^3}{3EI} + w_1\left[\frac{L(L-a_1)^3}{6EI} - \frac{(L-a_1)^4}{24EI}\right] \qquad (5)$$

$$\theta_0 = -F_\theta|_{x=L} = W\frac{L^2}{2EI} + w_1\frac{(L-a_1)^3}{3!EI}$$

$$M_0 = V_0 = 0$$

The desired deflection curve is then given by (1) and (2) as

$$v = v_0 - \theta_0 x + W\frac{x^3}{3!EI} + w_1\frac{\langle x-a_1\rangle^4}{4!EI}$$

with values of v_0 and θ_0 taken from (5).

EXAMPLE 2.2 Cantilevered Beam The slope, bending moment, and shear force equations of the beam of Fig. 2-6 are found from Eqs. (2.8*b*, *c*, *d*). The initial parameters of Eq. (5) of the previous example still apply. From the loading functions of Table 2-2,

$$\theta = \theta_0 + F_\theta = \theta_0 - W\frac{x^2}{2EI} - w_1\frac{\langle x-a_1\rangle^3}{3!EI}$$

$$M = F_M = -Wx - w_1\frac{\langle x-a_1\rangle^2}{2}$$

$$V = F_V = -W - w_1\langle x-a_1\rangle$$

EXAMPLE 2.3 Simply Supported Beam The deflection, slope, bending moment, and shear force relations (Eqs. 2.8) apply to boundary conditions other than those provided in Table 2-3. The initial parameters are found by applying the known boundary conditions to Eqs. (2.8). For a demonstration of the procedure, consider the familiar case in which both ends are simply supported, which is given in Table 2-3. The end conditions are $v|_{x=0} = M|_{x=0} = v|_{x=L} = M|_{x=L} = 0$. The first two conditions mean that $v_0 = M_0 = 0$. The final two conditions applied to Eqs. (2.8*a*) and (2.8*c*) give

$$v|_{x=L} = -\theta_0 L - V_0\frac{L^3}{6EI} + F_v|_{x=L} = 0$$

$$M|_{x=L} = V_0 L + F_M|_{x=L} = 0$$

We solve these two equations for θ_0 and V_0 and find

$$\theta_0 = \frac{F_v|_{x=L}}{L} + \frac{LF_M|_{x=L}}{6EI}$$

$$V_0 = -\frac{F_M|_{x=L}}{L}$$

These are the values given in Table 2-3.

2.4 General Uniform Beams

If the effects on a uniform beam of a compressive axial force P and an elastic foundation of modulus k (force/length^2) are taken into account, Eqs. (2.7) become

$$\begin{aligned} EI\frac{d^4v}{dx^4} + P\frac{d^2v}{dx^2} + kv &= w \\ EI\frac{d^3v}{dx^3} + P\frac{dv}{dx} &= -V \\ EI\frac{d^2v}{dx^2} &= -M \\ \frac{dv}{dx} &= -\theta \end{aligned} \tag{2.9}$$

These relations are appropriate for a beam with a tensile axial load if P is replaced by $-P$.

The solution of these equations is

Deflection:	$v = v_0(e_1 + \zeta e_3) - \theta_0 e_2 - M_0\frac{e_3}{EI} - V_0\frac{e_4}{EI} + F_v$	(2.10*a*)
Slope:	$\theta = v_0\lambda e_4 + \theta_0 e_1 + M_0\frac{e_2}{EI} + V_0\frac{e_3}{EI} F_\theta$	(2.10*b*)
Bending moment:	$M = v_0\lambda EIe_3 + \theta_0 EIe_0 + M_0 e_1 + V_0 e_2 + F_M$	(2.10*c*)
Shear force:	$V = v_0\lambda EI(e_2 + \zeta e_4) - \theta_0\lambda EIe_3 - M_0\lambda e_4 + V_0(e_1 + \zeta e_3) + F_V$	(2.10*d*)

The λ, ζ and e_i are defined in Table 2-4.

The F_v, F_θ, F_M, F_V are loading functions given in Table 2-5. If there is more than one load on a beam, F_v, F_θ, F_M, F_V functions are formed by adding the terms given in Table 2-5 for each load. In this table

$$e_i\langle x - a\rangle = \begin{cases} 0 & \text{if } x < a \\ e_i(x-a) & \text{if } x \geqslant a \end{cases} \tag{2.11}$$

TABLE 2-4 Values of e_i in Eqs. (2.10)

Ordinary Beam	Beam with Compressive Axial Force P	Beam with Tensile Axial Force P	Beam on Elastic (Winkler) Foundation, k (force/length^2)
$\zeta = \lambda = 0$	$\lambda = 0, \quad \alpha^2 = P/EI = \zeta$	$\lambda = 0, \quad \alpha^2 = P/EI, \quad \zeta = \alpha^2$	$\lambda = k/EI, \quad \zeta = 0, \quad \beta = (k/4EI)^{1/4}$
$e_0 = 0$	$e_0 = -\alpha \sin \alpha x$	$e_0 = \alpha \sinh \alpha x$	$e_0 = -\lambda e_4$
$e_1 = 1$	$e_1 = \cos \alpha x$	$e_1 = \cosh \alpha x$	$e_1 = \cosh \beta x \cos \beta x$
$e_2 = x$	$e_2 = \frac{1}{\alpha} \sin \alpha x$	$e_2 = \frac{1}{\alpha} \sinh \alpha x$	$e_2 = \frac{1}{2\beta}(\cosh \beta x \sin \beta x + \sinh \beta x \cos \beta x)$
$e_3 = \frac{x^2}{2}$	$e_3 = \frac{1}{\alpha^2}(1 - \cos \alpha x)$	$e_3 = \frac{1}{\alpha^2}(\cosh \alpha x - 1)$	$e_3 = \frac{1}{2\beta^2}(\sinh \beta x \sin \beta x)$
$e_4 = \frac{x^3}{6}$	$e_4 = \frac{1}{\alpha^3}(\alpha x - \sin \alpha x)$	$e_4 = \frac{1}{\alpha^3}(\sinh \alpha x - \alpha x)$	$e_4 = \frac{1}{4\beta^3}(\cosh \beta x \sin \beta x - \sinh \beta x \cos \beta x)$
$e_5 = \frac{x^4}{24}$	$e_5 = \frac{1}{\alpha^4}\left(\frac{\alpha^2 x^2}{2} + \cos \alpha x - 1\right)$	$e_5 = \frac{1}{\alpha^4}\left(-\frac{\alpha^2 x^2}{2} + \cosh \alpha x - 1\right)$	$e_5 = \frac{1 - e_1}{\lambda}$
$e_6 = \frac{x^5}{120}$	$e_6 = \frac{1}{\alpha^5}\left(\frac{\alpha^3 x^3}{6} + \sin \alpha x - \alpha x\right)$	$e_6 = \frac{1}{\alpha^5}\left(-\frac{\alpha^3 x^3}{6} + \sinh \alpha x - \alpha x\right)$	$e_6 = \frac{x - e_2}{\lambda}$

TABLE 2-5 Loading Functions F_v, F_θ, F_M, F_V for General Uniform Beams (Eqs. 2.10)

	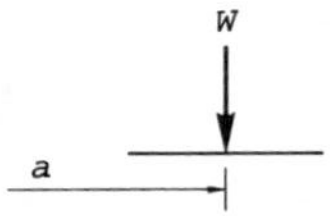 	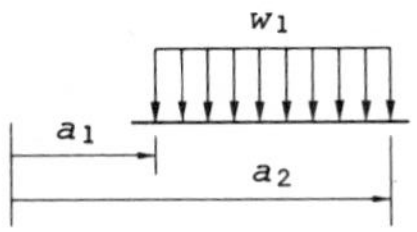 	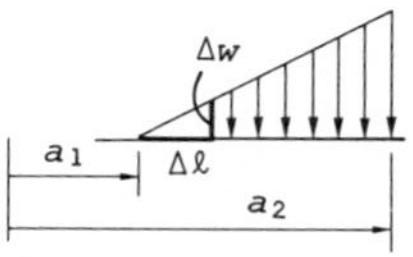 	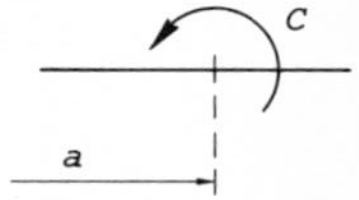
$F_v(x)$	$W \dfrac{e_4\langle x - a\rangle}{EI}$	$w_1 \dfrac{e_5\langle x - a_1\rangle - e_5\langle x - a_2\rangle}{EI}$	$\dfrac{\Delta w}{\Delta \ell} \dfrac{e_6\langle x - a_1\rangle - e_6\langle x - a_2\rangle}{EI}$ $- \dfrac{\Delta w}{\Delta \ell} \dfrac{(a_2 - a_1)}{EI} e_5\langle x - a_2\rangle$	$C \dfrac{e_3\langle x - a\rangle}{EI}$
$F_\theta(x)$	$-W \dfrac{e_3\langle x - a\rangle}{EI}$	$-w_1 \dfrac{e_4\langle x - a_1\rangle - e_4\langle x - a_2\rangle}{EI}$	$- \dfrac{\Delta w}{\Delta \ell} \dfrac{e_5\langle x - a_1\rangle - e_5\langle x - a_2\rangle}{EI}$ $+ \dfrac{\Delta w}{\Delta \ell} \dfrac{(a_2 - a_1)}{EI} e_4\langle x - a_2\rangle$	$-C \dfrac{e_2\langle x - a\rangle}{EI}$
$F_M(x)$	$-We_2\langle x - a\rangle$	$-w_1(e_3\langle x - a_1\rangle - e_3\langle x - a_2\rangle)$	$- \dfrac{\Delta w}{\Delta \ell} (e_4\langle x - a_1\rangle - e_4\langle x - a_2\rangle)$ $+ \dfrac{\Delta w}{\Delta \ell} (a_2 - a_1)\ e_3\langle x - a_2\rangle$	$-Ce_1\langle x - a\rangle$
$F_V(x)$	$-W(e_1\langle x - a\rangle$ $+ \zeta e_3\langle x - a\rangle)$	$-w_1 \big[(e_2\langle x - a_1\rangle - e_2\langle x - a_2\rangle)$ $-\zeta(e_4\langle x - a_1\rangle - e_4\langle x - a_2\rangle)\big]$	$- \dfrac{\Delta w}{\Delta \ell} (e_3\langle x - a_1\rangle + \zeta e_5\langle x - a_1\rangle$ $-e_3\langle x - a_2\rangle - \zeta e_5\langle x - a_2\rangle)$ $+ \dfrac{\Delta w}{\Delta \ell} (a_2 - a_1)(e_2\langle x - a_2\rangle + \zeta e_4\langle x - a_2\rangle)$	$C\lambda e_4\langle x - a\rangle$

For example, suppose $e_1 = \cos \alpha x$ in Table 2-5. Then

$$e_1\langle x-a\rangle = \cos \alpha \langle x-a\rangle = \langle x-a\rangle^0 \cos \alpha(x-a)$$
$$= \begin{cases} 0 & \text{if} \quad x < a \\ \cos \alpha(x-a) & \text{if} \quad x \geqslant a \end{cases}$$

This definition applies also to $e_i = 1$, that is, if $e_i = 1$ in Table 2-5, then

$$e_i\langle x-a\rangle = \langle x-a\rangle^0 = \begin{cases} 0 & \text{if} \quad x < a \\ 1 & \text{if} \quad x \geqslant a \end{cases} \qquad (2.12)$$

TABLE 2-6 Initial Parameters for General Uniform Beams (Eq. 2.10)

Right End → Left End ↓	1. Pinned 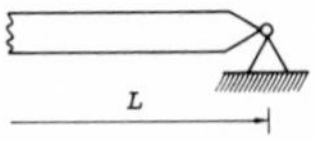	2. Fixed 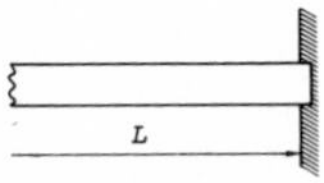
1. Pinned 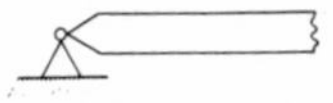 $v_0 = 0,\ M_0 = 0$	$\theta_0 = [-e_2 F_V - (e_4/EI)F_M]/\nabla$ $V_0 = (EIe_0 F_V + e_2 F_M)/\nabla$ $\nabla = e_0 e_4 - e_2^2$	$\theta_0 = [-e_3 F_v - (e_4/EI)F_\theta]/\nabla$ $V_0 = (EIe_1 F_v + EIe_2 F_\theta)/\nabla$ $\nabla = e_1 e_4 - e_2 e_3$
2. Fixed 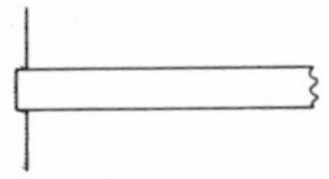 $v_0 = 0,\ \theta_0 = 0$	$M_0 = (-EIe_2 F_V - e_4 F_M)/\nabla$ $V_0 = (EIe_1 F_V + e_3 F_M)/\nabla$ $\nabla = e_1 e_4 - e_2 e_3$	$M_0 = (-EIe_3 F_v - EIe_4 F_\theta)/\nabla$ $V_0 = (EIe_2 F_v + EIe_3 F_\theta)/\nabla$ $\nabla = e_2 e_4 - e_3^2$
3. Free 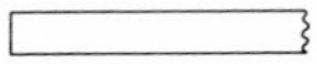 $M_0 = 0,\ V_0 = 0$	$v_0 = [-e_0 F_V - (e_2/EI)F_M]/\nabla$ $\theta_0 = \{\lambda e_3 F_V - [(e_1 + \zeta e_3)/EI]F_M\}/\nabla$ $\nabla = e_0(e_1 + \zeta e_3) + \lambda e_2 e_3$	$v_0 = (-e_1 F_v - e_2 F_\theta)/\nabla$ $\theta_0 = [\lambda e_4 F_v - (e_1 + \zeta e_3)F_\theta]/\nabla$ $\nabla = e_1(e_1 + \zeta e_3) + \lambda e_2 e_4$
4. Guided 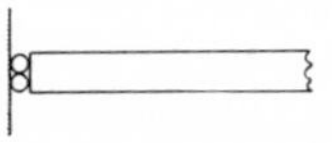 $\theta_0 = 0,\ V_0 = 0$	$v_0 = [-e_1 F_V - (e_3/EI)F_M]/\nabla$ $M_0 = [\lambda EIe_3 F_V - (e_1 + \zeta e_3)F_M]/\nabla$ $\nabla = e_1(e_1 + \zeta e_3) + \lambda e_3^2$	$v_0 = (-e_2 F_v - e_3 F_\theta)/\nabla$ $M_0 = [\lambda EIe_4 F_v - EI(e_1 + \zeta e_3)F_\theta]/\nabla$ $\nabla = e_2(e_1 + \zeta e_3) + \lambda e_3 e_4$
5. Partially Fixed 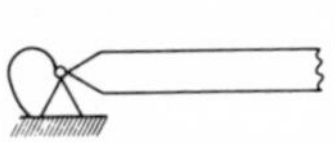 $v_0 = 0,\ M_0 = k_1^* \theta_0$	$\theta_0 = [-e_2 F_V - (e_4/EI)F_M]/\nabla$ $V_0 = (A_1 F_V + A_2 F_M)/\nabla$ $\nabla = A_1 e_4/EI - A_2 e_2$	$\theta_0 = [-(e_3/EI)F_v - (e_4/EI)F_\theta]/\nabla$ $V_0 = (A_3 F_v + A_2 F_\theta)/\nabla$ $\nabla = A_3 e_4/EI - A_2 e_3/EI$

Definitions: $F_V = F_V\big|_{x=L}$, $F_\theta = F_\theta\big|_{x=L}$, $F_M = F_M\big|_{x=L}$, $F_v = F_v\big|_{x=L}$

$e_i = e_i\big|_{x=L}$, $i = 0, 1, 2, 3, 4, 5, 6$

The initial values v_0, θ_0, M_0, V_0 (values of v, θ, M, V at the left end of the beam, $x = 0$) are provided in Table 2-6.

EXAMPLE 2.4 Beam on Elastic Foundation To illustrate the use of the general uniform-beam solution of Eqs. (2.10), consider the beam of Fig. 2-7, which lies on an elastic foundation. The response variables are given by Eqs. (2.10), with the loading functions for the distributed and concentrated forces given in Table 2-5, the e_i functions given by Table 2-4, and the initial parameters v_0, θ_0, M_0, V_0 taken from Table 2-6.

3. Free 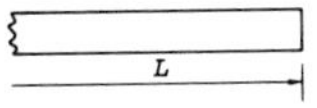	4. Guided 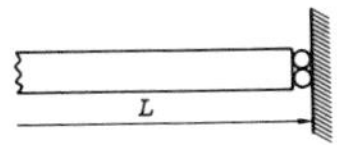	5. Partially Fixed 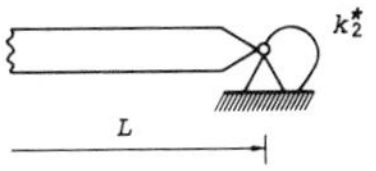
$\theta_0 = \{-[(e_1 + \zeta e_3)/EI]F_M + (e_2/EI)F_V\}/\nabla$ $V_0 = (-\lambda e_3 F_M - e_0 F_V)/\nabla$ $\nabla = e_0(e_1 + \zeta e_3) + \lambda e_2 e_3$	$\theta_0 = [-(e_1 + \zeta e_3)F_\theta + (e_3/EI)F_V]/\nabla$ $V_0 = (-\lambda EI e_3 F_\theta - e_1 F_V)/\nabla$ $\nabla = e_1(e_1 + \zeta e_3) + \lambda e_3^2$	$\theta_0 = (A_5 F_V + A_9 e_4)/\nabla$ $V_0 = (EIA_6 F_V - EIA_9 e_2)/\nabla$ $\nabla = A_5 e_2 + A_6 e_4$
$M_0 = [-(e_1 + \zeta e_3)F_M + e_2 F_V]/\nabla$ $V_0 = (-\lambda e_2 F_M - e_1 F_V)/\nabla$ $\nabla = e_1(e_1 + \zeta e_3) + \lambda e_2^2$	$M_0 = [-(e_1 + \zeta e_3)EIF_\theta + e_3 F_V]/\nabla$ $V_0 = (-\lambda EI e_2 F_\theta - e_2 F_V)/\nabla$ $\nabla = e_2(e_1 + \zeta e_3) + \lambda e_2 e_3$	$M_0 = (EIA_5 F_V + EIA_9 e_4)/\nabla$ $V_0 = (-EIA_7 F_V - EIA_9 e_3)/\nabla$ $\nabla = A_5 e_3 - A_7 e_4$
$v_0 = [(e_3/EI)F_M + (e_0/\lambda EI)F_V]/\nabla$ $\theta_0 = \{[(e_2 + \zeta e_4)/EI]F_M + (e_3/EI)F_V\}/\nabla$ $\nabla = -\lambda e_3^2 - e_0(e_2 + \zeta e_4)$	$v_0 = [e_3 F_\theta + (e_1/\lambda EI)F_V]/\nabla$ $\theta_0 = [(e_2 + \zeta e_4)F_\theta - (e_4/EI)F_V]/\nabla$ $\nabla = -\lambda e_3 e_4 - e_1(e_2 + \zeta e_4)$	$v_0 = (-A_6 F_V + A_9 e_2)/\nabla$ $\theta_0 = [\lambda A_8 F_V - A_9(e_1 + \zeta e_3)]/\nabla$ $\nabla = A_6(e_1 + \zeta e_3) + A_8 e_2$
$v_0 = [(e_2/EI)F_M - (e_1/\lambda EI)F_V]/\nabla$ $M_0 = [(e_2 + \zeta e_4)F_M - e_3 F_V]/\nabla$ $\nabla = -\lambda e_2 e_3 - e_1(e_2 + \zeta e_4)$	$v_0 = [e_2 F_\theta + (e_2/\lambda EI)F_V]/\nabla$ $M_0 = [EI(e_2 + \zeta e_4)F_\theta - e_4 F_V]/\nabla$ $\nabla = -\lambda e_2 e_4 - e_2(e_2 + \zeta e_4)$	$v_0 = (A_7 F_V + A_9 e_3)/\nabla$ $M_0 = [\lambda EIA_8 F_V + EIA_9(e_1 + \zeta e_3)]/\nabla$ $\nabla = \lambda A_8 e_3 - A_7(e_1 + \zeta e_3)$
$\theta_0 = [-(e_1 + \zeta e_3)F_M + e_2 F_V]/\nabla$ $V_0 = (-\lambda A_4 F_M - A_1 F_V)/\nabla$ $\nabla = A_1(e_1 + \zeta e_3) + \lambda A_4 e_2$	$\theta_0 = [-(e_1 + \zeta e_3)F_\theta + (e_3/EI)F_V]/\nabla$ $V_0 = (-\lambda A_4 F_\theta - A_3 F_V)/\nabla$ $\nabla = A_3(e_1 + \zeta e_3) + \lambda A_4 e_3/EI$	$\theta_0 = [A_5 F_V + A_9 e_4]/\nabla$ $V_0 = [EI(A_1 - k_2^* A_3)F_V - EIA_2 A_9]/\nabla$ $\nabla = A_2 A_5 + (A_1 - k_2^* A_3)e_4$

$A_1 = EIe_0 + k_1^* e_1$	$A_4 = EIe_3 + k_1^* e_2$	$A_7 = k_2^* e_2 - EIe_1$
$A_2 = e_2 + k_1^* e_3/EI$	$A_5 = k_2^* e_3 - EIe_2$	$A_8 = EIe_3 - k_2^* e_4$
$A_3 = e_1 + k_1^* e_2/EI$	$A_6 = EIe_0 - k_2^* e_1$	$A_9 = k_2^* F_\theta - F_M$

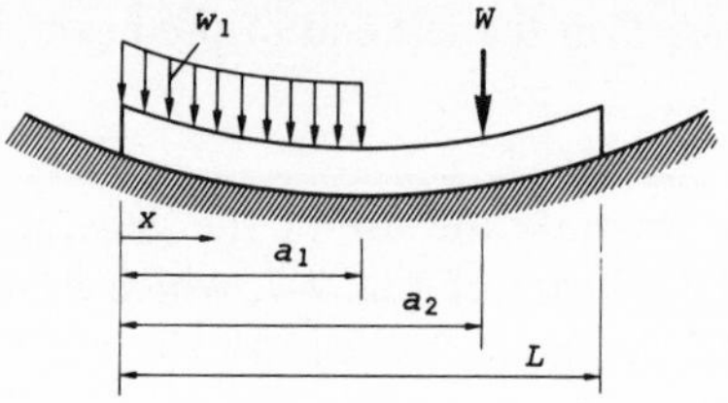

Fig. 2-7

Suppose only the deflection is desired. The initial values M_0 and V_0 can be set equal to zero since this is a free-free beam (Table 2-6). Then

$$v = v_0 e_1 - \theta_0 e_2 - M_0 \frac{e_3}{EI} - V_0 \frac{e_4}{EI} + F_v$$

$$= v_0 e_1 - \theta_0 e_2 + w_1 \left(\frac{e_5\langle x-0\rangle - e_5\langle x-a_1\rangle}{EI} \right) + W e_4 \langle x - a_2 \rangle$$

$$= v_0(\cosh \beta x \cos \beta x) - \theta_0 \frac{1}{2\beta}(\cosh \beta x \sin \beta x + \sinh \beta x \cos \beta x)$$

$$+ \frac{w_1}{EI} \left\{ \frac{1 - \cosh \beta x \cos \beta x}{k/EI} - \left[\frac{1 - \cosh \beta\langle x-a_1\rangle \cos \beta\langle x-a_1\rangle}{k/EI} \right] \right\}$$

$$+ W \frac{1}{4\beta^3}(\cosh \beta\langle x-a_2\rangle \sin \beta\langle x-a_2\rangle - \sinh \beta\langle x-a_2\rangle \cos \beta\langle x-a_2\rangle)$$

From Table 2-6,

$$v_0 = \left(\frac{e_3}{EI} F_M + \frac{e_0}{\lambda EI} F_V \right)_{x=L} \Big/ \nabla$$

$$\theta_0 = \left(\frac{e_2}{EI} F_M + \frac{e_3}{EI} F_V \right)_{x=L} \Big/ \nabla$$

$$\Delta = (-\lambda e_3^2 - e_0 e_2)_{x=L}$$

where, from Table 2-5,

$$F_M|_{x=L} = -We_2(L-a_2) - w_1[e_3(L) - e_3(L-a_1)]$$

$$F_V|_{x=L} = -We_1(L-a_2) - w_1[e_2(L) - e_2(L-a_1)]$$

and from Table 2-4,

$$e_0(L) = -\beta(\cosh \beta L \sin \beta L - \sinh \beta L \cos \beta L)$$

$$e_2(L) = \frac{1}{2\beta}(\cosh \beta L \sin \beta L + \sinh \beta L \cos \beta L)$$

$$e_3(L) = \frac{1}{2\beta^2} \sinh \beta L \sin \beta L$$

$$e_1(L-a_2) = \cosh \beta(L-a_2) \cos \beta(L-a_2)$$

$$e_2(L-a_i) = \frac{1}{2\beta}\left[\cosh \beta(L-a_i) \sin \beta(L-a_i) + \sinh \beta(L-a_i) \cos \beta(L-a_i)\right]$$

$$a_i = a_1 \quad \text{and} \quad a_2$$

$$e_3(L-a_1) = \frac{1}{2\beta^2} \sinh \beta(L-a_1) \sin \beta(L-a_1)$$

The deflection curve is now fully defined.

2.5 Column Buckling Loads

The buckling loads P_{cr} for columns of uniform cross section are given in Table 2-7. These are the Euler-theory unstable loads for elastic bars of length L.

2.6 Natural Frequencies

The natural frequencies ω_n, $n = 1, 2, \ldots,$ for the bending vibrations of uniform beams are given in Table 2-7.

B. COMPLEX BEAMS

2.7 The Transfer Matrix Method

In Part A, beam problems were solved using Eqs. (2.8). These equations can be written in the matrix form

$$\begin{bmatrix} v \\ \theta \\ M \\ V \end{bmatrix} = \begin{bmatrix} 1 & -x & -\dfrac{x^2}{2EI} & -\dfrac{x^3}{6EI} \\ 0 & 1 & \dfrac{x}{EI} & \dfrac{x^2}{2EI} \\ 0 & 0 & 1 & x \\ 0 & 0 & 0 & 1 \end{bmatrix} \begin{bmatrix} v_0 \\ \theta_0 \\ M_0 \\ V_0 \end{bmatrix} + \begin{bmatrix} F_v \\ F_\theta \\ F_M \\ F_V \end{bmatrix} \tag{2.13}$$

or

$$\begin{bmatrix} v \\ \theta \\ M \\ V \\ 1 \end{bmatrix} = \begin{bmatrix} 1 & -x & -\dfrac{x^2}{2EI} & -\dfrac{x^3}{6EI} & F_v \\ 0 & 1 & \dfrac{x}{EI} & \dfrac{x^2}{2EI} & F_\theta \\ 0 & 0 & 1 & x & F_M \\ 0 & 0 & 0 & 1 & F_V \\ 0 & 0 & 0 & 0 & 1 \end{bmatrix} \begin{bmatrix} v_0 \\ \theta_0 \\ M_0 \\ V_0 \\ 1 \end{bmatrix} \tag{2.14}$$

This 5×5 matrix is called the (extended) transfer matrix between the state variables at $x = 0$ and the state variables at a larger x. For a single-span uniform beam, this matrix equation is just as efficient as the simultaneous equations of Eq. (2.8). But for complex beams, for example, a beam with more than one span, variable cross-sectional properties, spring supports, and concentrated masses, the simultaneous equations become unwieldy. However, the transfer matrix of Eq.

TABLE 2-7 Natural Frequencies and Critical Axial Loads for Uniform Bars

Boundary Conditions	Frequency $\omega_n = (\beta_n L)^2 \sqrt{EI/\rho L^4}$	Critical Load P_{cr}
1. Fixed-free	$\beta_1 L = 1.8751$ $\beta_2 L = 4.6941$ $\beta_3 L = 7.8548$ $\beta_4 L = 10.9955$ .... $\beta_n L \approx (2n - 1)\pi/2$, n large	$\frac{\pi^2 EI}{4L^2}$
2. Hinged-hinged	$\beta_n L = n\pi$	$\pi^2 EI/L^2$
3. Fixed-hinged	$\beta_1 L = 3.9266$ $\beta_2 L = 7.0686$ $\beta_3 L = 10.2102$ $\beta_4 L = 13.3518$ .... $\beta_n L \approx (2n + 1)\pi/2$, n large	$20.19 \frac{EI}{L^2}$
4. Fixed-fixed	$\beta_1 L = 4.7300$ $\beta_2 L = 7.8532$ $\beta_3 L = 10.9956$ $\beta_4 L = 14.1372$ .... $\beta_n L \approx (2n + 1)\pi/2$, n large	$\frac{4\pi^2 EI}{L^2}$
5. Guided-guided	$\beta_n L = n\pi$	0
6. Free-free[1]	same as fixed-fixed	0
7. Free-hinged[1]	same as fixed-hinged	0
8. Hinged-guided	$\beta_n L = (2n - 1)\pi/2$	$\pi^2 EI/(4L^2)$
9. Fixed-guided	$\beta_1 L = 2.3650$ $\beta_2 L = 5.4978$ $\beta_3 L = 8.6394$ $\beta_4 L = 11.7810$ .... $\beta_n L \approx (4n - 1)\pi/4$, n large	$\frac{\pi^2 EI}{L^2}$
10. Free-guided[1]	same as fixed-guided	0

[1]This beam has a rigid body motion associated with the frequency = 0.

(2.14), properly manipulated, will lead efficiently to the response of a complex beam.

The details of the transfer matrix method, a so-called *mixed method*, are available in many sources, including Refs. 2.1 and 2.2. Also, the method is explained in some detail in the example problems of this and the following chapter.

2.8 Notation for Complex Beams

$A_s = Ak_s$	Equivalent shear area, A = cross-sectional area, k_s = shear form factor. See Table 2-1 or use a computer program such as BEAMSTRESS for k_s.
c	Moment intensity (force · length/length)
C	Concentrated applied moment (force · length)
$d = -1$	for vibrating beam
$d = 1$	for rotating shaft with whirl in the same angular direction as rotation
$d = -3$	for rotating shaft with whirl in the opposite direction as rotation (no eccentricity of mass)
G	Shear modulus of elasticity (force/length^2)
$I_T = \rho r^2$	Rotary inertia, transverse or diametrical mass moment of inertia per unit length (mass · length).
k	Winkler (elastic) foundation modulus (force/length^2)
k^*	Rotary foundation modulus (force · length/length)
k_i	Extension spring constant (force/length)
k_i^*	Rotary spring constant (force · length/radian)
ℓ	Length of segment, span of transfer matrix
$M_T = \int_A E\alpha\, \Delta T\, z\, dA$	Thermal moment
$r = r_y$	Radius of gyration of cross-sectional area about y axis (Table 2-1)
r_p	Polar radius of gyration (Table 2-1)
t	Time
α	Coefficient of thermal expansion (length/length · degree)
ΔT	Temperature change (degrees), that is, the temperature rise with respect to the reference temperature
D_V	Shear stiffness for composite beam
c^*	Rotary viscous damping coefficient (force · time, mass · length/time)
c'	External or viscous damping coefficient [force · time/length^2, mass/(time · length)]
c_ρ	Proportional viscous damping coefficient (1/time). If c' and c^* are chosen to be proportional to the mass, that is, $c' = c_\rho \rho$ and $c^* = c_\rho \rho r^2$, then c_ρ is the constant of proportionality.
c_i	Discrete extensional damping (dashpot) constant (force · time/length)
c_i^*	Discrete rotary damping constant (force · time · length)
M_i	Concentrated mass (mass)
I_{Ti}	Rotary inertia, transverse or diametrical mass moment of inertia of concentrated mass at station i (mass · length^2). This can be calculated as $I_{Ti} = \Delta a\, \rho r^2$, where Δa is the length of beam lumped at station i.
P_{Li}	Lumped axial force (force · length). This can be calculated as $P_{Li} = \Delta a\, P$, where Δa is a small segment of beam with axial force P.
c_1	Magnitude of distributed moment, uniform in x direction (force · length/length)
M_{T1}	Magnitude of distributed thermal moment, uniform in x direction.

$\dfrac{\Delta c}{\Delta \ell}$ Gradient of distributed moment, linearly varying in x direction (force · length/length^2)

$\dfrac{\Delta M_T}{\Delta \ell}$ Gradient of distributed thermal moment, linearly varying in x direction

Ω Frequency of steady-state forces and responses (radians/time)

ν Poisson's ratio

$\mathbf{U}_i$ Field matrix of the ith segment

$\overline{\mathbf{U}}_i$ Point matrix at $x = a_i$

The notation for transfer matrices is:

$$\mathbf{U}_i = \begin{bmatrix} U_{vv} & U_{v\theta} & U_{vM} & U_{vV} & F_v \\ U_{\theta v} & U_{\theta\theta} & U_{\theta M} & U_{\theta V} & F_\theta \\ U_{Mv} & U_{M\theta} & U_{MM} & U_{MV} & F_M \\ U_{Vv} & U_{V\theta} & U_{VM} & U_{VV} & F_V \\ 0 & 0 & 0 & 0 & 1 \end{bmatrix} \tag{2.15}$$

2.9 Differential Equations

The fundamental equations of motion in first-order form for the bending of a Timoshenko beam are

$$\frac{\partial v}{\partial x} = -\theta + \frac{V}{GA_s} \tag{2.16a}$$

$$\frac{\partial \theta}{\partial x} = \frac{M}{EI} + \frac{M_T}{EI} \tag{2.16b}$$

$$\frac{\partial M}{\partial x} = V + (k^* - P)\theta - d\rho r_y^2 \frac{\partial^2 \theta}{\partial t^2} - c(x, t) \tag{2.16c}$$

$$\frac{\partial V}{\partial x} = kv + \rho \frac{\partial^2 v}{\partial t^2} - w(x, t) \tag{2.16d}$$

with $d = -1$. In addition to bending, the Timoshenko beam includes the effects of shear deformation and rotary inertia. The expressions are reduced to those for a Rayleigh beam (bending, rotary inertia) by setting $1/GA_s = 0$, for a shear beam (bending, shear deformation) by setting $\rho r_y^2\, \partial^2\theta/\partial t^2 = 0$, and for an Euler-Bernoulli beam (bending) by setting $1/GA_s = \rho r_y^2\, \partial^2\theta/\partial t^2 = 0$. Equations (2.16) are appropriate for a beam with a tensile axial force if P is replaced by $-P$. In their present form they apply to beams with a compressive axial force P. In higher-order form, the equations become

$$w(x, t) = -\frac{\partial}{\partial x}\left[GA_s\left(\frac{\partial v}{\partial x} + \theta\right)\right] + kv + \rho \frac{\partial^2 v}{\partial t^2} \tag{2.17a}$$

$$c(x, t) = -\frac{\partial}{\partial x}\left(EI \frac{\partial \theta}{\partial x}\right) + GA_s \frac{\partial v}{\partial x} + (GA_s + k^* - P)\theta + \frac{\partial}{\partial x} M_T - d\rho r_y^2 \frac{\partial^2 \theta}{\partial t^2} \tag{2.17b}$$

$$V = GA_s\left(\frac{\partial v}{\partial x} + \theta\right) \tag{2.17c}$$

$$M = EI \frac{\partial \theta}{\partial x} - M_T \tag{2.17d}$$

For a rotating shaft, d is set equal to 1 or -3 for whirl in the same or the opposite direction, respectively, as rotation. The rotating shaft treated here has a circular cross section so that $r^2 = r_y^2 = r_z^2 = r_p^2/2$.

A more familiar form of these equations for the Euler-Bernoulli beam is

$$\frac{\partial^2}{\partial x^2} EI \frac{\partial^2 v}{\partial x^2} + \frac{\partial}{\partial x} P \frac{\partial v}{\partial x} + kv + \rho \frac{\partial^2 v}{\partial t^2} = w(x, t) \tag{2.18a}$$

$$V = -\frac{\partial}{\partial x} EI \frac{\partial^2 v}{\partial x^2} - P \frac{\partial v}{\partial x} \tag{2.18b}$$

$$M = -EI \frac{\partial^2 v}{\partial x^2} \tag{2.18c}$$

$$\theta = -\frac{\partial v}{\partial x} \tag{2.18d}$$

Composite Beams The centroid of a composite section (Fig. 2-8) is located at

$$\bar{y}^* = \frac{1}{A^*} \int y \, dA^* = \frac{1}{A^*} \sum_i \bar{y}_{0i} A_i^*, \qquad \bar{z}^* = \frac{1}{A^*} \int z \, dA^* = \frac{1}{A^*} \sum_i \bar{z}_{0i} A_i^* \tag{2.19}$$

where

$$dA^* = E \, dA / E_r, \qquad A_i^* = E_i A_i / E_r, \qquad A^* = \sum E_i A_i / E_r \tag{2.20}$$

and E_r is an arbitrary reference modulus that can be chosen to regulate the magnitude of the numbers being manipulated. If it is set equal to E, then for homogeneous beams, these so-called *modulus-weighted properties* will reduce to the ordinary geometric properties. The moments of inertia with respect to the y_0, z_0 axes are

$$I_{y_0}^* = \int_A z_0^2 \, dA^* = \sum_{i=1}^n \frac{E_i}{E_r} I_{y_{0i}} = \sum_{i=1}^n \frac{E_i}{E_r} \left(\bar{I}_{y_i} + \bar{z}_{0i}^2 A_i \right), \; I_{z_0}^* = \sum_{i=1}^n \frac{E_i}{E_r} \left(\bar{I}_{z_i} + \bar{y}_{0i}^2 A_i \right)$$

$$I_{y_0 z_0}^* = \sum_{i=1}^n \frac{E_i}{E_r} \left(\bar{I}_{y_i z_i} + \bar{y}_{0i} \bar{z}_{0i} A_i \right) \tag{2.21}$$

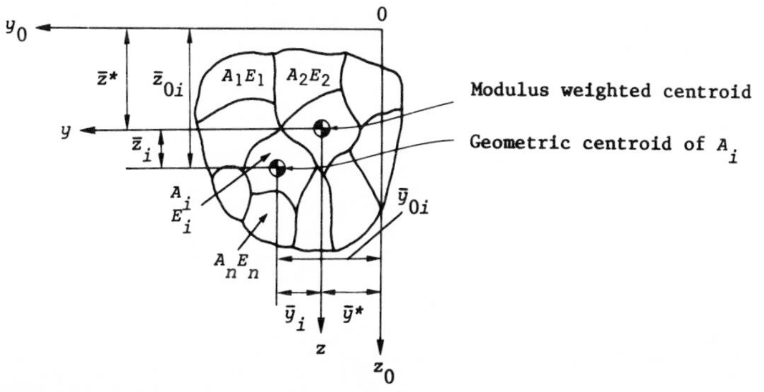

Fig. 2-8 Cross section of a composite beam.

where $I_{y_{0i}}$ is the moment of inertia of area A_i about the y_0 axis. $\bar{I}_{y_i}$, $\bar{I}_{z_i}$, $\bar{I}_{y_i z_i}$ are moments of inertia about the centroidal axes of area A_i. With respect to the modulus-weighted centroidal axes y, z, the moments of inertia are

$$I_y^* = I_{y_0}^* - \bar{z}^{*2}A^*, \; I_z^* = I_{z_0}^* - \bar{y}^{*2}A^*, \; I_{yz}^* = I_{y_0 z_0}^* - \bar{y}^*\bar{z}^*A^* \qquad (2.22)$$

To use the differential equations in Eqs. (2.16), (2.17), or (2.18), make the substitutions

Isotropic	Composite
EI	$E_r I_y^*$
GA_s	D_V (shear stiffness)
ρ	$\sum \rho_i$
r_y^2	$\dfrac{\Sigma A_i^* r_{yi}^2}{A^*}$

Unsymmetrical Bars The equations of motion of bars whose cross sections are not symmetrical about the z axis differ somewhat from those given above. For the case of $P = \rho = 1/GA_s = 0$, Eqs. (2.16) apply for a composite unsymmetrical beam if Eqs. (2.16*a*) and (2.16*b*) are replaced by

$$\frac{\partial v}{\partial x} = -\theta + \frac{1}{D_V}(\alpha_{zz}V_z - \alpha_{yz}V_y) \qquad (2.23)$$

$$\frac{\partial \theta}{\partial x} = \frac{(M_y + M_{Ty})I_z}{E_r(I_y I_z - I_{yz}^2)} + \frac{(M_z + M_{Tz})I_{yz}}{E_r(I_y I_z - I_{yz}^2)} \qquad (2.24)$$

where $M_y = M$, $M_{Ty} = M_T$, M_z is the internal moment about the z axis, and

$$M_{Ty} = \int_A E\alpha \, \Delta T \, z \, dA \qquad (2.25)$$

Also, $V_z = V$, V_y is the shear force in the y direction, and α_{zz} and α_{yz} are shear deformation coefficients.

2.10 Transfer Matrices for Uniform Segments

The transfer matrices for several commonly occurring beam segments of length ℓ are provided in Tables 2-8 to 2-12, These are called *field matrices* since they apply for segments of finite length. In these tables, the loads shown in Fig. 2-9 are taken into account.

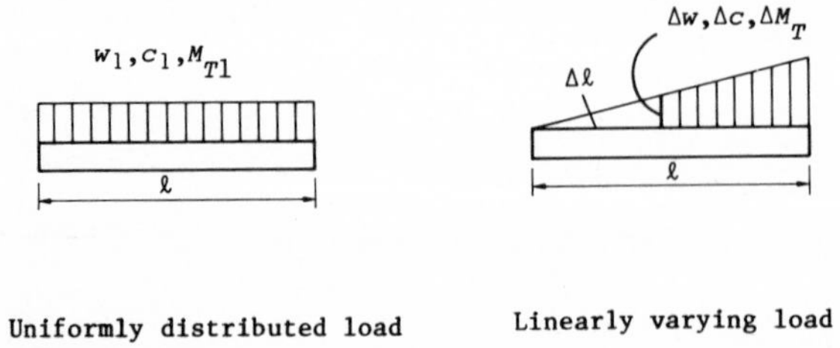

Fig. 2-9

TABLE 2-8 Massless Euler-Bernoulli Beam with Effect of Shear Deformation

Definition:

The subscript s on A_s has been dropped, i.e. $A = A_s$.

$$
\mathbf{U}_i = \begin{bmatrix}
1 & -\ell & -\dfrac{\ell^2}{2EI} & -\dfrac{\ell^3}{6EI} + \dfrac{\ell}{GA} & F_V \\
0 & 1 & \dfrac{\ell}{EI} & \dfrac{\ell^2}{2EI} & F_\theta \\
0 & 0 & 1 & \ell & F_M \\
0 & 0 & 0 & 1 & F_V \\
0 & 0 & 0 & 0 & 1
\end{bmatrix}
$$

$$F_V = w_1\left(\frac{\ell^4}{24EI} - \frac{\ell^2}{2GA}\right) + \frac{\Delta w}{\Delta \ell}\left(\frac{\ell^5}{120EI} - \frac{\ell^3}{6GA}\right) + c_1\frac{\ell^3}{6EI} + \frac{\Delta c}{\Delta \ell}\frac{\ell^4}{24EI}$$

$$F_\theta = -w_1\frac{\ell^3}{6EI} - \frac{\Delta w}{\Delta \ell}\frac{\ell^4}{24EI} - c_1\frac{\ell^2}{2EI} - \frac{\Delta c}{\Delta \ell}\frac{\ell^3}{6EI}$$

$$F_M = -w_1\frac{\ell^2}{2} - \frac{\Delta w}{\Delta \ell}\frac{\ell^3}{6} - c_1\ell - \frac{\Delta c}{\Delta \ell}\frac{\ell^2}{2}$$

$$F_V = -w_1\ell - \frac{\Delta w}{\Delta \ell}\frac{\ell^2}{2}$$

2.11 General Transfer Matrix for Uniform Segments

The general solution of the equations of motion of Eqs. (2.16) for a uniform beam segment is given in Table 2-13. This transfer matrix reduces for special cases to the matrices of Tables 2-8 to 2-12. To use Table 2-13, the engineer determines the relative magnitudes of various beam parameters and then selects the proper table entry. Column 5 of the definitions of e_i along with case 6 or 7 of the final page of definitions can be used for all beams regardless of the magnitude of the beam parameters, but this practice will frequently lead to elements involving complex numbers in the transfer matrix. Accordingly, computations with column 5 of the e_i definitions of Table 2-13 will require computers with complex arithmetic capabilities.

EXAMPLE 2.5 Use of Column 5 of Table 2-13 We wish to demonstrate that column 5 of the e_i definitions of Table 2-13 can be used for any beam. Consider the term e_1. From column 5 of the second page and case 6 of the third page,

$$e_1 = \frac{1}{b^2 - a^2}(b^2 \cos bx - a^2 \cos ax) \tag{1}$$

$$a^2 = \tfrac{1}{2}(\zeta - \eta) - \sqrt{\tfrac{1}{4}(\zeta + \eta)^2 - \lambda} \qquad b^2 = \tfrac{1}{2}(\zeta - \eta) + \sqrt{\tfrac{1}{4}(\zeta + \eta)^2 - \lambda} \tag{2}$$

TABLE 2-9 Euler-Bernoulli Beam with Axial Load and Effect of Shear Deformation

Definitions:

The subscript s on A_s has been dropped, i.e. $A = A_s$.

(a) For Compressive Axial Force P

$$\alpha = \sqrt{P/EI}\ , \quad d_1 = -\alpha^2$$

$$s = \frac{\sin \alpha \ell}{\alpha}\ , \quad c = \cos \alpha \ell$$

(b) For Tensile Axial Force P

$$\alpha = \sqrt{-P/EI}\ , \quad d_1 = \alpha^2$$

$$s = \frac{\sinh \alpha \ell}{\alpha}\ , \quad c = \cosh \alpha \ell$$

$$\mathbf{U}_i = \begin{bmatrix} 1 & -s & \dfrac{1-c}{d_1 EI} & \dfrac{\ell - s}{d_1 EI} + \dfrac{\ell}{GA} & F_V \\ 0 & c & \dfrac{s}{EI} & -\dfrac{1-c}{d_1 EI} & F_\theta \\ 0 & EIsd_1 & c & s & F_M \\ 0 & 0 & 0 & 1 & F_V \\ 0 & 0 & 0 & 0 & 1 \end{bmatrix}$$

$$F_V = w_1 \left[\frac{1}{d_1 EI} \left(\frac{c-1}{d_1} - \frac{\ell^2}{2} \right) - \frac{\ell^2}{2GA} \right] + \frac{\Delta w}{\Delta \ell} \left[\frac{1}{d_1 EI} \left(\frac{s - \ell}{d_1} - \frac{\ell^3}{6} \right) - \frac{\ell^3}{6GA} \right] - c_1 \frac{\ell - s}{d_1 EI} - \frac{\Delta c}{\Delta \ell} \frac{1}{d_1 EI} \left(\frac{\ell^2}{2} + \frac{1-c}{d_1} \right)$$

$$- M_{T1} \frac{c-1}{d_1 EI} + \frac{\Delta M_T}{\Delta \ell} \frac{1}{EI} \frac{\ell - s}{d_1}$$

$$F_\theta = w_1 \frac{\ell - s}{d_1 EI} + \frac{\Delta w}{\Delta \ell} \frac{1}{d_1 EI} \left(\frac{\ell^2}{2} + \frac{1-c}{d_1} \right) - c_1 \frac{c-1}{d_1 EI} - \frac{\Delta c}{\Delta \ell} \frac{s - \ell}{d_1 EI} + M_{T1} \frac{s}{EI} + \frac{\Delta M_T}{\Delta \ell} \frac{c-1}{d_1 EI}$$

$$F_M = w_1 \frac{1-c}{d_1^2} + \frac{\Delta w}{\Delta \ell} \frac{\ell - s}{d_1} - c_1 s - \frac{\Delta c}{\Delta \ell} \frac{c-1}{d_1} + M_{T1} (c-1) + \frac{\Delta M_T}{\Delta \ell} (s - \ell)$$

$$F_V = -w_1 \ell - \frac{\Delta w}{\Delta \ell} \frac{\ell^2}{2}$$

TABLE 2-10 Massless Euler-Bernoulli Beam on Elastic Foundation

Definitions:

$\lambda = k/EI, \quad \beta^4 = \lambda/4$

$e_0 = -\beta(\cosh \beta\ell \ \sin \beta\ell - \sinh \beta\ell \ \cos \beta\ell)$

$e_1 = \cosh \beta\ell \ \cos \beta\ell$

$e_2 = \frac{1}{2\beta} (\cosh \beta\ell \ \sin \beta\ell + \sinh \beta\ell \ \cos \beta\ell)$

$e_3 = \frac{1}{2\beta^2} \sinh \beta\ell \ \sin \beta\ell$

$e_4 = \frac{1}{4\beta^3} (\cosh \beta\ell \ \sin \beta\ell - \sinh \beta\ell \ \cos \beta\ell)$

$$\mathbf{U}_i = \begin{bmatrix} e_1 & -e_2 & -e_3/EI & -e_4/EI & F_v \\ \lambda e_4 & e_1 & e_2/EI & e_3/EI & F_\theta \\ \lambda EI e_3 & EI e_0 & e_1 & e_2 & F_M \\ \lambda EI e_2 & -\lambda EI e_3 & -\lambda e_4 & e_1 & F_V \\ 0 & 0 & 0 & 0 & 1 \end{bmatrix}$$

$$F_v = w_1 \frac{1 - e_1}{k} + \frac{\Delta w}{\Delta \ell} \frac{\ell - e_2}{k} + c_1 \frac{e_4}{EI} + \frac{\Delta c}{\Delta \ell} \frac{1 - e_1}{k} - \left(M_{T1} e_3 + \frac{\Delta M_T}{\Delta \ell} e_4\right)/EI$$

$$F_\theta = -w_1 \frac{e_4}{EI} - \frac{\Delta w}{\Delta \ell} \frac{1 - e_1}{k} - c_1 \frac{e_3}{EI} - \frac{\Delta c}{\Delta \ell} \frac{e_4}{EI} + \left(M_{T1} e_2 + \frac{\Delta M_T}{\Delta \ell} e_3\right)/EI$$

$$F_M = -w_1 e_3 - \frac{\Delta w}{\Delta \ell} e_4 - c_1 e_2 - \frac{\Delta c}{\Delta \ell} e_3 + M_{T1} (e_1 - 1) + \frac{\Delta M_T}{\Delta \ell} (e_2 - \ell)$$

$$F_V = -w_1 e_2 - \frac{\Delta w}{\Delta \ell} e_3 + c_1(1 - e_1) + \frac{\Delta c}{\Delta \ell} (\ell - e_2) - M_{T1} \lambda e_4 + \frac{\Delta M_T}{\Delta \ell} (e_1 - 1)$$

TABLE 2-11 Euler-Bernoulli Beam with Mass and Elastic Foundation

Definitions:

(a) Without Elastic Foundation

$$\beta^4 = -\lambda = \rho\omega^2/EI$$

(b) With Elastic Foundation, $k<\rho\omega^2$

$$\beta^4 = -\lambda = (\rho\omega^2 - k)/EI$$

If $k>\rho\omega^2$, use Table 2-10 with $\beta^4 = \frac{\lambda}{4} = \frac{k - \rho\omega^2}{4EI}$

$$e_0 = \frac{\beta}{2}\,(\sinh \beta\ell - \sin \beta\ell)$$

$$e_1 = \frac{1}{2}\,(\cosh \beta\ell + \cos \beta\ell)$$

$$e_2 = \frac{1}{2\beta}\,(\sinh \beta\ell + \sin \beta\ell)$$

$$e_3 = \frac{1}{2\beta^2}\,(\cosh \beta\ell - \cos \beta\ell)$$

$$e_4 = \frac{1}{2\beta^3}\,(\sinh \beta\ell - \sin \beta\ell)$$

$$\mathbf{U}_i = \begin{bmatrix} e_1 & -e_2 & -e_3/EI & -e_4/EI & F_V \\ \lambda e_4 & e_1 & e_2/EI & e_3/EI & F_\theta \\ \lambda EI e_3 & EI e_0 & e_1 & e_2 & F_M \\ \lambda EI e_2 & -\lambda EI e_3 & -\lambda e_4 & e_1 & F_V \\ 0 & 0 & 0 & 0 & 1 \end{bmatrix}$$

$$F_V = w_1 \frac{1 - e_1}{\lambda EI} - \frac{\Delta w}{\Delta\ell}\frac{1}{EI}\frac{\ell - e_2}{\lambda} + c_1 \frac{e_4}{EI} + \frac{\Delta c}{\Delta\ell}\frac{1 - e_1}{\lambda EI} - \left(M_{T1}\, e_3 + \frac{\Delta M_T}{\Delta\ell}\, e_4\right)/EI$$

$$F_\theta = -w_1 \frac{e_4}{EI} + \frac{\Delta w}{\Delta\ell}\frac{1 - e_1}{\lambda EI} - c_1 \frac{e_3}{EI} - \frac{\Delta c}{\Delta\ell}\frac{e_4}{EI} + \left(M_{T1}\, e_2 + \frac{\Delta M_T}{\Delta\ell}\, e_3\right)/EI$$

$$F_M = -w_1 e_3 - \frac{\Delta w}{\Delta\ell}\, e_4 - c_1 e_2 - \frac{\Delta c}{\Delta\ell}\, e_3 + M_{T1}\,(e_1 - 1) + \frac{\Delta M_T}{\Delta\ell}\,(e_2 - \ell)$$

$$F_V = -w_1 e_2 - \frac{\Delta w}{\Delta\ell}\, e_3 - c_1(e_1 - 1) - \frac{\Delta c}{\Delta\ell}\,(e_2 - \ell) - M_{T1}\,\lambda e_4 + \frac{\Delta M_T}{\Delta\ell}\,(e_1 - 1)$$

TABLE 2-12 Rigid Beam

Definitions:

(a) Massless with No Foundations

$$\rho = k = k^* = 0$$

(b) Vibrating Beam with No Foundations

$$k = k^* = 0, \quad d = -1$$

(c) Vibrating Beam with Foundations and Rotary Inertia

$$d = -1$$

(d) Rotating Shaft

$$d = \begin{cases} 1 & \text{whirl in same angular direction as rotation} \\ -3 & \text{whirl in opposite direction as rotation} \end{cases}$$

$$d_1 = \frac{\ell^3(\rho\omega^2 - k)}{6} + \ell(k^* + dr^2\rho\omega^2)$$

$$\mathbf{U}_i = \left[\begin{array}{cc|c|c|c|c} 1 & -\ell & 0 & 0 & F_V \\ \hline 0 & 1 & 0 & 0 & F_\theta \\ \hline \frac{\ell^2(k - \rho\omega^2)}{2} & d_1 & 1 & \ell & F_M \\ \hline \ell(k - \rho\omega^2) & \frac{\ell^2(\rho\omega^2 - k)}{2} & 0 & 1 & F_V \\ \hline 0 & 0 & 0 & 0 & 1 \end{array}\right]$$

$$F_V = 0$$

$$F_\theta = 0$$

$$F_M = -w_1 \frac{\ell^2}{2} - \frac{\Delta w}{\Delta \ell}\frac{\ell^3}{3!} - c_1\ell - \frac{\Delta c}{\Delta \ell}\frac{\ell^2}{2}$$

$$F_V = -w_1\ell - \frac{\Delta w}{\Delta \ell}\frac{\ell^2}{2}$$

TABLE 2-13 General Beam Solution

Definitions:

$\lambda = (k - \rho\omega^2)/EI$

$\eta = (k - \rho\omega^2)/GA_s$

$\zeta = (P - d\rho r^2\omega^2 - k^*)/EI$

$d = \begin{cases} -1 & \text{vibrating beam} \\ 1 & \text{rotating shaft with whirl in same angular direction as rotation} \\ -3 & \text{rotating shaft with whirl in opposite angular direction as rotation} \end{cases}$

Note: The column 5 definitions on the following page along with case 6 or 7 of the final page of this table apply for any magnitude of λ, ζ, η. However, usually the transfer matrix elements will then be complex quantities, and computations will require computers with complex arithmetic capabilities.

$$\mathbf{U}_i = \begin{bmatrix} e_1 + \zeta e_3 & -e_2 & -e_3/EI & -e_4/EI + (e_2 + \zeta e_4)/GA_s & F_V \\ \lambda e_4 & e_1 - \eta e_3 & (e_2 - \eta e_4)/EI & e_3/EI & F_\theta \\ \lambda EI e_3 & EI(e_0 - \eta e_2) & e_1 - \eta e_3 & e_2 & F_M \\ \lambda EI(e_2 + \zeta e_4) & -\lambda EI e_3 & -\lambda e_4 & e_1 + \zeta e_3 & F_V \\ 0 & 0 & 0 & 0 & 1 \end{bmatrix}$$

$$F_v = \left(w_1 e_5 + \frac{\Delta w}{\Delta \ell} e_6 + c_1 e_4 + \frac{\Delta c}{\Delta \ell} e_5\right)/EI - w_1(e_3 + \zeta e_5)/GA_s - \frac{\Delta w}{\Delta \ell}(e_4 + \zeta e_6)/GA_s - \left(M_{T1} e_3 + \frac{\Delta M_T}{\Delta \ell} e_4\right)/EI$$

$$F_\theta = -\left[w_1 e_4 + \frac{\Delta w}{\Delta \ell} e_5 + c_1(e_3 - \eta e_5) + \frac{\Delta c}{\Delta \ell}(e_4 - \eta e_6)\right]/EI + \left[M_{T1}(e_2 - \eta e_4) + \frac{\Delta M_T}{\Delta \ell}(e_3 - \eta e_5)\right]/EI$$

$$F_M = -w_1 e_3 - \frac{\Delta w}{\Delta \ell} e_4 - c_1(e_2 - \eta e_4) - \frac{\Delta c}{\Delta \ell}(e_3 - \eta e_5) + M_{T1}(e_1 - 1 - \eta e_3) + \frac{\Delta M_T}{\Delta \ell}(e_2 - \ell - \eta e_4)$$

$$F_V = -w_1(e_2 + \zeta e_4) - \frac{\Delta w}{\Delta \ell}(e_3 + \zeta e_5) + \lambda\left(c_1 e_5 + \frac{\Delta c}{\Delta \ell} e_6\right) - M_{T1}\lambda e_4 - \frac{\Delta M_T}{\Delta \ell}\lambda e_5$$

To use this general transfer matrix, follow the steps:

1. Calculate the three parameters λ, ζ, η. If shear deformation is not to be considered, set $1/GA_s = 0$.
2. Compare the magnitude of these parameters and look up the appropriate e_i functions on the following page.
3. Substitute these e_i expressions in the general transfer matrix above.

	1. $\lambda < 0$	$\lambda = 0,\ \lambda - \zeta\eta = 0$		$\lambda > 0,\ \lambda - \zeta\eta > 0$		
		2. $\zeta = \eta = 0$	3. $\eta = 0,\ \zeta \neq 0$	4. $\lambda - \zeta\eta = \frac{1}{4}(\zeta - \eta)^2$	5. $\lambda - \zeta\eta < \frac{1}{4}(\zeta - \eta)^2,\ \zeta - \eta \neq 0$	6. $\lambda - \zeta\eta > \frac{1}{4}(\zeta - \eta)^2,$
e_0	$\frac{1}{g}(a^3C - b^3D)$	0	$-\zeta B$	$-\frac{\zeta - \eta}{4}(3C + A\ell)$	$-\frac{1}{g}(b^3D - a^3C)$	$-(\lambda - \zeta\eta)e_4 - (\zeta - \eta)e_2$
e_1	$\frac{1}{g}(a^2A + b^2B)$	1	A	$\frac{1}{2}(2A - B\ell)$	$\frac{p}{g}(b^2B - a^2A)$	$AB - \frac{b^2 - a^2}{2ab}CD$
e_2	$\frac{1}{g}(aC + bD)$	ℓ	B	$\frac{1}{2}(C + A\ell)$	$\frac{p}{g}(bD - aC)$	$\frac{1}{2ab}(aAD + bBC)$
e_3	$\frac{1}{g}(A - B)$	$\frac{\ell^2}{2}$	$\frac{1}{\zeta}(1 - A)$	$\frac{C\ell}{2}$	$\frac{1}{g}(A - B)$	$\frac{1}{2ab}CD$
e_4	$\frac{1}{g}\left(\frac{C}{a} - \frac{D}{b}\right)$	$\frac{\ell^3}{6}$	$\frac{1}{\zeta}(\ell - B)$	$\frac{2}{(\zeta - \eta)}(C - A\ell)$	$\frac{1}{g}\left(\frac{C}{a} - \frac{D}{b}\right)$	$\frac{1}{2(a^2 + b^2)}\left(\frac{AD}{b} - \frac{BC}{a}\right)$
e_5	$\frac{1}{g}\left(\frac{A}{a^2} + \frac{B}{b^2}\right) - \frac{1}{a^2b^2}$	$\frac{\ell^4}{24}$	$\frac{1}{\zeta}\left(\frac{\ell^2}{2} - e_3\right)$	$\frac{2}{(\zeta - \eta)^2}(-2A - B\ell + 2)$	$\frac{p}{g}\left(\frac{B}{b^2} - \frac{A}{a^2}\right) + \frac{1}{a^2b^2}$	$\frac{1 - e_1}{\lambda - \zeta\eta} - \frac{\zeta - \eta}{\lambda - \zeta\eta}e_3$
e_6	$\frac{1}{g}\left(\frac{C}{a^3} + \frac{D}{b^3}\right) - \frac{\ell}{a^2b^2}$	$\frac{\ell^5}{120}$	$\frac{1}{\zeta}\left(\frac{\ell^3}{6} - e_4\right)$	$\frac{2}{(\zeta - \eta)^2}(-3C + A\ell + 2\ell)$	$\frac{p}{g}\left(\frac{D}{b^3} - \frac{C}{a^3}\right) + \frac{\ell}{a^2b^2}$	$\frac{\ell - e_2}{\lambda - \zeta\eta} - \frac{\zeta - \eta}{\lambda - \zeta\eta}e_4$

TABLE 2-13 General Beam Solution (*Continued*)

$\lambda < 0$	$\lambda = 0,\ \lambda - \zeta\eta = 0$	$\lambda > 0,\ \lambda - \zeta\eta > 0$		
		$\lambda - \zeta\eta = \frac{1}{4}(\zeta - \eta)^2$	$\lambda - \zeta\eta < \frac{1}{4}(\zeta - \eta)^2,\ \zeta - \eta \neq 0$	$\lambda - \zeta\eta > \frac{1}{4}(\zeta - \eta)^2$
1. $A = \cosh a\ell,\quad B = \cos b\ell$ $C = \sinh a\ell,\quad D = \sin b\ell$ $g = a^2 + b^2$ $a^2 = \sqrt{\beta^4 + \frac{1}{4}(\zeta + \eta)^2} - \frac{1}{2}(\zeta - \eta)$ $b^2 = \sqrt{\beta^4 + \frac{1}{4}(\zeta + \eta)^2} + \frac{1}{2}(\zeta + \eta)$ $\beta^4 = -\lambda$	**2.** $\zeta > 0:\ \alpha^2 = \zeta$ $A = \cos \alpha\ell$ $B = (\sin \alpha\ell)/\alpha$ **3.** $\zeta < 0:\ \alpha^2 = -\zeta$ $A = \cosh \alpha\ell$ $B = (\sinh \alpha\ell)/\alpha$	**4.** $\zeta - \eta > 0:\ \beta^2 = \frac{1}{2}(\zeta - \eta)$ $A = \cos \beta\ell,\quad B = \beta \sin \beta\ell$ $C = (\sin \beta\ell)/\beta$ **5.** $\zeta - \eta < 0:\ \beta^2 = -\frac{1}{2}(\zeta - \eta)$ $A = \cosh \beta\ell,\quad B = -\beta \sinh \beta\ell$ $C = (\sinh \beta\ell)/\beta$	**6.** $\zeta - \eta > 0:\ g = b^2 - a^2,\quad p = 1$ $A = \cos a\ell,\quad B = \cos b\ell$ $C = \sin a\ell,\quad D = \sin b\ell$ $a^2 = \frac{1}{2}(\zeta - \eta) - \sqrt{\frac{1}{4}(\zeta + \eta)^2 - \lambda}$ $b^2 = \frac{1}{2}(\zeta - \eta) + \sqrt{\frac{1}{4}(\zeta + \eta)^2 - \lambda}$ **7.** $\zeta - \eta < 0:\ g = a^2 - b^2,\quad p = -1$ $A = \cosh a\ell,\quad B = \cosh b\ell$ $C = \sinh a\ell,\quad D = \sinh b\ell$ $a^2 = -\frac{1}{2}(\zeta - \eta) + \sqrt{\frac{1}{4}(\zeta + \eta)^2 - \lambda}$ $b^2 = -\frac{1}{2}(\zeta - \eta) - \sqrt{\frac{1}{4}(\zeta + \eta)^2 - \lambda}$	**8.** $A = \cosh a\ell,\quad B = \cos b\ell$ $C = \sinh a\ell,\quad D = \sin b\ell$ $a^2 = \frac{1}{2}\sqrt{\lambda - \zeta\eta} - \frac{1}{4}(\zeta - \eta)$ $b^2 = \frac{1}{2}\sqrt{\lambda - \zeta\eta} + \frac{1}{4}(\zeta - \eta)$

Consider the cases:

1. Simple beam with axial force. Here $\lambda = 0 = \eta$, $\zeta = P/EI$. Then $a = 0$, $b^2 = \zeta$ and $e_1 = \cos bx$. This is the correct value for e_1 as given in column 3 of the e_i definitions of Table 2-13.

2. Simple beam. In this case $\zeta = \eta = \lambda = 0$. Thus $a = b = 0$ and (1) is indefinite. However, the value of e_1 can be found by taking the limit as $a \to b$ and $a \to 0$, $b \to 0$. To take the first limit, set $b = a + \Delta$ and

$$\lim_{\Delta \to 0} \frac{(a+\Delta)^2 \cos(a+\Delta)x - a^2 \cos ax}{(a+\Delta)^2 - a^2} = \cos ax - \frac{ax}{2} \sin ax$$

Finally,

$$e_1 = \lim_{a \to 0} \left(\cos ax - \frac{ax}{2} \sin ax \right) = 1$$

which is the correct value given in column 2 of the e_i definitions of Table 2-13.

The generality of column 5 of the e_i definitions of Table 2-13 has thus been demonstrated. However, a^2 and b^2 will in general be quantities containing complex numbers and must be treated accordingly. Also, as shown above, it is necessary to go to the limit if column 5 is to accommodate certain cases. This is a particularly delicate operation on the computer. In general, it is much simpler to determine the relative magnitudes of λ, ζ, η and enter the appropriate columns in Table 2-13.

2.12 Point Matrices

The transfer matrices that take into account concentrated (point) occurrences are listed as *point matrices* in Tables 2-14 and 2-15.

2.13 Transfer Matrices for Beams of Variable Cross Section

The transfer matrices for beam segments of length ℓ with variable cross sections are listed in Tables 2-16 and 2-17. The loading notation is shown in Fig. 2-9.

2.14 Transfer Matrices for Semi-Infinite and Infinite Beams

The transfer matrices for semi-infinite and infinite beams on an elastic foundation are given in Table 2-18. $\overline{U}_i^L$ is a point matrix used if the beam is infinite to the left, while $\overline{U}_i^R$ is employed for a beam infinite to the right. See Examples 2.13, 2.14, and 2.15 for the details on using these matrices. The theory underlying this approach is given in Reference 2.5.

TABLE 2-14 Point Matrix at $x = a_i$

Definitions:

(a) Concentrated Applied Transverse Force

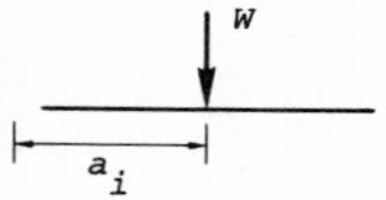

(b) Concentrated Applied Moment

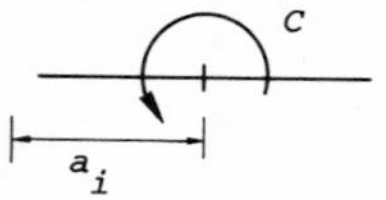

(c) Abrupt Change in Slope

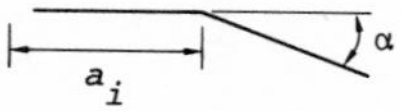

(d) Abrupt Change in Beam Axis

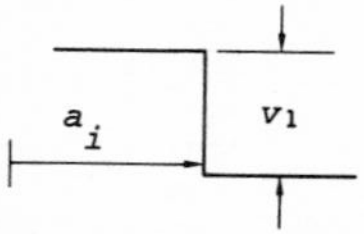

(e) Linear Hinge (shear release with spring)

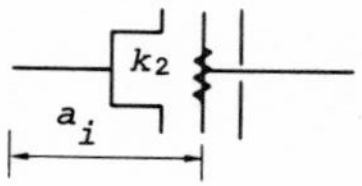

(f) Rotary Hinge (moment release with spring)

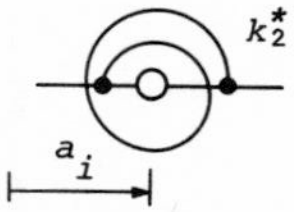

(g) Concentrated Mass

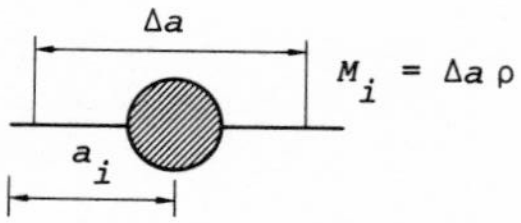

If rotary inertia is to be included, $I_{pi} = \Delta a \, \rho r^2$
d is defined in the notation.

$$
\overline{\mathbf{U}}_i = \begin{bmatrix} 1 & 0 & 0 & \frac{1}{k_2} & v_1 \\ 0 & 1 & \frac{1}{k_2^*} & 0 & -\alpha \\ 0 & P_{Li} + k_1^* + dI_{pi}\omega^2 & 1 & 0 & -C \\ k_1 - M_i\omega^2 & 0 & 0 & 1 & -W \\ 0 & 0 & 0 & 0 & 1 \end{bmatrix}
$$

(h) Lumped Axial Force

$P_{Li} = \Delta a \, P$

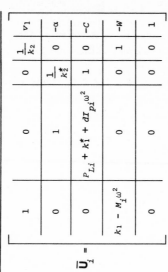

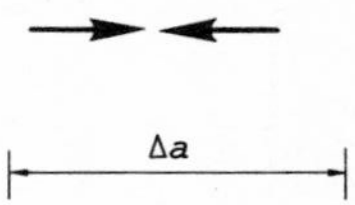

Substitute $-P$ for P if force is tensile.

For a distributed axial force p (force/length), P is found using the condition of equilibrium $\Sigma F_x = 0$.

(i) Branches and Attached Masses

	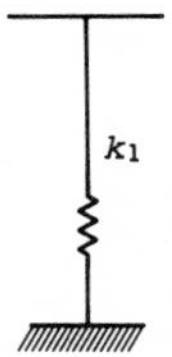	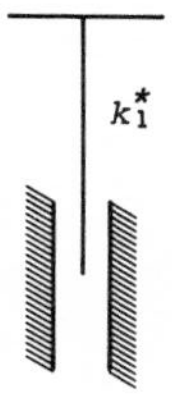	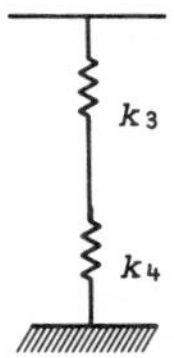	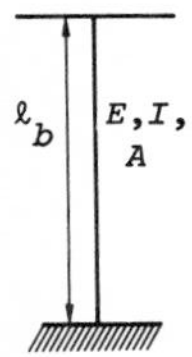	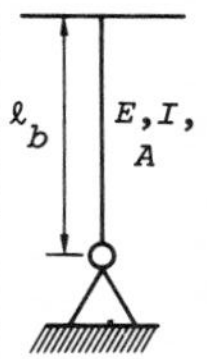	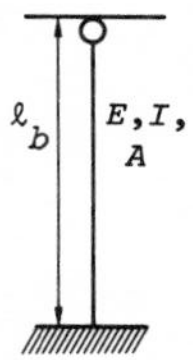	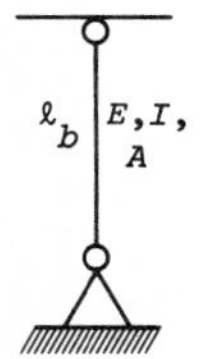	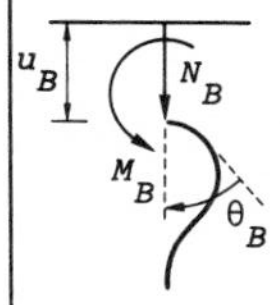 Arbitrary Branch B	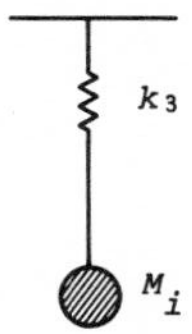	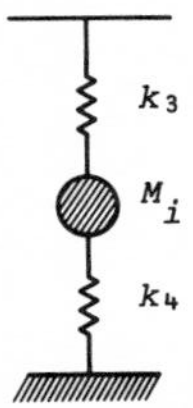
k_1	k_1	0	$\frac{k_3 k_4}{k_3 + k_4}$	$\frac{EA}{\ell_b}$	$\frac{EA}{\ell_b}$	$\frac{EA}{\ell_b}$	$\frac{EA}{\ell_b}$	$\frac{u_B(a_i)}{N_B(a_i)}$	$-\frac{k_3 M_i \omega^2}{k_3 - M_i \omega^2}$	$\frac{k_3\left(k_4 - M_i \omega^2\right)}{k_3 + k_4 - M_i \omega^2}$
k_1^*	0	k_1^*	0	$\frac{4EI}{\ell_b}$	$\frac{3EI}{\ell_b}$	0	0	$\frac{\theta_B(a_i)}{M_B(a_i)}$	0	0

Units: k_1, k_2, k_3, k_4 are force/length and k_1^*, k_2^* are force-length/length.

TABLE 2-15 Point Matrix for In-span Indeterminates at $x = a_i$

Definitions:

(a) Rigid Support

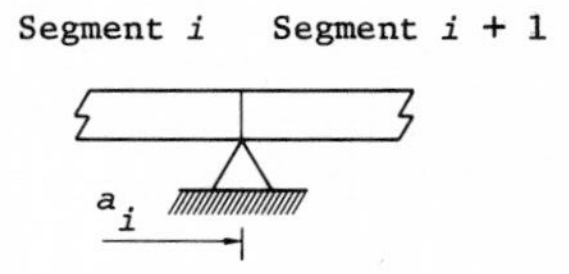

Nonzero k_{ij}:

$k_{Vj} = -U_{s_m j}/U_{s_m V}$, $j = v, \theta, M, V, F$

$$\bar{\mathbf{U}}_i = \begin{bmatrix} 1 + k_{vv} & k_{v\theta} & k_{vM} & k_{vV} & k_{vF} \\ k_{\theta v} & 1 + k_{\theta\theta} & k_{\theta M} & k_{\theta V} & k_{\theta F} \\ k_{Mv} & k_{M\theta} & 1 + k_{MM} & k_{MV} & k_{MF} \\ k_{Vv} & k_{V\theta} & k_{VM} & 1 + k_{VV} & k_{VF} \\ 0 & 0 & 0 & 0 & 1 \end{bmatrix}$$

(b) Moment Release

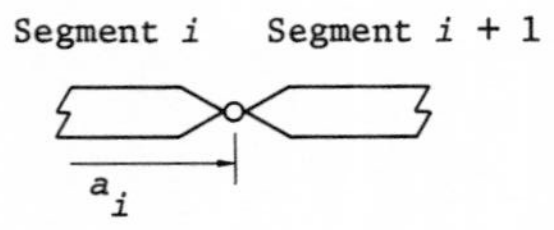

Nonzero k_{ij}:

$k_{\theta j} = -U_{s_m j}/U_{s_m \theta}$, $j = v, \theta, M, V, F$

(c) Angle Guide

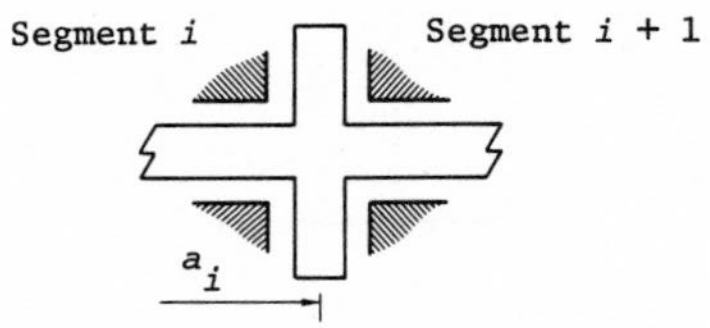

Nonzero k_{ij}:

$k_{Mj} = -U_{s_m j}/U_{s_m M}$, $j = v, \theta, M, V, F$

(d) Shear Release

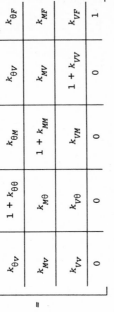

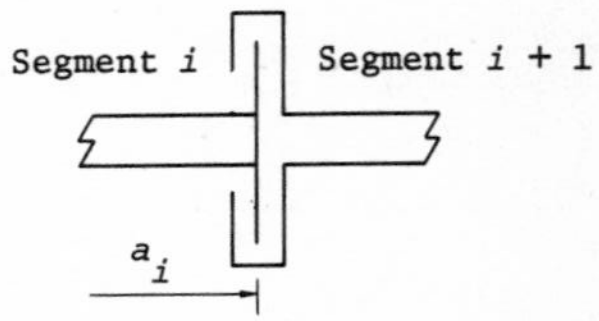

Nonzero k_{ij}:

$k_{Vj} = -U_{s_m j}/U_{s_m V}$, $j = v, \theta, M, V, F$

U_{iF} is defined to be F_i. s_m is v, θ, M, or V whichever is zero at the next in-span indeterminate to the right of $x = a_i$. If no in-span indeterminate occurs to the right of a_i, then s_m is one of the state variables (v, θ, M, or V) that is zero at the right end of the beam. All U_{ij} are the transfer matrix elements for the segment between $x = a_i$ and the next in-span indeterminate or the right end, as appropriate.

TABLE 2-16 Massless Euler-Bernoulli Beam with Variable Moment of Inertia

$$I = I_i(c_1 - c_2\ell)^n$$

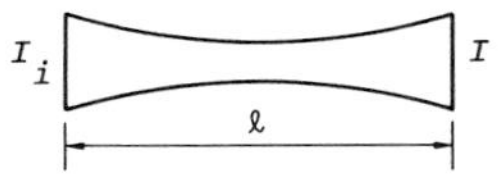

$$d_1 = c_1 - c_2\ell$$

$$d_2 = \ln c_1 - \ln d_1$$

$$\mathbf{U}_i = \begin{bmatrix} 1 & -\ell & U_{vM} & U_{vV} & F_V \\ 0 & 1 & U_{\theta M} & U_{\theta V} & F_\theta \\ 0 & 0 & 1 & \ell & F_M \\ 0 & 0 & 0 & 1 & F_V \\ 0 & 0 & 0 & 0 & 1 \end{bmatrix}$$

	$n = 1$	$n = 2$	$n = 3$	$n > 3$
U_{vM}	$\dfrac{d_1 d_2 - c_2\ell}{EI_i c_2^2}$	$\dfrac{c_2\ell - c_1 d_2}{EI_i c_1 c_2^2}$	$-\dfrac{\ell^2}{2EI_i c_1^2 d_1}$	$\dfrac{d_1^{n-2}[d_1 + (n-1)c_2\ell] - c_1^{n-1}}{(n-1)(n-2)EI_i c_1^{n-1} c_2^2 d_1^{n-2}}$
U_{vV}	$\dfrac{c_2^2\ell^2 + 2c_1 d_1 - 2c_1 c_2\ell}{2EI_i c_2^3}$	$\dfrac{2c_2\ell - (c_1 + d_1)d_2}{EI_i c_2^3}$	$\dfrac{c_2^2\ell^2 - 2c_1 c_2\ell + 2c_1 d_1 d_2}{2EI_i c_1 c_2^3 d_1}$	$\dfrac{c_1^{n-2}[2c_1 - (n-1)c_2\ell] - d_1^{n-2}[2c_1 + (n-3)c_2\ell]}{(n-1)(n-2)(n-3)EI_i c_1^{n-2} c_2^3 d_1^{n-2}}$
$U_{\theta M}$	$\dfrac{d_2}{EI_i c_2}$	$\dfrac{\ell}{EI_i c_1 d_1}$	$\dfrac{2c_1\ell - c_2\ell^2}{2EI_i c_1^2 d_1^2}$	$\dfrac{c_1^{n-1} - d_1^{n-1}}{(n-1)EI_i c_1^{n-1} c_2 d_1^{n-1}}$
$U_{\theta V}$	$\dfrac{c_1 d_2 - c_2\ell}{EI_i c_2^2}$	$\dfrac{c_2\ell - d_1 d_2}{EI_i c_2^2 d_1}$	$\dfrac{\ell^2}{2EI_i c_1 d_1^2}$	$\dfrac{d_1^{n-1} - (n-1)c_1^{n-2}d_1 + (n-2)c_1^{n-1}}{(n-1)(n-2)EI_i c_1^{n-2} c_2^2 d_1^{n-1}}$

$$F_M = -\frac{w_1\ell^2}{2} - \frac{\Delta w}{\Delta\ell}\frac{\ell^3}{3!}$$

$$F_V = -w_1\ell - \frac{\Delta w}{\Delta\ell}\frac{\ell^2}{2}$$

TABLE 2-16 Massless Euler-Bernoulli Beam with Variable Moment of Inertia (*Continued*)

	F_V	F_θ
$n = 1$	$\frac{w_1}{2EI_i c_2^4}\left(-\frac{c_2^3}{3}\ell^3 + \frac{3}{2}c_1c_2^2\ell^2 - c_1^2c_2\ell + c_1d_1^2d_2\right)$ $+ \frac{\Delta w/\Delta\ell}{2EI_i c_2^5}\left(-\frac{1}{12}c_2^4\ell^4 + \frac{11}{18}c_1c_2^3\ell^3 - \frac{5}{6}c_1^2c_2^2\ell^2\right.$ $\left.+ \frac{1}{3}c_1^3c_2\ell - \frac{1}{3}c_1d_1^3d_2\right)$	$\frac{w_1}{2EI_i c_2^3}(c_2^2\ell^2 + 2c_1d_1d_2 - 2c_1c_2\ell)$ $+ \frac{\Delta w/\Delta\ell}{2EI_i c_2^4}\left(\frac{1}{3}c_2^3\ell^3 - \frac{3}{2}c_1c_2^2\ell^2 + c_1^2c_2\ell - c_1d_1^2d_2\right)$
$n = 2$	$\frac{w_1}{EI_i c_2^4}\left[-\frac{5}{4}c_2^2\ell^2 + \frac{3}{2}c_1c_2\ell - (c_1 + \frac{1}{2}d_1)d_1d_2\right]$ $+ \frac{\Delta w/\Delta\ell}{EI_i c_2^5}\left[-\frac{17}{36}c_2^3\ell^3 + \frac{7}{6}c_1c_2^2\ell^2 - \frac{2}{3}c_1^2c_2\ell + \frac{1}{6}(3c_1 + d_1)d_1^2d_2\right]$	$\frac{w_1}{EI_i c_2^3}[2c_2\ell - (c_1 + d_1)d_2]$ $+ \frac{\Delta w/\Delta\ell}{2EI_i c_2^4}\left[\frac{5}{2}c_2^2\ell^2 - 3c_1c_2\ell + (2c_1 + d_1)d_1d_2\right]$
$n = 3$	$\frac{w_1}{2EI_i c_1c_2^4}\left[\frac{1}{2}c_2^2\ell^2 - 3c_1c_2\ell + (c_1^2 + 2c_1d_1)d_2\right]$ $+ \frac{\Delta w/\Delta\ell}{2EI_i c_1c_2^5}\left[\frac{1}{6}c_2^3\ell^3 - 2c_1c_2^2\ell^2 + 2c_1^2c_2\ell - c_1(c_1 + d_1)d_1d_2\right]$	$\frac{w_1}{2EI_i c_1c_2^3d_1}(c_2^2\ell^2 - 2c_1c_2\ell + 2c_1d_1d_2)$ $+ \frac{\Delta w/\Delta\ell}{2EI_i c_1c_2^4d_1}\left[\frac{1}{2}c_2^3\ell^3 + 3c_1^2c_2\ell - \frac{7}{2}c_1c_2^2\ell^2 - c_1(c_1 + 2d_1)d_1d_2\right]$

	F_V	F_θ
$n = 4$	$\frac{w_1}{6EI_i c_1^2 c_2^4}\left(\frac{1}{2} c_2^2 \ell^2 + 2c_1 c_2 \ell + \frac{c_2 \ell}{d_1} - 3c_1^2 d_2\right)$ $+ \frac{\Delta w/\Delta \ell}{6EI_i c_1^2 c_2^5}\left[\frac{1}{6} c_2^3 \ell^3 + c_1 c_2^2 \ell^2 - 4c_1^2 c_2 \ell + (c_1 + 3d_1)c_1^2 d_2\right]$	$\frac{w_1}{(n-1)(n-2)EI_i c_1^{n-2} c_2^3}\left[-c_2 \ell + \frac{(n-1)(c_1^{n-3} - d_1^{n-3})c_1}{(n-3)d_1^{n-3}} - \frac{(c_1^{n-2} - d_1^{n-2})c_1}{d_1^{n-2}}\right] + \frac{\Delta w/\Delta \ell}{6EI_i c_1^2 c_2^4}\left(-\frac{1}{2} c_2^2 \ell^2 - 3c_1 c_2 \ell - \frac{c_1 c_2^2 \ell^2}{d_1} + 3c_1^2 d_2\right)$
$n = 5$	$\frac{w_1}{(n-1)(n-2)(n-3)EI_i c_1^{n-2} c_2^4}\left[\frac{n-3}{2} c_2^2 \ell^2 + 2c_1 c_2 \ell + \frac{c_1^2(c_1^{n-3} - d_1^{n-3})}{d_1^{n-3}} - \frac{(n-1)(c_1^{n-4} - d_1^{n-4})c_1^2}{(n-4)d_1^{n-4}}\right]$ $+ \frac{\Delta w/\Delta \ell}{24EI_i c_1^3 c_2^5}\left[\frac{1}{3} c_2^3 \ell^3 + c_1 c_2^2 \ell^2 + \frac{(c_1 + 3d_1)c_1^2 c_2 \ell}{d_1} - 4c_1^3 d_2\right]$	For the w_1 term use the F_θ entry above in the $n = 4$ row. $\frac{\Delta w/\Delta \ell}{(n-1)(n-2)EI_i c_1^{n-2} c_2^4}\left[-\frac{1}{2} c_2^2 \ell^2 + \frac{(n-1)(c_1^{n-4} - d_1^{n-4})c_1^2}{(n-4)d_1^{n-4}} - \frac{(n-2)c_1 + (n-1)d_1}{n-3} \frac{(c_1^{n-3} - d_1^{n-3})c_1}{d_1^{n-3}} + \frac{(c_1^{n-2} - d_1^{n-2})c_1}{d_1^{n-3}}\right]$
$n \geq 6$	For the w_1 term use the F_V entry above in the $n = 5$ row. $\frac{\Delta w/\Delta \ell}{(n-1)(n-2)(n-3)EI_i c_1^{n-2} c_2^5}\left[\frac{n-3}{6} c_2^3 \ell^3 + c_1 c_2^2 \ell^2 - \frac{n-1}{n-5} \frac{(c_1^{n-5} - d_1^{n-5})c_1^3}{d_1^{n-5}} + \frac{(n-3)c_1 + (n-1)d_1}{n-4} \frac{(c_1^{n-4} - d_1^{n-4})c_1^2}{d_1^{n-4}} - \frac{(c_1^{n-3} - d_1^{n-3})c_1^2}{d_1^{n-3}}\right]$	For the w_1 term use the F_θ entry above in the $n = 4$ row. For the $\Delta w/\Delta \ell$ term use the F_θ entry above in the $n = 5$ row.

TABLE 2-17 Euler-Bernoulli Beam with Variable Cross Section and Compressive Axial Force

Definitions:

$$I = I_i(c_1 - c_2 \ell)^2$$

(a)

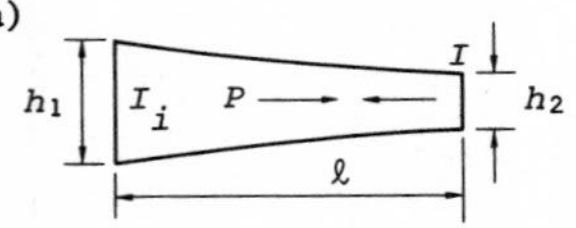

(b)

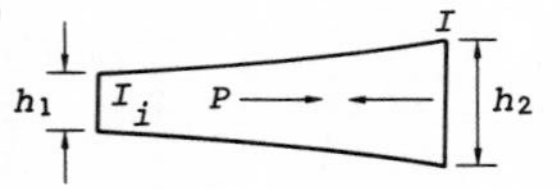

$c_1 = 1, \quad c_2 = \dfrac{h_2 - h_1}{h_1 \ell}$

$$\mathbf{U}_i = \left[\begin{array}{c|c|c|c|c} 1 & \dfrac{\sqrt{c_1}\, d_2 d_8}{\beta c_2} & d_{11} - \dfrac{d_2 d_8 d_{10}}{\sqrt{c_1}\, P} & -\dfrac{\sqrt{c_1}\, d_2 d_8}{\beta c_2 P} - \dfrac{\ell}{P} & F_V \\ \hline 0 & \dfrac{\sqrt{c_1}}{d_2}\left(d_9 - \dfrac{1}{2}\dfrac{d_8}{\beta}\right) & \dfrac{c_2}{\sqrt{c_1}\, d_2 P}\left(\dfrac{d_{12}}{2} + \dfrac{d_{13}}{d_7} + \dfrac{d_8 d_{10}}{2}\right) & \dfrac{1}{P}\left(1 - \dfrac{\sqrt{c_1}\, d_9}{d_2} + \dfrac{\sqrt{c_1}\, d_8}{2\beta d_2}\right) & F_\theta \\ \hline 0 & -\dfrac{P\sqrt{c_1 d_1}\, d_8}{\beta c_2} & \sqrt{\dfrac{d_1}{c_1}}\left(\dfrac{d_3}{d_5} + d_8 d_{10}\right) & \dfrac{\sqrt{c_1 d_1}\, d_8}{\beta c_2} & F_M \\ \hline 0 & 0 & 0 & 1 & F_V \\ \hline 0 & 0 & 0 & 0 & 1 \end{array}\right]$$

$\beta^2 = P/(c_2^2 E I_i) - \frac{1}{4}, \quad d_1 = c_1 - c_2 \ell, \quad d_2 = \sqrt{d_1}, \quad d_3 = \sin(\beta \ln d_1), \quad d_4 = \cos(\beta \ln d_1), \quad d_5 = \sin(\beta \ln c_1), \quad d_6 = \cos(\beta \ln c_1)$

$d_7 = \tan(\beta \ln c_1), \quad d_8 = d_4 d_5 - d_3 d_6, \quad d_9 = d_3 d_5 + d_4 d_6, \quad d_{10} = \dfrac{1}{2\beta} + \dfrac{1}{d_7}, \quad d_{11} = \dfrac{1 - d_2 d_3/(\sqrt{c_1}\, d_5)}{P}, \quad d_{12} = d_3/d_5 - d_9$

$d_{13} = \dfrac{d_4}{d_6} - d_9, \quad d_{14} = -\dfrac{\sqrt{d_1^3}\,(\frac{3}{2} d_3 - \beta d_4) - \sqrt{c_1^3}\,(\frac{3}{2} d_5 - \beta d_6)}{c_2(\frac{9}{4} + \beta^2)}, \quad d_{15} = -\dfrac{\sqrt{d_1}\,(\frac{1}{2} d_3 - \beta d_4) - \sqrt{c_1}\,(\frac{1}{2} d_5 - \beta d_6)}{c_2(\frac{1}{4} + \beta^2)},$

$$d_{16} = -\frac{1}{c_2^2}\left\{\frac{1}{\frac{25}{4} + \beta^2}\left[d_1^{5/2}\left(\frac{5}{2} d_3 - \beta d_4\right) - c_1^{5/2}\left(\frac{5}{2} d_5 - \beta d_6\right)\right] + c_1 c_2 d_{14}\right\}, \quad d_{17} = -\frac{1}{c_2^2}\left\{\frac{1}{\frac{9}{4} + \beta^2}\left[d_1^{3/2}\left(\frac{3}{2} d_3 - \beta d_4\right) - c_1^{3/2}\left(\frac{3}{2} d_5 - \beta d_6\right)\right] + c_1 c_2 d_{15}\right\}$$

$$d_{18} = -\frac{1}{c_2\left(\frac{9}{4} + \beta^2\right)}\left[d_1^{3/2}\left(\frac{3}{2} d_4 + \beta d_3\right) - c_1^{3/2}\left(\frac{3}{2} d_6 + \beta d_5\right)\right], \quad d_{19} = -\frac{1}{c_2\left(\frac{1}{4} + \beta^2\right)}\left[\sqrt{d_1}\left(\frac{1}{2} d_4 + \beta d_3\right) - \sqrt{c_1^3}\left(\frac{1}{2} d_6 + \beta d_5\right)\right]$$

$$d_{20} = -\frac{1}{c_2^2}\left\{\frac{1}{\frac{25}{4} + \beta^2}\left[d_1^{5/2}\left(\frac{5}{2} d_4 + \beta d_3\right) - c_1^{5/2}\left(\frac{5}{2} d_6 + \beta d_9\right)\right] + c_1 c_2 d_{18}\right\}, \quad d_{21} = -\frac{1}{c_2^2}\left\{\frac{1}{\frac{9}{4} + \beta^2}\left[d_1^{5/2}\left(\frac{3}{2} d_4 + \beta d_3\right) - c_1^{3/2}\left(\frac{3}{2} d_6 + \beta d_5\right)\right] + c_1 c_2 d_{19}\right\}$$

$$d_{22} = d_6 d_{14} - d_5 d_{18}, \quad d_{23} = d_6 d_{16} - d_5 d_{20}, \quad d_{24} = d_5 d_{15} + d_6 d_{19}, \quad d_{25} = d_5 d_{19} - d_6 d_{15}, \quad d_{26} = d_5 d_{17} + d_6 d_{21}, \quad d_{27} = d_5 d_{21} - d_6 d_{17}$$

$$F_V = -\left(w_1 + \frac{\Delta w}{\Delta \ell}\ell\right)\left(\frac{\sqrt{c_1}\, d_{22}}{\beta c_2 P} - \frac{\ell^2}{2P}\right) + \frac{\Delta w}{\Delta \ell}\left(\frac{\sqrt{c_1}\, d_{23}}{\beta c_2 P} - \frac{\ell^3}{3P}\right), \quad F_\theta = -\left(w_1 + \frac{\Delta w}{\Delta \ell}\ell\right)\frac{1}{P}\left(\ell - \sqrt{c_1}\, d_{24} + \frac{\sqrt{c_1}\, d_{25}}{2\beta}\right) + \frac{\Delta w}{\Delta \ell}\frac{1}{P}\left(\frac{\ell^2}{2} - \sqrt{c_1}\, d_{26} + \frac{\sqrt{c_1}}{2\beta} d_{23}\right)$$

$$F_M = \left(w_1 + \frac{\Delta w}{\Delta \ell}\ell\right)\frac{\sqrt{c_1}\, d_{22}}{\beta c_2} - \frac{\Delta w}{\Delta \ell}\frac{\sqrt{c_1}\, d_{23}}{\beta c_2}, \quad F_V = -\left(w_1 + \frac{\Delta w}{\Delta \ell}\ell\right)\ell + \frac{\Delta w}{\Delta \ell}\frac{\ell^2}{2}$$

TABLE 2-18 Semi-Infinite and Infinite Beams on Elastic Foundation

See Examples 2.13, 2.14, 2.15 for instructions on using these matrices. The point matrices $\overline{\mathbf{U}}_i^L$ and $\overline{\mathbf{U}}_i^R$ given here are not truly transfer matrices as the vector at $x = a_i$ contains arbitrary constants rather than state variables. However, the vector at other x coordinates is the state vector. As indicated in the example problems, all operations remain the same as those for the usual transfer matrices.

Definitions: $\lambda = (k - \rho\omega^2)/EI$, $\eta = (k - \rho\omega^2)/GA_s$, $\zeta = (P - d\rho r^2\omega^2 - k^*)/EI$. See Table 2-13 for the definition of d. Set $1/GA_s = 0$ for no shear deformation effects. $\beta = \dfrac{1}{EI} - \dfrac{\eta}{GA_s}$, $\gamma = \dfrac{1}{EI} - \dfrac{\zeta}{GA_s}$.

Beams Infinite to the Left

1. For $0 < \lambda - \zeta\eta < \frac{1}{4}(\zeta - \eta)^2$, $\zeta - \eta < 0$

$$\overline{\mathbf{U}}_i^L = \begin{bmatrix} 1 & 1 & 1 & 1 & 0 \\ -\dfrac{(\beta + a^2/GA_s)a}{\gamma} & -\dfrac{(\beta - b^2/GA_s)b}{\gamma} & \dfrac{(\beta + a^2/GA_s)a}{\gamma} & \dfrac{(\beta + b^2/GA_s)b}{\gamma} & 0 \\ EI(\eta - a^2) & EI(\eta - b^2) & EI(\eta - a^2) & EI(\eta - b^2) & 0 \\ \dfrac{(\eta - \zeta - a^2)a}{\gamma} & \dfrac{(\eta - \zeta - b^2)b}{\gamma} & \dfrac{(-\eta + \zeta + a^2)a}{\gamma} & \dfrac{(-\eta + \zeta + b^2)b}{\gamma} & 0 \\ 0 & 0 & 0 & 0 & 1 \end{bmatrix}$$

2. For $\lambda - \zeta\eta = \frac{1}{4}(\zeta - \eta)^2$, $\zeta - \eta < 0$

$$\overline{\mathbf{U}}_i^L = \begin{bmatrix} 1 & 0 & 1 & 0 & 0 \\ -\dfrac{(\beta + b^2/GA_s)b}{\gamma} & -\dfrac{\beta + 3b^2/GA_s}{\gamma} & \dfrac{(\beta + b^2/GA_s)b}{\gamma} & -\dfrac{\beta + 3b^2/GA_s}{\gamma} & 0 \\ EI(\eta - b^2) & -2EIb & EI(\eta - b^2) & 2EIb & 0 \\ \dfrac{(\eta - \zeta - b^2)b}{\gamma} & \dfrac{\eta - \zeta - 3b^2}{\gamma} & \dfrac{(-\eta + \zeta + b^2)b}{\gamma} & \dfrac{\eta - \zeta - 3b^2}{\gamma} & 0 \\ 0 & 0 & 0 & 0 & 1 \end{bmatrix}$$

3. For $\lambda - \zeta\eta > \frac{1}{4}(\zeta - \eta)^2$, $\zeta - \eta \gtreqless 0$

$$\overline{\mathbf{U}}_i^L = \begin{bmatrix} 1 & 0 & 1 & 0 & 0 \\ -\dfrac{a}{\gamma}(\beta + p/GA_s) & -\dfrac{b}{\gamma}(\beta - q/GA_s) & \dfrac{a}{\gamma}(\beta + p/GA_s) & -\dfrac{b}{\gamma}(\beta - q/GA_s) & 0 \\ EI(\eta - g) & -2EIab & EI(\eta - g) & 2EIab & 0 \\ \dfrac{a}{\gamma}(\eta - \zeta - p) & \dfrac{b}{\gamma}(\eta - \zeta + q) & -\dfrac{a}{\gamma}(\eta - \zeta - p) & \dfrac{b}{\gamma}(\eta - \zeta + q) & 0 \\ 0 & 0 & 0 & 0 & 1 \end{bmatrix}$$

Definitions(cont.)

Cases 1 and 4

$$a^2 = -\frac{1}{2}(\zeta - \eta) + \sqrt{\frac{1}{4}(\zeta - \eta)^2 - (\lambda - \zeta\eta)}$$

$$b^2 = -\frac{1}{2}(\zeta - \eta) - \sqrt{\frac{1}{4}(\zeta - \eta)^2 - (\lambda - \zeta\eta)}$$

Cases 2 and 5

$$b^2 = -\frac{1}{2}(\zeta - \eta)$$

Cases 3 and 6

$$a^2 = \frac{1}{2}\sqrt{\lambda - \zeta\eta} - \frac{1}{4}(\zeta - \eta), \quad g = a^2 - b^2, \quad h = a^2 + b^2$$

$$b^2 = \frac{1}{2}\sqrt{\lambda - \zeta\eta} + \frac{1}{4}(\zeta - \eta), \quad p = a^2 - 3b^2, \quad q = b^2 - 3a^2$$

Beams Infinite to the Right

4. For $0 < \lambda - \zeta\eta < \frac{1}{4}(\zeta - \eta)^2, \quad \zeta - \eta < 0$

$$\overline{\mathbf{U}}_i^R = \frac{1}{2(b^2 - a^2)}\begin{bmatrix} b^2 - \eta & \frac{1}{a}(-b^2 + \eta - \zeta) & \frac{1}{EI} & \frac{1}{a}(\beta + b^2/GA_s) & 0 \\ \eta - a^2 & \frac{1}{b}(a^2 - \eta + \zeta) & -\frac{1}{EI} & -\frac{1}{b}(\beta + a^2/GA_s) & 0 \\ b^2 - \eta & \frac{1}{a}(b^2 - \eta + \zeta) & \frac{1}{EI} & -\frac{1}{a}(\beta + b^2/GA_s) & 0 \\ \eta - a^2 & \frac{1}{b}(-a^2 + \eta - \zeta) & -\frac{1}{EI} & \frac{1}{b}(\beta + a^2/GA_s) & 0 \\ 0 & 0 & 0 & 0 & 1 \end{bmatrix}$$

5. For $\lambda - \zeta\eta = \frac{1}{4}(\zeta - \eta)^2, \quad \zeta - \eta < 0$

$$\overline{\mathbf{U}}_i^R = \begin{bmatrix} \frac{1}{2} & \frac{\eta - \zeta - 3b^2}{4b^3} & 0 & \frac{\beta + 3b^2/GA_s}{4b^3} & 0 \\ \frac{\eta - b^2}{4b} & -\frac{\eta - \zeta - b^2}{4b^2} & -\frac{1}{4EIb} & -\frac{\beta + b^2/GA_s}{4b^2} & 0 \\ \frac{1}{2} & -\frac{\eta - \zeta - 3b^2}{4b^3} & 0 & -\frac{\beta + 3b^2/GA_s}{4b^3} & 0 \\ -\frac{\eta - b^2}{4b} & -\frac{\eta - \zeta - b^2}{4b^2} & \frac{1}{4EIb} & -\frac{\beta + b^2/GA_s}{4b^2} & 0 \\ 0 & 0 & 0 & 0 & 1 \end{bmatrix}$$

6. For $\lambda - \zeta\eta > \frac{1}{4}(\zeta - \eta)^2, \quad \zeta - \eta \gtrless 0$

$$\overline{\mathbf{U}}_i^R = \begin{bmatrix} \frac{1}{2} & \frac{\eta - \zeta + q}{4ah} & 0 & \frac{\beta - q/GA_s}{4ah} & 0 \\ \frac{\eta - g}{4ab} & -\frac{\eta - \zeta - p}{4bh} & -\frac{1}{4EIab} & -\frac{\beta + p/GA_s}{4bh} & 0 \\ \frac{1}{2} & -\frac{\eta - \zeta + q}{4ah} & 0 & -\frac{\beta - q/GA_s}{4ah} & 0 \\ -\frac{\eta - g}{4ab} & -\frac{\eta - \zeta - p}{4bh} & \frac{1}{4EIab} & -\frac{\beta + p/GA_s}{4bh} & 0 \\ 0 & 0 & 0 & 0 & 1 \end{bmatrix}$$

2.15 Loading Functions

The loading functions are provided in the previous transfer matrix tables for most common loadings. These functions can be calculated for other loadings from the formulas:

$$F_v = -\int_0^\ell w(x) U_{vV}(\ell - x)\, dx - \int_0^\ell c(x) U_{vM}(\ell - x)\, dx + \int_0^\ell \frac{M_T(x)}{EI} U_{v\theta}(\ell - x)\, dx \tag{2.26a}$$

$$F_\theta = -\int_0^\ell w(x) U_{\theta V}(\ell - x)\, dx - \int_0^\ell c(x) U_{\theta M}(\ell - x)\, dx + \int_0^\ell \frac{M_T(x)}{EI} U_{\theta\theta}(\ell - x)\, dx \tag{2.26b}$$

$$F_M = -\int_0^\ell w(x) U_{MV}(\ell - x)\, dx - \int_0^\ell c(x) U_{MM}(\ell - x)\, dx + \int_0^\ell \frac{M_T(x)}{EI} U_{M\theta}(\ell - x)\, dx \tag{2.26c}$$

$$F_V = -\int_0^\ell w(x) U_{VV}(\ell - x)\, dx - \int_0^\ell c(x) U_{VM}(\ell - x)\, dx + \int_0^\ell \frac{M_T(x)}{EI} U_{V\theta}(\ell - x)\, dx \tag{2.26d}$$

where $w(x)$ is a general applied distributed force (force intensity), $c(x)$ is an arbitrary applied moment intensity, and M_T is the thermal moment due to a temperature change ΔT across the beam depth. The notation $U_{ij}(\ell - x)$ refers to the U_{ij} given in the transfer matrix tables with ℓ replaced by $\ell - x$. In each entry of Eqs. (2.26*a*, *b*, *c*, *d*), it is permissible to switch the $(\ell - x)$ dependency from the transfer matrix element to the loading variable, for example,

$$F_v = -\int_0^\ell w(x) U_{vV}(\ell - x)\, dx = -\int_0^\ell w(\ell - x) U_{vV}(x)\, dx \tag{2.26e}$$

where now $U_{ij}(x)$ is U_{ij} from the tables with ℓ replaced by x.

EXAMPLE 2.6 Calculation of Loading Functions Find the loading function F_v for a massless Euler-Bernoulli beam with no shear deformation. Suppose the loading is given by $w(\ell) = w_1 + (\Delta w / \Delta \ell)\ell$.

From Table 2-8, $U_{vV}(\ell) = -\ell^3/6EI$. Then, from Eq. (2.26*a*),

$$F_v = -\int_0^\ell w(x) U_{vV}(\ell - x)\, dx = -\int_0^\ell \left(w_1 + \frac{\Delta w}{\Delta \ell} x\right)\left[-\frac{(\ell - x)^3}{6EI}\right] dx$$

$$= w_1 \frac{\ell^4}{24EI} + \frac{\Delta w}{\Delta \ell} \frac{\ell^5}{120EI}$$

Alternatively, from the final integral in Eq. (2.26*e*),

$$F_v = -\int_0^\ell w(\ell - x) U_{vV}(x)\, dx = \int_0^\ell \left[w_1 + \frac{\Delta w}{\Delta \ell}(\ell - x) \right]\left(-\frac{x^3}{6EI} \right) dx$$

$$= w_1 \frac{\ell^4}{24EI} + \frac{\Delta w}{\Delta \ell} \frac{\ell^5}{120EI}$$

2.16 Static Response

The procedure for calculating the deflection, slope, bending moment, and shear force due to static loading is presented briefly here.

The vector

$$\mathbf{s} = \begin{bmatrix} v \\ \theta \\ M \\ V \\ 1 \end{bmatrix} \tag{2.27}$$

is composed of state variables and is called the *state vector*. The transfer matrix $\mathbf{U}_i$ of Eq. (2.15) serves to "transfer" the state variables from left (for example, $x = 0$) to right (for example, $x = \ell$) along the beam. That is,

$$\mathbf{s}_{x=\ell} = \mathbf{U}_i \mathbf{s}_{x=0} = \mathbf{U}_i \mathbf{s}_0 \tag{2.28}$$

Transfer matrices are presented in this chapter for virtually all physically plausible types of beams. These include field matrices $\mathbf{U}_j$ that represent a beam segment of finite length and point matrices $\overline{\mathbf{U}}_j$ for concentrated occurrences such as a concentrated force.

The state vector $\mathbf{s}$ at any point along the beam is found by progressive multiplication of the transfer matrices for all occurrences from left to right up to that point. That is, the state variables at any point j are given by

$$\mathbf{s}_j = \mathbf{U}_j \mathbf{U}_{j-1} \cdots \overline{\mathbf{U}}_k \cdots \mathbf{U}_2 \mathbf{U}_1 \mathbf{s}_0 \tag{2.29}$$

Here $\mathbf{U}_1$ is the transfer matrix for the first segment at the left end of the beam. Matrix $\mathbf{U}_2$ takes the second occurrence into account. Matrix $\overline{\mathbf{U}}_k$ accounts for a point occurrence at $x = a_k$, and so on. If there are n occurrences along the beam, the state variables at the right end of the beam become

$$\mathbf{s}_{x=L} = \mathbf{s}_n = \mathbf{U}_n \mathbf{U}_{n-1} \cdots \overline{\mathbf{U}}_k \cdots \mathbf{U}_2 \mathbf{U}_1 \mathbf{s}_0 = \mathbf{U}\mathbf{s}_0 \tag{2.30}$$

The initial parameters v_0, θ_0, M_0, V_0 composing $\mathbf{s}_0$ are found from the two boundary conditions that occur at each end of the beam.

A computer program for a beam usually consists of a processor that calls up stored transfer matrices as required and performs the matrix multiplications indicated in Eqs. (2.29) and (2.30). First the *overall* or *global* transfer matrix $\mathbf{U}$ of Eq. (2.30) must be formed in order to apply the boundary conditions to evaluate the initial state vector $\mathbf{s}_0$. Then, with $\mathbf{s}_0$ known, a second "sweep" along the member with Eq. (2.29) is required to compute and print the displacements and forces along the beam.

Summary of the Calculation Procedure

1. Model the beam system in terms of segments (*sections*) that connect locations (*stations*) of point occurrences (for example, applied concentrated forces) or abrupt changes (for example, a jump in the cross-sectional area).

2. Calculate the moment of inertia I and other constants such as the shear-corrected area or elastic foundation modulus for each section.

3. Calculate the elements of the transfer matrices for all the sections connecting stations. Compute the elements of the point matrices for the concentrated occurrences. The formulas of Tables 2-8 to 2-18 are used for these purposes.

4. Calculate the global transfer matrix by multiplying in sequence all transfer matrices from the left end to the right end of the beam. That is, calculate $\mathbf{U}$ of Eq. (2.30).

5. Evaluate the initial variables v_0, θ_0, M_0, V_0 of

$$\mathbf{s}_{x=0} = \mathbf{s}_0 = \begin{bmatrix} v \\ \theta \\ M \\ V \\ 1 \end{bmatrix}_0$$

by using the boundary conditions. This can be accomplished by eliminating the unnecessary rows and columns of Eq. (2.30) and solving the remaining equations. In the case of infinite beams, use the procedure outlined in Examples 2.13, 2.14, and 2.15.

6. Compute the deflection, slope, internal moment, and shear force at all points of interest using Eq. (2.29).

7. Use the bending moment and shear force to find the stresses as given by Section 2.2.

EXAMPLE 2.7 Beam with Variable Cross Section Find the deflection, slope, moment, and shear force curves for the beam of variable cross section as shown in Fig. 2-10. A uniformly distributed load of magnitude w_1 is applied along the length of the beam. Neglect shear-deformation effects.

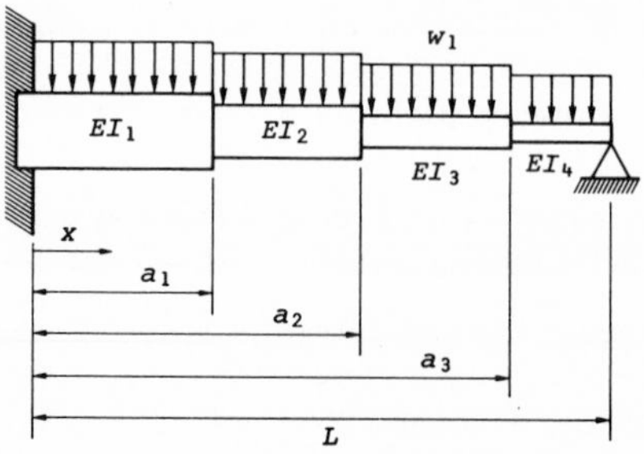

Fig. 2-10

If the boundaries are included, there are five locations of concentrated occurrences or abrupt changes, that is, there are five stations. There are four sections connecting these occurrences. The number of sections is always one less than the number of stations. The transfer matrix for each uniformly loaded section is given by Table 2-8 as

$$\mathbf{U}_i = \begin{bmatrix} 1 & -\ell_i & -\dfrac{\ell_i^2}{2EI_i} & -\dfrac{\ell_i^3}{3!EI_i} & \dfrac{w_1\ell_i^4}{4!EI_i} \\ 0 & 1 & \dfrac{\ell_i}{EI_i} & \dfrac{\ell_i^2}{2EI_i} & -\dfrac{w_1\ell_i^3}{3!EI_i} \\ 0 & 0 & 1 & \ell_i & -\dfrac{w_1\ell_i^2}{2} \\ 0 & 0 & 0 & 1 & -w_1\ell_i \\ 0 & 0 & 0 & 0 & 1 \end{bmatrix} \tag{1}$$

Here ℓ_i is the length of the ith segment, that is, $\ell_1 = a_1$ for $\mathbf{U}_1$, $\ell_2 = a_2 - a_1$ for $\mathbf{U}_2$, $\ell_3 = a_3 - a_2$ for $\mathbf{U}_3$, $\ell_4 = L - a_3$ for $\mathbf{U}_4$. If results between the ends of a segment are desired, then ℓ_i in $\mathbf{U}_j$ of Eq. (2.29) is chosen to be less than the length of the segment. The initial parameter vector $\mathbf{s}_0$ is determined from the boundary conditions at each end of the beam. These conditions are applied to $\mathbf{s}_0$ and $\mathbf{s}_{x=L}$ of Eq. (2.30), that is,

$$\mathbf{s}_{x=L} = \mathbf{s}_n = \mathbf{U}_4\mathbf{U}_3\mathbf{U}_2\mathbf{U}_1\mathbf{s}_0 = \mathbf{U}\mathbf{s}_0 \tag{2}$$

Since the left end is fixed, $v_0 = 0$ and $\theta_0 = 0$. The conditions at the simply supported right end are $v_{x=L} = 0$ and $M_{x=L} = 0$. These are applied to the first and third rows of (2). Equation (2), that is, $\mathbf{s}_{x=L} = \mathbf{U}\mathbf{s}_0$, can be written as

$$\begin{bmatrix} v = 0 \\ \theta \\ M = 0 \\ V \\ 1 \end{bmatrix}_{x=L} = \begin{bmatrix} \cdot & \cdot & U_{vM} & U_{vV} & F_v \\ \cdot & \cdot & \cdot & \cdot & \cdot \\ \cdot & \cdot & U_{MM} & U_{MV} & F_M \\ \cdot & \cdot & \cdot & \cdot & \cdot \\ \cdot & \cdot & 0 & 0 & 1 \end{bmatrix} \begin{bmatrix} v = 0 \\ \theta = 0 \\ M \\ V \\ 1 \end{bmatrix}_{x=0} \tag{3}$$

Cancel columns 1 and 2 because $v_0 = \theta_0 = 0$.

Cancel rows 2 and 4 because $\theta_{x=L}$ and $V_{x=L}$ are unknown.

where U_{kj} and F_k are the elements of $\mathbf{U}$ of (2). The equations $v_{x=L} = 0$, $M_{x=L} = 0$ are used to compute M_0, V_0, the remaining unknown initial parameters. Thus, from (3)

$$\begin{aligned} v_{x=L} &= 0 = M_0U_{vM} + V_0U_{vV} + F_v \\ M_{x=L} &= 0 = M_0U_{MM} + V_0U_{MV} + F_M \end{aligned} \tag{4}$$

Equations (4) are solved for M_0, V_0 giving

$$M_0 = (F_M U_{vV} - F_v U_{MV})|_{x=L}/\nabla$$
$$V_0 = (F_v U_{MM} - F_M U_{vM})|_{x=L}/\nabla \qquad (5)$$
$$\nabla = (U_{vM} U_{MV} - U_{MM} U_{vV})_{x=L}$$

Since $\mathbf{s}_0$ is now known, the variables v_0, θ_0, M_0, V_0 can be calculated by placing $v_0 = 0$, $\theta_0 = 0$ and (5) in Eq. (2.29). The results are then printed out at desired locations. For example, just to the left of each station,

$$\mathbf{s}_{x=a_1} = \mathbf{U}_1 \mathbf{s}_0$$
$$\mathbf{s}_{x=a_2} = \mathbf{U}_2 \mathbf{U}_1 \mathbf{s}_0$$
$$\mathbf{s}_{x=a_3} = \mathbf{U}_3 \mathbf{U}_2 \mathbf{U}_1 \mathbf{s}_0$$
$$\mathbf{s}_{x=a_4} = \mathbf{U}_4 \mathbf{U}_3 \mathbf{U}_2 \mathbf{U}_1 \mathbf{s}_0$$

Between stations results are computed by appropriately adjusting the coordinate in the transfer matrix for that section.

Formulas for the Initial Parameters The application of the boundary conditions to Eq. (2.30) to calculate v_0, θ_0, M_0, V_0 as described above can always be employed to determine the initial parameters for beams. Formulas to accomplish the same thing are listed in Table 2-19. In this table, values of v_0, θ_0, M_0, V_0 are listed according to boundary conditions. The U_{kj} and F_k in this table are elements of the overall or global transfer matrix extending from $x = 0$ to $x = L$, that is, the elements of $\mathbf{U}$ of Eq. (2.30). Note that no loadings are shown on the beams of Table 2-19. This is because all loadings should be incorporated in the solution by the loading functions of the transfer matrix tables or Eqs. (2.26).

Table 2-19 can be condensed into a single set of formulas. Let s_1, s_2, s_3, s_4 be the state variables (v, θ, M, V but not necessarily in this order) with initial values s_{10}, s_{20}, s_{30}, s_{40}. Two of the initial parameters are known by observation to be zero at the left end of the beam. Assume these are s_{10} and s_{20}. If s_k, s_j ($k, j = 1$, 2, 3, or 4, $k \neq j$) are the state variables that are zero at the right end, that is, $(s_k)_{x=L} = (s_j)_{x=L} = 0$, then the initial parameters are given by

$$s_{10} = 0$$
$$s_{20} = 0$$
$$s_{30} = \left(F_{s_j} U_{s_k s_4} - F_{s_k} U_{s_j s_4}\right)|_{x=L}/\nabla \qquad (2.31)$$
$$s_{40} = \left(U_{s_j s_3} F_{s_k} - U_{s_k s_3} F_{s_j}\right)|_{x=L}/\nabla$$
$$\nabla = \left(U_{s_k s_3} U_{s_j s_4} - U_{s_j s_3} U_{s_k s_4}\right)_{x=L}$$

where U_{kj} and F_k are elements of the transfer matrix extending from $x = 0$ to $x = L$, that is, the elements of $\mathbf{U}$ of Eq. (2.30).

EXAMPLE 2.8 Determination of Initial Parameters Use Eqs. (2.31) to find the initial parameters for the beam of Example 2.7 and Fig. 2-10.

TABLE 2-19 Initial Parameters for Beams with No In-span Supports

Left End ↓ / Right End →	1. Pinned	2. Fixed	3. Free	4. Guided
	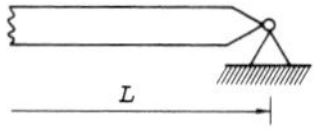	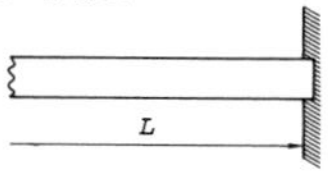 This column also applies to beams infinite to the right.	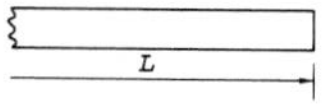	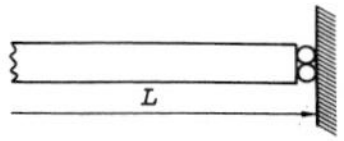
1. Pinned 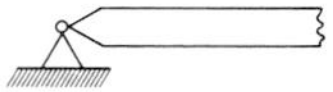 $v_0 = 0,\ M_0 = 0$	$\theta_0 = (F_M U_{vV} - F_v U_{MV})/\nabla$ $V_0 = (F_v U_{M\theta} - F_M U_{v\theta})/\nabla$ $\nabla = U_{v\theta} U_{MV} - U_{M\theta} U_{vV}$	$\theta_0 = (F_\theta U_{vV} - F_v U_{\theta V})/\nabla$ $V_0 = (F_v U_{\theta\theta} - F_\theta U_{v\theta})/\nabla$ $\nabla = U_{v\theta} U_{\theta V} - U_{\theta\theta} U_{vV}$	$\theta_0 = (F_V U_{MV} - F_M U_{VV})/\nabla$ $V_0 = (F_M U_{V\theta} - F_V U_{M\theta})/\nabla$ $\nabla = U_{M\theta} U_{VV} - U_{V\theta} U_{MV}$	$\theta_0 = (F_V U_{\theta V} - F_\theta U_{VV})/\nabla$ $V_0 = (F_\theta U_{V\theta} - F_V U_{\theta\theta})/\nabla$ $\nabla = U_{\theta\theta} U_{VV} - U_{\theta V} U_{V\theta}$
2. Fixed 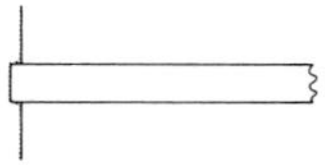 $v_0 = 0,\ \theta_0 = 0$	$M_0 = (F_M U_{vV} - F_v U_{MV})/\nabla$ $V_0 = (F_v U_{MM} - F_M U_{vM})/\nabla$ $\nabla = U_{vM} U_{MV} - U_{MM} U_{vV}$	$M_0 = (F_\theta U_{vV} - F_v U_{\theta V})/\nabla$ $V_0 = (F_v U_{\theta M} - F_\theta U_{vM})/\nabla$ $\nabla = U_{vM} U_{\theta V} - U_{\theta M} U_{vV}$	$M_0 = (F_V U_{MV} - F_M U_{VV})/\nabla$ $V_0 = (F_M U_{VM} - F_V U_{MM})/\nabla$ $\nabla = U_{MM} U_{VV} - U_{VM} U_{MV}$	$M_0 = (F_V U_{\theta V} - F_\theta U_{VV})/\nabla$ $V_0 = (F_\theta U_{VM} - F_V U_{\theta M})/\nabla$ $\nabla = U_{\theta M} U_{VV} - U_{\theta V} U_{VM}$
3. Free 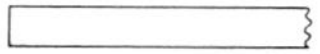 $M_0 = 0,\ V_0 = 0$ This row also applies to beams infinite to the left.	$v_0 = (F_M U_{v\theta} - F_v U_{M\theta})/\nabla$ $\theta_0 = (F_v U_{Mv} - F_M U_{vv})/\nabla$ $\nabla = U_{vv} U_{M\theta} - U_{Mv} U_{v\theta}$	*$v_0 = (F_\theta U_{v\theta} - F_v U_{\theta\theta})/\nabla$ $\theta_0 = (F_v U_{\theta v} - F_\theta U_{vv})/\nabla$ $\nabla = U_{vv} U_{\theta\theta} - U_{\theta v} U_{v\theta}$	$v_0 = (F_V U_{M\theta} - F_M U_{V\theta})/\nabla$ $\theta_0 = (F_M U_{Vv} - F_V U_{Mv})/\nabla$ $\nabla = U_{Mv} U_{V\theta} - U_{M\theta} U_{Vv}$	$v_0 = (F_V U_{\theta\theta} - F_\theta U_{V\theta})/\nabla$ $\theta_0 = (F_\theta U_{Vv} - F_V U_{\theta v})/\nabla$ $\nabla = U_{\theta v} U_{V\theta} - U_{\theta\theta} U_{Vv}$
4. Guided 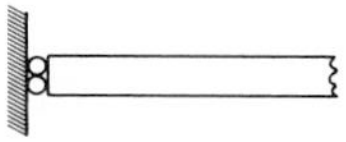 $\theta_0 = 0,\ V_0 = 0$	$v_0 = (F_M U_{vM} - F_v U_{MM})/\nabla$ $M_0 = (F_v U_{Mv} - F_M U_{vv})/\nabla$ $\nabla = U_{vv} U_{MM} - U_{Mv} U_{vM}$	$v_0 = (F_\theta U_{vM} - F_v U_{\theta M})/\nabla$ $M_0 = (F_v U_{\theta v} - F_\theta U_{vv})/\nabla$ $\nabla = U_{vv} U_{\theta M} - U_{vM} U_{\theta v}$	$v_0 = (F_V U_{MM} - F_M U_{VM})/\nabla$ $M_0 = (F_M U_{Vv} - F_V U_{Mv})/\nabla$ $\nabla = U_{Mv} U_{VM} - U_{MM} U_{Vv}$	$v_0 = (F_V U_{\theta M} - F_\theta U_{VM})/\nabla$ $M_0 = (F_\theta U_{Vv} - F_V U_{\theta v})/\nabla$ $\nabla = U_{\theta v} U_{VM} - U_{\theta M} U_{Vv}$

Definitions: $U_{kj} = U_{kj}\big|_{x=L}$, $k,\ j = v,\ \theta,\ M,\ V$; $F_k = F_k\big|_{x=L}$, $k = v,\ \theta,\ M,\ V$

*These entries apply to beams infinite in both axial directions.

Since the left end is fixed, $v_0 = 0 = s_{10}$, $\theta_0 = 0 = s_{20}$. This leaves $s_3 = M$, $s_4 = V$. Because of the simply supported right end, $M_{x=L} = (s_k)_{x=L} = 0$, $v_{x=L} = (s_j)_{x=L} = 0$. Then from Eq. (2.31),

$$v_0 = s_{10} = 0$$
$$\theta_0 = s_{20} = 0$$
$$M_0 = s_{30} = (F_v U_{MV} - F_M U_{vV})|_{x=L}/\nabla$$
$$V_0 = s_{40} = (U_{vM} F_M - U_{MM} F_v)|_{x=L}/\nabla$$
$$\nabla = (U_{MM} U_{vV} - U_{vM} U_{MV})_{x=L}$$

where U_{kj} and F_k are the elements of **U** of (2) of Example 2.7. These are the same results found in the previous example. These are also the formulas listed in Table 2-19 for a fixed-pinned beam.

Initial Parameters for Beams with In-Span Indeterminate Conditions If rigid supports, moment releases, shear releases, or angle guides (Fig. 2-11) occur along a beam, special consideration must be given to the evaluation of the initial parameters.

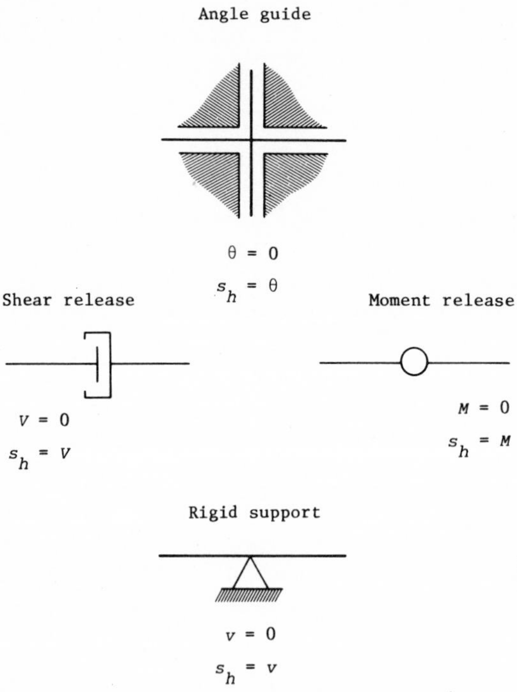

Fig. 2-11 In-span indeterminate conditions.

These will be referred to as in-span indeterminate conditions because there occurs at each such condition a variable of temporarily unknown magnitude. For example, at a rigid support the deflection is known (usually zero), while the reaction force is unknown. In modeling a real structure by beams with indeterminate in-span conditions, it is well to be aware that it is possible to create a kinematically unstable situation if certain combinations of conditions, for example, two shear releases, occur in sequence.

The procedure for incorporating in-span indeterminate conditions into a solution is rather complicated; a computer should be employed. The technique can be summarized as follows:

1. Calculate the transfer matrices for all individual sections.
2. Obtain subglobal matrices between the in-span indeterminate conditions.
3. Calculate the point matrices for all in-span indeterminate conditions using the formulas of Table 2-15. This step essentially evaluates the unknown variable at the in-span condition.
4. Insert these in-span condition point matrices in the sequence of transfer matrices and compute the global transfer matrix. Recall that the computation of the point matrix (step 3) next to the right end uses one of the boundary conditions at the right end. Therefore only the single unused boundary condition at the right end is available for use in evaluating the initial parameters. However, the known in-span condition next to the left end has not been employed and is available for use in finding the initial parameters.
5. Set up two simultaneous equations for the initial parameters that are not zero at the left end. The first one employs the global matrix and the unused boundary condition at the right end. The second one employs the subglobal matrix from the left end to the first in-span indeterminate condition. This equation satisfies the known condition at the in-span indeterminate next to the left end.
6. Solve the simultaneous equations. The initial parameters are now known and the response calculations proceed as before.

Formulas similar to Eqs. (2.31) are available for beams with in-span indeterminate conditions. Assume that the first in-span indeterminate occurring from left to right along the beam occurs at $x = a_1$. Designate the state variable that is zero at this in-span indeterminate by s_h. These zero-state variables are displayed in Fig. 2-11. Then the initial parameters are given by

$$
\begin{aligned}
s_{10} &= 0 \\
s_{20} &= 0 \\
s_{30} &= \left(F_{s_j} U_{s_h s_4} - F_{s_h} U_{s_j s_4}\right)/\nabla \\
s_{40} &= \left(U_{s_j s_3} F_{s_h} - U_{s_h s_3} F_{s_j}\right)/\nabla \\
\nabla &= U_{s_h s_3} U_{s_j s_4} - U_{s_j s_3} U_{s_h s_4}
\end{aligned}
\tag{2.32}
$$

where

- s_j is the state variable that (1) is zero at $x = L$ and (2) was not used as s_m in the point matrices for in-span indeterminates (Table 2-15)
- $U_{s_h j}$ ($j = s_3$ or s_4), F_{s_h} are the transfer matrix elements for the segment between $x = 0$ and $x = a_1$
- $U_{s_j k}$ ($k = s_3$ or s_4), F_{s_j} are the elements of the global transfer matrix extending from $x = 0$ to $x = L$

Miscellaneous Examples The following are a variety of examples that illustrate the procedures and formulas of this section.

EXAMPLE 2.9 Cantilevered Beam Find the deflection, slope, moment, and shear equations for a uniform beam fixed at the left end and free at the right end with a transverse concentrated force W acting at the right end (Fig. 2-12). This simple example demonstrates how to handle loadings at the boundaries.

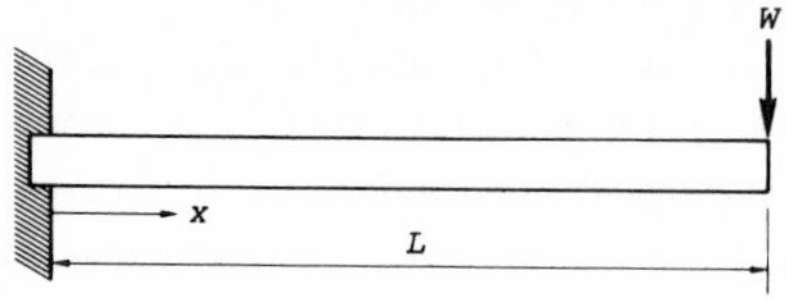

Fig. 2-12

No distinction is made between loadings at the ends and those occurring in-span; that is, all loadings along a beam are taken into account by field or point matrices. Thus the formulas of Table 2-19 or Eqs. (2.31), which are appropriate for homogeneous end conditions—that is, boundaries with no loading—suffice to determine the initial parameters ($\mathbf{s}_0$) for any beam.

For our problem, using the field matrix of Table 2-8, we find

$$\begin{bmatrix} v \\ \theta \\ M \\ V \\ 1 \end{bmatrix}_\ell = \begin{bmatrix} 1 & -\ell & -\dfrac{\ell^2}{2EI} & -\dfrac{\ell^3}{3!EI} & 0 \\ 0 & 1 & \dfrac{\ell}{EI} & \dfrac{\ell^2}{2EI} & 0 \\ 0 & 0 & 1 & \ell & 0 \\ 0 & 0 & 0 & 1 & 0 \\ 0 & 0 & 0 & 0 & 1 \end{bmatrix} \begin{bmatrix} v \\ \theta \\ M \\ V \\ 1 \end{bmatrix}_0 \qquad 0 \leqslant \ell < L \tag{1}$$

At $x = L$, the point matrix for a concentrated force W of Table 2-14 is introduced:

$$\begin{bmatrix} v \\ \theta \\ M \\ V \\ 1 \end{bmatrix}_L = \begin{bmatrix} 1 & 0 & 0 & 0 & 0 \\ 0 & 1 & 0 & 0 & 0 \\ 0 & 0 & 1 & 0 & 0 \\ 0 & 0 & 0 & 1 & -W \\ 0 & 0 & 0 & 0 & 1 \end{bmatrix} \begin{bmatrix} 1 & -L & -\dfrac{L^2}{2EI} & -\dfrac{L^3}{3!EI} & 0 \\ 0 & 1 & \dfrac{L}{EI} & \dfrac{L^2}{2EI} & 0 \\ 0 & 0 & 1 & L & 0 \\ 0 & 0 & 0 & 1 & 0 \\ 0 & 0 & 0 & 0 & 1 \end{bmatrix} \begin{bmatrix} v \\ \theta \\ M \\ V \\ 1 \end{bmatrix}_0$$

$$= \begin{bmatrix} 1 & -L & -\dfrac{L^2}{2EI} & -\dfrac{L^3}{3!EI} & 0 \\ 0 & 1 & \dfrac{L}{EI} & \dfrac{L^2}{2EI} & 0 \\ 0 & 0 & 1 & L & 0 \\ 0 & 0 & 0 & 1 & -W \\ 0 & 0 & 0 & 0 & 1 \end{bmatrix} \begin{bmatrix} v \\ \theta \\ M \\ V \\ 1 \end{bmatrix}_0 \tag{2}$$

Either the boundary conditions $v_{x=0} = 0$, $\theta_{x=0} = 0$, $M_{x=L} = 0$, $V_{x=L} = 0$ can be applied to (2) to evaluate v_0, θ_0, M_0, V_0, or use can be made of Table 2-19 or Eqs. (2.31). In the first case, (2) becomes

$$\begin{bmatrix} v \\ \theta \\ 0 \\ 0 \\ 1 \end{bmatrix}_L = \begin{bmatrix} 1 & -L & -\dfrac{L^2}{2EI} & -\dfrac{L^3}{3!EI} & 0 \\ 0 & 1 & \dfrac{L}{EI} & \dfrac{L^2}{2EI} & 0 \\ 0 & 0 & 1 & L & 0 \\ 0 & 0 & 0 & 1 & -W \\ 0 & 0 & 0 & 0 & 1 \end{bmatrix} \begin{bmatrix} 0 \\ 0 \\ M \\ V \\ 1 \end{bmatrix}_0 \tag{3}$$

If the first two columns and the first two rows are canceled, (3) reduces to two simultaneous equations for the two unknowns M_0, V_0. We find $0 = M_0 + V_0 L$, $0 = V_0 - W$ or $V_0 = W$ and $M_0 = -V_0 L = -WL$. With these values, the desired curves are expressed by (1) for any point along the beam.

EXAMPLE 2.10 Beam with Variable Axial Force and Elastic Support Find the deflection, slope, moment, and shear curves for the beam column of Fig. 2-13.

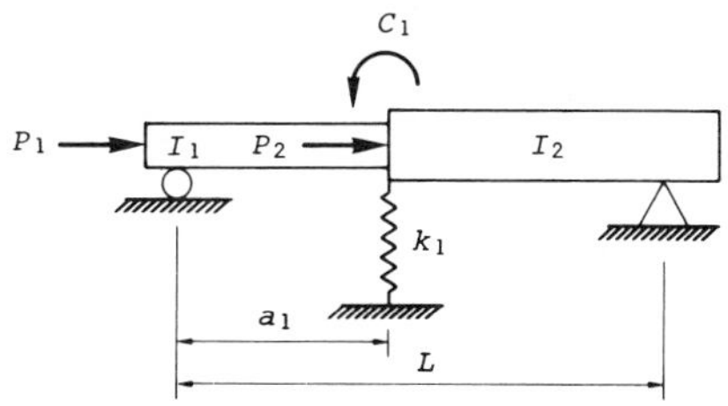

Fig. 2-13

The solution will involve three transfer matrices—two field matrices and one point matrix. The state variables at any location will be given by Eq. (2.29). Evaluation of $\mathbf{s}_0$ requires $\mathbf{U}$ of Eq. (2.30) to be established. We write

$$\mathbf{s}_{x=L} = \mathbf{U}_2 \overline{\mathbf{U}}_1 \mathbf{U}_1 \mathbf{s}_0 = \mathbf{U}\mathbf{s}_0 \tag{1}$$

where $\mathbf{U}_1$ is taken from Table 2-9 with $\ell = a_1$, $P = P_1$, $\alpha^2 = P_1/EI_1$, $I = I_1$, $w_1 = \Delta w = c_1 = \Delta c = 0$. From Table 2-14, the point matrix $\overline{\mathbf{U}}_1$ to account for C_1 and k_1 is given by

$$\overline{\mathbf{U}}_1 = \begin{bmatrix} 1 & 0 & 0 & 0 & 0 \\ 0 & 1 & 0 & 0 & 0 \\ 0 & 0 & 1 & 0 & -C_1 \\ k_1 & 0 & 0 & 1 & 0 \\ 0 & 0 & 0 & 0 & 1 \end{bmatrix}$$

Finally $\mathbf{U}_2$ is formed from Table 2-9 where $\ell = L - a_1$, $P = P_1 + P_2$, $\alpha^2 = (P_1 + P_2)/EI_2$, $I = I_2$, $w_1 = \Delta w = c_1 = \Delta c = 0$. Note that P is taken as the actual axial force in the segment as found from axial equilibrium requirements and not as P_2, the force applied at one side of the segment. These same transfer matrices, using an adjusted ℓ, are employed in setting up Eq. (2.29) if results between the outer supports and the spring are desired.

The vector $\mathbf{s}_0$ of Eq. (2.29) is found by applying the boundary conditions to (1). This has been formalized in Table 2-19 or Eq. (2.31). Consider the use of Eq. (2.31). Since the left-end conditions are $v_{x=0} = M_{x=0}$, s_1 and s_2 are chosen as v and M. Then $s_3 = \theta$, $s_4 = V$. Also, the right-end conditions are $v_{x=L} = M_{x=L} = 0$ so that $s_k = v$, $s_j = M$. Then from Eq. (2.31),

$$s_{10} = v_0 = 0 \qquad s_{20} = M_0 = 0 \qquad s_{30} = \theta_0 = (F_M U_{vV} - F_v U_{MV})|_{x=L}/\nabla$$

$$s_{40} = V_0 = (U_{M\theta} F_v - U_{v\theta} F_M)|_{x=L}/\nabla \qquad \nabla = (U_{v\theta} U_{MV} - U_{M\theta} U_{vV})|_{x=L}$$

where these transfer matrix elements are taken from $\mathbf{U}$ of (1).

EXAMPLE 2.11 Continuous Beam Set up the transfer matrix solution for the responses of the beam of Fig. 2-14 which has a rigid support at $x = a_1$.

A beam with such in-span supports requires special attention, especially in applying the boundary conditions. The solution employs three transfer matrices —a field matrix for each of the two segments and a point matrix for the rigid (unyielding) support. To calculate the desired variables along the beam with Eq. (2.29), it is necessary to evaluate $\mathbf{s}_0$ so that the boundary conditions are satisfied. This entails use of Eq. (2.30),

$$\mathbf{s}_{x=L} = \mathbf{U}_2 \overline{\mathbf{U}}_1 \mathbf{U}_1 \mathbf{s}_0 = \mathbf{U}\mathbf{s}_0 \tag{1}$$

The field matrix $\mathbf{U}_1$ is given by Table 2-8 with $\ell = a_1$ and $\Delta w = c_1 = \Delta c = 0$. Matrix $\mathbf{U}_2$ is also taken from Table 2-8. In this case $\ell = L - a_1$ and $\Delta w = c_1 = \Delta c = 0$. The point matrix $\overline{\mathbf{U}}_1$ for the rigid support is provided in Table 2-15. Since no in-span support occurs to the right of the support at $x = a_1$, s_m in $\overline{\mathbf{U}}_1$ should be assigned to be v or θ because these are the state variables that are zero at $x = L$. Choose $s_m = v$. We find

$$\overline{\mathbf{U}}_1 = \begin{bmatrix} 1 & 0 & 0 & 0 & 0 \\ 0 & 1 & 0 & 0 & 0 \\ 0 & 0 & 1 & 0 & 0 \\ -\dfrac{U_{vv}}{U_{vV}} & -\dfrac{U_{v\theta}}{U_{vV}} & -\dfrac{U_{vM}}{U_{vV}} & 0 & -\dfrac{U_{vF}}{U_{vV}} \\ 0 & 0 & 0 & 0 & 1 \end{bmatrix}$$

$$= \begin{bmatrix} 1 & 0 & 0 & 0 & 0 \\ 0 & 1 & 0 & 0 & 0 \\ 0 & 0 & 1 & 0 & 0 \\ \dfrac{3!EI}{(L-a_1)^3} & -\dfrac{3!EI}{(L-a_1)^2} & -\dfrac{3EI}{(L-a_1)} & 0 & \dfrac{w_1(L-a_1)}{4} \\ 0 & 0 & 0 & 0 & 1 \end{bmatrix} \tag{2}$$

where $U_{vV} = -(L - a_1)^3/3!EI$. All elements of (2) are taken from $\mathbf{U}_2$, that is, from the transfer matrix extending from the support at $x = a_1$ to the next support (the right end of the beam for this example).

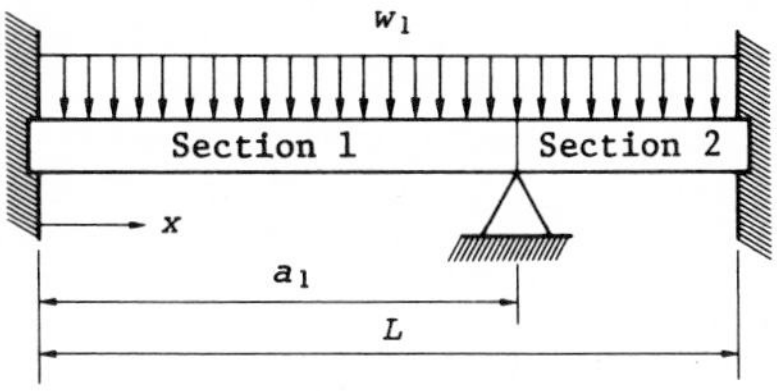

Fig. 2-14

Since $v_{x=0} = 0$, and $\theta_{x=0} = 0$, two initial parameters are known, that is, $v_0 = \theta_0 = 0$. The other two initial parameters are found from other available conditions. Recall that $v_{x=L} = 0$ was employed in establishing the point matrix (2). Thus, only the condition $\theta_{x=L} = 0$ remains at the right end. The second unused condition is $v_{x=a_1} = 0$. These two conditions provide the two simultaneous equations

$$\begin{aligned} \theta_{x=L} &= 0 = M_0 U_{\theta M} + V_0 U_{\theta V} + F_\theta \\ v_{x=a_1} &= 0 = M_0 U^*_{vM} + V_0 U^*_{vV} + F^*_v \end{aligned} \tag{3}$$

where U^*_{vM}, U^*_{VV}, F^*_v are taken from $\mathbf{U}_1$ and $U_{\theta M}$, $U_{\theta V}$, F_θ are elements of $\mathbf{U}$ of (1).

Solution of (3) gives

$$M_0 = (F_\theta U^*_{vV} - F^*_v U_{\theta V})/\nabla = \left[F_\theta \left(-\frac{a_1^3}{6EI} \right) - \frac{w_1 a_1^4}{24EI} U_{\theta V} \right] \Big/ \nabla$$

$$V_0 = (U_{\theta M} F^*_v - U^*_{vM} F_\theta)/\nabla = \left[U_{\theta M} \left(-\frac{w_1 a_1^4}{24EI} \right) - \left(-\frac{a_1^2}{2EI} \right) F_\theta \right] \Big/ \nabla \tag{4}$$

$$\nabla = U^*_{vM} U_{\theta V} - U_{\theta M} U^*_{vV} = -\frac{a_1^2}{2EI} U_{\theta V} - U_{\theta M} \left(-\frac{a_1^3}{6EI} \right)$$

Equations (4) can also be obtained from Eqs. (2.32). Since v and θ are zero at the left end, s_1 and s_2 are chosen to be v and θ. Then $s_3 = M$, $s_4 = V$. Also, $s_h = v$. By definition s_j is the state variable that is zero at $x = L$ and that was not used as s_m. Since s_m was set equal to v, we choose $s_j = \theta$. Then Eq. (2.32) gives $s_{10} = v_0 = 0$, $s_{20} = \theta_0 = 0$ and $s_{30} = M_0$, $s_{40} = V_0$ are the same as (4).

EXAMPLE 2.12 Thermally Loaded Beam Find the displacements, internal forces, and stresses caused by a thermal loading on a beam of rectangular cross section with dimensions b and h. The beam is fixed at one end and pinned at the other. The temperature change through the beam varies as $\Delta T = \Delta T_1 (2z/h + 1)^2/4$, where z is measured from the centroid of the cross section, positive downwards.

First the thermal moment M_T is calculated. Since M_T will not vary along the beam axis, let $M_T = M_{T1}$. Then the solution proceeds in the same manner as for a mechanically loaded beam.

For the given temperature change, we find

$$M_T = \int_A E\alpha\, \Delta T\, zb\, dz = \frac{E\alpha\, \Delta T_1\, b}{4} \int_{-h/2}^{h/2} \left(2\frac{z}{h} + 1\right)^2 z\, dz = \frac{E\alpha I\, \Delta T_1}{h} = M_{T1} \tag{1}$$

with $I = bh^3/12$. The transfer matrix can be taken from Table 2-13 with $\lambda = \lambda - \zeta\eta = \zeta = \eta = 0$. With $\ell = x$, this gives

$$\mathbf{U} = \begin{bmatrix} 1 & -x & -\dfrac{x^2}{2EI} & -\dfrac{x^3}{6EI} & -\dfrac{\alpha\, \Delta T_1}{2h}x^2 \\ 0 & 1 & \dfrac{x}{EI} & \dfrac{x^2}{2EI} & \dfrac{\alpha\, \Delta T_1}{h}x \\ 0 & 0 & 1 & x & 0 \\ 0 & 0 & 0 & 1 & 0 \\ 0 & 0 & 0 & 0 & 1 \end{bmatrix} \tag{2}$$

The initial parameters are listed in Table 2-19 as

$$\begin{aligned} &v_0 = \theta_0 = 0, \\ &M_0 = (F_M U_{vV} - F_v U_{MV})|_{x=L}/\nabla \qquad V_0 = (F_v U_{MM} - F_M U_{vM})|_{x=L}/\nabla \\ &\nabla = (U_{vM} U_{MV} - U_{MM} U_{vV})|_{x=L} \end{aligned} \tag{3}$$

Substitution of the elements of (2) into (3) leads to

$$\nabla = -L^3/3EI \qquad M_0 = -3EI\alpha\, \Delta T_1/2h \qquad V_0 = 3EI\alpha\, \Delta T_1/2hL \tag{4}$$

Finally, with $v_0 = \theta_0 = 0$, the deflection and internal forces are given by

$$\begin{aligned} v &= M_0 U_{vM} + V_0 U_{vV} + F_v = \frac{\alpha\, \Delta T_1\, x^2}{4h}\left(1 - \frac{x}{L}\right) \\ M &= M_0 U_{MM} + V_0 U_{MV} + F_M = \frac{3EI\alpha\, \Delta T_1}{2h}\left(\frac{x}{L} - 1\right) \\ V &= M_0 U_{VM} + V_0 U_{VV} + F_V = \frac{3EI\alpha\, \Delta T_1}{2hL} \end{aligned} \tag{5}$$

If thermal effects are included, the normal stress is given by

$$\sigma = \frac{P_T}{A} + \frac{(M + M_T)z}{I} - E\alpha\, \Delta T \tag{6}$$

where

$$P_T = \int_A E\alpha\, \Delta T\, dA = \frac{E\alpha\, \Delta T_1\, b}{4} \int_{-h/2}^{h/2} \left(2\frac{z}{h} + 1\right)^2 dz = \frac{E\alpha A\, \Delta T_1}{3} \tag{7}$$

with $A = bh$. Then, from (6),

$$\sigma = \frac{E\alpha\,\Delta T_1}{3} + \frac{1}{I}\left[\frac{3EI\alpha\,\Delta T_1}{2h}\left(\frac{x}{L} - 1\right) + \frac{E\alpha I\,\Delta T_1}{h}\right]z - E\alpha\,\frac{\Delta T_1}{4}\left(\frac{2z}{h} + 1\right)^2$$

$$= E\alpha\,\Delta T_1\left[\frac{1}{3} + \frac{3z}{2h}\left(\frac{x}{L} - \frac{1}{3}\right) - \frac{1}{4}\left(\frac{2z}{h} + 1\right)^2\right] \qquad (8)$$

EXAMPLE 2.13 Beam of Infinite Length to the Right Find the deflection, slope, bending moment, and shear force for the semi-infinite beam on the elastic foundation of Fig. 2-15. The beam is very long in the positive x direction.

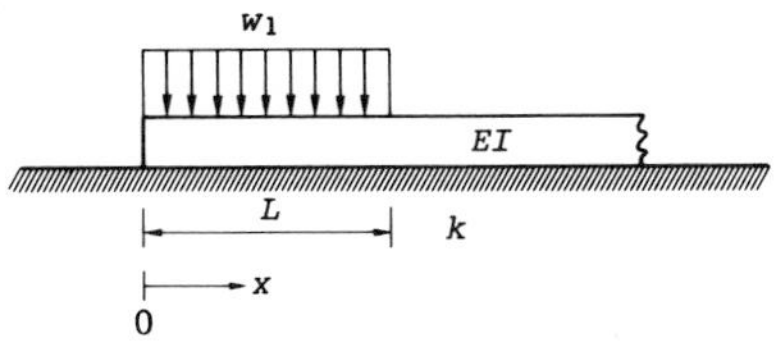

Fig. 2-15

The usual analyses of long beams on elastic foundations normally encounter numerical difficulties. These difficulties are overcome by using the point matrices of Table 2-18. These point matrices are employed during the calculation of the initial parameters. Once the initial parameters have been evaluated, the usual transfer-matrix analysis applies.

The boundary conditions at the left end are $M_{x=0} = 0$ and $V_{x=0} = 0$. Thus, $M_0 = V_0 = 0$. The point matrix $\overline{\mathbf{U}}_i^R$ of Table 2-18 is used to find v_0, θ_0. *Application of* $\overline{\mathbf{U}}_i^R = \overline{\mathbf{U}}_1^R$ *at* $x = L$ *converts the state vector into a vector for which the first two elements must be zero for a beam that is infinite to the right.* For the beam of Fig. 2-15 we write

$$\mathbf{s}_{x=L} = \overline{\mathbf{U}}_1^R \mathbf{U}_1|_{x=L}\mathbf{s}_0 \qquad (1)$$

where $\mathbf{U}_1|_{x=L}$ is taken from Table 2-10 with $\ell = L$ and $\overline{\mathbf{U}}_1^R$ is the point matrix of Table 2-18. Although the symbol $\mathbf{s}$ for a state vector is used, $\mathbf{s}_{x=L}$ is not actually a state vector because its elements are arbitrary constants, not the state variables v, θ, M, V. However, as we shall see, in practice $\mathbf{s}_{x=L}$ of (1) is treated the same as a state vector. Consider now the selection of $\overline{\mathbf{U}}_1^R$ from Table 2-18. For the static response of a beam on elastic foundation, $\lambda = k/EI$, $\zeta = \eta = 0$. Thus, use case 6 with $a = b$, $g = 0$, $h = 2b^2$, $p = q = -2b^2$, $\beta = \gamma = 1/EI$.

Since $M_0 = V_0 = 0$ and the first two elements of $\mathbf{s}_{x=L}$ are zero, (1) becomes

$$\begin{bmatrix} 0 \\ 0 \\ C \\ D \\ 1 \end{bmatrix} = \overline{\mathbf{U}}_1^R \mathbf{U}_1 \begin{bmatrix} v \\ \theta \\ 0 \\ 0 \\ 1 \end{bmatrix}_0 \qquad (2)$$

C, D are arbitrary constants that play no role in the analysis. Equation (2) can be solved for v_0, θ_0. The initial parameters can also be computed using the proper entries in Table 2.19, that is, for our beam, the second column, third row. We find

$$v_0 = \frac{w_1}{4EIb^4}\left[1 - e^{-bL}(\cos bL - \sin bL)\right], \qquad \theta_0 = \frac{w_1}{2EIb^3} e^{-bL} \sin bL \quad (3)$$

Now that v_0, θ_0, M_0, V_0 are known, the usual relation, Eq. (2.29), is employed for computing the state variables, even beyond the final occurrence along the beam. That is,

$$\mathbf{s}_{x \leqslant L} = \mathbf{U}_1 \mathbf{s}_0, \qquad \mathbf{s}_{x \geqslant L} = \mathbf{U}_2 \mathbf{U}_1 \mathbf{s}_0 \quad (4)$$

where $\mathbf{U}_2$ is taken from Table 2-10 with $w_1 = 0$ and ℓ equal to any coordinate of interest. Substitution of (3) in (4) gives the deflection for $x \leqslant L$

$$v = \frac{w_1}{8EIb^4}\{2 - e^{-b(L-x)} \cos b(L-x) + e^{-b(L+x)}[2 \sin bL \cos bx - \cos b(L-x)]\}$$

For $x \geqslant L$

$$v = \frac{w_1}{8EIb^4}\{[e^{-b(x-L)} - e^{-b(x+L)}]\cos b(x-L) + 2e^{-b(x+L)} \sin bL \cos bx\} \quad (5)$$

The slope, bending moment, and shear force also follow from the same manipulations.

This procedure applies to an infinite beam to the right on an elastic foundation of any complexity, that is, with multiple loading or with variable foundation modulus or cross section. Simply insert $\overline{\mathbf{U}}_1^R$ after the final occurrence in moving to the right along the beam. $\overline{\mathbf{U}}_1^R$ employs the characteristics—for example, E, I, A_s—of the beam extending to infinity to the right of the last occurrence and creates a state vector in which the first two elements are zero. The initial parameters are evaluated, as indicated above in (2), by establishing a relationhip at the final occurrence.

EXAMPLE 2.14 Beam of Infinite Length to the Left Develop the equations for the response of the semi-infinite beam of Fig. 2-16. This beam is very long to the left.

Moving from left to right, select the location of the first occurrence along the beam as the origin. Then set up the expression

$$\mathbf{s}_{x=L} = \mathbf{U}\overline{\mathbf{U}}_1^L \mathbf{s}_0 \quad (1)$$

where $\mathbf{U}$ is the usual global transfer matrix from $x = 0$ to $x = L$, taking into account the change of cross section, the distributed load, and the concentrated force W. $\overline{\mathbf{U}}_{\mathrm{I}}^{L}$ is the point matrix of Table 2-18. $\overline{\mathbf{U}}_{\mathrm{I}}^{L}$ utilizes the properties of the beam to the left of the first occurrence.

In the case of the beam of Fig. 2-16, $\overline{\mathbf{U}}_{\mathrm{I}}^{L}$ is located in Table 2-18 by noting that $\lambda = k_1/EI_1$, $\zeta = \eta = 0$. This means that $\overline{\mathbf{U}}_{\mathrm{I}}^{L}$ is the point matrix taken from case 3 with $a = b$, $g = 0$, $h = 2b^2$, $p = q = -2b^2$, and $\beta = \gamma = 1/EI_1$.

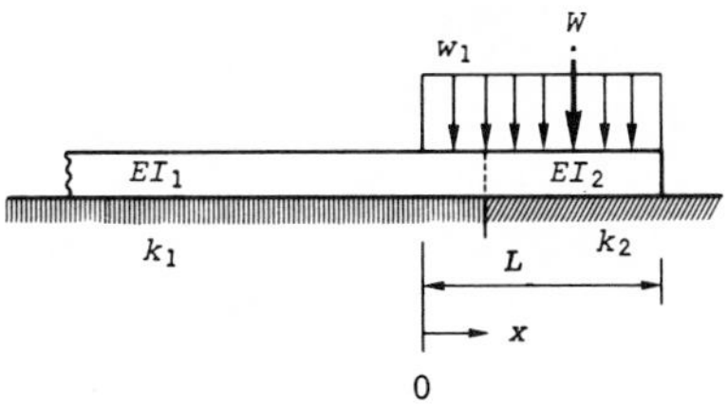

Fig. 2-16

The apparent boundary conditions for the beam of Fig. 2-16 are $M_{x=L} = V_{x=L} = 0$. These conditions inserted in (1) will help in establishing $\mathbf{s}_0$. *The point matrix* $\overline{\mathbf{U}}_{\mathrm{I}}^{L}$ *adjusts* $\mathbf{s}_0$ *so that its third and fourth elements are zero.* Although the initial vector is written $\mathbf{s}_0$, it is not actually the initial state vector. It is an artificial vector established as a convenience for beams that are infinite to the left. Using these conditions, (1) becomes

$$\begin{bmatrix} v \\ \theta \\ M = 0 \\ V = 0 \\ 1 \end{bmatrix}_{x=L} = \mathbf{U}\mathbf{U}_{\mathrm{I}}^{L}\mathbf{s}_0 = \mathbf{U}\overline{\mathbf{U}}_{\mathrm{I}}^{L} \begin{bmatrix} A \\ B \\ C = 0 \\ D = 0 \\ 1 \end{bmatrix}_{x=0} \tag{2}$$

A and B are used in $\mathbf{s}_0$ rather than v_0, θ_0 since the elements of $\mathbf{s}_0$ are not actual state variables. Equation (2) provides two equations from which A, B can be evaluated. These same variables can be found with the entry in the third row, third column of Table 2-19. Once A, B are evaluated, the state variables are determined from

$$\mathbf{s}_{0\leqslant x\leqslant L} = \mathbf{U}_n\mathbf{U}_{n-1} \cdots \mathbf{U}_2\mathbf{U}_1\overline{\mathbf{U}}_{\mathrm{I}}^{L} \begin{bmatrix} A \\ B \\ C = 0 \\ D = 0 \\ 1 \end{bmatrix} \tag{3}$$

where $\mathbf{U}_1$, $\mathbf{U}_2$, . . . are the usual transfer matrices that progressively account for occurrences to the right of $x = 0$.

EXAMPLE 2.15 Infinite Beam Set up the equations for deflection, slope, moment, and shear in the infinite beam of Fig. 2-17. This beam is long in both axial directions.

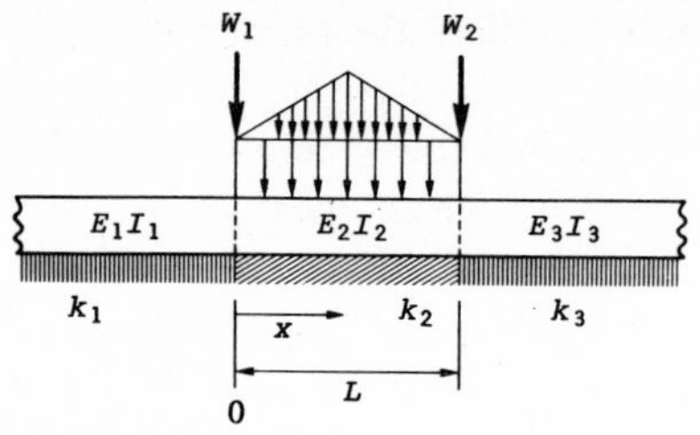

Fig. 2-17

Note that a coordinate system is established for which the origin is situated at the first occurrence in moving from left to right, and $x = L$ is at the final occurrence. This problem is solved in a way similar to the two previous examples. In order to find the initial parameters, we write

$$\mathbf{s}_{x=L} = \overline{\mathbf{U}}_2^R \mathbf{U}|_{x=L} \overline{\mathbf{U}}_1^L \mathbf{s}_0$$

where $\mathbf{U}|_{x=L}$ is a global transfer matrix taking all occurrences between $x = 0$ and $x = L$ into account. In this case, neither $\mathbf{s}_{x=L}$ nor $\mathbf{s}_0$ are state vectors. The third and fourth elements in $\mathbf{s}_0$ are zero as are the first two elements in $\mathbf{s}_{x=L}$. Thus, A, B are computed from

$$\begin{bmatrix} 0 \\ 0 \\ C \\ D \\ 1 \end{bmatrix} = \overline{\mathbf{U}}_2^R \mathbf{U} \overline{\mathbf{U}}_1^L \begin{bmatrix} A \\ B \\ 0 \\ 0 \\ 1 \end{bmatrix} \tag{1}$$

This is equivalent to using the third row, second column of Table 2-19.

As in the previous two examples, the properties of the beam of Fig. 2-17 dictate that the point matrices $\overline{\mathbf{U}}_1^L$ and $\overline{\mathbf{U}}_2^R$ are taken from case 3 and 6, respectively, of Table 2-18. For $\overline{\mathbf{U}}_1^L$, use $\lambda = k_1/E_1I_1$, $\zeta = \eta = g = 0$, $a = b$, $h = 2b^2$, $p = q = -2b^2$, $\beta = \gamma = 1/E_1I_1$. For $\overline{\mathbf{U}}_1^R$, use $\lambda = k_3/E_3I_3$, $\zeta = \eta = 0 = g = 0$, $a = b$, $h = 2b^2$, $p = q = -2b^2$, $\beta = \gamma = 1/E_3I_3$.

After computing the unknowns A, B, the state vector at any $x \leqslant L$ is found from Eq. (3) of Example 2.14. For $x \leqslant L$, use

$$\mathbf{s}_{0 \leqslant x \leqslant L} = \mathbf{U} \overline{\mathbf{U}}_1^L \begin{bmatrix} A \\ B \\ 0 \\ 0 \\ 1 \end{bmatrix}$$

For $x > L$, use

$$\mathbf{s}_{x>L} = \mathbf{U}_2 \mathbf{U}|_{x=L} \overline{\mathbf{U}}_1^L \begin{bmatrix} A \\ B \\ 0 \\ 0 \\ 1 \end{bmatrix}$$

where U_2 is the ordinary transfer matrix that applies for the section to the right of $x = L$.

2.17 Stability

The critical axial load is the smallest root of the characteristic equation resulting from the evaluation of the initial parameters. If this characteristic equation is represented by ∇, then the buckling load P_{cr} is the lowest value of P in ∇ that makes ∇ equal to zero.

The procedure for calculating the critical load and corresponding mode shape involves many of the same steps used in a static analysis. The model employed for a stability analysis requires the inclusion of the axial force. This force can be modeled as being continuously distributed or it can be lumped at particular locations along the beam. The lumped force representation enjoys certain computational advantages (Ref. 2.6). For a distributed force model use Table 2-9 or the general transfer matrix of Table 2-13. The point matrix of Table 2-14 is employed for the lumped force idealization.

Summary of the Calculation Procedure

1. Model the beam in terms of sections, as in the static analysis.

2. Calculate the same constants, for example, I, as in the static analysis. In addition, if the axial force varies along the beam, use axial equilibrium requirements to compute the axial force in each section. Frequently, the magnitudes of the axial forces in each section are known to remain in a prescribed ratio relative to a nominal value. Then the stability analysis is employed to find the nominal value. If a lumped model is to be used, the magnitude of the concentrated forces P_{Li} must be determined.

3, 4. Steps 3 and 4 are the same as in the case of static analysis. The fifth column terms in the transfer matrices, that is, the loading functions F_v, F_θ, F_M, F_V, can be ignored during the calcuation of P_{cr} and the corresponding mode shape.

5. Instead of the boundary conditions providing the initial parameters v_0, θ_0, M_0, V_0, they lead to the characteristic function ∇. The lowest root of $\nabla = 0$ is the critical load P_{cr}. For simple beams these roots can be found analytically. For nonsimple beams numerical root-finding techniques must be employed.

6. The mode shape is calculated as in step 6 for a static analysis. In the calculation, one of the initial parameters (v_0, θ_0, M_0, or V_0) that is not known to be zero at the left end should be assigned a value such as 1. For example, if the beam is fixed at the left end, $v_0 = 0$, $\theta_0 = 0$ and M_0 or V_0 should be set equal to one. This is consistent with the definition of the magnitude of a mode shape as having an unknown amplitude. The boundary conditions permit the remaining unknown initial parameter to be found in terms of the one that is set equal to 1. Now, use Eq. (2.29) to calculate the mode shapes for v, θ, M and V.

Formulas for the Characteristic Equation The critical force P_{cr} of a beam is that value of P that makes ∇ of Table 2-19 or Eq. (2.31) or (2.32), as appropriate, equal to zero. This ∇ is formed from the elements of the global transfer matrix **U** of Eq. (2.30). It is seen in these equations that the initial parameters and, hence, the response variables experience unrestrained growth when ∇ is zero. For complex beams it is necessary to evaluate the critical force from $\nabla = 0$ with a computational root-finding technique. This, in general, is a time-consuming process.

EXAMPLE 2.16 Buckling Load for a Pinned Column Determine the critical axial force in a pinned-pinned column.

The transfer matrix for a beam with compressive axial load P is given in Table 2-9. Set $1/GA_s = 0$ if shear deformation effects are to be neglected. Apply the condition of simply supported ends ($v_0 = M_0 = 0$, $v_{x=L} = M_{x=L} = 0$) to the transfer matrix and find

$$\begin{aligned} \theta_0 U_{v\theta}|_{x=L} + V_0 U_{vV}|_{x=L} &= 0 \\ \theta_0 U_{M\theta}|_{x=L} + V_0 U_{MV}|_{x=L} &= 0 \end{aligned} \tag{1}$$

where the loading functions have been ignored. The condition for nontrivial solutions to exist for these homogeneous equations is that the determinant of the initial parameters is zero. Thus, we establish the characteristic equation

$$\nabla = (U_{v\theta} U_{MV} - U_{vV} U_{M\theta})_{x=L} = 0 \tag{2}$$

From Table 2-9,

$$\nabla = -\frac{\sin\alpha L}{\alpha}\frac{\sin\alpha L}{\alpha} - \frac{\alpha L - \sin\alpha L}{\alpha^3 EI}\alpha^2 EI\frac{\sin\alpha L}{\alpha} = -\frac{L}{\alpha}\sin\alpha L = 0 \tag{3}$$

with $\alpha^2 = P/EI$. From (3)

$$\sin\alpha L = 0 \qquad \text{or} \qquad \alpha L = \sqrt{\frac{P}{EI}}\,L = \pi$$

where only the smallest meaningful root αL of $\sin\alpha L = 0$ is of interest for instability. Thus

$$P_{cr} = \frac{EI\pi^2}{L^2} \tag{4}$$

The same result is obtained from Table 2-19 or Eqs. (2.31), since these both give the same ∇ as (2).

The mode shape for displacement requires knowledge of the initial parameters. From (1), we obtain

$$V_0 = -\theta_0 (U_{v\theta}/U_{vV})_{x=L} \qquad \text{or} \qquad V_0 = -\theta_0 (U_{M\theta}/U_{MV})_{x=L} \tag{5}$$

Note from $\nabla = 0$, of (2), that these two expressions for V_0 are equivalent, that is, $(U_{v\theta}/U_{vV})_{x=L} = (U_{M\theta}/U_{MV})_{x=L}$. Then, from Table 2-9,

$$\begin{aligned} V_0 &= -\theta_0 (U_{v\theta}/U_{vV})_{x=L} = -\theta_0 \left(-\frac{\sin\alpha L}{\alpha} \right) \Big/ \left(-\frac{\alpha L - \sin\alpha L}{\alpha^3 EI} \right) \Bigg|_{\alpha L = \pi} \\ &= -\theta_0 (0) \Big/ \left(-\frac{\pi}{\alpha^3 EI} \right) = 0 \end{aligned} \tag{6}$$

Since a mode shape describes only a shape and not a precise amplitude, we can assign a value to θ_0. Let $\theta_0 = 1$. Then the initial parameters are $v_0 = M_0 = 0$, $\theta_0 = 1$, $V_0 = 0$. Return to the transfer matrix of Table 2-9 to find that for $\ell = x$,

$$v = -\theta_0 \frac{\sin \alpha x}{\alpha} = -\frac{\sin \alpha x}{\alpha} \tag{7}$$

where $\alpha = \sqrt{P_{cr}/EI}$.

EXAMPLE 2.17 Stepped Column Find the critical load of the stepped column of Fig. 2-18.

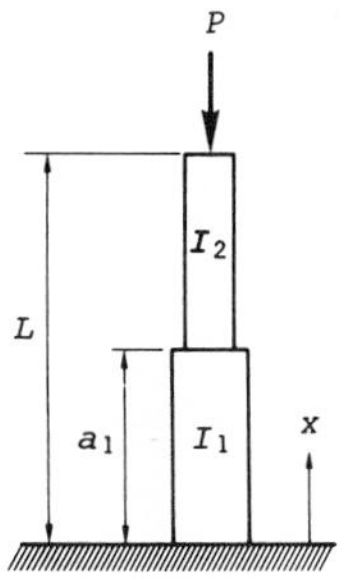

Fig. 2-18

The boundary conditions are applied to $\mathbf{s}_0$ and $\mathbf{s}_{x=L}$ of Eq. (2.30), that is,

$$\mathbf{s}_{x=L} = \mathbf{U}_2\mathbf{U}_1\mathbf{s}_0 = \mathbf{U}\mathbf{s}_0 \tag{1}$$

where, from Table 2-9 for no shear deformation effects,

$$\mathbf{U}_i = \begin{bmatrix} 1 & -\dfrac{\sin \alpha_i \ell_i}{\alpha_i} & \dfrac{\cos \alpha_i \ell_i - 1}{P_i} & \dfrac{\sin \alpha_i \ell_i}{\alpha_i P_i} - \dfrac{\ell_i}{P_i} \\ 0 & \cos \alpha_i \ell_i & \dfrac{\alpha_1 \sin \alpha_i \ell_i}{P_i} & \dfrac{1 - \cos \alpha_i \ell_i}{P_i} \\ 0 & -\dfrac{P_i \sin \alpha_i \ell_i}{\alpha_i} & \cos \alpha_i \ell_i & \dfrac{\sin \alpha_i \ell_i}{\alpha_i} \\ 0 & 0 & 0 & 1 \end{bmatrix} \tag{2}$$

with $\alpha_i^2 = P/EI_i$, $\ell_1 = a_1$ for $\mathbf{U}_1$, $\ell_2 = L - a_1$ for $\mathbf{U}_2$. In (1), $\mathbf{U}$ is the global transfer matrix spanning $x = 0$ to $x = L$. Since the left (lower) end is fixed, $v_0 = \theta_0 = 0$. Because of the free right (upper) end, $M_{x=L} = V_{x=L} = 0$. These conditions can be applied to (1) directly, or Table 2-19 (Eqs. 2.31) can be used. We choose the latter approach. In the notation of Eqs. (2.31), $v_0 = s_{10} = 0$, $\theta_0 = s_{20} = 0$, $s_3 = M$, $s_4 = V$, $(s_k)_{x=L} = M_{x=L} = 0$, $(s_j)_{x=L} = V_{x=L} = 0$. From Eqs. (2.31),

$$\begin{aligned} M_0 = s_{30} &= (F_V U_{MV} - F_M U_{VV})_{x=L}/\nabla \\ V_0 = s_{40} &= (U_{VM} F_M - U_{MM} F_V)_{x=L}/\nabla \\ \nabla &= (U_{MM} U_{VV} - U_{VM} U_{MV})_{x=L} \end{aligned} \tag{3}$$

where U_{kj}, F_k are the elements of the global transfer matrix $\mathbf{U}$ of (1). The constants M_0, V_0 will become large if ∇ is zero. This is the criterion for stability. Upon carrying out the matrix multiplication indicated in (1) and placing the resulting global transfer matrix elements in (3), we find

$$\nabla = -\frac{\alpha_2}{\alpha_1} \sin \alpha_1 a_1 \sin\left[\alpha_2(L - a_1)\right] + \cos \alpha_1 a_1 \cos\left[\alpha_2(L - a_1)\right] \tag{4}$$

If we set $\nabla = 0$,

$$\frac{\alpha_1}{\alpha_2} - \tan \alpha_1 a_1 \tan\left[\alpha_2(L - a_1)\right] = 0 \tag{5}$$

The lowest value of P that satisfies (5) is P_{cr}. Except for special cases, for example, $a_1 = L - a_1$, this value is found by a numerical root-finding technique. In fact, in problems more complex than this example, no attempt should be made to analytically form the characteristic equation (5). The whole procedure should be implemented numerically. A computer program can be written that evaluates ∇ numerically for each trial P in the search routine. The value of P should be incremented until ∇ changes sign. Then one of several available root-finding techniques can be used to close in on P_{cr}.

EXAMPLE 2.18 Variable Axial Force To find the condition of instability caused by the axial force of the beam of Fig. 2-19, we can use many of the relations of Example 2.17. The required overall transfer matrix is $\mathbf{U} = \mathbf{U}_2\mathbf{U}_1$ of Eq. (1), where $\mathbf{U}_i$ is given by Eq. (2) with $P = P_1$, $\alpha_1^2 = P_1/EI_1$, $\ell_1 = a_1$ for $\mathbf{U}_1$ and $P = P_1 + P_1^*$ (the axial force in the second section is $P_1 + P_1^*$, not just P_1^*), $\alpha_2^2 = (P_1 + P_1^*)/EI_2$, $\ell_2 = L - a_1$ for $\mathbf{U}_2$. The beam is simply supported at both ends; hence, in the notation for Eq. (2.31), $s_{10} = v_0 = 0$, $s_{20} = M_0 = 0$, $s_3 = \theta$, $s_4 = V$, $(s_k)_{x=L} = v_{x=L} = 0$, $(s_j)_{x=L} = M_{x=L} = 0$. The critical load is found by setting $\nabla = 0$. From Table 2-19 or Eq. (2.31), $\nabla = (U_{v\theta}U_{MV} - U_{M\theta}U_{vV})_{x=L}$, in which the elements of the overall transfer matrix $\mathbf{U}$ are substituted. The relation $\nabla = 0$ leads to

$$\frac{1}{\alpha_1^2} - \frac{\alpha_1^2 LEI_1/P_1^* + a_1}{\alpha_1 \tan \alpha_1 a_1} - \frac{I_1/I_2}{\alpha_2^2} - \frac{\alpha_2^2 LEI_1/P_1^* - (L - a_1)I_1/I_2}{\alpha_2 \tan \alpha_2(L - a_1)} = 0 \tag{1}$$

Combinations of P_1 and P_1^* that satisfy this equation define the critical condition. Usually some additional information is given relating P_1 and P_1^*, for example, $P_1^* = cP_1$ where c is a known constant. Then (1) is a relationship for a single unknown, the lowest value of which provides the critical loading situation.

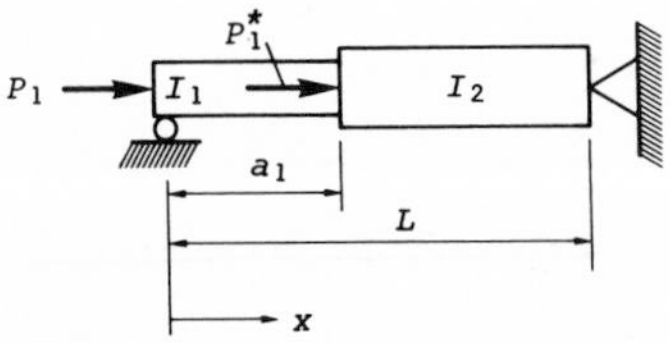

Fig. 2-19

EXAMPLE 2.19 Lumped Force Model Suppose the critical load of the column in Fig. 2-18 is to be found with a lumped force model rather than the usual representation of Example 2.17. Lumping an axial force for stability is similar to the lumping of mass for free dynamics. There are a variety of methods of selecting the discretization increment size. In general, the more lumps, the closer the calculated critical load will be to the theoretically correct force. The technique of lumping is quite simple. Calculate the force P in the beam along an increment of length Δa. Then let $P_{Li} = P\,\Delta a$ and use the point matrix of Table 2-14.

The lumped forces are connected by forceless sections represented by the transfer matrix of Table 2-8. Computationally, the lumped force model is frequently easier to handle than the usual continuous representation. In addition, for complex beams with variable axial forces, especially those with forces that vary continuously along the beam, it may be difficult to set up the solution unless the lumped force model is employed.

Assume the force in the column of Fig. 2-18 is to be discretized in four lumps to each side of the change in cross section. Let the interval over which the force P is lumped be denoted by Δa. Then $P_{Li} = P\,\Delta a$ and the point matrix (Table 2-14) for each lump is

$$\overline{\mathbf{U}}_i = \begin{bmatrix} 1 & 0 & 0 & 0 & 0 \\ 0 & 1 & 0 & 0 & 0 \\ 0 & P_{Li} & 1 & 0 & 0 \\ 0 & 0 & 0 & 1 & 0 \\ 0 & 0 & 0 & 0 & 1 \end{bmatrix} \tag{1}$$

Between each lumped station will be a segment of length ℓ with transfer matrix (Table 2-8)

$$\mathbf{U}_i = \begin{bmatrix} 1 & -\ell & -\dfrac{\ell^2}{2EI} & -\dfrac{\ell^3}{6EI} & 0 \\ 0 & 1 & \dfrac{\ell}{EI} & \dfrac{\ell^2}{2EI} & 0 \\ 0 & 0 & 1 & \ell & 0 \\ 0 & 0 & 0 & 1 & 0 \\ 0 & 0 & 0 & 0 & 1 \end{bmatrix} \qquad I = I_1 \quad \text{or} \quad I_2 \tag{2}$$

The global transfer matrix is formed as

$$\mathbf{s}_{x=L} = \mathbf{U}_{10}\overline{\mathbf{U}}_8\mathbf{U}_9\overline{\mathbf{U}}_7\mathbf{U}_8 \cdots \overline{\mathbf{U}}_1\mathbf{U}_1\mathbf{s}_0 = \mathbf{U}\mathbf{s}_0 \tag{3}$$

The boundary conditions still give the characteristic equation

$$\nabla = (U_{MM}U_{VV} - U_{VM}U_{MV})_{x=L} = 0 \tag{4}$$

Note that if all lumped forces P_{Li} are equal, (4) is a polynomial in P_{Li} rather than a transcendental function of P as found in Eq. (4) of Example 2.17. Herein lies the computational advantage. It is simpler to find the roots of a polynomial than the roots of a transcendental function.

The characteristic equation (4) is searched for P_{Li} by a numerical root-finding technique. Finally P is determined as $P_{Li}/\Delta a$. In many instances it may not be convenient to set all P_{Li} equal to each other (for $i = 1, 2, 3, \ldots$). For such cases, in practice, there will be a known proportionality between each P_{Li} and some nominal value. This is similar to $P_1^* = cP_1$ of Example 2.18. The nominal value becomes the single unknown in Eq. (4) that will be found with a root-finding technique.

2.18 Free Dynamic Response-Natural Frequencies

The natural frequencies ω_n are the roots of the characteristic equation resulting from the evaluation of the initial parameters. If this characteristic equation is represented by ∇, then the natural frequencies ω_n are those values of ω in ∇ that make ∇ equal to zero.

The procedure for calculating frequencies is essentially the same as that for stability analysis. The search for a critical axial load is replaced by a search for the frequencies ω_n. The model employed for free dynamic analyses requires the inclusion of mass. The mass can be modeled as being continuously distributed or it can be lumped at particular locations along the beam axis.

A free dynamic analysis follows the step-by-step stability calculation procedure outlined in the previous section, with the mass-related properties being computed and inserted in the analysis in step 2. The rest of the procedure remains essentially the same.

Formulas for the Characteristic Equations The natural frequencies ω_n, $n = 1, 2, \ldots$, of a beam are those values of ω that make ∇ of Table 2-19 or Eq. (2.31) [or (2.32)] equal to zero. This ∇ is formed from the elements of the global transfer matrix **U** of Eq. (2.30). For complex beams it is necessary to evaluate the frequencies from $\nabla = 0$ with a computational root-finding technique. Root-searching procedures are, in general, time-consuming.

The mode shapes for deflection, slope, bending moment, and shear forces are given by Eq. (2.29) using initial parameters defined by

$$\begin{aligned} s_{10} &= 0 \\ s_{20} &= 0 \\ s_{30} &= s_{30} \\ s_{40} &= -s_{30} U_{s_k s_3} / U_{s_k s_4} \end{aligned} \tag{2.33}$$

where the definitions of Eqs. (2.31) still apply. The variable s_{30} is assigned to be the initial parameter that can be given any value. Usually it is given the value of 1. The unspecified initial parameter is to be expected since the mode shapes describe only a shape of a state variable and not a precise amplitude. The fifth-column terms in the transfer matrices, that is, the loading functions F_v, F_θ, F_M, F_V, can be ignored during the calculation of natural frequencies and mode shapes. If there are in-span indeterminates, then the initial parameters are given by

$$\begin{aligned} s_{10} &= 0 \\ s_{20} &= 0 \\ s_{30} &= s_{30} \\ s_{40} &= -s_{30} U_{s_j s_3} / U_{s_j s_4} \end{aligned} \tag{2.34}$$

where the definitions of Eq. (2.32) apply.

EXAMPLE 2.20 Uniform Cantilevered Beam Find the frequencies and mode shapes of a uniform beam fixed on the left end and free on the right end.

The natural frequencies are found as the roots of $\nabla = 0$, where ∇ is the characteristic equation found by applying the boundary conditions to Eq. (2.30) or ∇ can be taken from Table (2-19) or Eqs. (2.31). The end conditions are $v_{x=0} = v_0 = 0$, $\theta_{x=0} = \theta_0 = 0$, $M_{x=L} = 0$, $V_{x=L} = 0$. The conditions applied to Eqs. (2.30) give

$$\nabla = (U_{MM} U_{VV} - U_{VM} U_{MV})_{x=L} \tag{1}$$

If Eqs. (2.31) are used, set $s_1 = v$, $s_2 = \theta$. The remaining state variables are M and V, therefore $s_3 = M$, $s_4 = V$. The conditions at the right end are $M_{x=L} = V_{x=L} = 0$. Thus, we define $s_k = M$, $s_j = V$. Then, from Eqs. (2.31), $s_k = M$, $s_j = V$ and

$$\nabla = U_{s_k s_3} U_{s_j s_4} - U_{s_j s_3} U_{s_k s_4} = (U_{MM} U_{VV} - U_{VM} U_{MV})_{x=L} \tag{2}$$

Take the elements U_{kj} from the transfer matrix of Table 2-11 with $\ell = L$. We find

$$\begin{aligned} \nabla &= \frac{1}{2}(\cosh \beta L + \cos \beta L) \frac{1}{2}(\cosh \beta L + \cos \beta L) \\ &\quad - \frac{\beta}{2}(\sinh \beta L - \sin \beta L) \frac{1}{2\beta}(\sinh \beta L + \sin \beta L) \\ &= \frac{1}{2} \cos \beta L \cosh \beta L + \frac{1}{2} = 0 \end{aligned}$$

The values of ω, or, equivalently, values of βL, that make $\cos \beta L \cosh \beta L + 1$ equal to zero are desired.

These are

$$\begin{aligned} \beta_1 L &= 1.875 \\ \beta_2 L &= 4.694 \\ &\cdots\cdots\cdots \end{aligned}$$

and the frequencies are found from $\omega_n = (\beta_n L)^2 \sqrt{EI/\rho L^4}$.

The mode shapes can be developed in a manner similar to that used in Example 2.16 for stability. Or Eq. (2.33) can be employed. For the latter approach, the initial parameters required for the mode shapes are given by

$$s_{10} = v_0 = 0$$

$$s_{20} = \theta_0 = 0$$

$$s_{30} = M_0$$

$$s_{40} = V_0 = -s_{30}\frac{U_{s_k s_3}}{U_{s_k s_4}} = -M_0\frac{U_{VM}}{U_{VV}} = -M_0\frac{-\frac{\beta}{2}(\sinh\beta L - \sin\beta L)}{\frac{1}{2}(\cosh\beta L + \cos\beta L)}$$

These values are inserted in Eq. (2.29). For example, in the case of the deflection

$$v = M_0 U_{vM} + V_0 U_{vV} = M_0\left[U_{vM}(\ell) - U_{vV}(\ell)\frac{U_{VM}(L)}{U_{VV}(L)}\right]$$

$$= \frac{M_0}{2\beta^2 EI}\left[\cosh\beta\ell - \cos\beta\ell - (\sinh\beta\ell - \sin\beta\ell)\frac{\sinh\beta L - \sin\beta L}{\cosh\beta L + \cos\beta L}\right] \qquad (3)$$

where ℓ can assume any value between $x = 0$ and $x = L$. A different mode shape is provided by (3) for each frequency ω_n. The parameter M_0 is the mode shape amplitude of unknown magnitude which is often set equal to one.

EXAMPLE 2.21 Beam on Elastic Supports As an example of a complex beam, consider the shaft of Fig. 2-20*a* that is resting on elastic supports.

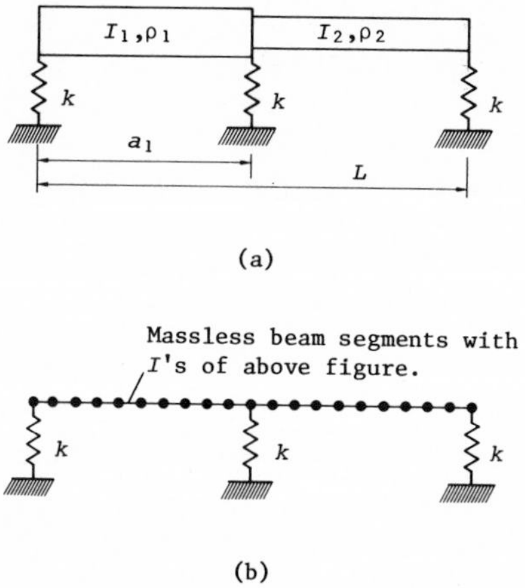

Fig. 2-20

The mode shapes are given by Eq. (2.29) as

$$\begin{aligned} \mathbf{s} &= \mathbf{U}_1\overline{\mathbf{U}}_1\mathbf{s}_0 \qquad 0 < x < a_1 \\ \mathbf{s} &= \mathbf{U}_2\overline{\mathbf{U}}_2\mathbf{U}_1\overline{\mathbf{U}}_1\mathbf{s}_0 \qquad a_1 < x < L \end{aligned} \qquad (1)$$

and at the right end, $x = L$, from Eq. (2.30)

$$\mathbf{s}_{x=L} = \overline{\mathbf{U}}_3\mathbf{U}_2\overline{\mathbf{U}}_2\mathbf{U}_1\overline{\mathbf{U}}_1\mathbf{s}_0 = \mathbf{U}\mathbf{s}_0 \qquad (2)$$

where, from Table 2.14,

$$\overline{\mathbf{U}}_3 = \overline{\mathbf{U}}_2 = \overline{\mathbf{U}}_1 = \begin{bmatrix} 1 & 0 & 0 & 0 \\ 0 & 1 & 0 & 0 \\ 0 & 0 & 1 & 0 \\ k & 0 & 0 & 1 \end{bmatrix} \tag{3}$$

and $\mathbf{U}_1$, $\mathbf{U}_2$ are the usual transfer matrices that can be selected from the general transfer matrix of Table 2-13 according to the beam model under consideration, for example, segments with bending, shear deformation, and/or rotary inertia. Note that the fifth row and fifth column of (3) have been dropped since they are not needed for free-dynamics analyses.

If desired, a lumped-mass model (Fig. 2-20*b*) can be employed for the segments between the spring supports. The $\mathbf{U}_1$ and $\mathbf{U}_2$ are formed by progressive multiplication of transfer matrices for a massless segment (Table 2-8) and a point mass (Table 2-14).

The ends of the beam are considered to be free in spite of the end spring supports. This is the case because such supports are incorporated into the overall transfer matrix and hence need not be taken into account in the boundary conditions. From the definitions associated with Eqs. (2.31), we set $s_1 = M$, $s_2 = V$ since $M_{x=0} = V_{x=0} = 0$. Then $s_3 = v$, $s_4 = \theta$. As a result of $M_{x=L} = V_{x=L} = 0$, s_k and s_j are set equal to M and V. Then, from Eqs. (2.33),

$$\begin{aligned} s_{10} &= M_0 = 0 \\ s_{20} &= V_0 = 0 \\ s_{30} &= v_0 \\ s_{40} &= -v_0 \left.\frac{U_{Mv}}{U_{M\theta}}\right|_{x=L} \end{aligned} \tag{4}$$

and from Eq. (2.31)

$$\nabla = (U_{Mv}U_{V\theta} - U_{M\theta}U_{Vv})_{x=L} \tag{5}$$

where the U_{kj} terms in (4) and (5) are the elements of $\mathbf{U}$ of (2). The frequencies are the roots of $\nabla = 0$. The mode shapes are found by inserting these frequencies and the initial parameters of (4) into (1).

Polynomial Characteristic Equation for Lumped-Mass Models It is of interest that the global transfer matrix $\mathbf{U}$ and hence ∇ can be expressed in terms of a polynomial in ω^2 if a lumped-mass model is used. For example, consider a segment of beam with mass M_k at $x = a_k$ and a massless section of length ℓ extending to $x = a_{k+1}$. The transfer matrix for this combined segment can be arranged as

$$\mathbf{U}_{k+1}\overline{\mathbf{U}}_k = \omega^2 \mathbf{S}_k + \mathbf{R}_k$$

where, for an Euler-Bernoulli beam, $\mathbf{U}_{k+1}$ is a transfer matrix from Table 2-8 for the massless segment and $\overline{\mathbf{U}}_k$ is a point matrix from Table 2-14 for the point mass. Upon rearrangement it is seen that $\mathbf{S}_k$ and $\mathbf{R}_k$ take the form

$$\mathbf{R}_k = \begin{bmatrix} 1 & -\ell & -\dfrac{\ell^2}{2EI} & -\dfrac{\ell^3}{3!EI} \\ 0 & 1 & \dfrac{\ell}{EI} & \dfrac{\ell^2}{2EI} \\ 0 & 0 & 1 & \ell \\ 0 & 0 & 0 & 1 \end{bmatrix} \qquad \mathbf{S}_k = M_k \begin{bmatrix} -\dfrac{\ell^3}{3!EI} & 0 & 0 & 0 \\ -\dfrac{\ell^2}{2EI} & 0 & 0 & 0 \\ -\ell & 0 & 0 & 0 \\ 1 & 0 & 0 & 0 \end{bmatrix}$$

For n such segments, the global transfer matrix may be written as

$$\mathbf{U} = (\omega^2\mathbf{S}_n + \mathbf{R}_n)(\omega^2\mathbf{S}_{n-1} + \mathbf{R}_{n-1}) \cdots (\omega^2\mathbf{S}_1 + \mathbf{R}_1)$$
$$= \mathbf{A}_0 + \mathbf{A}_1\omega^2 + \mathbf{A}_2\omega^4 + \cdots + \mathbf{A}_n\omega^{2n}$$

where the $\mathbf{A}_j$ are 4×4 matrices resulting from the multiplication of the $(\omega^2\mathbf{S}_k + \mathbf{R}_k)$ factors. Although a general expression for $\mathbf{A}_j$ is readily derived, this serves no useful purpose because it is a simple matter with the computer to progressively multiply these factors and keep track of the coefficients. Similarly, the frequency determinant ∇ then becomes an easily identified polynomial. The frequency equation $\nabla = 0$ is now eligible for solution by standard polynomial-root-solving routines. In contrast to the technique explained previously for continuous-mass systems, it is not necessary to redevelop the global transfer matrix for each new trial ω^2. The commonly available polynomial-root-finding routines require only the coefficients of the polynomial as input. A single analysis of the structural member (formation of the global transfer matrix) suffices to provide these coefficients. Furthermore, most polynomial software packages are quite reliable and do not require estimates of the eigenvalues being computed. Thus, the polynomial approach appears to avoid the sometimes hazardous, time-consuming repetitive analyses, and the need for root estimates of the usual approach. The procedure applies as well for the complex frequencies occurring in damped systems. It is also useful for various stability problems, including the usual buckling problem and rotating shaft whirl instability. The drawback to the polynomial construction scheme, other than the fact that it applies only to lumped-parameter systems, is the possibility of encountering numerical instabilities for complex members or if very high frequencies are being sought.

EXAMPLE 2.22 Continuous Beam Suppose the springs of the beam of Fig. 2-20 are so stiff that the beam can be considered to be rigidly supported at the springs. We consider then the natural frequencies of the beam of Fig. 2-21.

The mode shapes are found from

$$\begin{aligned} \mathbf{s} &= \mathbf{U}_1\mathbf{s}_0 & 0 \leqslant x < a_1 \\ \mathbf{s} &= \mathbf{U}_2\overline{\mathbf{U}}_2\mathbf{U}_1\mathbf{s}_0 & a_1 < x \leqslant L \end{aligned} \tag{1}$$

where $\mathbf{U}_1$ and $\mathbf{U}_2$ are the field matrices covering the two sections. $\overline{\mathbf{U}}_2$ is the point matrix at $x = a_1$ that accounts for the rigid in-span support.

Equations (2.34) and (2.32) with $s_1 = v$, $s_2 = M$, $s_3 = \theta$, $s_4 = V$, $s_j = M$, $s_h = v$, give

$$s_{10} = v_0 = 0, \quad s_{20} = M_0 = 0, \quad s_{30} = \theta_0, \quad s_{40} = V_0 = -\theta_0(U_{M\theta}/U_{MV})_{x=L} \tag{2}$$

$$\nabla = U_{s_h s_3}U_{s_j s_4} - U_{s_j s_3}U_{s_h s_4} = U_{v\theta}U_{MV} - U_{M\theta}U_{vV} \tag{3}$$

where $U_{s_h s_3} = U_{v\theta}$ and $U_{s_h s_4} = U_{vV}$ are elements of the transfer matrix spanning $a_1 < x \leqslant L$. $U_{M\theta}$ and U_{MV} are elements of the global transfer matrix for the whole beam. Since s_j, a zero boundary condition at $x = L$, was chosen to be M, $v_{x=L} = 0$ is available for establishing $\overline{\mathbf{U}}_2$; that is, $s_m = v$. From Table 2-15

$$\overline{\mathbf{U}}_2 = \begin{bmatrix} 1 & 0 & 0 & 0 \\ 0 & 1 & 0 & 0 \\ 0 & 0 & 1 & 0 \\ k_{Vv} & k_{V\theta} & k_{VM} & 1 + k_{VV} \end{bmatrix}$$

$$\text{with} \quad k_{Vv} = -\frac{U_{vv}}{U_{vV}} \qquad k_{V\theta} = -\frac{U_{v\theta}}{U_{vV}} \qquad k_{VM} = -\frac{U_{vM}}{U_{vV}} \qquad k_{VV} = -1$$

where these elements are for the transfer matrix of the segment between $x = a_1$ and $x = L$.

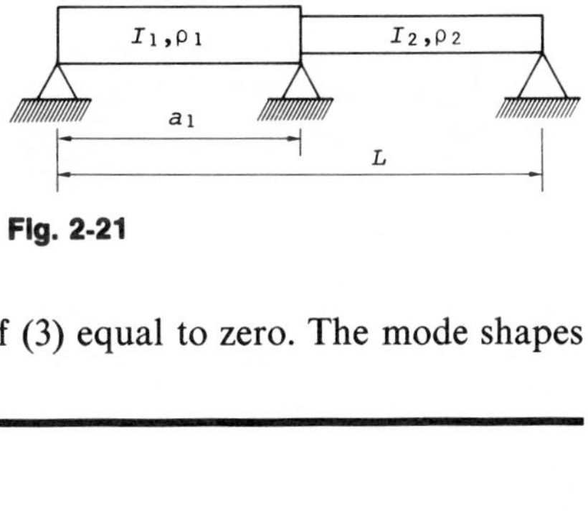

Fig. 2-21

The frequencies are found by setting ∇ of (3) equal to zero. The mode shapes are identified by placing (2) in (1).

2.19 Forced Harmonic Motion

If the forcing functions vary as $\sin \Omega t$ or $\cos \Omega t$, where Ω is the frequency of the loading, the state variables v, θ, M, and V also vary as $\sin \Omega t$ or $\cos \Omega t$. The spatial distribution of v, θ, M, and V are found by setting up a static solution using those transfer matrices containing ω. In these transfer matrices, ω should be replaced by Ω. The solution procedure is exactly the same as solving for the static response, since time need not enter the analysis. The initial parameters v_0, θ_0, M_0 and V_0 are given by Table 2-19 [Eqs. (2.31)] and Eqs. (2.32).

2.20 Dynamic Response Due to Arbitrary Loading

In terms of a modal solution, the time-dependent state variables resulting from arbitrary dynamic loading are expressed by

Displacement method:

$$\begin{aligned} v(x,t) &= \sum_n A_n(t) v_n(x) \\ \theta(x,t) &= \sum_n A_n(t) \theta_n(x) \\ M(x,t) &= \sum_n A_n(t) M_n(x) \\ V(x,t) &= \sum_n A_n(t) V_n(x) \end{aligned} \tag{2.35a}$$

Acceleration method:

$$\begin{aligned} v(x,t) &= v_s(x,t) + \sum_n B_n(t) v_n(x) \\ \theta(x,t) &= \theta_s(x,t) + \sum_n B_n(t) \theta_n(x) \\ M(x,t) &= M_s(x,t) + \sum_n B_n(t) M_n(x) \\ V(x,t) &= V_s(x,t) + \sum_n B_n(t) V_n(x) \end{aligned} \tag{2.35b}$$

where

$$A_n(t) = \frac{\eta_n(t)}{N_n} \qquad B_n(t) = \frac{\xi_n(t)}{N_n} \tag{2.36}$$

These relations provide the complete dynamic response (Ref. 2.7). That is, these are the solutions to Eqs. (2.16).

Two normal-mode solution forms—the *displacement* and *acceleration* methods—are presented here. Both techniques accept arbitrary loading, time-dependent boundary conditions, and in-span conditions. The approaches differ in that the displacement method employs only the natural-mode information in constructing a solution, whereas the acceleration method takes advantage of the static response in addition to natural modes and frequencies. Both methods involve a summation over the natural mode shapes. Of the two approaches, the acceleration method usually exhibits superior convergence properties for the time period during which the loading is being applied.

No Damping or Proportional Damping For viscous damping, the terms $c' \, \partial v/\partial t$ and $c^* \, \partial \theta/\partial t$ should be added to the right-hand sides of Eqs. (2.16*d*) and (2.16*c*), respectively. If the damping is "proportional," then $c' = c_\rho \rho$ and $c^* = c_\rho \rho r^2$. Set $c' = 0$ or $c^* = 0$ for no linear or rotary viscous damping, respectively.

In Eqs. (2.35*b*) the terms with the subscript s are static solutions, as given in Section 2.16, that are determined for the applied loading at each point in time. The terms v_n, θ_n, M_n, V_n are the mode shapes, that is, v, θ, M, V with $\omega = \omega_n$,

$n = 1, 2, \ldots,$ as explained in Section 2.18. The quantities in Eqs. (2.36) are defined as

$$N_n = \int_0^L \rho\left(v_n^2 + r^2\theta_n^2\right) dx \tag{2.37a}$$

$$\eta_n(t) = e^{-\zeta_n\omega_n t}\left[\cos\alpha_n t + \frac{\zeta_n\omega_n}{\alpha_n}\sin\alpha_n t\right]\eta_n(0) + e^{-\zeta_n\omega_n t}\,\frac{\sin\alpha_n t}{\alpha_n}\,\frac{\partial\eta_n}{\partial t}(0)$$
$$+ \int_0^t f_n(\tau)e^{-\zeta_n\omega_n(t-\tau)}\,\frac{\sin\alpha_n(t-\tau)}{\alpha_n}\,d\tau \tag{2.37b}$$

where $\alpha_n = \omega_n\sqrt{1-\zeta_n^2}$, $\zeta_n = c_\rho/2\omega_n$. $\xi_n(t)$ is taken from $\eta_n(t)$ by replacing

$$\eta_n(0) \quad \text{by} \quad \eta_n(0) - \frac{f_n(0)}{\omega_n^2}$$

$$\frac{\partial\eta_n}{\partial t}(0) \quad \text{by} \quad \frac{\partial\eta_n}{\partial t}(0) - \frac{1}{\omega_n^2}\frac{\partial f_n}{\partial t}(0)$$

$$f_n(\tau) \quad \text{by} \quad -\frac{1}{\omega_n^2}\left(\frac{\partial^2}{\partial\tau^2} + c_\rho\frac{\partial}{\partial\tau}\right)f_n(\tau)$$

If $\zeta_n > 1$, replace sin by sinh, cos by cosh, and $\sqrt{1-\zeta_n^2}$ by $\sqrt{\zeta_n^2 - 1}$. For zero viscous damping, $\zeta_n = 0$.

Set $r^2 = 0$ if rotary inertia effects are not to be taken into account.

$$\eta_n(0) = \int_0^L \rho\left[v(x,0)v_n + r^2\theta(x,0)\theta_n\right] dx \tag{2.38}$$

$$f_n(t) = \int_0^L \left[w(x,t)v_n + c(x,t)\theta_n + \frac{M_T(x,t)}{EI}M_n\right] dx + h_n(a_k,t) \tag{2.39a}$$

In the case of the acceleration method, $f_n(t)$ can alternatively be expressed as

$$f_n(t) = \omega_n^2\int_0^L \left[\rho v_s(x,t)v_n + \rho r^2\theta_s(x,t)\theta_n\right] dx \tag{2.39b}$$

The function $h_n(a_k, t)$ accounts for nonhomogeneous displacement $v(a_k, t)$ or rotation $\theta(a_k, t)$ conditions at $x = a_k$, such as those produced by supports or prescribed time-dependent displacements located in-span or on the boundary:

$$h_n(a_k,t) = -v(a_k,t)\Delta V_n(a_k) - \theta(a_k,t)\Delta M_n(a_k) \tag{2.40a}$$

$$\Delta V_n(a_k) = V_n(a_k^-) - V_n(a_k^+)$$
$$\Delta M_n(a_k) = M_n(a_k^-) - M_n(a_k^+)$$

$+(-)$ means just to the right (left) of $x = a_k$

If the right end ($x = L$) has a nonzero (for example, time-dependent) displacement, then Eq. (2.40a) reduces for $a_k = L$ to

$$h_n(L,t) = -v(L,t)V_n(L) - \theta(L,t)M_n(L) \tag{2.40b}$$

If the left end ($x = 0$) has a nonhomogeneous displacement, then Eq. (2.40*a*) reduces for $a_k = 0$ to

$$h_n(0, t) = +v(0, t)V_n(0) + \theta(0, t)M_n(0) \tag{2.40c}$$

EXAMPLE 2.23 Zero Initial Conditions The response formulas given above accept arbitrary loading, prescribed displacements, initial position, and initial velocity conditions. Suppose for a beam with no rotary inertia $v(x, 0)$ and $\partial v(x, 0)/\partial t$ are zero. Then, if there are no prescribed displacements and $w(x, t)$ is the only loading, the responses are given by the above formulas with Eq. (2.37*b*) reducing to

$$\eta_n(t) = \int_0^t \int_0^L w(x, \tau)v_n \, dx \, \frac{\sin \omega_n(t - \tau) \, d\tau}{\omega_n} \tag{1}$$

EXAMPLE 2.24 Prescribed Initial Conditions Assume the beam in the previous example is released from the prescribed initial displacement $v(x, 0) = g_1(x)$ and velocity $\partial v(x, 0)/\partial t = g_2(x)$. If $w(x, t) = 0$ then

$$\eta_n(t) = \cos \omega_n t \int_0^L \rho g_1(x)v_n(x) \, dx + \frac{\sin \omega_n t}{\omega_n} \int_0^L \rho g_2(x)v_n(x) \, dx \tag{1}$$

EXAMPLE 2.25 Time-Independent Loading In the case of a loading that is not time-dependent, the loading "rides" the member; then $w(x, t) = w(x)$, and Eq. (1) of Example 2.23 reduces to

$$\eta_n(t) = \frac{(1 - \cos \omega_n t)}{\omega_n^2} \int_0^L w(x)v_n(x) \, dx \tag{1}$$

for zero initial conditions. It is interesting that although the loading is independent of time, the response of the beam appears to be time-dependent. We can resolve this apparent paradox by noting that the assumption of zero initial conditions is not valid for a truly time-independent problem since the loading is on the beam at $t = 0$, and hence there is an initial displacement. In the case of a statics problem, the acceleration, $\partial^2 v(x, t)/\partial t^2$, is zero and as a result $\partial^2 \eta_n(t)/\partial t^2 = 0$. Then, it can be shown that $\eta_n(t)$ is given by

$$\eta_n(t) = \frac{1}{\omega_n^2} \int_0^L w(x)v_n(x) \, dx \tag{2}$$

rather than by (1). The complete static solution becomes

$$v(x) = \sum_{n=1} \frac{v_n(x) \int_0^L w(x)v_n(x) \, dx}{\omega_n^2 \int_0^L \rho v_n v_n \, dx} \tag{3}$$

We see from (3) that if the mode shapes are available, it is possible to compute the static response. However, since the development of each mode shape involves the application of the transfer matrix method, it is simpler to compute the static response as in Section 2.16 with a single application of the transfer matrix method. Also, the usual static analysis of Section 2.16 avoids the summations and integration of (3).

EXAMPLE 2.26 Steady-State Loading If the loading is harmonic, say $w(x, t) = w(x) \cos \Omega t$, then Eq. (1) of Example 2.23 becomes

$$\eta_n(t) = \frac{\cos \Omega t - \cos \omega_n t}{\omega_n^2 - \Omega^2} \int_0^L w(x) v_n(x)\, dx \tag{1}$$

As is to be expected, $\eta_n(t)$ and the response variables increase with time as Ω approaches one of the natural frequencies ω_n. Thus, as $\Omega \to \omega_n$,

$$\eta_n(t) = \frac{t}{2} \frac{\sin \omega_n t}{\omega_n} \int_0^L w(x) v_n(x)\, dx \tag{2}$$

EXAMPLE 2.27 Moving Load Assume a load of velocity v^* moves across the beam. Then

$$w(x, t) = w(t) \langle x - v^* t \rangle^{-1} \tag{1}$$

and, from the delta function integration property of Eq. (1.5), Eq. (1) of Example 2.23 reduces to

$$\eta_n(t) = \int_0^t w(\tau) v_n(v^* \tau) \frac{\sin \omega_n (t - \tau)}{\omega_n}\, d\tau \tag{2}$$

while the load remains on the beam. The effect on the motion of the mass of the load is not taken into account.

EXAMPLE 2.28 Impacting Mass Drop a mass M_1 on a member and assume it impacts on the beam at $x = a_1$ with velocity v^*. Treat the free-motion problem as though the mass were an integral part of the beam; that is, find the mode shapes for an undamped beam with the mass M_1 built into it. The initial conditions become

$$v(x, 0) = 0 \qquad \frac{\partial v}{\partial t}(x, 0) = v^* \langle x - a_1 \rangle^{-1} \tag{1}$$

and $\eta_n(t)$ of Eq. (1) of Example 2.24 takes the form

$$\eta_n(t) = \frac{\sin \omega_n t}{\omega_n} M_1 v^* w_n(a_1) \tag{2}$$

where the integration property of Eq. (1.5) for the delta function $\langle x - a_1 \rangle^{-1}$ has been used. It is assumed here that relative to the mass dropped on the beam, the mass of the beam under the point of impact is negligible.

EXAMPLE 2.29 Solution Details for the Transient Response of a Beam

Displacement Method Consider the simple problem of a cantilever beam with a time-dependent force on the free end. Suppose the beam can be represented by a lumped-mass model with a single mass at the free end (Fig. 2-22). We choose this model for demonstrating the procedure for a simple solution, not for realistically approximating a practical beam.

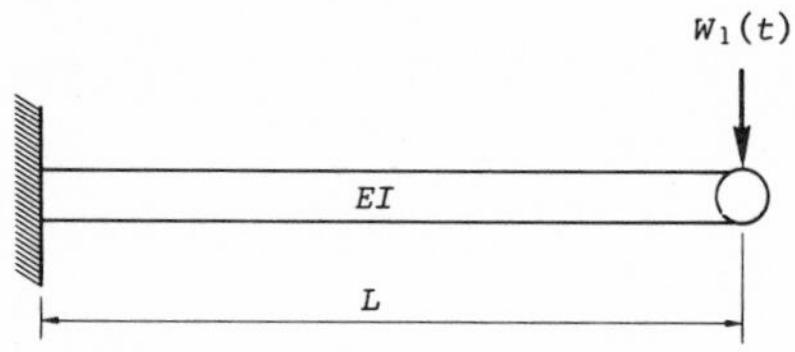

Fig. 2-22

A forced dynamic solution begins with a study of the free motion of the beam. From Tables 2-8 and 2-14 and Eqs. (2.30), the displacements and forces are given by

$$\begin{bmatrix} v \\ \theta \\ M \\ V \end{bmatrix}_\ell = \begin{bmatrix} 1 & -\ell & -\dfrac{\ell^2}{2EI} & -\dfrac{\ell^3}{3!EI} \\ 0 & 1 & \dfrac{\ell}{EI} & \dfrac{\ell^2}{2EI} \\ 0 & 0 & 1 & \ell \\ 0 & 0 & 0 & 1 \end{bmatrix} \begin{bmatrix} v \\ \theta \\ M \\ V \end{bmatrix}_0 \tag{1}$$

where ℓ can take any value from $x = 0$ up to $x < L$. At $x = L$,

$$\begin{bmatrix} v \\ \theta \\ M \\ V \end{bmatrix}_L = \begin{bmatrix} 1 & 0 & 0 & 0 \\ 0 & 1 & 0 & 0 \\ 0 & 0 & 1 & 0 \\ -M_1^*\omega_n^2 & 0 & 0 & 1 \end{bmatrix} \begin{bmatrix} 1 & -L & -\dfrac{L^2}{2EI} & -\dfrac{L^3}{3!EI} \\ 0 & 1 & \dfrac{L}{EI} & \dfrac{L^2}{2EI} \\ 0 & 0 & 1 & L \\ 0 & 0 & 0 & 1 \end{bmatrix} \begin{bmatrix} v \\ \theta \\ M \\ V \end{bmatrix}_0$$

$$= \begin{bmatrix} 1 & -L & -\dfrac{L^2}{2EI} & -\dfrac{L^3}{3!EI} \\ 0 & 1 & \dfrac{L}{EI} & \dfrac{L^2}{2EI} \\ 0 & 0 & 1 & L \\ -M_1^*\omega_n^2 & LM_1^*\omega_n^2 & \dfrac{L^2M_1^*\omega_n^2}{2EI} & 1 + \dfrac{L^3M_1^*\omega_n^2}{3!EI} \end{bmatrix} \begin{bmatrix} v \\ \theta \\ M \\ V \end{bmatrix}_0 \tag{2}$$

The frequencies are found by setting $\nabla = 0$, where ∇ is selected from Table 2-19 [or Eqs. (2.31)] for a fixed-free beam. Thus

$$\nabla = (U_{MM}U_{VV} - U_{MV}U_{VM})_{x=L} = 0 \tag{3}$$

Substitution of the elements of (2) into (3) gives

$$\nabla = 1\left(1 + \frac{L^3M_1^*\omega_n^2}{3!EI}\right) - L\frac{L^2M_1^*\omega_n^2}{2EI} = 1 - \frac{L^3M_1^*\omega_n^2}{3EI} = 0$$

or

$$\omega_1^2 = \frac{3EI}{L^3M_1^*} \tag{4}$$

There is only a single frequency for this single degree of freedom model.

Insertion of the frequency of (4) and the initial parameters of Eqs. (2.33) in (1) and (2) gives for $0 \leqslant x < L$

$$\begin{aligned} v_1 &= M_0\left(-\frac{x^2}{2EI} + \frac{x^3}{3!EIL}\right) \\ \theta_1 &= M_0\left(\frac{x}{EI} - \frac{x^2}{2EIL}\right) \\ M_1 &= M_0\left(1 - \frac{x}{L}\right) \\ V_1 &= -M_0\left(\frac{1}{L}\right) \end{aligned} \tag{5}$$

where we have replaced ℓ by x.

At $x = L$,

$$v_1 = -M_0\left(\frac{1}{\omega_1^2LM_1^*}\right) \qquad \theta_1 = M_0\frac{L}{2EI}$$

$$M_1 = 0 \qquad V_1 = 0 \tag{6}$$

The displacement-method response to the time-varying force $W_1(t)$ is given by Eqs. (2.35a). Since there is a single natural frequency ($\omega_n = \omega_1$), these expressions reduce to

$$\begin{aligned} v(x,t) &= A_1(t)v_1(x) \\ \theta(x,t) &= A_1(t)\theta_1(x) \\ M(x,t) &= A_1(t)M_1(x) \\ V(x,t) &= A_1(t)V_1(x) \end{aligned} \tag{7}$$

where from Eqs. (2.36) and (2.37) for zero damping and no rotary inertia,

$$A_1(t) = \left[\int_0^L \rho v(x,0)v_1\,dx\cos\omega_1 t + \int_0^L \rho\frac{\partial v}{\partial t}(x,0)v_1\,dx\frac{\sin\omega_1 t}{\omega_1}\right.$$

$$\left. + \int_0^t\int_0^L w(x,\tau)v_1\,dx\frac{\sin\omega_1(t-\tau)}{\omega_1}\,d\tau\right] \Big/ \int_0^L \rho v_1^2\,dx$$

Since $\rho = M_1^* \langle x - L \rangle^{-1}$, $w(x,t) = W_1(t)\langle x - L\rangle^{-1}$,

$$A_1(t) = \left[M_1^* v(L,0) v_1(L) \cos \omega_1 t + M_1^* \frac{\partial v}{\partial t}(L,0) v_1(L) \frac{\sin \omega_1 t}{\omega_1} + \int_0^t W_1(\tau) v_1(L) \frac{\sin \omega_1 (t-\tau)}{\omega_1}\, d\tau \right] \Big/ M_1^* v_1^2(L)$$

$$= \frac{v(L,0)}{v_1(L)} \cos \omega_1 t + \frac{\partial v}{\partial t}(L,0) \frac{\sin \omega_1 t}{\omega_1 v_1(L)} + \frac{1}{M_1^* v_1(L)} \int_0^t W_1(\tau) \frac{\sin \omega_1 (t-\tau)}{\omega_1}\, d\tau$$

where the singularity function integral of Eq. (1.5) has been used. The mode shape equations (6) give

$$A_1(t) = -v_1(L,0) \frac{M_1^* \omega_1^2 L}{M_0} \cos \omega_1 t - \frac{\partial v}{\partial t}(L,0) \frac{M_1^* \omega_1 L}{M_0} \sin \omega_1 t - \frac{\omega_1 L}{M_0} \int_0^t W_1(\tau) \sin \omega_1 (t-\tau)\, d\tau \tag{8}$$

From (5), (6), (7), and (8), the final form of the response is

$$\left.\begin{aligned}
v(x,t) &= \left(\frac{x^2}{2EI} - \frac{x^3}{3!\,EIL} \right) \omega_1 L \int_0^t W_1(\tau) \sin \omega_1 (t-\tau)\, d\tau \\
\theta(x,t) &= \left(\frac{x^2}{2EIL} - \frac{x}{EI} \right) \omega_1 L \int_0^t W_1(\tau) \sin \omega_1 (t-\tau)\, d\tau \\
M(x,t) &= \left(\frac{x}{L} - 1 \right) \omega_1 L \int_0^t W_1(\tau) \sin \omega_1 (t-\tau)\, d\tau \\
V(x,t) &= \omega_1 \int_0^t W_1(\tau) \sin \omega_1 (t-\tau)\, d\tau
\end{aligned}\right\} \quad 0 \leqslant x < L$$

and for the mass at $x = L$,

$$\begin{aligned}
v(L,t) &= \frac{1}{\omega_1 M_1^*} \int_0^t W_1(\tau) \sin \omega_1 (t-\tau)\, d\tau \\
\theta(L,t) &= -\frac{\omega_1 L^2}{2EI} \int_0^t W_1(\tau) \sin \omega_1 (t-\tau)\, d\tau \\
M(L,t) &= 0 \\
V(L,t) &= 0
\end{aligned}$$

where it has been assumed that the initial displacement $v(L,0)$ and velocity $\partial v\,(L,0)/\partial t$ of the mass are zero.

Acceleration Method The acceleration form of the transient response may be used as an alternative solution which for many loadings W_1 will converge more rapidly than the previous displacement solution. From Eq. (2.35*b*)

$$\begin{aligned} v(x,t) &= v_s(x,t) + B_1(t)v_1(x) \\ \theta(x,t) &= \theta_s(x,t) + B_1(t)\theta_1(x) \\ M(x,t) &= M_s(x,t) + B_1(t)M_1(x) \\ V(x,t) &= V_s(x,t) + B_1(t)V_1(x) \end{aligned} \tag{9}$$

where $v_s(x, t)$ and so forth are the static solutions determined as described in Section 2.16 for the load $W_1(t)$ evaluated at each point in time of interest. The temporal coefficient $B_1(t)$ is given by Eqs. (2.36), (2.37), and (2.38) as

$$\begin{aligned} B_1(t) = \Bigg\{ &\cos\omega_1 t \int_0^L \left[\rho v(x,0)v_1 - \frac{w(x,0)}{\omega_1^2} v_1 \right] dx \\ &+ \frac{\sin\omega_1 t}{\omega_1} \int_0^L \left[\rho \frac{\partial v}{\partial t}(x,0)v_1 - \frac{1}{\omega_1^2} \frac{\partial w(x,0)}{\partial t} v_1 \right] dx \\ &- \frac{1}{\omega_1^2} \int_0^t \int_0^L \frac{\partial^2 w(x,\tau)}{\partial \tau^2} v_1\, dx \frac{\sin\omega_1(t-\tau)}{\omega_1} d\tau \Bigg\} \Bigg/ \int_0^L \rho v_1^2\, dx \end{aligned} \tag{10}$$

Again, $\rho = M_1^* \langle x - L \rangle^{-1}$ and $w(x, t) = W_1(t)\langle x - L \rangle^{-1}$. Make use of the mode shapes (6) to find

$$\begin{aligned} B_1(t) = &\left[\frac{v(L,o)}{v_1(L)} - \frac{W_1(0)}{\omega_1^2 M_1^* v_1(L)} \right] \cos\omega_1 t \\ &+ \left[\frac{\partial v}{\partial t}(L,0) \frac{1}{v_1(L)} - \frac{\partial W_1}{\partial t}(0) \frac{1}{\omega_1^2 M_1^* v_1(L)} \right] \frac{\sin\omega_1 t}{\omega_1} \\ &- \frac{1}{\omega_1^2 M_1^* v_1(L)} \int_0^t \frac{\partial^2 W_1(\tau)}{\partial \tau^2} \frac{\sin\omega_1(t-\tau)}{\omega_1} d\tau \end{aligned} \tag{11}$$

$$\begin{aligned} B_1(t) = &\left[-v_1(L,0) \frac{M_1^* \omega_1^2 L}{M_0} + W_1(0) \frac{L}{M_0} \right] \cos\omega_1 t \\ &+ \left[-\frac{\partial v}{\partial t}(L,0) \frac{M_1^* \omega_1^2 L}{M_0} + \frac{\partial W_1(0)}{\partial t} \frac{L}{M_0} \right] \frac{\sin\omega_1 t}{\omega_1} \\ &+ \frac{L}{\omega_1 M_0} \int_0^t \frac{\partial^2 W_1(\tau)}{\partial \tau^2} \sin\omega_1(t-\tau)\, d\tau \end{aligned} \tag{12}$$

for no viscous damping or rotary inertia. The static portion of the solution given in (9) can be found from Section 2.16. In the cantilevered beam example in that section (Example 2.9), it was found that the initial parameters were given by $V_0 = W_1(t)$ and $M_0 = -LW_1(t)$. Note that $W_1(t)$ is evaluated for each step in time. From the transfer matrix in Eq. (1) of Example 2.9 for the static state variables, and from Eqs. (5), (6), (9), and (12) of the present example, the final acceleration form of the response, for $0 \leqslant x < L$, is given by

$$v(x, t) = \frac{x^2 L W_1(t)}{2EI} - \frac{x^3 W_1(t)}{3!EI} + \left(\frac{-x^2}{2EI} + \frac{x^3}{3!EIL} \right) \frac{L}{\omega_1} \int_0^t \frac{\partial^2 W_1(\tau)}{\partial \tau^2} \sin \omega_1 (t-\tau)\, d\tau$$

$$\theta(x, t) = -\frac{xLW_1(t)}{EI} + \frac{x^2 W_1(t)}{2EI} + \left(\frac{x}{EI} - \frac{x^2}{2EIL} \right) \frac{L}{\omega_1} \int_0^t \frac{\partial^2 W_1(\tau)}{\partial \tau^2} \sin \omega_1 (t-\tau)\, d\tau$$

$$M(x, t) = -LW_1(t) + xW_1(t) + (1 - x/L) \frac{L}{\omega_1} \int_0^t \frac{\partial^2 W_1(\tau)}{\partial \tau^2} \sin \omega_1 (t-\tau)\, d\tau$$

$$V(x, t) = W_1(t) - \frac{1}{\omega_1} \int_0^t \frac{\partial^2 W_1(\tau)}{\partial \tau^2} \sin \omega_1 (t-\tau)\, d\tau$$

For the mass at $x = L$, use Eq. (3) in Example 2.9 for the static-state variables to give

$$v(L, t) = \frac{L^3}{2EI} W_1(t) - \frac{L^3 W_1(t)}{3!EI} - \frac{1}{\omega_1^3 M_1^*} \int_0^t \frac{\partial^2 W_1(\tau)}{\partial \tau^2} \sin \omega_1 (t-\tau)\, d\tau$$

$$\theta(L, t) = -\frac{L^2}{EI} W_1(t) + \frac{L^2 W_1(t)}{2EI} + \frac{L}{2EI\omega_1} \int_0^t \frac{\partial^2 W_1(\tau)}{\partial \tau^2} \sin \omega_1 (t-\tau)\, d\tau$$

$$M(L, t) = 0$$

$$V(L, t) = 0$$

where it has been assumed that all initial conditions are set equal to zero.

EXAMPLE 2.30 Time-Dependent Applied Displacements The response of a beam due to time-varying prescribed displacements deserves special attention. For zero damping, initial conditions, temperature change, and time-varying slope, Eqs. (2.37*b*) and (2.39*a*) reduce to

$$\eta_n(t) = \int_0^t \left[\int_0^L w(x, \tau) v_n(x)\, dx - v(a_1, \tau)\, \Delta V_n(a_1) \right] \frac{\sin \omega_n (t-\tau)\, d\tau}{\omega_n} \tag{1}$$

where the in-span time-varying displacement is assumed to occur at $x = a_1$.

Consider a beam where the left-hand boundary is displaced with the function $v(0, t)$ (Fig. 2-23a) and the right-hand boundary is fixed. As indicated in Eq. (2.40c), (1) for this case reduces to

$$\eta_n(t) = \int_0^t \left[\int_0^L w(x, \tau) v_n(x) \, dx + v(0, t) V_n(0) \right] \frac{\sin \omega_n(t - \tau) \, d\tau}{\omega_n} \tag{2}$$

The frequencies and mode shapes are determined for the problem in which the time-dependent displacement conditions are set equal to zero. These are referred to as the "homogeneous" boundary conditions. In the case of the beam of Fig. 2-23a, the frequencies and mode shapes of Fig. 2-23b are employed in (2) and Eqs. (2.35).

Suppose a support is placed at $x = a_1$ of this beam (Fig. 2-24a). Then the mode shapes required for the solution are found from the beam of Fig. (2-24b). The function $\eta_n(t)$ is given by (1).

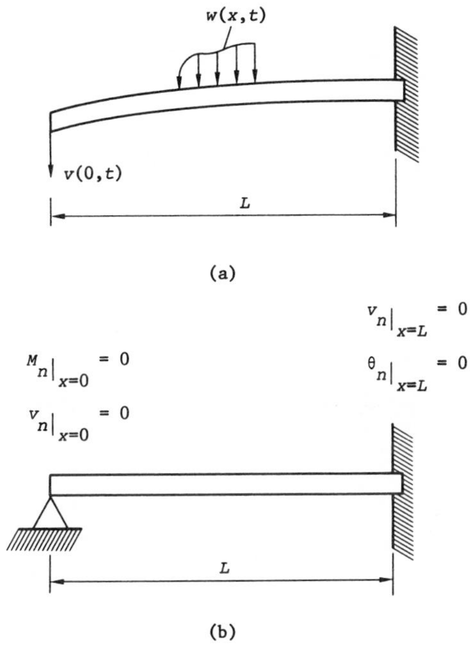

Fig. 2-23

Nonproportional Damping The formulas of this chapter need considerable adjustment in order to make them applicable to materials with nonproportional viscous damping. As noted above, Eqs. (2.16d) and (2.16c) should be supplemented by the appropriate viscous damping terms on the right-hand sides. Replace the frequency ω_n by s_n, where s_n is a complex number. The real part of s_n is commonly referred to as the *damping exponent*, whereas the imaginary part of s_n is called the frequency of the damped free vibration. Also, the mode shapes and some of the transfer matrices are complex functions. Finding the complex mode shapes and frequencies is considerably more difficult than the computation of the corresponding undamped quantities.

Equations (2.35) and (2.36) remain valid but now the static solutions refer to

$$v_s(x, t) = -\sum_n \frac{v_n(x) f_n(t)}{s_n N_n} \tag{2.41}$$

where $v_n(x)$ and N_n are the mode shape and norm associated with the complex eigenvalue s_n. Similar expansions apply for the slope, moment, and shear.

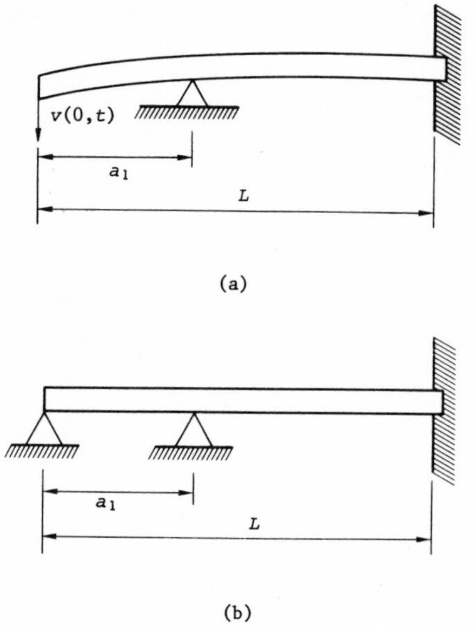

Fig. 2-24

The definitions of the previous subsection must be adjusted so that

$$N_n = 2s_n \int_0^L (\rho v_n^2 + \rho r^2 \theta_n^2)\, dx + \int_0^L (c' v_n^2 + c^* \theta_n^2)\, dx \tag{2.42a}$$

$$\eta_n(t) = e^{s_n t}\eta_n(0) + \int_0^t e^{s_n(t-\tau)} f_n(\tau)\, d\tau \tag{2.42b}$$

$$\xi_n(t) = e^{s_n t}\left[\frac{1}{s_n^2}\frac{\partial f_n}{\partial t}(0) + \frac{1}{s_n} f_n(0) + \eta_n(0)\right] - \frac{1}{s_n^2}\frac{\partial f_n}{\partial t}(t)$$
$$+ \frac{1}{s_n^2}\int_0^t e^{s_n(t-\tau)} \frac{\partial^2 f_n}{\partial \tau^2}(\tau)\, d\tau \tag{2.43}$$

$$\eta_n(0) = \int_0^L \Big[\rho s_n v(x,0) v_n(x) + \rho s_n r^2 \theta(x,0)\theta_n(x) + c' v(x,0) v_n(x)$$
$$+ c^*\theta(x,0)\theta_n(x) + \rho \frac{\partial v}{\partial t}(x,0)\, v_n(x) + \rho r^2 \frac{\partial \theta}{\partial t}(x,0)\theta_n(x)\Big]\, dx \tag{2.44}$$

In the case of the acceleration method, $f_n(t)$ can be alternatively expressed as

$$f_n(t) = -s_n \int_0^L \left[c' v_s(x, t) v_n + c^* \theta_s(x, t) \theta_n \right] dx$$

$$- s_n^2 \int_0^L \left[\rho v_s(x, t) v_n + \rho r^2 \theta_s(x, t) \theta_n \right] dx \tag{2.45}$$

The computational difficulties often encountered in nonproportional damping can be eased if the damping occurs in discrete form or if a lumped-parameter model can be employed. For example, if the damping can be lumped into discrete dashpots and if the mass can be lumped at particular locations along the beam, then the complex quantities appear only in point matrices and the complex frequency search is relatively simple to construct. In particular, the polynomial form of the characteristic equation, which was discussed in Section 2.18, can be formed. For discrete damping with extensional dashpot constant c_i (force · time/length) and rotary dashpot constant c_i^* (force · time/length) at $x = a_i$ use the point matrix of Table 2-14 for $x = a_i$ with k_1 replaced by $k_1 + sc_i$ and ω^2 by $-s^2$ for extensional damping. Also replace k_1^* by $k_1^* + sc_i^*$ if discrete rotary damping is present. Furthermore, in N_n set

$$\int_0^L \left(c' v_n^2 + c^* \theta_n^2 \right) dx = \sum_i c_i v_n^2(a_i) + \sum_i c_i^* \theta_n^2(a_i) \tag{2.46}$$

and in $\eta_n(0)$ set

$$\int_0^L \left[c' v(x, 0) v_n(x) + c^* \theta(x, 0) \theta_n(x) \right] dx$$

$$= \sum_i c_i v(a_i, 0) v_n(a_i) + \sum_i c_i^* \theta(a_i, 0) \theta_n(a_i) \tag{2.47}$$

Voigt-Kelvin Material The formulas of this chapter are also applicable to a beam of Voigt-Kelvin material. In the equations of motion, Eqs. (2.16), replace E by $E(1 + \epsilon\, \partial/\partial t)$ where ϵ is the Voigt-Kelvin damping coefficient and where similar behavior in dilatation and shear is assumed. If the beam is resting on a proportional-damped foundation, retain the $c_\rho \rho\, \partial v/\partial t$ term on the right-hand side of Eq. (2.16*d*) and the $c_\rho \rho r^2\, \partial\theta/\partial t$ on the right-hand side of Eq. (2.16*c*). The response formulas of this section apply if c_ρ is replaced by $c_\rho + \epsilon\omega_n^2$ and ζ_n is redefined as $\zeta_n = c_\rho/(2\omega_n) + \epsilon\omega_n/2$. Continue to use the same undamped mode shapes employed in a previous subsection and the corresponding nth undamped frequency ω_n.

Base Motion through Flexible Support System The transient response of a beam, which is supported by N in-span flexible supports or bearings subjected to a ground motion, can be determined by modifying the previous beam formulation. The model for this flexible support system is shown in Fig. (2-25), where the additional notation for a support (bearing) at $x = a_i$ is defined as follows:

k_1, c_1	Discrete stiffness and nonproportional damping constants, respectively, between the beam and discrete support mass
k_2, c_2	Discrete stiffness and nonproportional damping constants, respectively, between the discrete support mass and the support base

$\overline{M}$	Discrete support mass
$\bar{v}(a_i, t)$	Vertical deflection of the support mass
$g(a_i, t)$, $\dot{g}(a_i, t)$	Base deflection and velocity, respectively; $\dot{g}$ is defined as $\partial g/\partial t$

Note that all of the above quantities occur at $x = a_i$, the location of the ith in-span support, and in general all these quantities will be different for each support.

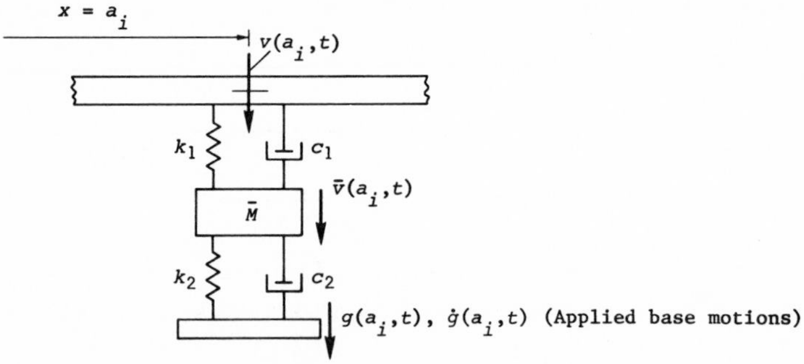

Fig. 2-25 Model of flexible support with ground motion.

The equations of motion are also modified as follows:

1. Include in Eqs. (2.16) an equation of motion corresponding to the discrete support masses, that is,

$$\sum_{i=1}^{N}\left\{\overline{M}\,\frac{\partial^2\bar{v}}{\partial t^2}(a_i, t) + k_1\left[\bar{v}(a_i, t) - v(a_i, t)\right] + c_1\left[\frac{\partial\bar{v}(a_i, t)}{\partial t} - \frac{\partial v(a_i, t)}{\partial t}\right]\right.$$

$$\left. + k_2\left[\bar{v}(a_i, t) - g(a_i, t)\right] + c_2\left[\frac{\partial\bar{v}}{\partial t}(a_i, t) - \frac{\partial g(a_i, t)}{\partial t}\right]\right\} = 0 \qquad (2.48)$$

2. Add the following term to the right hand side of Eq. (2.16d):

$$\sum_{i=1}^{N}\left\{k_1\left[v(a_i, t) - \bar{v}(a_i, t)\right] + c_1\left[\frac{\partial v}{\partial t}(a_i, t) - \frac{\partial\bar{v}}{\partial t}(a_i, t)\right]\right\} \qquad (2.49)$$

The complex frequency corresponding to damped free motion may be found with transfer matrices. The effect of the ith support is taken into account by insertion of a point matrix, which includes the effective support dynamic stiffness, when transfer matrices are multiplied down the beam. This complex support impedance point matrix is given by

$$\overline{\mathbf{U}}_i = \begin{bmatrix} 1 & 0 & 0 & 0 \\ 0 & 1 & 0 & 0 \\ 0 & 0 & 1 & 0 \\ -Z_i & 0 & 0 & 1 \end{bmatrix} \qquad (2.50)$$

where

$$Z_i = \frac{(k_1 + sc_1)(k_2 + sc_2 + s^2\overline{M})}{(k_1 + sc_1) + (k_2 + sc_2 + s^2\overline{M})} \tag{2.51}$$

and where the subscript i emphasizes that Z_i depends on c_1, c_2, k_1, k_2, $\overline{M}$ which are different, in general, for each ith support. The complex frequencies $s = s_n$ are then found by applying the boundary conditions as explained in Section 2.18. Finally, to find the mode shape $\bar{v}(a_i, t)$ corresponding to the new degree of freedom, the following relation may be used

$$\bar{v}_n(a_i) = \left(\frac{k_1 + s_n c_1}{k_1 + s_n c_1 + k_2 + s_n c_2 + s_n^2 \overline{M}} \right) v_n(a_i) \tag{2.52}$$

where $\bar{v}_n(a_i)$ is the nth modal deflection of the discrete mass at the ith bearing.

The transient response of a beam with N in-span flexible supports may be found by redefining N_n, $\eta_n(0)$, and $f_n(t)$. As a simple model, consider first no damping either in the N bearings or in the beam, that is, $c' = c^* = c_1 = c_2 = 0$. The frequencies and mode shapes will be real rather than complex. Equations (2.37*a*), (2.38), (2.39*a*), and (2.39*b*) are replaced by

$$N_n = \int_0^L \rho(v_n^2 + r^2\theta_n^2)\, dx + \sum_{i=1}^{N} \overline{M}\left[\bar{v}_n(a_i)\right]^2 \tag{2.53a}$$

$$\eta_n(0) = \int_0^L \rho\left[v(x,0)v_n + r^2\theta(x,0)\theta_n\right] dx + \sum_{i=1}^{N} \overline{M}\bar{v}(a_i, 0)\bar{v}_n(a_i) \tag{2.53b}$$

$$f_n(t) = \int_0^L \left[w(x,t)v_n + c(x,t)\theta_n + \frac{M_T(x,t)}{EI} M_n \right] dx + \sum_{i=1}^{N} \left[k_2 g(a_i, t) + c_2 \dot{g}(a_i, t)\right]\bar{v}_n(a_i) + h_n(a_k, t) \tag{2.53c}$$

In the case of the acceleration method, $f_n(t)$ can alternatively be expressed as

$$f_n(t) = \omega_n^2 \int_0^L \left[\rho v_s(x,t)v_n + \rho r^2\theta_s(x,t)\theta_n\right] dx + \omega_n^2 \sum_{i=1}^{N} \overline{M}\bar{v}_s(a_i, t)\bar{v}_n(a_i) \tag{2.53d}$$

where ω_n is real and $\bar{v}_s(a_i, t)$ is the static deflection of the ith discrete support mass at each point in time of the applied loading.

For nonproportional damping in the model, Eqs. (2.42a) and (2.44) are replaced by

$$N_n = 2s_n\left[\int_0^L \left(\rho v_n^2 + \rho r_y^2\theta_n^2\right) dx + \sum_{i=1}^{N} \overline{M}\bar{v}_n^2(a_i, t)\right] + \int_0^L \left(c'v_n^2 + c^*\theta_n^2\right) dx$$

$$+ \sum_{i=1}^{N} \left[c_1 v_n^2(a_i) - 2v_n(a_i)c_1\bar{v}_n(a_i) + (c_1 + c_2)\bar{v}_n^2(a_i)\right] \tag{2.54}$$

$$\eta_n(0) = \int_0^L \Bigg[\rho s_n v(x, 0)v_n(x) + \rho s_n r^2\theta(x, 0)\theta_n(x) + c'v(x, 0)v_n(x)$$

$$+ c^*\theta(x, 0)\theta_n(x) + \rho \frac{\partial v}{\partial t}(x, 0)\, v_n(x) + \rho r^2 \frac{\partial \theta}{\partial t}(x, 0)\, \theta_n(x)\Bigg] dx$$

$$+ \sum_{i=1}^{N} \Bigg\{\overline{M}\left[s_n\bar{v}(a_i, 0)\bar{v}_n(a_i) + \frac{\partial \bar{v}}{\partial t}(a_i, 0)\, \bar{v}_n(a_i)\right]$$

$$+ c_1 v_n(a_i)v(a_i, 0) - c_1 v_n(a_i)\bar{v}(a_i, 0)$$

$$- c_1\bar{v}_n(a_i)v(a_i, 0) + (c_1 + c_2)\bar{v}_n(a_i)\bar{v}(a_i, 0)\Bigg\} \tag{2.55}$$

And for the acceleration method, $f_n(t)$ becomes

$$f_n(t) = -s_n\int_0^L \left[c'v_s(x, t)v_n + c^*\theta_s(x, t)\theta_n\right] dx$$

$$- s_n^2\int_0^L \left[\rho v_s(x, t)v_n + \rho r^2\theta_s(x, t)\theta_n\right] dx$$

$$- \sum_{i=1}^{N} \Big\{s_n\big[c_1 v_n(a_i)v_s(a_i, t) - c_1 v_n(a_i)\bar{v}_s(a_i, t) - c_1\bar{v}_n(a_i)v_s(a_i, t)$$

$$+ (c_1 + c_2)\bar{v}_n(a_i)\bar{v}_s(a_i, t)\big] + s_n^2\overline{M}\bar{v}_s(a_i, t)\bar{v}_n(a_i)\Big\} \tag{2.56}$$

C. COMPUTER PROGRAMS AND EXAMPLES

2.21 Benchmark Examples

Complicated beam problems should be solved with the assistance of a computer program. The following examples are provided as benchmark examples against which a reader's own program can be checked. The computer program BEAMRESPONSE was used for these examples. BEAMRESPONSE calculates the deflection, slope, bending moment, and shear force of a beam for static and steady-state conditions, the critical load and mode shape for stability, and the natural frequencies and mode shapes for transverse vibrations. The beam can be formed of segments of different geometric or material properties with any mechanical or thermal loading, in-span supports, foundations, and boundary conditions. The user can include any or all of bending, shear deformation, and rotary inertia effects.

EXAMPLE 2.31 Beam on Elastic Foundation Find the deflection, slope, bending moment, and shear force along the statically loaded beam of Fig. 2-26. Let $E = 30(10^6)\ \text{lb/in}^2$, $I = 25\ \text{in}^4$.

The results are given in Fig. 2-27.

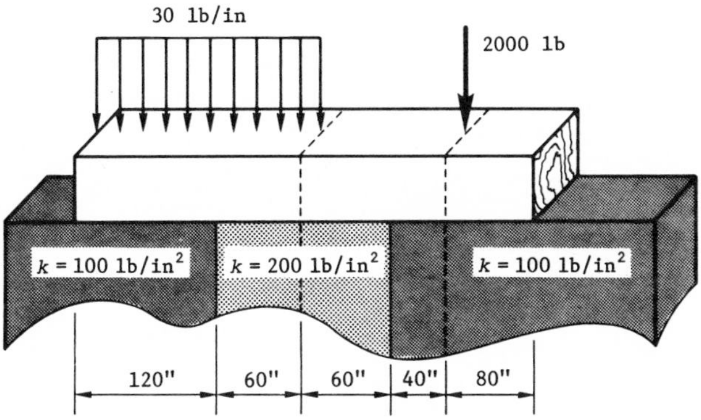

Fig. 2-26 Example 2.31.

AXIAL LOCATION	DEFLECTION	SLOPE	MOMENT	SHEAR
0.	3.4725E-01	1.1517E-03	0.	0.
1.2000E+02	1.8706E-01	1.6015E-03	-6.6306E+02	-3.2840E+02
1.8000E+02	1.0568E-01	8.9350E-04	-1.7976E+04	-4.1400E+02
2.4000E+02	1.0094E-01	-6.0534E-04	-8.2454E+03	7.3046E+02
2.8000E+02	1.2206E-01	-1.1331E-04	2.9720E+04	-8.1720E+02
3.6000E+02	7.0515E-02	8.8436E-04	0.	-4.3656E-11

Fig. 2-27 Partial output for Example 2.31.

EXAMPLE 2.32 Instability of Beam with Variable Axial Load Calculate the value of P_0 that would make the beam of Fig. 2-28 unstable. Note that P_1, P_2, P_3, P_4 remain in proportion to the nominal value P_0 throughout the loading process.

The unstable load is computed to be $P_0 = 8.97$ MN.

$E = 200\ GN/m^2 = 200(10^9)\ N/m^2$, $I_1 = 43\,500\ cm^4 = 4.35(10^{-4})m^4$, $I_2 = 6.5(10^{-4})m^4$

$I_3 = I_4 = 9.3(10^{-4})m^4$, $I_5 = 1.12(10^{-3})m^4$

$P_1/P_0 = 0.967$, $P_2/P_0 = 0.917$, $P_3/P_0 = 1.05$, $P_4/P_0 = 2.15$

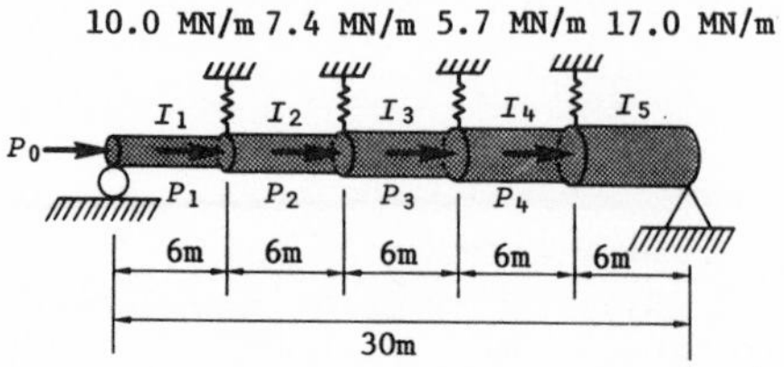

Fig. 2-28 Example 2.32.

EXAMPLE 2.33 Natural Frequencies of Continuous Beam Find the first two natural frequencies of the beam of Fig. 2-29. Here $E = 30(10^6)\ lb/in^2$, $I_1 = 24{,}800\ in^4$, $I_2 = 6200\ in^4$, $\rho_1 = 0.4\ lb \cdot sec^2/in^2$, $\rho_2 = 0.2\ lb \cdot sec^2/in^2$, $A_1 = 545.53\ in^2$, $A_2 = 272.76\ in^2$. Include the effects of shear deformation. The cross-sectional areas given have been adjusted by the shear shape factor.

The frequencies are found to be 42.1 and 134.7 Hz.

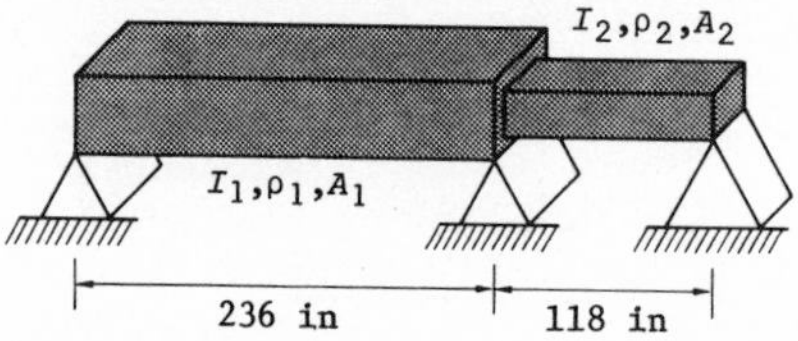

Fig. 2-29 Example 2.33.

EXAMPLE 2.34 Beam with Springs and Complicated Loading Find the deflection, slope, bending moment, and shear force in the axially loaded beam of Fig. 2-30. The right end is guided. The total axial load is required in the transfer matrices for each section; for example, the compressive load in section 2 is 13,300 + 8900 = +22,200 N. The output is shown in Fig. 2-31.

$E = 200\ GN/m^2 = 200(10^9)\ N/m^2$, $I_1 = 125\ cm^4 = 1.25(10^{-6})m^4$,

$I_2 = 2.5(10^{-6})m^4$

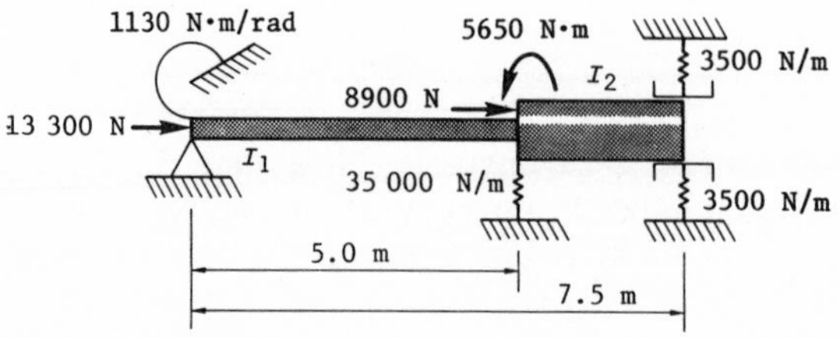

Fig. 2-30 Example 2.34.

AXIAL LOCATION	DEFLECTION	SLOPE	MOMENT	SHEAR
0.	0.	-.7135E-02	-.8063E+01	.4612E+03
.1250E+01	.8223E-02	-.5449E-02	.6778E+03	.4612E+03
.2500E+01	.1224E-01	-.4505E-03	.1308E+04	.4612E+03
.3750E+01	.8137E-02	.7448E-02	.1830E+04	.4612E+03
.5000E+01	-.7320E-02	.1759E-01	-.3449E+04	.2050E+03
.5625E+01	-.1696E-01	.1322E-01	-.3535E+04	.2050E+03
.6250E+01	-.2383E-01	.8781E-02	-.3560E+04	.2050E+03
.6875E+01	-.2793E-01	.4348E-02	-.3523E+04	.2050E+03
.7500E+01	-.2929E-01	0.	-.3425E+04	.2050E+03

Fig. 2-31 Partial output for Example 2.34.

EXAMPLE 2.35 Thermal Loading Find the deflection, slope, bending moment, and shear force at the midpoint of a thermally loaded uniform beam on an elastic foundation. This steel beam has a moment of inertia of 1000 in^4, is 100 in long, and rests on an elastic foundation of modulus 100,000 lb/in^2. The ends are simply supported. A uniform thermal moment of 1000 lb · in. is applied along the beam.

The deflection is $8.4(10^{-6})$ in, the slope is zero, the moment is -969.3 lb · in, and the shear force is zero.

EXAMPLE 2.36 Frequency Analysis of a Drilling Platform Find the first three natural frequencies of the drilling platform of Fig. 1-4. Considerable discussion of the modeling of this structure was given in Chap. 1. This is modeled as a beam column with rotary in-span supports. When beam parameters are selected to represent the structure well, the length and moments of inertia of the legs, platform, and truss cause no problems. The lengths are measured directly from the design drawings and the moments of inertia are calculated by standard methods.

When the span length of the members is determined, care must be taken to make sure that the relative positions of the forces and masses are correct. We can determine the locations of the idealized lumped masses and supporting springs for the platform and lower truss from the neutral-axis concepts. For greater accuracy, we choose to model the middle truss as two springs. Using $k^* = 6EI/\ell^2$, $E = 4.32(10^6)$ kip/ft^2, we calculate the spring constants shown in Fig. 2.32.

In some frame-analysis problems, the finite length of very stiff joints can be neglected without substantially affecting the results. However, this approximation should be used with caution, especially for dynamic problems. Very stiff joints can be treated as rigid or as possessing moments of inertia of large magnitudes. For this beam we choose the values shown in Fig. 2-32.

The mass of the structure must include the weights of structural components and the water surrounding the structure. Handbooks on hydrodynamics will provide the value of the added (so-called virtual) mass due to water. The figures given in Fig. 2-32 are for one-fourth of the structure. The mass at the bottom has no effect on the results and is taken to be zero.

The frequencies are computed to be 0.3164, 0.8315, 2.6307 Hz.

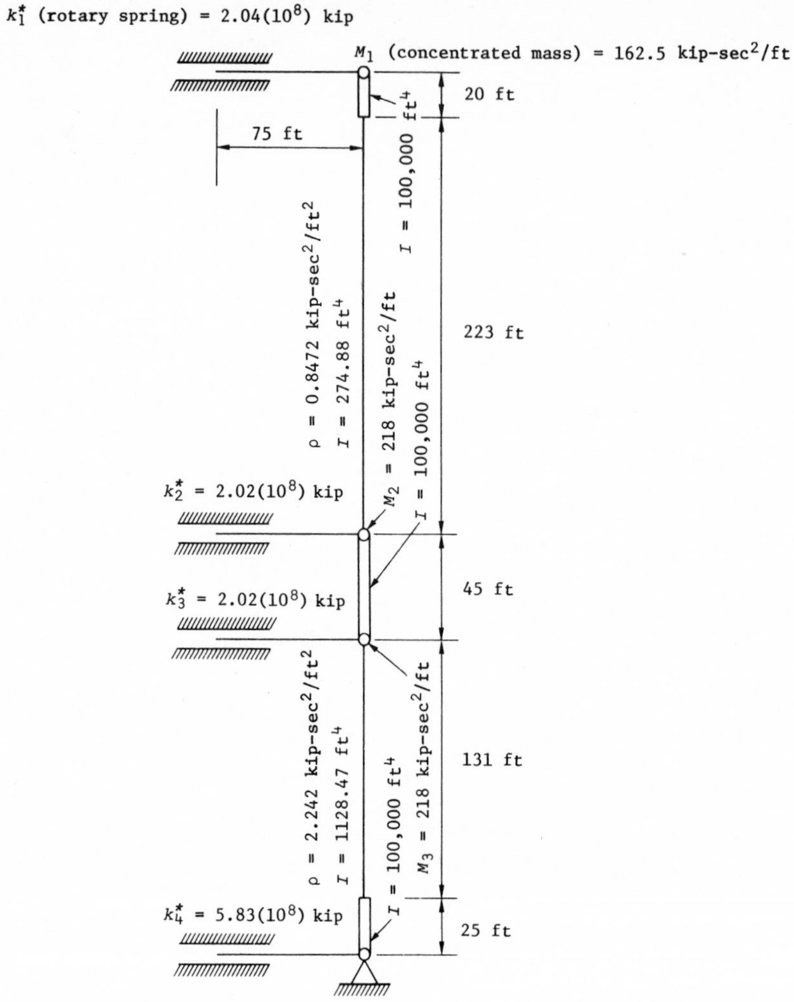

Fig. 2-32 Example 2.36.

EXAMPLE 2.37 Beam on Stiff Elastic Foundation Consider a uniform beam of length 10 in on an elastic foundation of modulus of $4.4(10^4)$ lb/in^2. The modulus of elasticity is $4.8(10^6)$ lb/in^2 and the moment of inertia is $2(10^{-5})$ in^4. A load of 100 lb is placed on the left end of this free-free beam. Find the deflections and slopes at the ends. This beam, which is a quite flexible beam on a stiff foundation, is selected because—unless special care is taken—numerical inaccuracy problems may be encountered. The Riccati transformation discussed in Appendix 2 was used to stabilize these calculations.

The deflection and slope at the left end are $1.487(10^{-2})$ in and $4.866(10^{-2})$ radians and at the right end are $1.283(10^{-16})$ in and $1.159(10^{-15})$ radians.

EXAMPLE 2.38 Column with Continuously Varying Axial Force Suppose a column is subjected to a linearly distributed axial force $p_x(L - x)$. Find the value of p_x that buckles the column.

The critical loads (Ref. 2.6) for various boundary conditions are: hinged-hinged, $(p_x)_{cr} = 18.5\ EI/L^3$; fixed-hinged, $(p_x)_{cr} = 52.5\ EI/L^3$; fixed-fixed, $(p_x)_{cr} = 74.7\ EI/L^3$; fixed-free, $(p_x)_{cr} = 7.8\ EI/L^3$.

REFERENCES

2.1 Pilkey, W. D., and Pilkey, O. H.: *Mechanics of Solids*, Quantum Publishers, New York, 1974.

2.2 Pestel, E., and Leckie, F.: *Matrix Methods in Elastomechanics*, McGraw-Hill, New York, 1963.

2.3 Hopkins, R. B.: *Design Analysis of Shafts and Beams*, McGraw-Hill, New York, 1970.

2.4 Cowper, G. R.: "The Shear Coefficient in Timoshenko's Beam Theory," *Trans. ASME*, ser. E., vol. 33, pp. 355–340, 1966.

2.5 Pilkey, W. D., Chang, P. Y., and Hsu, D.: "Complex Infinite Beams on an Elastic Foundation," *J. of the Struct. Div., Proc. Am. Soc. Civ. Eng.*, vol. 103, 1977.

2.6 Pilkey, W. D., and O'Connor, K. J.: "Lumped Parameter Model for Stability Analysis, " *J. Struct. Div., Proc. Am. Soc. Civ. Eng.*, vol. 99, pp. 1702 – 1707, 1973.

2.7 Pilkey, W. D.: "Normal Mode Solutions for the Dynamic Response of Structural Members with Time Dependent In-Span Conditions," *J. Acoust. Soc. Am.*, vol. 44, pp. 1675–1678, 1968.

THREE

Torsional Systems

This chapter treats the static twisting motion and torsional vibration of bars, including shafting and geared disk systems. Response formulas for simple bars and fundamental stress formulas are handled in Part A. The formulas for more complex bars are provided in Part B. Part C contains some examples of problems solved with computer programs. These examples include static and dynamic analyses of bars, as well as the calculation of torsional constants and torsional shear stresses on cross sections of any shape.

The formulas of this chapter apply to torsionally loaded members without restraints against warping. Shear stresses and twisting angles for members restrained from warping can be found with the formulas of Chap. 5 (Torsion of Thin-Walled Beams).

A. STRESS FORMULAS AND SIMPLE BARS

3.1 Notation and Conventions

ϕ	Angle of twist, rotation (radians)
T	Twisting moment, torque (force · length)
G	Shear modulus of elasticity (force/length^2)
J	Torsional constant (length^4). In the case of a circular cross section, J is the polar moment of inertia I_x of the cross-sectional area with respect to the axis of the bar. Table 3-1 provides values of J for several shapes.

L	Length of bar
T_1, T_i	Concentrated applied torque (force · length)
m_x	Distributed torque, twisting moment intensity (force · length/length)
m_{x1}	Magnitude of distributed torque that is uniform in x direction (force · length/length)
τ	Torsional shear stress (force/length^2)
q	Shear flow (force/length)
t	Thickness of thin-walled section (length)
A^*	Area enclosed by thin-walled section (length^2)
ρ^*	Mass per unit volume (mass/length^3)
ω	Natural frequency (radian/second)

Positive angle of twist ϕ and torque T are shown in Fig. 3-1.

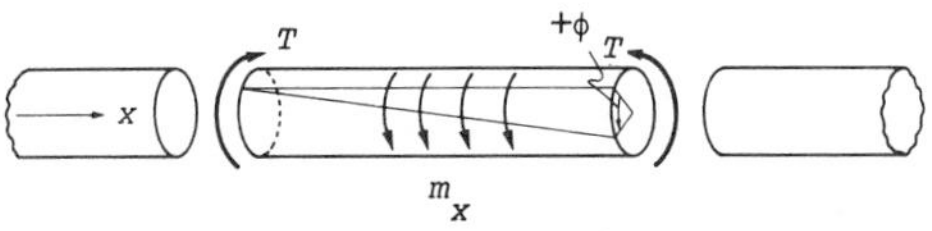

Fig. 3-1 Positive angle of twist ϕ and torque T.

3.2 Torsional Stresses

The tables of this chapter provide the internal twisting moment at any point along a bar. Once the twisting moment is known, the torsional stresses can be calculated from the stress formulas given in this section.

Circular Cross Section The torsional shear stress τ on a cross section of a circular shaft, either solid or hollow, is (Fig. 3-2*a*)

$$\tau = \frac{Tr}{J} \tag{3.1}$$

where r is the radial distance from the longitudinal axis (x) through the center of the cross section. For circular cross sections $J = I_x$, where I_x is the polar moment of inertia of the cross section about the central axis of a shaft. J is given by

$$J = I_x = \int_A r^2 \, dA = \int_A (z^2 + y^2) \, dA = I_y + I_z \tag{3.2}$$

where A is the cross-sectional area. Values of J for noncircular sections are provided in Table 3-1.

Thick Noncircular Sections The peak stresses for several common thick noncircular sections are listed in Table 3-1.

Hollow Thin-Walled Shaft The torsional stress in a hollow thin-walled shaft is assumed to be uniformly distributed through the thickness t (Fig.3-2*b*). This stress is

$$\tau = \frac{q}{t} = \frac{T}{2A^*t} \tag{3.3}$$

where q is the shear flow and A* is the area enclosed by the middle line of the wall.

Thin-Walled Open Sections The torsional stress in a thin-walled open section is assumed to vary linearly across the thickness of a cross section and acts in a direction parallel to the edges (Fig. 3-2c). The stresses are equal in magnitude and opposite in direction at the two edges of a thin wall. At the edges where the maximum stresses occur

$$\tau = \frac{Tt}{J} \tag{3.4a}$$

where t is the thickness at the location at which the stress is being calculated. The torsional constant is

$$J = \frac{\alpha}{3} \sum_{i=1}^{m} b_i t_i^3 \tag{3.4b}$$

where m is the number of straight or curved segments of thickness t_i and width or height b_i composing the section. The constant α is a shape factor. If no information concerning α is available, then set $\alpha = 1$. Values of J and α for some thin-walled sections are given in Table 3-1.

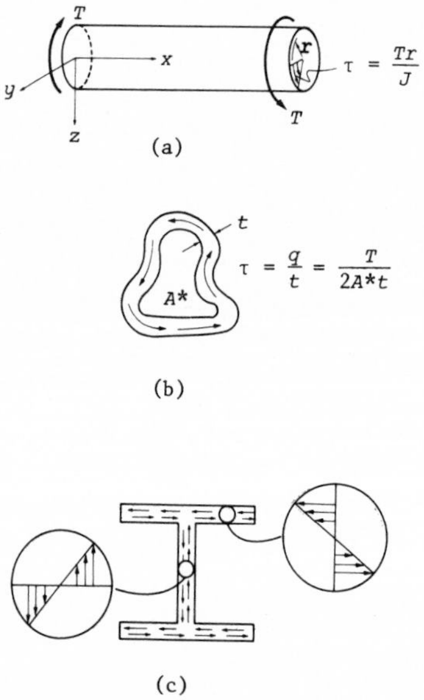

Fig. 3-2

3.3 Angle of Twist

One end of a shaft of length L of constant cross section twists

$$\phi = \frac{TL}{GJ} \tag{3.5}$$

radians with respect to the other end. The torsional constant J is listed for many shapes in Table 3-1.

TABLE 3-1 Formulas for Stress, Torsional Constant J, and Other Torsion-Related Constants

Thick Noncircular Sections - The maximum stress for the cross section occurs at point i

1. Ellipse

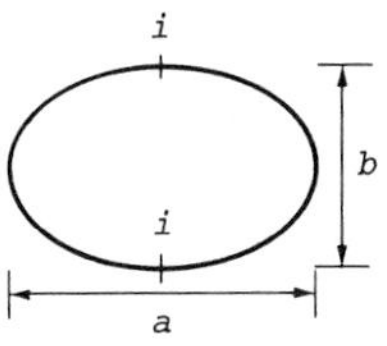

$$J = \frac{\pi a^3 b^3}{16(a^2 + b^2)}$$

$$\tau = \frac{16T}{\pi a b^2}$$

2. Hollow Ellipse

$k = a_i/a = b_i/b$

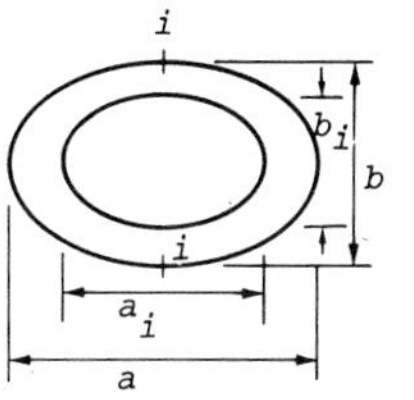

$$J = \frac{\pi a^3 b^3}{16(a^2 + b^2)}(1 - k^4)$$

$$\tau = \frac{16T}{\pi a b^2(1 - k^4)}$$

3.

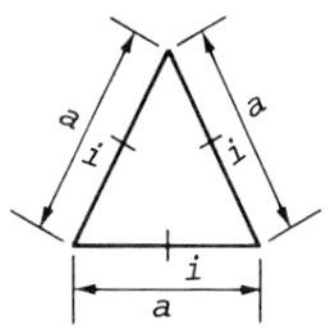

$$J = \frac{a^4\sqrt{3}}{80}$$

$$\tau = \frac{20T}{a^3}$$

4.

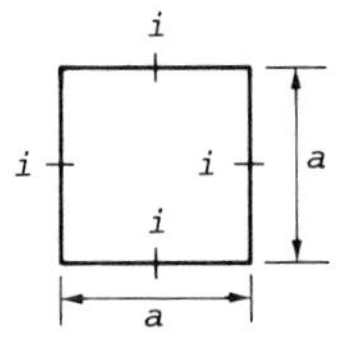

$$J = 0.1406a^4$$

$$\tau = \frac{4.81T}{a^3}$$

5.

$a > b$

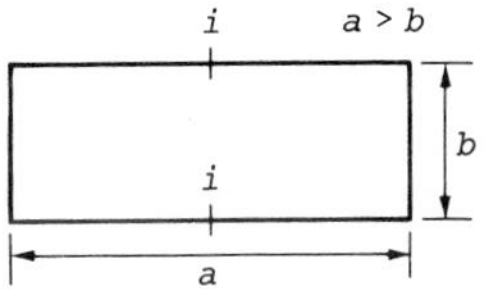

$$J = \frac{ab^3}{3}\left(1 - 0.630\frac{b}{a} + 0.052\frac{b^5}{a^5}\right)$$

$$\tau = \frac{3T}{ab^2\left(1 - 0.630\frac{b}{a} + 0.250\frac{b^2}{a^2}\right)}$$

6.

Hollow Thin-Walled Sections

A^* = area enclosed by the middle line of wall

S = entire length of middle line of wall

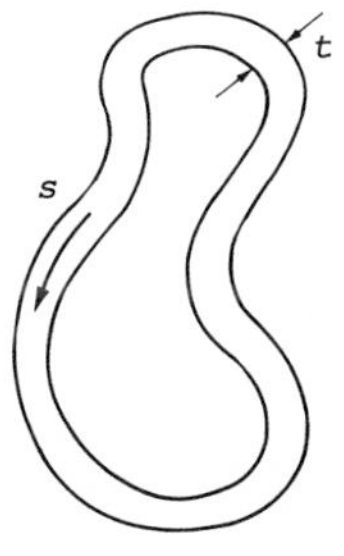

$$J = \frac{4A^{*2}}{\int_0^S (1/t)\, ds}$$

For constant t:

$$J = \frac{4A^{*2}t}{S}$$

TABLE 3-1 Formulas for Stress, Torsional Constant *J*, and Other Torsion-Related Constants (*Continued*)

Circular Cross Sections

7. Solid

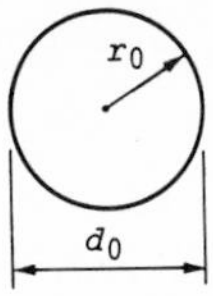

$$J = \frac{\pi r_0^4}{2} = \frac{\pi d_0^4}{32}$$

8. Hollow

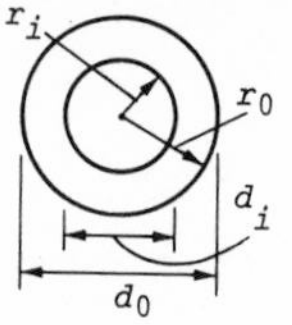

$$J = \frac{\pi}{2}(r_0^4 - r_i^4) = \frac{\pi}{32}(d_0^4 - d_i^4)$$

9. Very Thin

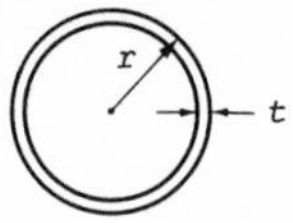

$$J \approx 2\pi r^3 t$$

Thin-Walled Open Sections (Ref. 3.1, 5.6)

e locates the shear center S.
ω is the sectorial coordinate (Ref. 5.6).
Γ is the warping constant.
Q_ω is the first sectorial moment (Ref. 5.6).

See Chapter 5 for more complete definitions of these parameters.

10. Any Open Section

$$J = \frac{\alpha}{3}\sum_{i=1}^{m} b_i t_i^3$$

m is the number of the straight or curved segments of thickness t_i and width or height b_i comprising the section. Set $\alpha = 1$ except as designated otherwise.

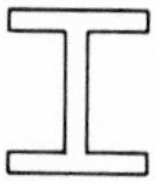

$\alpha = 1.31$

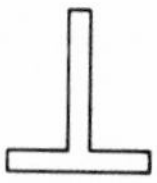

$\alpha = 1.12$

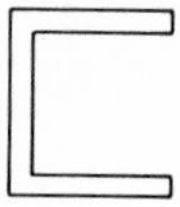

$\alpha = 1.12$

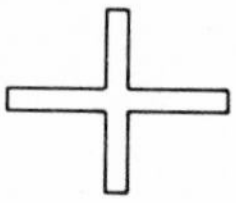

$\alpha = 1.17$

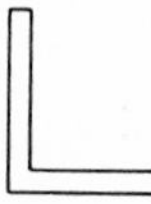

$\alpha = 1$

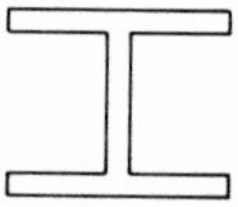

$\alpha = 1.29$

$\alpha = 1.12$

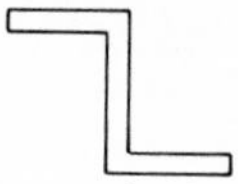

$\alpha = 1.12$

11.

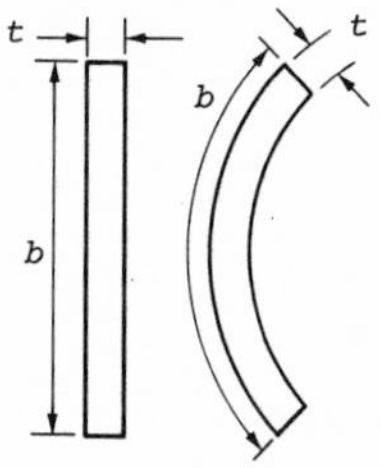

$$J = \frac{bt^3}{3}$$

12.

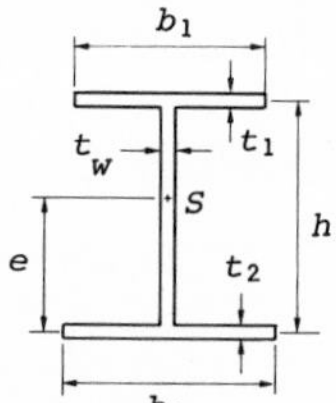

Ref. 5.4

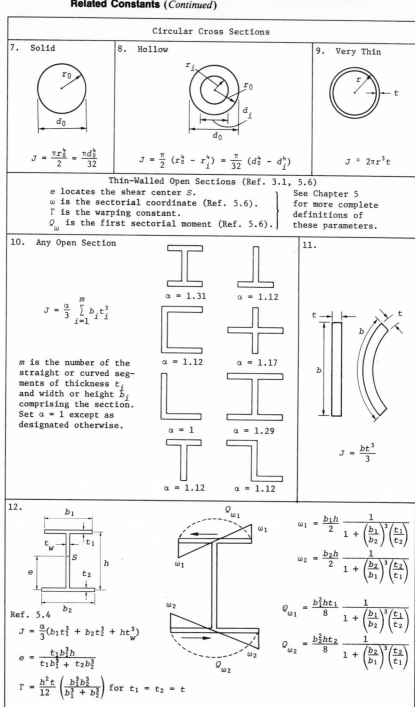

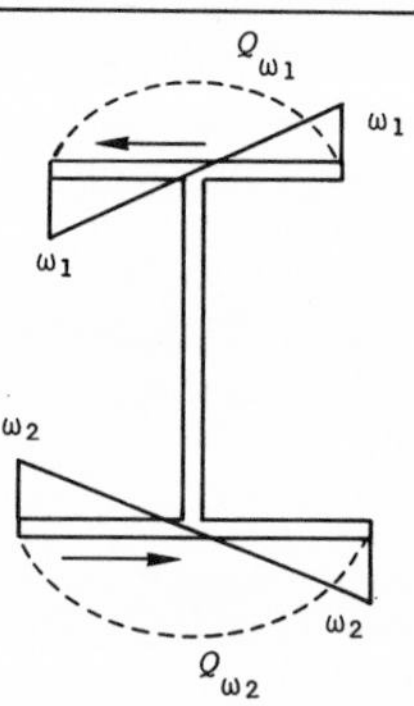

$$J = \frac{\alpha}{3}(b_1 t_1^3 + b_2 t_2^3 + h t_w^3)$$

$$e = \frac{t_1 b_1^3 h}{t_1 b_1^3 + t_2 b_2^3}$$

$$\Gamma = \frac{h^2 t}{12}\left(\frac{b_1^3 b_2^3}{b_1^3 + b_2^3}\right) \quad \text{for } t_1 = t_2 = t$$

$$\omega_1 = \frac{b_1 h}{2}\,\frac{1}{1 + \left(\frac{b_1}{b_2}\right)^3\left(\frac{t_1}{t_2}\right)}$$

$$\omega_2 = \frac{b_2 h}{2}\,\frac{1}{1 + \left(\frac{b_2}{b_1}\right)^3\left(\frac{t_2}{t_1}\right)}$$

$$Q_{\omega 1} = \frac{b_1^2 h t_1}{8}\,\frac{1}{1 + \left(\frac{b_1}{b_2}\right)^3\left(\frac{t_1}{t_2}\right)}$$

$$Q_{\omega 2} = \frac{b_2^2 h t_2}{8}\,\frac{1}{1 + \left(\frac{b_2}{b_1}\right)^3\left(\frac{t_2}{t_1}\right)}$$

TABLE 3-1 Formulas for Stress, Torsional Constant J, and Other Torsion-Related Constants (*Continued*)

13.

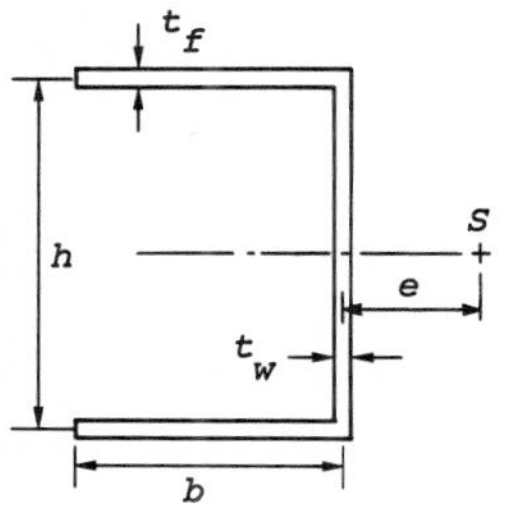

Ref. 5.2 $J = \frac{\alpha}{3}(2bt_f^3 + ht_w^3)$

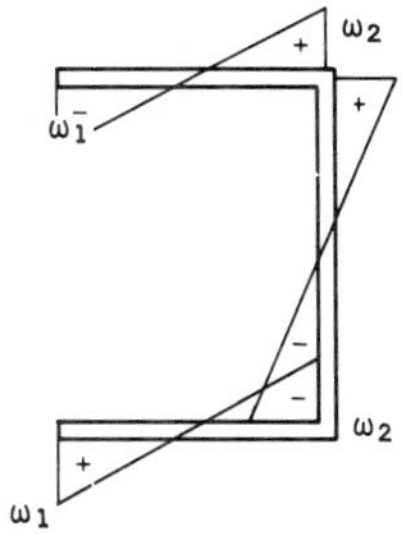

$$e = \frac{3t_f b^2}{6bt_f + ht_w}$$

$$\omega_1 = \frac{bh}{2}\,\frac{3 + \left(\frac{h}{b}\right)\left(\frac{t_w}{t_f}\right)}{6 + \left(\frac{h}{b}\right)\left(\frac{t_w}{t_f}\right)}$$

$$\omega_2 = \frac{bh}{2}\,\frac{1}{2 + \frac{1}{3}\left(\frac{h}{b}\right)\left(\frac{t_w}{t_f}\right)}$$

$$\Gamma = \frac{b^3h^2t_f}{12}\,\frac{2ht_w + 3bt_f}{ht_w + 6bt_f}$$

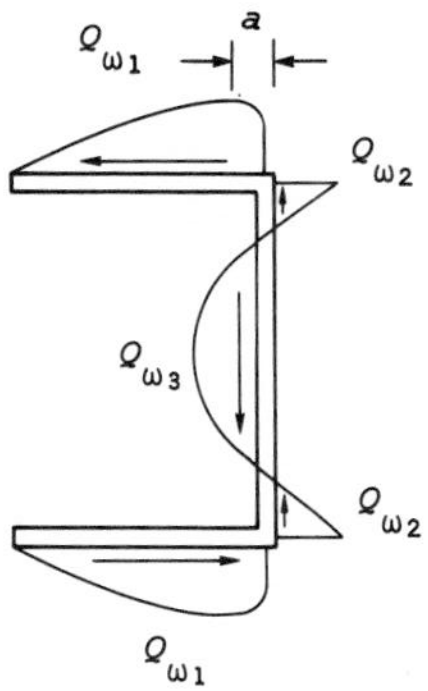

$$Q_{\omega_1} = \frac{b^2ht}{4}\left[\frac{3 + \left(\frac{h}{b}\right)\left(\frac{t_w}{t_f}\right)}{6 + \left(\frac{h}{b}\right)\left(\frac{t_w}{t_f}\right)}\right]^2$$

$$Q_{\omega_2} = \frac{b^2ht}{4}\,\frac{1}{1 + \left(\frac{b}{h}\right)\left(\frac{t_f}{t_w}\right)}$$

$$Q_{\omega_3} = -\frac{b^2ht}{8}\,\frac{1}{1 + 6\left(\frac{b}{h}\right)\left(\frac{t_f}{t_w}\right)}$$

$$a = \frac{b}{2 + \frac{1}{3}\left(\frac{h}{b}\right)\left(\frac{t_w}{t_f}\right)}$$

14.

Ref. 5.6

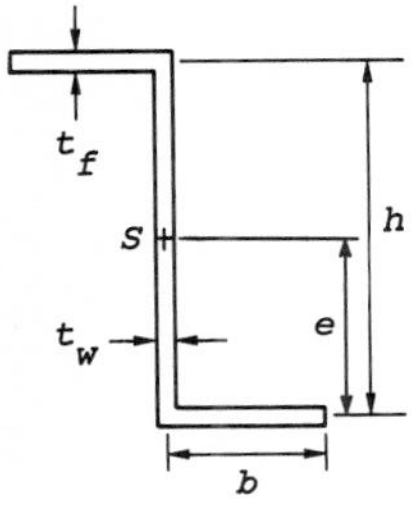

$$e = \frac{h}{2}$$

$$\Gamma = \frac{b^3h^2t_f}{12}\left[\frac{bt_f + 2ht_w}{2bt_f + ht_w}\right]$$

$J = \frac{\alpha}{3}(2bt_f^3 + ht_w^3)$

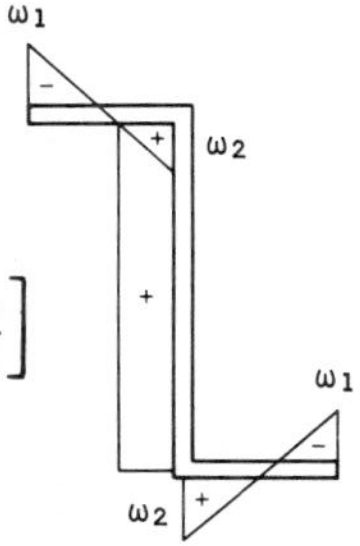

$$\omega_1 = \frac{bh}{2}\left(\frac{\frac{t_f}{t_w} + \frac{h}{b}}{2\,\frac{t_f}{t_w} + \frac{h}{b}}\right)$$

$$\omega_2 = \frac{b^2ht_f}{2(2bt_f + ht_w)}$$

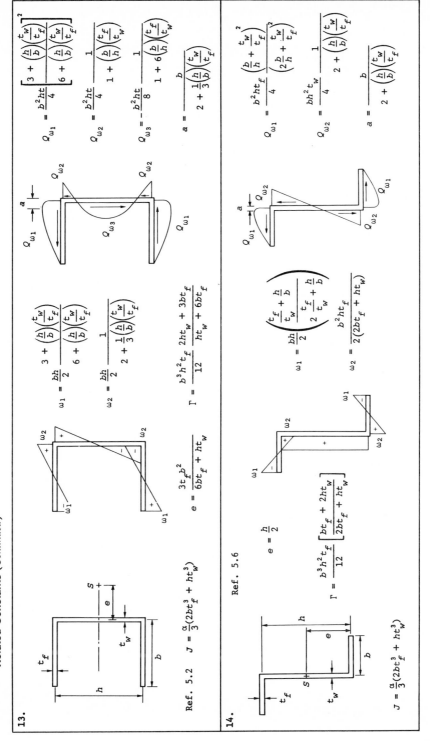

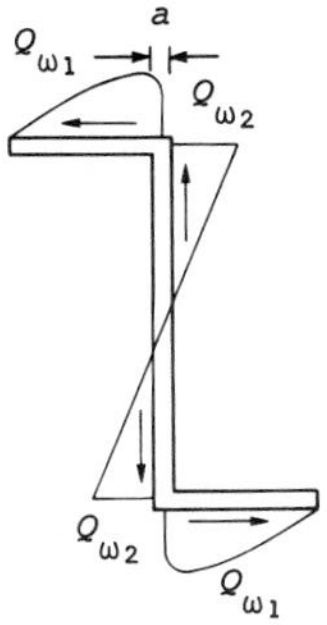

$$Q_{\omega_1} = \frac{b^2ht_f}{4}\,\frac{\left(\frac{b}{h} + \frac{t_w}{t_f}\right)^2}{\left(2\frac{b}{h} + \frac{t_w}{t_f}\right)^2}$$

$$Q_{\omega_2} = \frac{bh^2t_w}{4}\,\frac{1}{2 + \left(\frac{h}{b}\right)\left(\frac{t_w}{t_f}\right)}$$

$$a = \frac{b}{2 + \left(\frac{h}{b}\right)\left(\frac{t_w}{t_f}\right)}$$

TABLE 3-1 Formulas for Stress, Torsional Constant J, and Other Torsion-Related Constants (*Continued*)

Structural Shapes (Ref. 3.2)	
15. 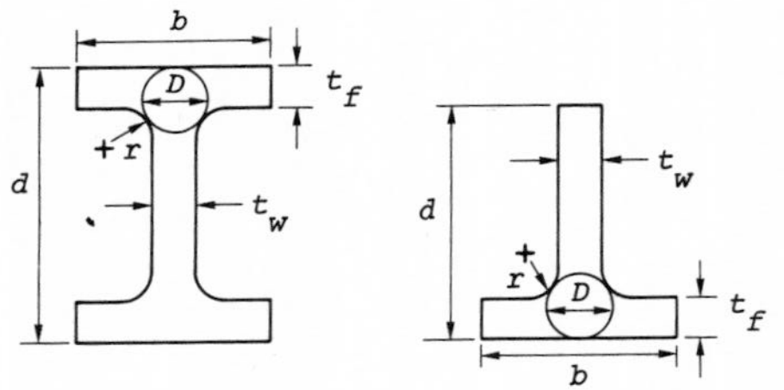	$J_I = \frac{2}{3}bt_f^3 + \frac{1}{3}(d - 2t_f)\,t_w^3 + 2\alpha D^4 - 0.420\,t_f^4$ $J_\perp = \frac{1}{3}bt_f^3 + \frac{1}{3}(d - t_f)\,t_w^3 + \alpha D^4 - 0.210\,t_f^4 - 0.105\,t_w^4$ $D = \dfrac{(t_f + r)^2 + t_w\left(r + \frac{t_w}{4}\right)}{2r + t_f}$ $\alpha = -0.0420 + 0.2204\,\frac{t_w}{t_f} + 0.1355\,\frac{r}{t_f} - 0.0865\,\frac{t_w r}{t_f^2} - 0.0725\left(\frac{t_w}{t_f}\right)^2$
16. 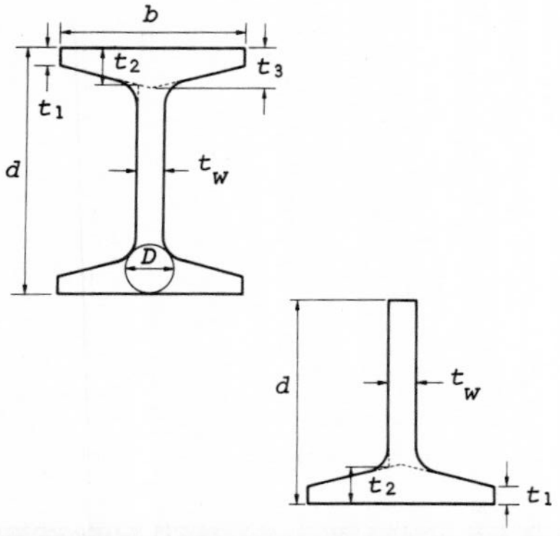	$J_I = \frac{1}{6}(b - t_w)(t_1 + t_2)(t_1^2 + t_2^2) + \frac{2}{3}t_w t_2^3 + \frac{1}{3}(d - 2t_2)\,t_w^3 + 2\alpha D^4 - 4Vt_1^4$ $J_\perp = \frac{1}{12}(b - t_w)(t_1 + t_2)(t_1^2 + t_2^2) + \frac{1}{3}t_w t_2^3 + \frac{1}{3}(d - t_2)\,t_w^3 + \alpha D^4 - 2Vt_1^4 - 0.105\,t_w^4$ $D = \dfrac{(F + t_3)^2 + t_w\left(r + \frac{t_w}{4}\right)}{F + r + t_3}$ $\alpha = -0.0836 + 0.2536\,\frac{t_w}{t_2} + 0.1268\,\frac{r}{t_2} - 0.0806\,\frac{t_w r}{t_2^2} - 0.0858\left(\frac{t_w}{t_2}\right)^2$ $F = rS\left(\sqrt{\frac{1}{S^2} + 1} - 1 - \frac{t_w}{2r}\right)$ $V = 0.10504 + 0.10000S + 0.08480S^2 + 0.06746S^3 + 0.05153S^4$ $S = \dfrac{2(t_2 - t_1)}{(b - t_w)}$ = slope of the flange

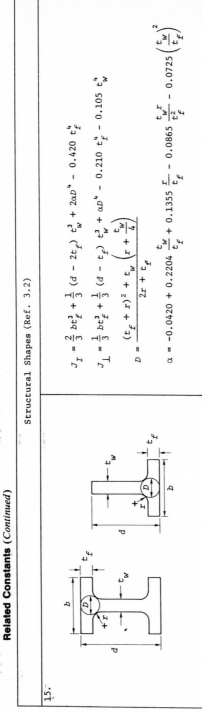

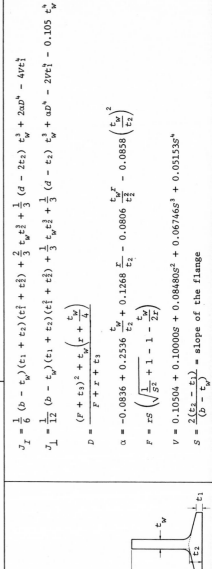

17.

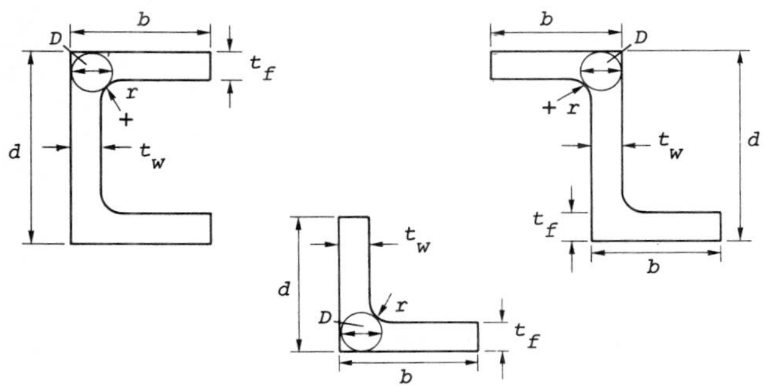

$$J_{\text{[}} = J_{\text{]}} = \frac{2}{3} b t_f^3 + \frac{1}{3} (d - 2t_f)\ t_w^3 + 2\alpha D^4 - 0.420\ t_f^4$$

$$J_{\text{L}} = \frac{1}{3} b t_f^3 + \frac{1}{3} (d - t_f)\ t_w^3 + \alpha D^4 - 0.210\ t_f^4 - 0.105\ t_w^4$$

$$D = 2[(3r + t_w + t_f) - \sqrt{2(2r + t_w)(2r + t_f)}]$$

$$\alpha = -0.0908 + 0.2621 \frac{t_w}{t_f} + 0.1231 \frac{r}{t_f} - 0.0752 \frac{t_w r}{t_f^2} - 0.0945 \left(\frac{t_w}{t_f}\right)^2$$

18.

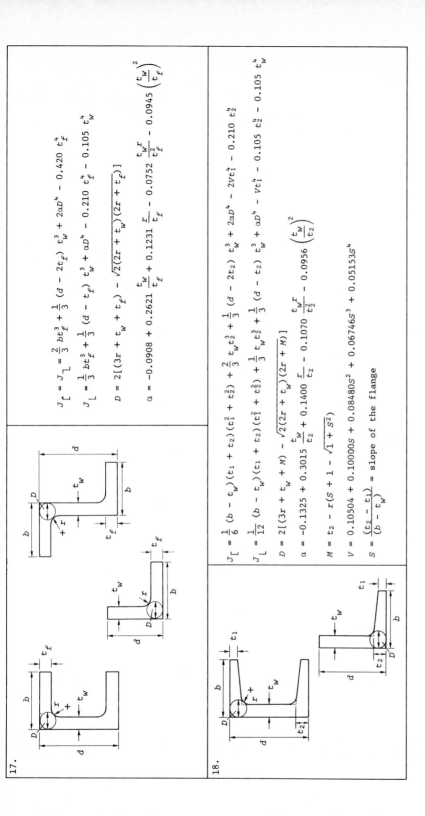

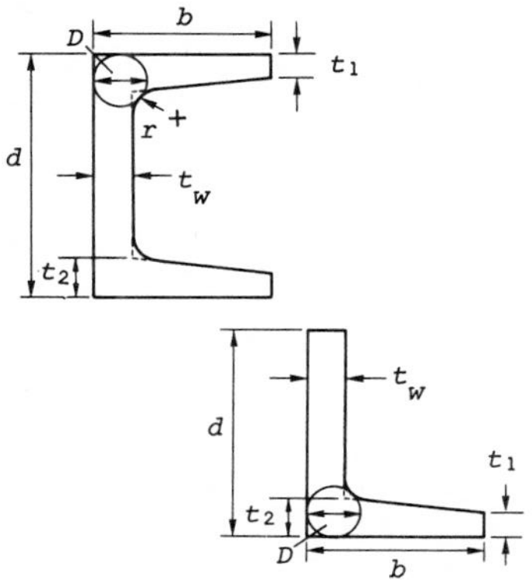

$$J_{\text{[}} = \frac{1}{6} (b - t_w)(t_1 + t_2)(t_1^2 + t_2^2) + \frac{2}{3} t_w t_2^3 + \frac{1}{3} (d - 2t_2)\ t_w^3 + 2\alpha D^4 - 2Vt_1^4 - 0.210\ t_2^4$$

$$J_{\text{L}} = \frac{1}{12} (b - t_w)(t_1 + t_2)(t_1^2 + t_2^2) + \frac{1}{3} t_w t_2^3 + \frac{1}{3} (d - t_2)\ t_w^3 + \alpha D^4 - Vt_1^4 - 0.105\ t_2^4 - 0.105\ t_w^4$$

$$D = 2[(3r + t_w + M) - \sqrt{2(2r + t_w)(2r + M)}]$$

$$\alpha = -0.1325 + 0.3015 \frac{t_w}{t_2} + 0.1400 \frac{r}{t_2} - 0.1070 \frac{t_w r}{t_2^2} - 0.0956 \left(\frac{t_w}{t_2}\right)^2$$

$$M = t_2 - r(S + 1 - \sqrt{1 + S^2})$$

$$V = 0.10504 + 0.10000S + 0.08480S^2 + 0.06746S^3 + 0.05153S^4$$

$$S = \frac{(t_2 - t_1)}{(b - t_w)} = \text{slope of the flange}$$

The differential equations equivalent to Eq. (3.5) are:

$$\frac{d\phi}{dx} = \frac{T}{GJ} \qquad \frac{dT}{dx} = -m_x \tag{3.6}$$

These relations are integrated to provide the angle of twist and the torque.

Angle of twist:	$\phi = \phi_0 + T_0 \dfrac{x}{GJ} + F_\phi$	(3.7*a*)
Torque:	$T = T_0 + F_T$	(3.7*b*)

The F_ϕ, F_T are loading functions given in Table 3-2. If there is more than one load on a bar, the F_ϕ, F_T functions are formed by adding the terms given in Table 3-2 for each load. The initial parameters ϕ_0, T_0 (values of ϕ, T at the left end of the bar, $x = 0$) are provided in Table 3-3.

TABLE 3-2 Loading Functions $F_\phi(x)$, $F_T(x)$ for Bar of Constant Cross Section (Eqs. 3.7)

	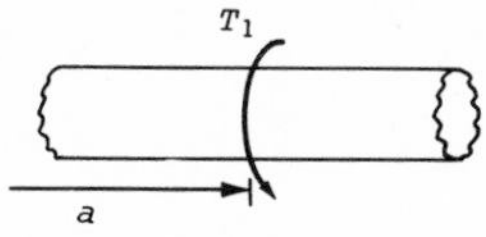	Uniformly Distributed Torque 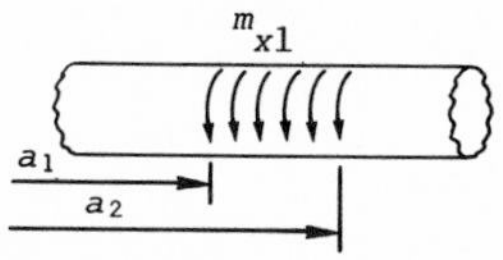
$F_\phi(x)$	$\dfrac{-T_1<x-a>}{GJ}$	$-\dfrac{m_{x1}}{2GJ}(<x-a_1>^2 - <x-a_2>^2)$
$F_T(x)$	$-T_1<x-a>^0$	$-m_{x1}(<x-a_1> - <x-a_2>)$

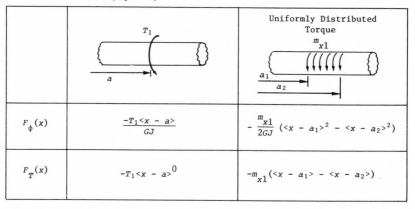

3.4 Natural Frequencies

The natural frequencies ω_n, $n = 1, 2, \ldots,$ for the twisting of uniform shafts are given by

$$\omega_n = \frac{n\pi}{L}\sqrt{\frac{G}{\rho^*}} \tag{3.8}$$

with $\beta^2 = \omega^2\rho^*/G$, where

$n = 1, 2, \ldots$ for fixed-fixed ends
$n = \frac{1}{2}, \frac{3}{2}, \ldots$ for fixed-free ends
$n = 0, 1, 2, \ldots$ for free-free ends

TABLE 3-3 Initial Parameters for Bar of Constant Cross Section (Eqs. 3.7)

Right End → / Left End ↓	1. Fixed 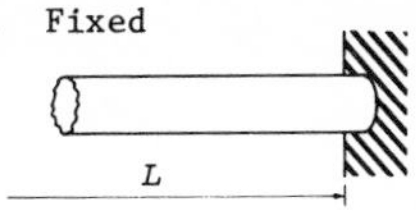	2. Free 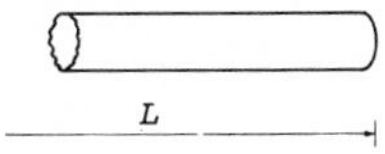
1. Fixed 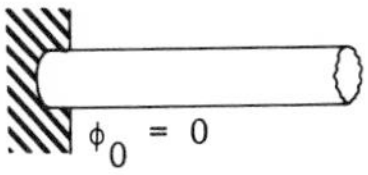 $\phi_0 = 0$	$T_0 = -\frac{GJ}{L}(F_\phi)_{x=L}$	$T_0 = -(F_T)_{x=L}$
2. Free 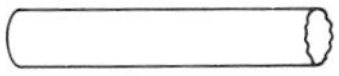 $T_0 = 0$	$\phi_0 = -(F_\phi)_{x=L}$	Kinematically Unstable*

* A kinematically unstable bar can only be analyzed for its natural frequencies.

B. COMPLEX BARS

3.5 Notation for Complex Bars

r_p Polar radius of gyration (length); r_p is the radius of gyration of the cross-sectional area with respect to the axis of the bar (Table 2-1).

ρ Mass per unit length (mass/length)

$I_p = \rho r_p^2$ Polar mass moment of inertia per unit length (mass-length). For a hollow circular section $I_p = \rho(r_{\text{outer}}^2 + r_{\text{inner}}^2)/2$.

I_{pi} Polar mass moment of inertia of concentrated mass at station i (mass · length^2). This can be calculated as $I_{pi} = \Delta a \rho r_p^2$ where Δa is the length of shaft lumped at station i. For a disk of mass M_i, $I_{pi} = M_i(r_{\text{outer}}^2 + r_{\text{inner}}^2)/2$.

$\dfrac{\Delta m_x}{\Delta \ell}$ Gradient of distributed torque, linearly varying in x direction (force · length/length^2)

k Torsional spring constant (force · length)

k_t Elastic foundation modulus (force · length/length)

ℓ Length of segment, span of transfer matrix

t Time

c Discrete torsional damping constant (mass · length^2/time, force · length^2 · time/length)

c_t External or viscous damping coefficient for continuous damping (mass · length/time, force · length · time/length)

c_ρ Proportional viscous damping coefficient (1/time). If the damping is chosen to be proportional to the mass, that is, $c_t = c_\rho \rho r_p^2$, then c_ρ is a constant of proportionality

$\mathbf{U}_i$ Field matrix of the ith segment

$\overline{\mathbf{U}}_i$ Point matrix at $x = a_i$

The notation for transfer matrices is:

$$\mathbf{U}_i = \begin{bmatrix} U_{\phi\phi} & U_{\phi T} & F_\phi \\ U_{T\phi} & U_{TT} & F_T \\ 0 & 0 & 1 \end{bmatrix} \tag{3.9}$$

TABLE 3-4 Massless Bar

$\mathbf{U}_i =$

1	$\frac{\ell}{GJ}$	F_ϕ
0	1	F_T
0	0	1

$$F_\phi = -\frac{1}{GJ}\left(m_{x1}\frac{\ell^2}{2} + \frac{\Delta m_x}{\Delta \ell}\frac{\ell^3}{6}\right)$$

$$F_T = -\left(m_{x1}\ell + \frac{\Delta m_x}{\Delta \ell}\frac{\ell^2}{2}\right)$$

3.6 Differential Equations

The fundamental equations of motion in first-order form for the torsion of a shaft are

$$\frac{\partial \phi}{\partial x} = \frac{T}{GJ}$$
$$\frac{\partial T}{\partial x} = k_t\phi + \rho r_p^2 \frac{\partial^2 \phi}{\partial t^2} - m_x(x,t) \tag{3.10}$$

In higher-order form, these equations become

$$m_x = -\frac{\partial}{\partial x}\left(GJ\frac{\partial \phi}{\partial x}\right) + k_t\phi + \rho r_p^2 \frac{\partial^2 \phi}{\partial t^2}$$
$$T = GJ\frac{\partial \phi}{\partial x} \tag{3.11}$$

3.7 Transfer Matrices for Uniform Segments

The transfer matrices for a shaft segment of length ℓ are provided in Tables 3-4 to 3-7. In these tables, uniformly distributed torques m_{x1} and linearly varying torques of gradient $\Delta m_x/\Delta \ell$ are taken into account.

3.8 Point Matrices

The transfer matrices that take into account concentrated (point) occurrences are listed in Tables 3-8 and 3-9.

TABLE 3-5 Massless Bar on Elastic Foundation

$\mathbf{U}_i =$

$\cosh \beta\ell$	$\frac{\sinh \beta\ell}{GJ\beta}$	F_ϕ
$GJ\beta \sinh \beta\ell$	$\cosh \beta\ell$	F_T
0	0	1

$$F_\phi = -\frac{m_{x1}}{GJ}\frac{(-1 + \cosh \beta\ell)}{\beta^2} - \frac{1}{GJ}\frac{\Delta m_x}{\Delta \ell}\frac{(-\beta\ell + \sinh \beta\ell)}{\beta^3}$$

$$F_T = -m_{x1}\frac{\sinh \beta\ell}{\beta} - \frac{\Delta m_x}{\Delta \ell}\frac{(-1 + \cosh \beta\ell)}{\beta^2}$$

Definition: $\beta^2 = k_t/GJ$

TABLE 3-6 Bar with Mass and Foundation

$$\mathbf{U}_i = \begin{bmatrix} 1 & 0 & F_\phi \\ \ell(k_t - \rho r_p^2 \omega^2) & 1 & F_T \\ 0 & 0 & 1 \end{bmatrix}$$

$F_\phi = 0$

$F_T = -m_{x1}\ell - \dfrac{\Delta m_x}{\Delta \ell}\dfrac{\ell^2}{2}$

Definitions:

(a) Massless with no foundation: $\rho = k_t = 0$

(b) Vibrating bar with no foundation: $k_t = 0$

(c) Vibrating bar with foundation: Give k_t and ρ their correct values

3.9 Loading Functions

The loading functions are provided in the previous transfer matrix tables for most common loadings. These functions can be calculated for other loadings from the formulas:

$$\begin{aligned} F_\phi &= -\int_0^\ell m_x(x) U_{\phi T}(\ell - x)\, dx = -\int_0^\ell m_x(\ell - x) U_{\phi T}(x)\, dx \\ F_T &= -\int_0^\ell m_x(x) U_{TT}(\ell - x)\, dx = -\int_0^\ell m_x(\ell - x) U_{TT}(x)\, dx \end{aligned} \tag{3.12}$$

where m_x is an arbitrary torque intensity. The notations $U_{ij}(\ell - x)$ and $U_{ij}(x)$ refer to the U_{ij} given in the transfer matrix tables with ℓ replaced by $\ell - x$ and x, respectively. Similar definitions apply for $m_x(\ell - x)$ and $m_x(x)$.

TABLE 3-7 Rigid Bar

$$\mathbf{U}_i = \begin{bmatrix} \cos\beta\ell & \dfrac{\sin\beta\ell}{GJ\beta} & F_\phi \\ -GJ\beta\sin\beta\ell & \cos\beta\ell & F_T \\ 0 & 0 & 1 \end{bmatrix}$$

$F_\phi = -m_{x1}\dfrac{1}{GJ\beta^2}(1 - \cos\beta\ell) - \dfrac{\Delta m_x}{\Delta\ell}\dfrac{1}{GJ\beta^3}(\beta\ell - \sin\beta\ell)$

$F_T = -m_{x1}\dfrac{1}{\beta}\sin\beta\ell - \dfrac{\Delta m_x}{\Delta\ell}\dfrac{1}{\beta^2}(1 - \cos\beta\ell)$

Definitions:

$\beta^2 = \dfrac{\omega^2 \rho r_p^2 - k_t}{GJ}$

$\dfrac{\rho r_p^2}{J} = \rho^*$ for circular cross section

If $\beta^2 < 0$, use Table 3-5 with $\beta^2 = \dfrac{k_t - \omega^2 \rho r_p^2}{GJ}$

TABLE 3-8 Point Matrices at $x = a_i$

Definitions: Care should be taken in using this table in that case (b) can not occur at the same axial location as cases (c) and (d). That is, based on physical reasoning either $1/k = 0$ or $k_b - I_{pi}\omega^2 = 0$ at a particular location. Multiply point matrices according to the sequence of occurrences.

$$\bar{\mathbf{U}}_i = \begin{bmatrix} 1 & 1/k & 0 \\ k_b - I_{pi}\omega^2 & 1 & -T_i \\ 0 & 0 & 1 \end{bmatrix}$$

(a) Concentrated Applied Torque

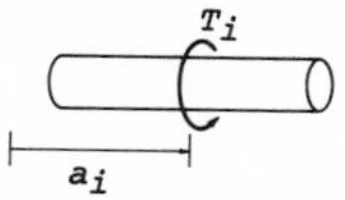

(b) Torsional Spring or Shaft Section

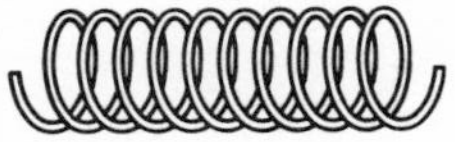

The spring constant k is equivalent to GJ/ℓ of Table 3-4 for a shaft being treated as a torsional spring

For a coil spring

$$k = \frac{Ed^4}{32ND}$$

E = Young's modulus
d = spring wire diameter
D = mean coil diameter
N = number of spring coils

(c) Branch System

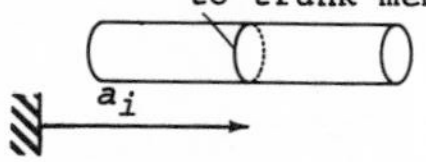

$$T|_{\text{branch at } x=a_i} = k_b\phi|_{\text{branch at } x=a_i}$$

(d) Disk

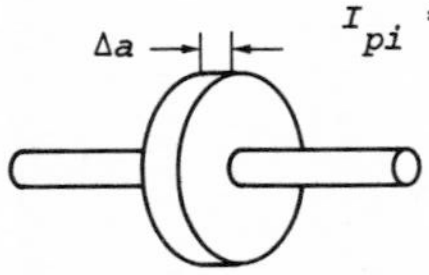

$$I_{pi} = \rho r_p^2\,\Delta a$$

For bar of circular cross section
$\rho r_p^2 = \rho^* J$

TABLE 3-9 Point Matrix for Disk System

$$\overline{\mathbf{U}}_i = \begin{bmatrix} (-1)^{h-g}\dfrac{r_g}{r_h} & 0 & 0 \\ U_{T\phi} & (-1)^{h-g}\dfrac{r_h}{r_g} & 0 \\ 0 & 0 & 1 \end{bmatrix}$$

Definitions:

(a) Geared Disk System

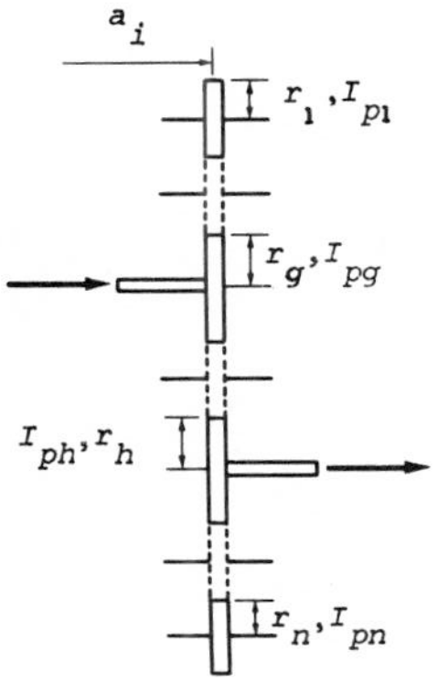

$$U_{T\phi} = -(-1)^{h-g}\frac{r_h}{r_g}\,\omega^2\sum_{i=1}^{n}\left(I_{pi}\,\frac{r_1^2}{r_i^2}\right)$$

(b) Geared Disk System with Branches

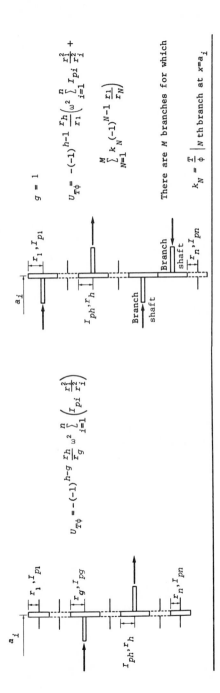

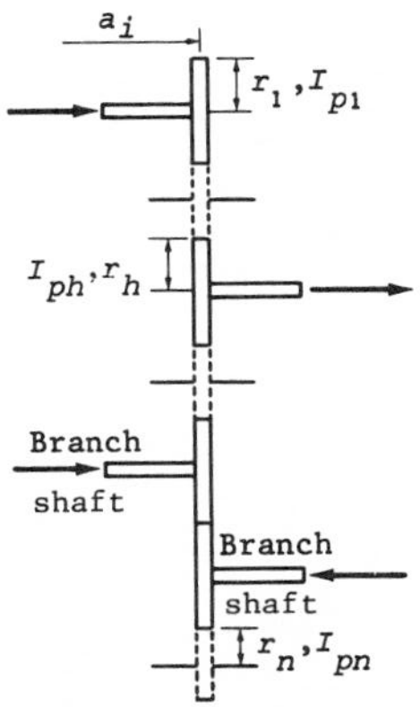

$g = 1$

$$U_{T\phi} = -(-1)^{h-1}\frac{r_h}{r_1}\left(\omega^2\sum_{i=1}^{n} I_{pi}\,\frac{r_1^2}{r_i^2} + \sum_{N=1}^{M} k_N(-1)^{N-1}\frac{r_1}{r_N}\right)$$

There are M branches for which

$$k_N = \frac{T}{\phi}\bigg|_{N\text{th branch at } x=a_i}$$

EXAMPLE 3.1 Loading Functions for a Bar on Elastic Foundation Find F_ϕ, F_T for the case of a bar on an elastic foundation for

$$m_x(\ell) = m_{x1} + \frac{\Delta m_x}{\Delta \ell}\,\ell$$

From Table 3-5,

$$U_{\phi T}(\ell) = \frac{\sinh \beta\ell}{GJ\beta} \qquad U_{TT}(\ell) = \cosh \beta\ell$$

Then, from Eq. (3.12),

$$\begin{aligned} F_\phi &= -\int_0^\ell \left[m_{x1} + \frac{\Delta m_x}{\Delta \ell}(\ell - x) \right] \frac{\sinh \beta x}{GJ\beta}\,dx \\ &= -\frac{1}{GJ\beta^2}\left[m_{x1}(\cosh \beta\ell - 1) + \frac{\Delta m_x}{\Delta \ell}\,\frac{\sinh \beta\ell - \beta\ell}{\beta} \right] \end{aligned}$$

$$\begin{aligned} F_T &= -\int_0^\ell \left[m_{x1} + \frac{\Delta m_x}{\Delta \ell}(\ell - x) \right] \cosh \beta x\,dx \\ &= -\left[m_{x1}\,\frac{\sinh \beta\ell}{\beta} + \frac{\Delta m_x}{\Delta \ell}\,\frac{\cosh \beta\ell - 1}{\beta^2} \right] \end{aligned}$$

3.10 Static Response

The procedure for calculating the angle of twist and internal torque resulting from static loading is the same as that given in Section 2.16 for the response of beams. In brief, the state vector

$$\mathbf{s} = \begin{bmatrix} \phi \\ T \\ 1 \end{bmatrix} \tag{3.13}$$

at any point along the bar is found by progressive multiplication of the transfer matrices for all occurrences from left to right up to that point. That is, the state variables at any point j are given by

$$\mathbf{s}_j = \mathbf{U}_j\mathbf{U}_{j-1} \cdots \overline{\mathbf{U}}_k \cdots \mathbf{U}_2\mathbf{U}_1\mathbf{s}_0 \tag{3.14}$$

where $\mathbf{s}_0 = \mathbf{s}_{x=0}$. If the right end of the bar occurs at station n, the state variables at the right end of the bar become

$$\mathbf{s}_{x=L} = \mathbf{s}_n = \mathbf{U}_n\mathbf{U}_{n-1} \cdots \overline{\mathbf{U}}_k \cdots \mathbf{U}_2\mathbf{U}_1\mathbf{s}_0 = \mathbf{U}\mathbf{s}_0 \tag{3.15}$$

The matrix **U** is the overall or global transfer matrix. The initial parameters ϕ_0, T_0 composing $\mathbf{s}_0$ are found from the boundary condition that occurs at each end of the bar.

Summary of the Calculation Procedure

1. Model the torsion system in terms of sections that connect locations (stations) of point occurrences, for example, applied concentrated torques, or abrupt changes in cross-sectional area.

2. Calculate the torsional constant J and other constants such as the modulus of elastic foundation for each section.

3. Calculate the elements of the transfer matrices for all the segments (sections) connecting the stations, that is, locations of point occurrences. Compute the elements of the point matrices for the concentrated occurrences. The formulas of Tables 3-4 to 3-9 are used for these purposes.

4. Calculate the global matrix by multiplying in sequence all transfer matrices from the left end to the right end of the bar. That is, calculate **U** of Eq. (3.15).

5. Evaluate the initial variables ϕ_0, T_0 of

$$\mathbf{s}_{x=0} = \mathbf{s}_0 = \begin{bmatrix} \phi \\ \mathrm{T} \\ 1 \end{bmatrix}_0$$

by using the boundary conditions. This is accomplished by eliminating the unnecessary rows and columns of Eq. (3.15) and solving the remaining equations.

6. Calculate the angle of twist and internal torque at all points of interest using Eq. (3.14).

7. Use the internal torque to calculate the stresses as given by the formulas of Section 3.2.

EXAMPLE 3.2 Static Analysis of Shaft Find the angles of twist and internal twisting moments along the shaft system of Fig. 3-3.

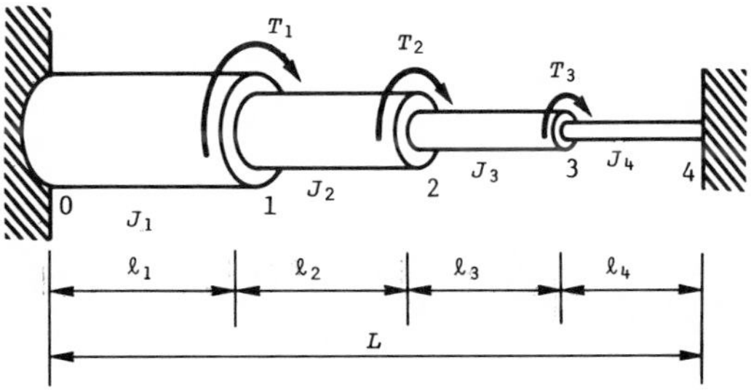

Fig. 3-3

If the boundaries are included, there are five locations of concentrated occurrences; that is, there are five stations. These occur at 0, 1, 2, 3, 4. There are four sections connecting these occurrences. The number of sections is always one less than the number of stations. The transfer matrices can now be formed. From Table 3-4 the transfer matrices for the four sections ($m = 1, 2, 3, 4$) are

$$\mathbf{U}_m = \begin{bmatrix} 1 & \dfrac{\ell_m}{GJ_m} & 0 \\ 0 & 1 & 0 \\ 0 & 0 & 1 \end{bmatrix} \qquad m = 1, 2, 3, 4 \tag{1}$$

or

$$\mathbf{U}_m = \begin{bmatrix} 1 & \dfrac{1}{k_m} & 0 \\ 0 & 1 & 0 \\ 0 & 0 & 1 \end{bmatrix} \qquad \text{with } k_m = \frac{GJ_m}{\ell_m} \tag{2}$$

From Table 3-8

$$\overline{\mathbf{U}}_i = \begin{bmatrix} 1 & 0 & 0 \\ 0 & 1 & T_i \\ 0 & 0 & 1 \end{bmatrix} \qquad i = 1, 2, 3 \tag{3}$$

for stations $i = 1, 2, 3$.

The global transfer matrix $\mathbf{U}$ is given by Eq. (3.15). In our case

$$\mathbf{U} = \mathbf{U}_4\overline{\mathbf{U}}_3\mathbf{U}_3\overline{\mathbf{U}}_2\mathbf{U}_2\overline{\mathbf{U}}_1\mathbf{U}_1$$

or

$$\mathbf{U} = \begin{bmatrix} 1 & \dfrac{1}{k_4} + \dfrac{1}{k_3} + \dfrac{1}{k_2} + \dfrac{1}{k_1} & \dfrac{T_3}{k_4} + \dfrac{T_2}{k_3} + \dfrac{T_1}{k_2} \\ 0 & 1 & T_3 + T_2 + T_1 \\ 0 & 0 & 1 \end{bmatrix} \tag{4}$$

The boundary conditions are now applied to (4) (Eq. 3.15) to calculate ϕ_0, T_0 of $\mathbf{s}_0$. The bar is fixed at each end so that $\phi_0 = 0$ and $\phi_4 = 0$. From (4),

$$\begin{bmatrix} \phi = 0 \\ T \\ 1 \end{bmatrix}_4 = \begin{bmatrix} & \dfrac{1}{k_4} + \dfrac{1}{k_3} + \dfrac{1}{k_2} + \dfrac{1}{k_1} & \dfrac{T_3}{k_4} + \dfrac{T_2}{k_3} + \dfrac{T_1}{k_2} \\ \hline & 0 & 1 \end{bmatrix} \begin{bmatrix} \phi = 0 \\ T \\ 1 \end{bmatrix}_0 \tag{5}$$

Cancel column 1 because $\phi_0 = 0$. Cancel row 2 because T_4 is unknown.

The equation $\phi_4 = 0$ is used to compute T_0, the remaining unknown initial condition. Thus, from (5)

$$\left(\frac{1}{k_4}+\frac{1}{k_3}+\frac{1}{k_2}+\frac{1}{k_1}\right)T_0=-\left(\frac{T_3}{k_4}+\frac{T_2}{k_3}+\frac{T_1}{k_2}\right)$$

$$T_0=-\frac{\dfrac{T_3}{k_4}+\dfrac{T_2}{k_3}+\dfrac{T_1}{k_2}}{\dfrac{1}{k_4}+\dfrac{1}{k_3}+\dfrac{1}{k_2}+\dfrac{1}{k_1}} \quad \text{or} \quad T_0=-\frac{\dfrac{T_3\ell_4}{GJ_4}+\dfrac{T_2\ell_3}{GJ_3}+\dfrac{T_1\ell_2}{GJ_2}}{\dfrac{\ell_4}{GJ_4}+\dfrac{\ell_3}{GJ_3}+\dfrac{\ell_2}{GJ_2}+\dfrac{\ell_1}{GJ_1}} \tag{6}$$

For the special case of $J_m = J$, for $m = 1, 2, 3, 4$, (6) reduces to

$$T_0=-\frac{T_3\ell_4+T_2\ell_3+T_1\ell_2}{\ell_4+\ell_3+\ell_2+\ell_1} \tag{7}$$

Since both initial parameters (ϕ_0, T_0) are now known, the angle of twist and twisting moment can be calculated along the shaft from Eq. (3.14). Thus

$$\begin{aligned}
\mathbf{s}_4 &= \mathbf{U}\mathbf{s}_0 = \mathbf{U}_4\overline{\mathbf{U}}_3\mathbf{U}_3\overline{\mathbf{U}}_2\mathbf{U}_2\overline{\mathbf{U}}_1\mathbf{U}_1\mathbf{s}_0\\
\mathbf{s}_{3+} &= \overline{\mathbf{U}}_3\mathbf{U}_3\overline{\mathbf{U}}_2\mathbf{U}_2\overline{\mathbf{U}}_1\mathbf{U}_1\mathbf{s}_0\\
\mathbf{s}_{3-} &= \mathbf{U}_3\overline{\mathbf{U}}_2\mathbf{U}_2\overline{\mathbf{U}}_1\mathbf{U}_1\mathbf{s}_0\\
\mathbf{s}_{2+} &= \overline{\mathbf{U}}_2\mathbf{U}_2\overline{\mathbf{U}}_1\mathbf{U}_1\mathbf{s}_0\\
\mathbf{s}_{2-} &= \mathbf{U}_2\overline{\mathbf{U}}_1\mathbf{U}_1\mathbf{s}_0\\
\mathbf{s}_{1+} &= \overline{\mathbf{U}}_1\mathbf{U}_1\mathbf{s}_0\\
\mathbf{s}_{1-} &= \mathbf{U}_1\mathbf{s}_0
\end{aligned} \tag{8}$$

Note that because of the concentrated torques at the stations, the state variables to the right and left of the stations differ. Values to the right of a station are indicated by +—for example, 3 +—while − indicates values to the left of a station.

Consider the values of ϕ and T given by (8) at stations 4, 1 +, and 1 −. For simplicity we use the special case of a uniform shaft, that is, $J_m = J$ for all m. Then, from (1), (3), and (8).

$$\mathbf{s}_4=\mathbf{U}\mathbf{s}_0=\begin{bmatrix}\phi\\ T\\ 1\end{bmatrix}_4=\begin{bmatrix}1 & \dfrac{\ell_4+\ell_3+\ell_2+\ell_1}{GJ} & \dfrac{T_3\ell_4+T_2\ell_3+T_1\ell_2}{GJ}\\ 0 & 1 & T_3+T_2+T_1\\ 0 & 0 & 1\end{bmatrix}\begin{bmatrix}\phi\\ T\\ 1\end{bmatrix}_0 \tag{9}$$

This gives

$$\phi_4=\frac{1}{GJ}\left[(\ell_4+\ell_3+\ell_2+\ell_1)T_0+T_3\ell_4+T_2\ell_3+T_1\ell_2\right] \tag{10}$$

Substituting (7) in (10), we find $\phi_4 = 0$. Also, from (9)

$$T_4 = T_0 + T_3 + T_2 + T_1 = -\frac{T_3\ell_4 + T_2\ell_3 + T_1\ell_2}{\ell_4 + \ell_3 + \ell_2 + \ell_1} + T_3 + T_2 + T_1$$

For the special case of $T_i = T$, $\ell_i = \ell$, $i = 1, 2, 3$ and $\ell_4 = \ell$ we have

$$T_0 = -\frac{3}{4}T \qquad T_4 = -\frac{3}{4}T + 3T = 2\frac{1}{4}T$$

Similarly, at station 1 − we have

$$\mathbf{s}_1 = \mathbf{U}_1\mathbf{s}_0 = \begin{bmatrix} \phi \\ T \\ 1 \end{bmatrix}_{1-} = \begin{bmatrix} 1 & \frac{\ell_1}{GJ} & 0 \\ 0 & 1 & 0 \\ 0 & 0 & 1 \end{bmatrix} \begin{bmatrix} \phi \\ T \\ 1 \end{bmatrix}_0 \quad \text{or} \quad \phi_{1-} = \frac{\ell_1}{GJ}T_0, \quad T_{1-} = T_0$$

and at station 1 +

$$\mathbf{s}_1 = \overline{\mathbf{U}}_1\mathbf{U}_1\mathbf{s}_0 = \begin{bmatrix} \phi \\ T \\ 1 \end{bmatrix}_{1+} = \begin{bmatrix} 1 & \frac{\ell_1}{GJ} & 0 \\ 0 & 1 & T_1 \\ 0 & 0 & 1 \end{bmatrix} \begin{bmatrix} \phi \\ T \\ 1 \end{bmatrix}_0 \quad \text{or} \quad \phi_{1+} = \frac{\ell_1}{GJ}T_0, \quad T_{1+} = T_1 + T_0$$

For some problems it may be desirable to calculate the state variables at points between stations. For example, suppose the state variables are sought at $\ell_2/2$ distance from station 1. First, set up the transfer matrix extending from station 1 to $1\frac{1}{2}$ (at $\ell_2/2$). From Table 3-4

$$\mathbf{U}_{1\frac{1}{2}} = \begin{bmatrix} 1 & \frac{\ell_2}{2GJ} & 0 \\ 0 & 1 & 0 \\ 0 & 0 & 1 \end{bmatrix}$$

Then calculate $\mathbf{s} = \mathbf{U}_{1\frac{1}{2}}\overline{\mathbf{U}}_1\mathbf{U}_1\mathbf{s}_0$ as

$$\begin{bmatrix} \phi \\ T \\ 1 \end{bmatrix}_{1\frac{1}{2}} = \begin{bmatrix} 1 & \frac{\ell_2}{2GJ} + \frac{\ell_1}{GJ} & \frac{T_1}{2GJ} \\ 0 & 1 & T_1 \\ 0 & 0 & 0 \end{bmatrix} \begin{bmatrix} \phi \\ T \\ 1 \end{bmatrix}_0$$

or

$$\phi_{1\frac{1}{2}} = \frac{1}{GJ}\left(\frac{\ell_2}{2} + \ell_1\right)T_0, \quad T_{1\frac{1}{2}} = T_0 + T_1$$

Formulas for the Initial Parameters The application of the boundary conditions to Eq. (3.15) to calculate ϕ_0, T_0 as described above suffices to determine the initial parameters for all shafts. Formulas to accomplish the same thing are listed in Table 3-10. In this table values of ϕ_0, T_0 are listed according to boundary conditions. The U_{kj} and F_k in this table are elements of the overall or global transfer matrix extending from $x = 0$ to $x = L$, that is, the elements of **U** of Eq. (3.15).

TABLE 3-10 Initial Parameters

Left End ↓ / Right End →	1. Fixed	2. Free
1. Fixed ($\phi_0 = 0$)	$T_0 = -(F_\phi)_{x=L}/\nabla$ $\nabla = (U_{\phi T})_{x=L}$	$T_0 = -(F_T)_{x=L}/\nabla$ $\nabla = (U_{TT})_{x=L}$
2. Free ($T_0 = 0$)	$\phi_0 = -(F_\phi)_{x=L}/\nabla$ $\nabla = (U_{\phi\phi})_{x=L}$	$\phi_0 = -(F_T)_{x=L}/\nabla$ $\nabla = (U_{T\phi})_{x=L}$

Table 3-10 can be condensed into a single set of formulas. Let s_1, s_2 be the state variables (ϕ, T but not necessarily in this order) with initial values s_{10}, s_{20}. One of the initial parameters is known by observation to be zero at the left end of the bar. Let this be s_{10}. If s_k, $k = 1$ or 2, is the state variable that is zero at the right end, that is, $(s_k)_{x=L} = 0$, then the initial parameters are given by

$$s_{10} = 0$$
$$s_{20} = -\left(F_{s_k}\right)_{x=L}/\nabla \tag{3.16}$$
$$\nabla = \left(U_{s_k s_2}\right)_{x=L}$$

where $U_{s_k s_2}$ and F_{s_k} are elements of the transfer matrix extending from $x = 0$ to $x = L$, that is, the elements of **U** of Eq. (3.15).

EXAMPLE 3.3 Twisting of an Embedded Bar Consider the problem of finding the twist and torque for the bar of Fig. 3-4.

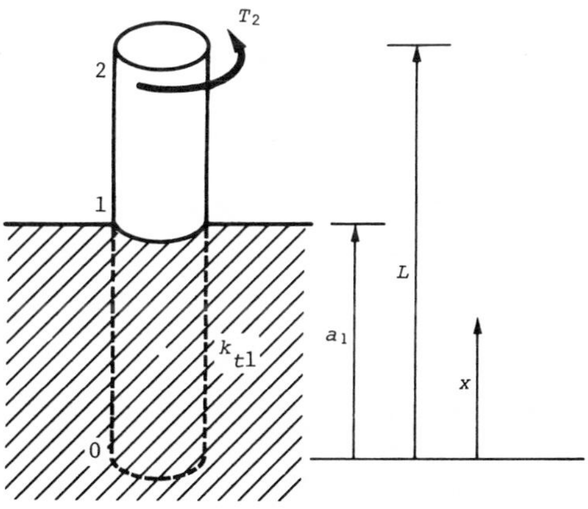

Fig. 3-4

The responses are given by (Eq. 3.14 and 3.15)

$$\mathbf{s}_{x=a_1} = \mathbf{U}_1\mathbf{s}_0 \tag{1}$$

$$\mathbf{s}_{x=L} = \overline{\mathbf{U}}_2\mathbf{U}_2\mathbf{U}_1\mathbf{s}_0 = \mathbf{U}\mathbf{s}_0 \tag{2}$$

where, from Tables 3-5, 3-4, and 3-8

$$\mathbf{U}_1 = \begin{bmatrix} \cosh \beta a_1 & \dfrac{\sinh \beta a_1}{GJ\beta} & 0 \\ GJ\beta \sinh \beta a_1 & \cosh \beta a_1 & 0 \\ 0 & 0 & 1 \end{bmatrix} \qquad \beta^2 = \frac{k_{t1}}{GJ}$$

$$\mathbf{U}_2 = \begin{bmatrix} 1 & \dfrac{(L - a_1)}{GJ} & 0 \\ 0 & 1 & 0 \\ 0 & 0 & 1 \end{bmatrix} \qquad \overline{\mathbf{U}}_2 = \begin{bmatrix} 1 & 0 & 0 \\ 0 & 1 & -T_2 \\ 0 & 0 & 1 \end{bmatrix} \tag{3}$$

The solution can be printed out at locations other than $x = a_1$, L of (1) and (2) if appropriate values of x are substituted in (1) and (2). For example,

$$\mathbf{s}_{x=\ell} = \mathbf{U}_1\mathbf{s}_0 \qquad 0 \leqslant \ell \leqslant a_1$$

where ℓ replaces a_1 in (3).

The initial parameters for this free-free bar are found in Table 3-10. These involve the global transfer matrix of (2). We find

$$\phi_0 = -(F_T)_{x=L}/\nabla = T_2/\nabla$$

$$T_0 = 0$$

$$\nabla = (U_{T\phi})_{x=L} = GJ\beta \sinh \beta a_1$$

and the solution is complete.

3.11 Free Dynamic Response—Natural Frequencies

The natural frequencies ω_n, $n = 1, 2, \ldots,$ are roots of the characteristic equation resulting from the evaluation of the initial parameters. If this characteristic equation is represented by ∇, then the natural frequencies ω_n are those values of ω in ∇ that make ∇ equal to zero.

The procedure for calculating frequencies and mode shapes involves many of the same steps used in a static analysis. The model employed for free dynamic analyses requires the inclusion of mass. As with other structural members, the mass can be modeled as being continuously distributed or it can be lumped at particular locations along the shaft.

Summary of Calculation Procedure

1. Model the shaft system in terms of sections, as in the case of a static analysis.

2. Calculate the same constants, for example, J, as in a static analysis. In addition, calculate the mass densities ρ and polar radii of gyration r_p for sections

in which the distributed mass is to be taken into account. For lumped-mass models, calculate the polar mass moment of inertia I_{pi} for each concentrated mass.

3, 4. Steps 3 and 4 are the same as in the case of a static analysis.

5. Instead of the boundary conditions providing the initial parameters ϕ_0, T_0, they lead to the characteristic function ∇. The roots of $\nabla = 0$ are the natural frequencies. For simple shafts these roots can be found analytically. For complex shafts, for example, systems with several sections or lumped masses, numerical root-finding techniques must be employed.

6. The mode shapes are calculated as in step 6 for a static analysis. In this calculation, the initial parameter (ϕ_0 or T_0) that is not known to be zero at the left end should be assigned a value, such as 1. For example, if the shaft is fixed at the left end, ϕ_0 is zero and T_0 should be set equal to 1. Then use Eq. (3.14) to calculate the mode shapes for the angle of twist and the internal torque.

EXAMPLE 3.4 Natural Frequencies of Torsional System Find the natural frequencies for the shaft system of Fig. 3-3. The static analysis of this shaft was treated in Example 3.2.

A typical lumped-mass model for such a shaft would have the mass concentrated at stations 1, 2, and 3. The corresponding polar mass moments of inertia would be I_{p1}, I_{p2}, and I_{p3}. For simplicity, we choose to set $I_{p1} = I_{p3} = 0$.

From Table 3-8 the point matrix for the lumped mass at station 2 is

$$\overline{\mathbf{U}}_2 = \begin{bmatrix} 1 & 0 & 0 \\ -I_{p2}\omega^2 & 1 & 0 \\ 0 & 0 & 1 \end{bmatrix} \tag{1}$$

The transfer matrices of Eq. (2) in Example 3.2 still remain valid for sections 1, 2, 3, and 4.

The global transfer matrix is given by Eq. (3.15) in the form

$$\mathbf{U} = \mathbf{U}_4\mathbf{U}_3\overline{\mathbf{U}}_2\mathbf{U}_2\mathbf{U}_1 \tag{2}$$

where $\overline{\mathbf{U}}_2$ is given by (1) and $\mathbf{U}_1$, $\mathbf{U}_2$, $\mathbf{U}_3$, and $\mathbf{U}_4$ are taken from $\mathbf{U}_m$, $m = 1, 2, 3, 4$ of Eq. (2), Example 3.2. We find

$$\mathbf{U} = \begin{bmatrix} 1 - I_{p2}\omega^2\left(\frac{1}{k_4} + \frac{1}{k_3}\right) & \frac{1}{k_4} + \frac{1}{k_3} + \frac{1}{k_2} + \frac{1}{k_1} - I_{p2}\omega^2\left(\frac{1}{k_2} + \frac{1}{k_1}\right)\left(\frac{1}{k_4} + \frac{1}{k_3}\right) & 0 \\ -I_{p2}\omega^2 & 1 - \left(\frac{1}{k_2} + \frac{1}{k_1}\right)I_{p2}\omega^2 & 0 \\ 0 & 0 & 1 \end{bmatrix} \tag{3}$$

The boundary conditions $\phi_0 = \phi_4 = 0$ applied to Eq. (3.15) appear as

$$\begin{bmatrix} \phi = 0 \\ T \\ 1 \end{bmatrix}_4 = \left[\begin{array}{c|cc} & \frac{1}{k_4} + \frac{1}{k_3} + \frac{1}{k_2} + \frac{1}{k_1} - I_{p2}\omega^2\left(\frac{1}{k_2} + \frac{1}{k_1}\right)\left(\frac{1}{k_4} + \frac{1}{k_3}\right) & 0 \\ \hline & 0 & 1 \end{array}\right] \begin{bmatrix} \phi = 0 \\ T \\ 1 \end{bmatrix}_0 \tag{4}$$

The first column and second row are cancelled because ϕ_0 is zero and T_4 is unknown, respectively. The only meaningful equation in (4) is the $\phi_4 = 0$ relationship. This appears as

$$\left[\frac{1}{k_4} + \frac{1}{k_3} + \frac{1}{k_2} + \frac{1}{k_1} - I_{p2}\omega^2\left(\frac{1}{k_2} + \frac{1}{k_1}\right)\left(\frac{1}{k_4} + \frac{1}{k_3}\right)\right]T_0 = 0 \tag{5}$$

Since T_0 is not zero, we have

$$\nabla = \left[\frac{1}{k_4} + \frac{1}{k_3} + \frac{1}{k_2} + \frac{1}{k_1} - I_{p2}\omega^2\left(\frac{1}{k_2} + \frac{1}{k_1}\right)\left(\frac{1}{k_4} + \frac{1}{k_3}\right)\right] = 0 \tag{6}$$

Since the term in brackets is the characteristic function, it is designated by ∇. Equation (6) gives the natural frequency

$$\omega^2 = \frac{\dfrac{1}{k_4} + \dfrac{1}{k_3} + \dfrac{1}{k_2} + \dfrac{1}{k_1}}{\left(\dfrac{1}{k_2} + \dfrac{1}{k_1}\right)\left(\dfrac{1}{k_4} + \dfrac{1}{k_3}\right)I_{p2}} \tag{7}$$

This reduces to

$$\omega^2 = \frac{\ell_4 + \ell_3 + \ell_2 + \ell_1}{(\ell_2 + \ell_1)(\ell_4 + \ell_3)} \frac{GJ}{I_{p2}}$$

if $J_m = J$ for all m.

The mode shapes are given by Eq. (3.14) with $\phi_0 = 0$, $T_0 = 1$.

Formulas for the Initial Parameters The procedure just described for calculating frequencies can be formalized somewhat by using Table 3-10 or, equivalently, Eq. (3.16). The natural frequencies ω_n ($n = 1, 2, \ldots,$) of a bar are those values of ω that make ∇ of Table 3-10 or Eqs. (3.16) equal to zero. This ∇ is formed by the elements of the global transfer matrix **U** of Eq. (3.15). For complex bars it is necessary to evaluate the frequencies from $\nabla = 0$ with a computational root-finding technique.

The mode shapes for the angle of twist and torque are given by Eq. (3.14) with initial parameters

$$\left.\begin{aligned}\phi_0 &= 0\\ T_0 &= T_0\end{aligned}\right\}\quad \text{for fixed left ends}$$
$$\left.\begin{aligned}\phi_0 &= \phi_0\\ T_0 &= 0\end{aligned}\right\}\quad \text{for free left ends} \tag{3.17}$$

Note that in each case one initial parameter remains unspecified. This is to be expected since the mode shapes describe only a shape of a state variable and not a precise amplitude. Frequently this unspecified initial parameter is set equal to unity. The third-column terms in the transfer matrices, that is, the loading functions F_ϕ, F_T, can be ignored during the calculation of natural frequencies and mode shapes.

In terms of the definitions of Eq. (3.16), the initial parameters of the mode shapes are defined by

$$\begin{aligned} s_{10} &= 0 \\ s_{20} &= s_{20} \end{aligned} \tag{3.18}$$

where s_{20} is assigned to be the initial parameter that can be given any value. Usually, s_{20} is taken to be unity.

EXAMPLE 3.5 Simple Lumped-Mass System Consider a simple lumped-mass system with a single mass located at the right end of a massless bar of length L. This system is chosen to replace a bar (GJ) of length L with continuously distributed mass. The left end of the bar is fixed, the right end with mass (I_{p1}) is free. The transfer matrix of Table 3-4 is appropriate for the bar, while the disk is represented by the point matrix of Table 3-8, with $k_b = 1/k = T_i = 0$. The product of the two matrices gives the overall transfer matrix of the system:

$$\mathbf{U} = \begin{bmatrix} 1 & \dfrac{L}{GJ} & 0 \\ -\omega_n^2 I_{pi} & 1 - \dfrac{\omega_n^2 I_{pi} L}{GJ} & 0 \\ 0 & 0 & 1 \end{bmatrix} \tag{1}$$

Often the massless bar is considered as a spring of constant $k = GJ/L$. Then the point matrix of Table 3-8, with $k_b = I_{Pi} = T_i = 0$, replaces the field matrix of Table 3-4. Note that in such cases where a shaft is modeled as a torsional spring of constant k, the length of the spring does not enter the problem. However, it is customary during computations to assign an axial coordinate (x) to locate the mass so that the computer output for this mass is easily identified.

The natural frequencies are given by setting ∇ of Table 3-10 equal to zero. Using (1), we find

$$\nabla = (U_{TT})_{x=L} = 1 - \frac{\omega_n^2 I_p L}{GJ} = 0 \qquad \text{or} \qquad \omega_n^2 = \omega_1^2 = \frac{GJ}{I_p L}$$

a result given in many elementary vibration texts.

EXAMPLE 3.6 Gear Systems Find the natural frequencies of the geared system of Fig. 3-5. We assign locations $x = 0$, a_1, a_2 to the disks. These distances serve only to locate the disks for identification purposes, since the axial distances between the disks do not affect the solution. We find

$$\begin{bmatrix} \phi \\ T \end{bmatrix}_{a_2} = \bar{\mathbf{U}}_2 \mathbf{U}_2 \bar{\mathbf{U}}_1 \mathbf{U}_1 \bar{\mathbf{U}}_0 \begin{bmatrix} \phi \\ T \end{bmatrix}_0 = \mathbf{U} \begin{bmatrix} \phi \\ T \end{bmatrix}_0 \tag{1}$$

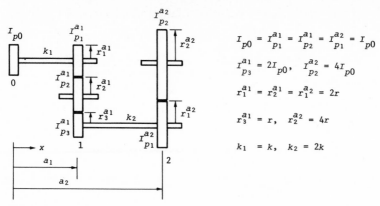

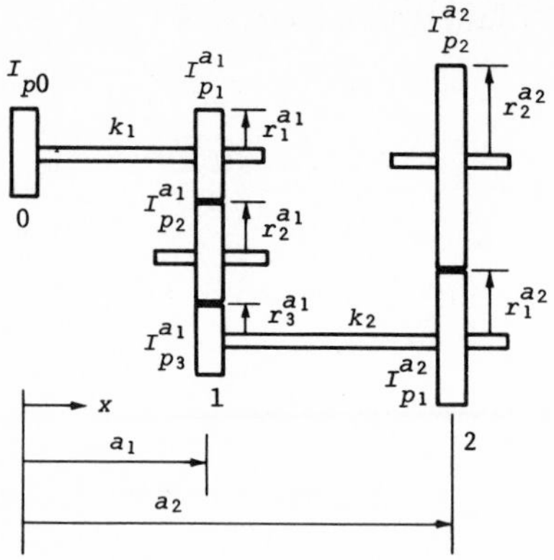

$$I_{p0} = I^{a_1}_{p1} = I^{a_1}_{p2} = I^{a_2}_{p1} = I_{p0}$$

$$I^{a_1}_{p3} = 2I_{p0}, \quad I^{a_2}_{p2} = 4I_{p0}$$

$$r^{a_1}_1 = r^{a_1}_2 = r^{a_2}_1 = 2r$$

$$r^{a_1}_3 = r, \quad r^{a_2}_2 = 4r$$

$$k_1 = k, \quad k_2 = 2k$$

Fig. 3-5

where from Table 3-8

$$\overline{\mathbf{U}}_0 = \begin{bmatrix} 1 & 0 \\ -\omega^2 I_{p0} & 1 \end{bmatrix} \qquad \mathbf{U}_1 = \begin{bmatrix} 1 & \dfrac{1}{k} \\ 0 & 1 \end{bmatrix} \qquad \mathbf{U}_2 = \begin{bmatrix} 1 & \dfrac{1}{2k} \\ 0 & 1 \end{bmatrix}$$

and from Table 3-9

$$\overline{\mathbf{U}}_1 = \begin{bmatrix} \dfrac{r^{a_1}_1}{r^{a_1}_3} & 0 \\ -\dfrac{r^{a_1}_3}{r^{a_1}_1}\omega^2\left[I^{a_1}_{p1}\left(\dfrac{r^{a_1}_1}{r^{a_1}_1}\right)^2 + I^{a_1}_{p2}\left(\dfrac{r^{a_1}_1}{r^{a_1}_2}\right)^2 + I^{a_1}_{p3}\left(\dfrac{r^{a_1}_1}{r^{a_1}_3}\right)^2\right] & \dfrac{r^{a_1}_3}{r^{a_1}_1} \end{bmatrix}$$

$$= \begin{bmatrix} 2 & 0 \\ -5\omega^2 I_{p0} & \frac{1}{2} \end{bmatrix}$$

$$\overline{\mathbf{U}}_2 = \begin{bmatrix} -\dfrac{r^{a_2}_1}{r^{a_2}_2} & 0 \\ \dfrac{r^{a_2}_2}{r^{a_2}_1}\omega^2\left[I^{a_2}_{p1}\left(\dfrac{r^{a_2}_1}{r^{a_2}_1}\right)^2 + I^{a_2}_{p2}\left(\dfrac{r^{a_2}_1}{r^{a_2}_2}\right)^2\right] & -\dfrac{r^{a_2}_2}{r^{a_2}_1} \end{bmatrix} = \begin{bmatrix} -\frac{1}{2} & 0 \\ 4\omega^2 I_{p0} & -2 \end{bmatrix}$$

Note that the third columns and third rows of the transfer matrices of the transfer matrix tables have been dropped since they are not needed.

The frequencies are found by setting ∇ of Table 3-10 for free-free boundary conditions equal to zero. The element $U_{T\phi}$ in ∇ belongs to **U** of (1). After the necessary matrix multiplications are performed, we find

$$\nabla = (U_{T\phi})_{x=L} = 0$$

$$= \omega^2 I_{p0}\left[8\left(1 - \frac{\omega^2 I_{p0}}{k}\right) - \frac{10\omega^2 I_{p0}}{k}\left(1 - \frac{\omega^2 I_{p0}}{k}\right) - \frac{\omega^2 I_{p0}}{k} + 10\left(1 - \frac{\omega^2 I_{p0}}{k}\right) + 1\right] \tag{2}$$

The roots of (2) are calculated to be

$$\omega_1^2 = 0 \qquad \omega_2^2 = k/I_{p0} \qquad \omega_3^2 = 1.9k/I_{p0}$$

The first frequency corresponds to rigid-body shaft rotation without torsion.

EXAMPLE 3.7 System with Branch A branched system requires the determination of an equivalent spring constant k_b that is introduced into the analysis with the transfer matrix of Table 3-8. For the system of Fig. 3-6*a*, the overall transfer

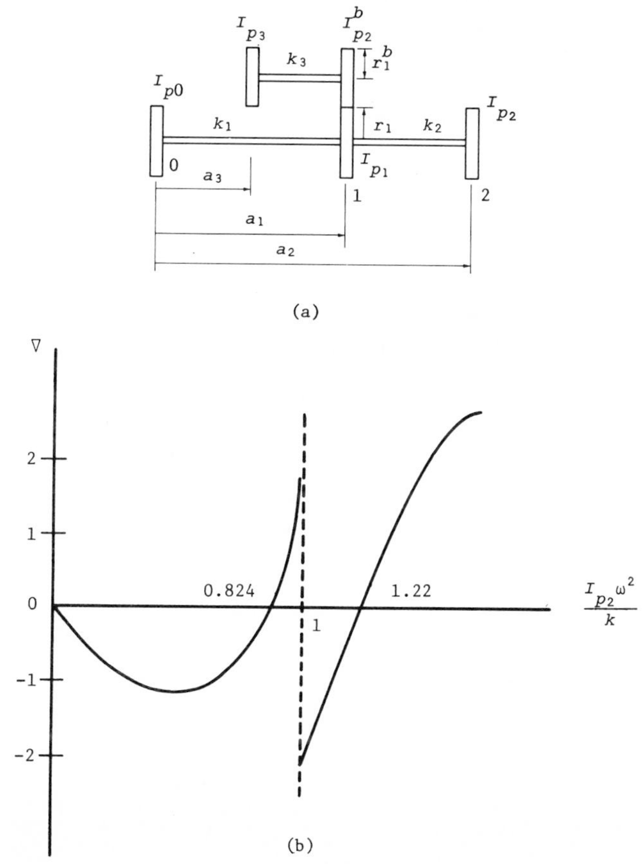

Fig. 3-6

matrix is given by

$$\begin{bmatrix} \phi \\ T \end{bmatrix}_{a_2} = \overline{\mathbf{U}}_2\mathbf{U}_2\overline{\mathbf{U}}_1\mathbf{U}_1\overline{\mathbf{U}}_0\begin{bmatrix} \phi \\ T \end{bmatrix}_0 = \mathbf{U}\begin{bmatrix} \phi \\ T \end{bmatrix}_0 \tag{1}$$

where, from Table 3-8,

$$\underset{i=0,2}{\overline{\mathbf{U}}_i = \begin{bmatrix} 1 & 0 \\ -I_{pi}\omega^2 & 1 \end{bmatrix}} \qquad \underset{j=1,2}{\mathbf{U}_j = \begin{bmatrix} 1 & 1/k_j \\ 0 & 1 \end{bmatrix}} \qquad \overline{\mathbf{U}}_1 = \begin{bmatrix} 1 & 0 \\ k_b & 1 \end{bmatrix}$$

The branch coefficient k_b is the constant of proportionality between the twisting moment and the angular displacement of the branch at the connection between the branch and the trunk systems; that is,

$$T|_{\text{branch at } a_1} = k_b\phi|_{\text{branch at } a_1}$$

Assume the effect of the mass of the gears is negligible, that is, $I_{p1} = I^b_{p1} = 0$. Then for the branch

$$\begin{bmatrix} \phi \\ \\ T \end{bmatrix}_{a_1} = \begin{bmatrix} -\dfrac{r_1}{r^b_1} & 0 \\ 0 & -\dfrac{r^b_1}{r_1} \end{bmatrix}\begin{bmatrix} 1 & \dfrac{1}{k_3} \\ 0 & 1 \end{bmatrix}\begin{bmatrix} 1 & 0 \\ -I_{p3}\omega^2 & 1 \end{bmatrix}\begin{bmatrix} \phi \\ \\ T \end{bmatrix}_{a_3}$$

$$= \begin{bmatrix} 2\left(\dfrac{I_{p3}\omega^2}{k_3} - 1\right) & -\dfrac{2}{k_3} \\ \dfrac{I_{p3}\omega^2}{2} & -\dfrac{1}{2} \end{bmatrix}\begin{bmatrix} \phi \\ \\ T \end{bmatrix}_{a_3} \tag{2}$$

where r^b_1 has been set equal to $2r_1$. Since the left end of the branch is free, $T|_{a_3} = 0$. Substitute $T|_{a_3} = 0$ in the right-hand vector of (2) and eliminate $\phi|_{a_3}$ to find

$$T|_{a_1(\text{branch})} = \frac{1}{4}\,\frac{I_{p3}\omega^2}{\dfrac{I_{p3}\omega^2}{k_3} - 1}\,\phi|_{a_1(\text{branch})}$$

Thus

$$k_b = \frac{1}{4}\,\frac{I_{p3}\omega^2}{\dfrac{I_{p3}\omega^2}{k_3} - 1}$$

The frequencies are found for this free-free system by setting $\nabla = U_{T\phi}$ (Table 3-10) equal to zero. $U_{T\phi}$ is taken from (1). If $k_3 = k_1 = k_2/2 = k$, $I_{p0}/2 = I_{p3} = I_{p2} = I_p$, the plot of ∇ appears as in Fig. 3-6*b*. For specified values of I_p and k, the frequencies can be taken from Fig. 3-6*b* at positions where $\nabla = 0$. The discontinuity in the curve is of considerable interest because this type of occurrence can cause major problems in computational implementation of a solution for branch systems. It corresponds to the natural frequency of the branch. Note that the branch spring constant k_b becomes inordinately large at this frequency.

3.12 Forced Harmonic Motion

If the forcing functions vary as sin Ωt or cos Ωt, where Ω is the frequency of the loading, the state variables ϕ and T also vary as sin Ωt or cos Ωt. The spatial distribution of ϕ and T are found by replacing ω in the appropriate transfer matrices by Ω. The solution procedure is the same as for the static response, since time need not enter the analysis. The initial parameters are given by Table 3-10 or Eqs. (3.16).

3.13 Dynamic Response Due to Arbitrary Loading

In terms of a modal solution, the time-dependent state variables resulting from arbitrary dynamic loading are expressed by

Displacement method:

$$\phi(x,t) = \sum_n A_n(t)\phi_n(x) \qquad T(x,t) = \sum_n A_n(t)T_n(x) \tag{3.19a}$$

Acceleration method:

$$\phi(x,t) = \phi_s(x,t) + \sum_n B_n(t)\phi_n(x) \qquad T(x,t) = T_s(x,t) + \sum_n B_n(t)T_n(x) \tag{3.19b}$$

where

$$A_n(t) = \frac{\eta_n(t)}{N_n} \qquad B_n(t) = \frac{\xi_n(t)}{N_n} \tag{3.20}$$

These relations provide the complete dynamic response. That is, these are the solutions to Eqs. (3.10).

No Damping or Proportional Damping for a Continuous Member For viscous damping, a term $c_t\, \partial\phi/\partial t$ should be added to the right-hand side of the second equation in Eqs. (3.10) when a torsional member is described with damping continuously distributed along the axial direction. If the damping is "proportional," then $c_t = c_\rho \rho r_p^2$. Set $c_t = 0$ for no viscous damping.

In Eqs. (3.19*b*) the terms with the subscript s are static solutions that are determined for the applied loading at each point in time. These static solutions can be taken from Section 3.10. ϕ_n, T_n are the mode shapes, that is, ϕ, T with

$\omega = \omega_n$, $n = 1, 2, \ldots,$ as explained in Section 3.11. The quantities in Eq. (3.20) are defined as

$$N_n = \int_0^L \rho r_p^2 \phi_n^2 dx \tag{3.21}$$

$$\eta_n(t) = e^{-\zeta_n \omega_n t}\left(\cos \alpha_n t + \frac{\zeta_n \omega_n}{\alpha_n} \sin \alpha_n t\right)\eta_n(0) + e^{-\zeta_n \omega_n t} \frac{\sin \alpha_n t}{\alpha_n} \frac{\partial \eta_n}{\partial t}(0)$$

$$+ \int_0^t f_n(\tau) e^{-\zeta_n \omega_n (t-\tau)} \frac{\sin \alpha_n (t-\tau)}{\alpha_n} d\tau \tag{3.22}$$

$\xi_n(t)$ is taken from $\eta_n(t)$ by replacing

$$\eta_n(0) \quad \text{by} \quad \eta_n(0) - \frac{f_n(0)}{\omega_n^2} \qquad \frac{\partial \eta_n}{\partial t}(0) \quad \text{by} \quad \frac{\partial \eta_n}{\partial t}(0) - \frac{1}{\omega_n^2}\frac{\partial f_n}{\partial t}(0)$$

$$f_n(\tau) \quad \text{by} \quad -\frac{1}{\omega_n^2}\left(\frac{\partial^2}{\partial \tau^2} + c_\rho \frac{\partial}{\partial \tau}\right) f_n(\tau)$$

In Eq. (3.22),

$$\alpha_n = \omega_n \sqrt{1 - \zeta_n^2} \qquad \zeta_n = \frac{c_\rho}{2\omega_n}$$

If $\zeta_n > 1$, replace sin by sinh, cos by cosh, and $\sqrt{1 - \zeta_n^2}$ by $\sqrt{\zeta_n^2 - 1}$, $\zeta_n = 0$ for zero viscous damping.

$$\eta_n(0) = \int_0^L \rho r_p^2 \phi(x, 0)\phi_n \, dx \tag{3.23}$$

$$f_n(t) = \int_0^L m_x(x, t)\phi_n \, dx + \phi(0, t)T_n(0) - \phi(L, t)T_n(L) \tag{3.24}$$

The function $m_x(x,t)$ is the applied torque intensity; $\phi(0, t)$ and $\phi(L, t)$ are the applied time-dependent angles of twist at the left and right ends of the bar, respectively. In the case of the acceleration method, $f_n(t)$ can alternatively be expressed as

$$f_n(t) = \omega_n^2 \int_0^L \rho r_p^2 \phi_s(x, t)\phi_n(x) \, dx \tag{3.25}$$

Suppose an undamped bar is connected to the ground with a spring of stiffness k_1 at the left end or k_L at the right end. Angular ground motion $\phi(0, t)$ at $x = 0$ may be included by adding $k_1\phi(0, t)\phi_n(a_1)$ to $f_n(t)$, where $x = a_1$ is the coordinate of the connection of the spring to the bar. For prescribed ground motion $\phi(L, t)$ at $x = L$, add $k_L\phi(L, t)\phi_n(a_L)$ to $f_n(t)$, where a_L is the coordinate of the connection of the spring to the bar.

No Damping or Proportional Damping for Discrete Members When a lumped-mass approximation to the continuous torsion member is desired, some adjustment to the previous formulation must be made. In order to define proportional damping for discrete members, it is helpful to define viscous damping and stiffness matrices $[c]$ and $[k]$.

To establish $[c]$ in terms of the dashpots in the discrete model, suppose that the discrete mass lumps are numbered consecutively beginning with 1 at the far left-hand mass. Then assign dashpot and stiffness values of c_{i+1} and k_{i+1} to the field between mass stations i and $i+1$ (Fig. 3-7a). This means that if the left-hand boundary is fixed, the first left-hand field is designated by c_1 and k_1 as shown in Fig. 3-7b, whereas if a mass is located at a free left-hand boundary, c_1 and k_1 will be zero and the first field is assigned the values c_2 and k_2 (Fig. 3-7c). The matrix $[c]$ is defined as

$$[c] = \begin{bmatrix} c_1 + c_2 & -c_2 & 0 & \cdot & & & \cdot & \cdot & 0 \\ -c_2 & c_2 + c_3 & -c_3 & 0 & & & \cdot & \cdot & 0 \\ 0 & -c_3 & & c_i + c_{i+1} & -c_{i+1} & 0 & \cdot & \cdot & 0 \\ \cdot & 0 & & -c_{i+1} & & & & & \cdot \\ \cdot & \cdot & & 0 & & & & & \cdot \\ \cdot & \cdot & & \cdot & & & & & -c_L \\ 0 & 0 & & 0 & & & & -c_L & c_L + c_{L+1} \end{bmatrix} \tag{3.26}$$

where L is the number of mass lumps. A similar definition exists for the stiffness matrix $[k]$.

Proportional damping is defined by the relation

$$c_{ij} = c_p I_{pi} \delta_{ij} + c_k k_{ij} \tag{3.27}$$

where I_{pi} is the mass moment of inertia of the lumped mass at station i, c_{ij} and k_{ij} are the elements of the damping and stiffness matrices, and

$$\delta_{ij} = \begin{cases} 0 & \text{if} \quad i \neq j \\ 1 & \text{if} \quad i = j \end{cases} \tag{3.28}$$

Set $c_{ij} = 0$ for no viscous damping. The constant of proportionality c_k is a proportional viscous damping coefficient that serves the same role for stiffness as c_p does for mass.

Equations (3.19) and (3.20) remain valid for discrete torsional systems with no damping or proportional damping. Some of the quantities in Eqs. (3.19) and (3.20), however, are redefined as

$$N_n = \sum_{i=1}^{L} I_{pi} [\phi_n(a_i)]^2 \tag{3.29}$$

$$\eta_n(0) = \sum_{i=1}^{L} I_{pi} \phi(a_i, 0) \phi_n(a_i) \tag{3.30}$$

$$f_n(t) = \sum_{i=1}^{L} T_{a_i} \phi_n(a_i) \tag{3.31}$$

or, in the case of the acceleration method, $f_n(t)$ can be expressed as

$$f_n(t) = \omega_n^2 \sum_{i=1}^{L} I_{pi}\phi_s(a_i, t)\phi_n(a_i) \tag{3.32}$$

where a_i is the axial location of the discrete masses and T_{a_i} is the applied torque at mass location a_i.

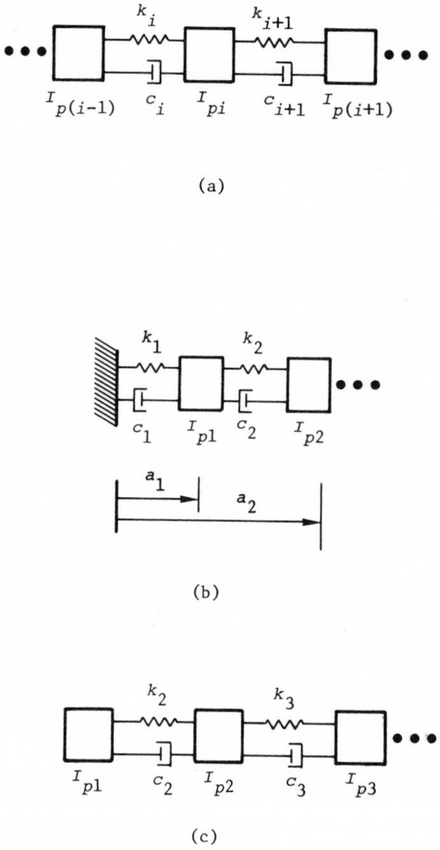

Fig. 3-7 Notation for dashpot and stiffness values used in the discrete model. (*a*) General labeling. (*b*) Left-hand end fixed. (*c*) Left-hand end free.

Transfer matrices may be used for this discrete model as was done for a continuous member when the free-motion undamped response was calculated. The point matrix for each lumped mass is given in Table 3-8. The transfer matrix for a discrete stiffness is given by the spring term in Table 3-8.

For angular base motion of a system with fixed left end ($x = 0$), the terms $k_1\phi(0,t)\phi_n(a_1) + c_1\, \partial\phi(0,t)/\partial t\ \phi_n(a_1)$ should be added to $f_n(t)$, wherein $\phi(0, t)$ and $\partial\phi(0, t)/\partial t$ are the applied angular base displacement and velocity. The first mass is located at $x = a_1$. If the $x = L$ end is fixed and is subjected to angular base motion, then add the terms $k_{L+1}\phi(L, t)\phi_n(a_L) + c_{L+1}\, \partial\phi(L,t)/\partial t\ \phi_n(a_L)$ to $f_n(t)$ where the final mass is located at $x = a_L$.

Nonproportional Damping for Continuous Members The formulas of this chapter need considerable adjustment in order to make them applicable to continuous members made of materials with nonproportional viscous damping. As noted above, Eqs. (3.10) should be supplemented by the term $c_t\,\partial\phi/\partial t$ on the right-hand side. Replace the frequency ω_n by s_n, where s_n is a complex number. The real part of s_n is commonly referred to as the damping exponent, whereas the imaginary part of s_n is called the frequency of the damped free vibration. The mode shape associated with s_n can now also be complex.

Equations (3.19) and (3.20) remain valid, but now the static solutions refer to

$$\phi_s(x,t) = -\sum_{n=1} \frac{\phi_n(x) f_n(t)}{s_n N_n} \tag{3.33}$$

where $\phi_n(x)$, N_n are now the mode shape and norm associated with the complex eigenvalue s_n.

The definitions of this section must be adjusted so that

$$N_n = 2s_n \int_0^L \rho r_p^2 \phi_n^2\,dx + \int_0^L c_t \phi_n^2\,dx \tag{3.34}$$

$$\eta_n(t) = e^{s_n t}\eta_n(0) + \int_0^t e^{s_n(t-\tau)} f_n(\tau)\,d\tau \tag{3.35}$$

$$\xi_n(t) = e^{s_n t}\left[\frac{1}{s_n^2}\frac{\partial f_n}{\partial t}(0) + \frac{1}{s_n} f_n(0) + \eta_n(0)\right] - \frac{1}{s_n^2}\frac{\partial f_n}{\partial t}(t) + \frac{1}{s_n^2}\int_0^t e^{s_n(t-\tau)}\frac{\partial^2 f_n}{\partial \tau^2}(\tau)\,d\tau \tag{3.36}$$

$$\eta_n(0) = \int_0^L \left[c_t\phi(x,0)\phi_n(x) + \rho r_p^2 \frac{\partial\phi(x,0)}{\partial t}\phi_n(x) + s_n\rho r_p^2\phi(x,0)\phi_n(x)\right] dx \tag{3.37}$$

and, in the case of the acceleration method, $f_n(t)$ can be expressed as

$$f_n(t) = -s_n \int_0^L \phi_s(x,t) c_t \phi_n(x)\,dx - s_n^2 \int_0^L \phi_s(x,t)\rho r_p^2 \phi_n(x)\,dx \tag{3.38}$$

Angular applied base motion is handled as in the proportional damping case.

Nonproportional Damping for Discrete Models The frequencies and mode shapes for a discrete model are obtained with the point matrix of Table 3-8. For inclusion of discrete damping with constant c_i at $x = a_i$, use this point matrix with $1/k$ replaced by $1/(k_i + sc_i)$ and $-I_{pi}\omega^2$ by $I_{pi}s^2$.

To find the transient response, use

$$N_n = 2s_n \sum_{i=1}^{L} I_{pi}\phi_n^2(a_i) + \boldsymbol{\phi}_n^T[c]\boldsymbol{\phi}_n \tag{3.39}$$

where $\boldsymbol{\phi}_n$ is a column vector made up of the free damped angular displacement of each discrete mass, that is,

$$\boldsymbol{\phi}_n = \begin{bmatrix} \phi_n(a_1) \\ \phi_n(a_2) \\ \vdots \\ \phi_n(a_i) \\ \vdots \\ \phi_n(a_L) \end{bmatrix} \tag{3.40}$$

and $\boldsymbol{\phi}_n^T$ is the corresponding row vector

$$\boldsymbol{\phi}_n^T = [\phi_n(a_1)\ \phi_n(a_2) \cdots \phi_n(a_L)] \tag{3.41}$$

With this notation, the following two changes are also needed

$$\eta_n(0) = \sum_{i=1}^{L} \left[\phi_n(a_i) I_{pi} \frac{\partial \phi_n}{\partial t}(a_i, 0) + s_n \phi_n(a_i) I_{pi} \phi(a_i, 0) \right] + \boldsymbol{\phi}_n^T [c] \boldsymbol{\phi}(0) \tag{3.42}$$

where $\boldsymbol{\phi}(0)$ is a vector of the initial angular displacements at each discrete mass. And in the case of the acceleration method, $f_n(t)$ can be expressed as

$$f_n(t) = -s_n \boldsymbol{\phi}_s(t)^T [c] \boldsymbol{\phi}_n - s_n^2 \sum_{i=1}^{L} [\phi_s(a_i, t) I_{pi} \phi_n(a_i)] \tag{3.43}$$

where $\phi_s(a_i, t)$ is the static angular displacement at each discrete mass and $\boldsymbol{\phi}_s(t)$ is the corresponding vector, both evaluated for each point in time of the applied loading. Use Eq. (3.33) to compute $\phi_s(a_i, t)$.

The $f_n(t)$ for angular applied base motion is selected as in the case for proportional damping.

Voigt-Kelvin Material for Continuous Member The formulas of this chapter are applicable to a continuous bar of Voigt-Kelvin material. In the equations of motion, Eqs. (3.10), replace G by $G(1 + \epsilon\, \partial/\partial t)$ where ϵ is the Voigt-Kelvin damping coefficient, and where similar behavior in dilatation and shear is assumed. Also, add a term $c_\rho r_p^2\, \partial\phi/\partial t$ to the right-hand side of the second equation in Eqs. (3.10) if the bar is proportionally damped in the twist direction. The response formulas of this section apply if c_ρ is replaced by $(c_\rho + \epsilon\omega_n^2)$ and ζ_n is defined as $\zeta_n = c_\rho/(2\omega_n) + \epsilon\omega_n/2$. Continue to use the same undamped mode shape employed above, where ω_n is the corresponding nth undamped frequency.

C. COMPUTER PROGRAMS AND EXAMPLES

3.14 Stresses and Torsional Constant

Torsional stresses are difficult to compute for many cross-sectional shapes. Computers can be used to great advantage in such calculations. For a bar of arbitrary cross section, the computer program BEAMSTRESS can be employed for such calculations, as well as for the evaluation of the torsional constant. Chapter 14 contains several examples of the calculation of torsional stresses and constants by BEAMSTRESS.

3.15 Benchmark Examples for Twisting Response

Complicated torsion problems should be solved with the assistance of a computer program. The following examples are provided as benchmark examples against which a reader's own program can be checked. The computer program TWIST was used for these examples. For static and steady-state torsional loads, TWIST calculates the angle of twist and the twisting moment of a shaft. It also computes the natural frequencies and mode shapes of torsional vibration. The torsional system can be formed of segments with any loading, gears, branches, foundations, and boundary conditions.

EXAMPLE 3.8 Static Response of an Imbedded Bar Find the angle of twist and internal torque along the shaft of Fig. 3-8, which is partially imbedded in an elastic foundation. Set $J = 1.57$ in^4, $G = 12(10^6)$ psi, $k_t = 200$ lb · in/in, and $m_x = 20$ lb · in/in.

Some of the computed results are given in Fig. 3-9.

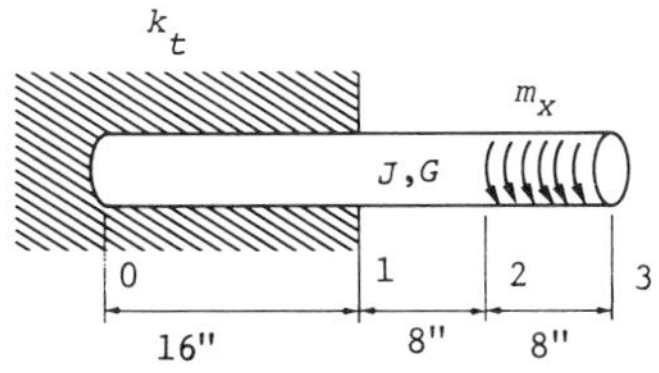

Fig. 3-8 Example 3.8.

AXIAL LOCATION	ANGLE	TWISTING MOMENT
0.	4.99774E-02	0.
5.33333E+00	4.99849E-02	5.33119E+01
1.06667E+01	5.00075E-02	1.06640E+02
1.60000E+01	5.00453E-02	1.60000E+02
1.86667E+01	5.00679E-02	1.60000E+02
2.13333E+01	5.00906E-02	1.60000E+02
2.40000E+01	5.01132E-02	1.60000E+02
2.66667E+01	5.01321E-02	1.06667E+02
2.93333E+01	5.01434E-02	5.33333E+01
3.20000E+01	5.01472E-02	-9.09495E-13

Fig. 3-9 Partial output for example 3.8.

EXAMPLE 3.9 Natural Frequencies of a Shaft Find the first two natural frequencies and mode shapes of the shaft of Fig. 3-10. Let $I_{p0} = I_{p1} = I_{p2} = I_{p3} = 1$ N · m · sec^2, $I_{p4} = 2$ N · m · sec^2, $k_1 = k_2 = k_3 = 0.17(10)^6$ N · m/rad and $k_4 = 0.23(10)^6$ N · m/rad.

They are found to be 36.37 and 75.52 Hz.

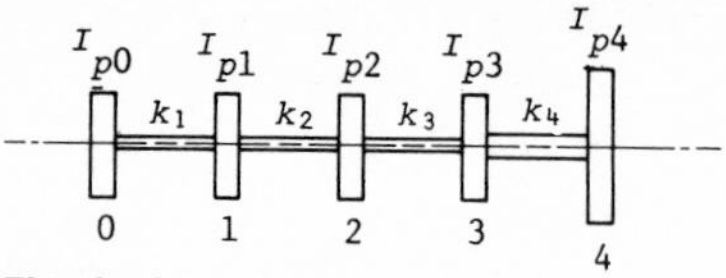

Fig. 3-10 Example 3.9.

EXAMPLE 3.10 A Geared Torsional System Find the first two natural frequencies of the geared system of Fig. 3-5. Let $I_{p0} = 10$ lb · in. · sec^2, $r = 1$ in, $k = 10^5$ lb · in/rad, $a_2 = 2a_1$.

The frequencies are calculated to be 15.91 and 21.93 Hz.

REFERENCES

3.1 Pörschmann, H.: *Bautechnische Berechnungstafeln für Ingenieure*, Teubner, Leibzig, 1971.

3.2 Kollbrunner, C. F., and Basler, K.: *Torsion in Structures* (translated from the German edition by E. C. Glauser), Springer-Verlag, New York, 1969.

FOUR

Extensional Systems

This chapter considers the static axial (longitudinal) extension and longitudinal vibration of bars and spring mass systems. Response formulas for simple bars and fundamental stress formulas are treated in Part A. The formulas for more complex bars are provided in Part B. Part C contains some examples of problems solved by a computer program.

The equations of motion, and hence the solutions, for small longitudinal (axial) extension of strings are precisely the same as those for the extension of bars. Consequently, the formulas and tables of this chapter apply for problems concerning the extension of strings. This chapter is also appropriate for the small-displacement theory of transverse motion of strings. In this case, the transverse displacement and transverse force of the string replace the axial displacement and axial force, respectively, of the bar. In addition, the prescribed axial force in the string is substituted for EA for the extension bar of this chapter. Also the string's transverse applied loadings replace the bar's axial applied loadings.

A. STRESS FORMULAS AND SIMPLE BARS

4.1 Notation and Conventions

u, Δ	Axial displacement (length)
P	Axial force (force)
E	Modulus of elasticity (force/length^2)

A	Cross-sectional area (length^2)
L	Length of bar (length)
k	Spring constant (force/length)
P_1, P_i	Concentrated applied axial force
p_x	Distributed axial force, loading intensity (force/length)
p_{x1}	Magnitude of distributed axial force that is uniform in the x direction (force/length)
$\sigma = \sigma_x$	Axial stress (force/length^2)
α	Coefficient of thermal expansion (length/length · degree)
ΔT	Change in temperature (degree), that is, the temperature rise with respect to the reference temperature.
ρ	Mass per unit length (mass/length)
ρ^*	Mass per unit volume (mass/length^3)
ω	Natural frequency (radian/second)

Positive displacement u and force P are shown in Fig. 4-1.

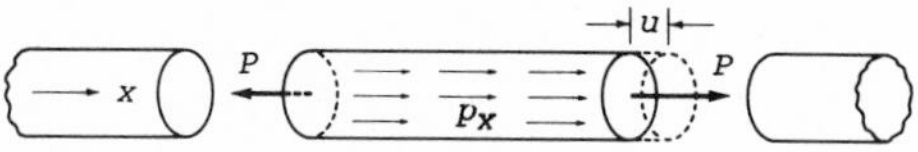

Fig. 4-1 An extension bar.

4.2 Normal Stress

The tables of this chapter provide the internal axial force at any point along a bar. The normal stresses can be calculated from the stress formula

$$\sigma = \frac{P}{A} \tag{4.1}$$

This normal stress is uniformly distributed over the cross section.

4.3 Extension

A bar of length L of constant cross section extends

$$\Delta = \frac{PL}{AE} \tag{4.2}$$

For a spring of spring constant k or a bar treated as a spring of stiffness $k = AE/L$, the extension is given by

$$\Delta = \frac{P}{k} \tag{4.3}$$

The differential equations equivalent to Eq. (4.2) are

$$\frac{du}{dx} = \frac{P}{AE} \tag{4.4a}$$

$$\frac{dP}{dx} = -p_x \tag{4.4b}$$

If thermal effects are taken into account, Eqs. (4.2) and (4.4a) become

$$\Delta = \frac{PL}{AE} + \alpha(\Delta T)L \qquad \text{and} \qquad \frac{du}{dx} = \frac{P}{AE} + \alpha\,\Delta T \tag{4.5}$$

where ΔT is the change in temperature (degrees) and α is the coefficient of linear thermal expansion.

The differential relations are integrated to provide the extension and axial force

Extension:	$u = u_0 + P_0 \dfrac{x}{EA} + F_u$	(4.6a)
Axial force:	$P = P_0 + F_P$	(4.6b)

F_u and F_P are loading functions given in Table 4-1. If there is more than one load on a bar, the F_u, F_P functions are formed by adding the terms given in Table 4-1 for each load. The initial parameters u_0, P_0 (values of u, P at the left end of the bar, $x = 0$) are provided in Table 4-2.

EXAMPLE 4.1 Column with Fixed Ends Find the displacements, axial force, and stress in the bar of Fig. 4-2.

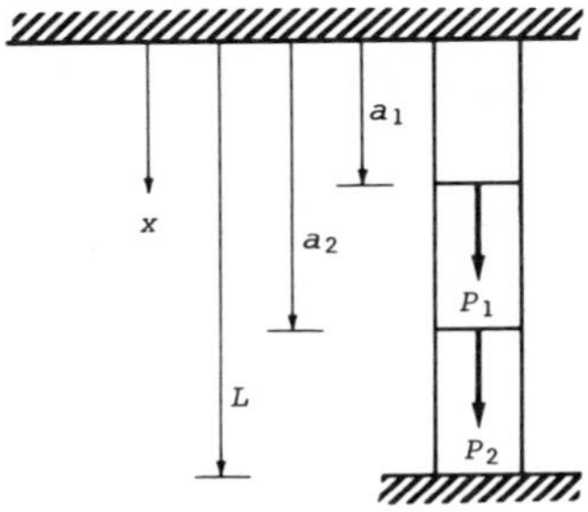

Fig. 4-2

The displacement u and axial force P are given by Eqs. (4.6) with loading functions F_u, F_P taken from Table 4-1 as

$$F_u = -\frac{P_1\langle x - a_1\rangle}{EA} - \frac{P_2\langle x - a_2\rangle}{EA}$$
$$F_P = -P_1\langle x - a_1\rangle^0 - P_2\langle x - a_2\rangle^0 \qquad (1)$$

From Table 4-2, the initial parameters are given by

$$u_0 = 0 \qquad P_0 = \frac{-EA}{L} F_{u\,|\,x=L} = \frac{P_1(L - a_1)}{L} + \frac{P_2(L - a_2)}{L} \qquad (2)$$

Substitution of (1) and (2) in Eqs. (4.6) gives the extension and axial force along the bar:

$$u = \frac{P_1(L - a_1) + P_2(L - a_2)}{L} \frac{x}{EA} - \frac{P_1\langle x - a_1\rangle + P_2\langle x - a_2\rangle}{EA}$$

$$P = \frac{1}{L}\left[P_1(L - a_1) + P_2(L - a_2)\right] - P_1\langle x - a_1\rangle^0 - P_2\langle x - a_2\rangle^0$$

The stress is found from Eq. (4.1) as $\sigma = P/A$.

TABLE 4-1 Loading Functions $F_u(x)$, $F_p(x)$ for Bar of Constant Cross Section (Eqs. 4.6)

	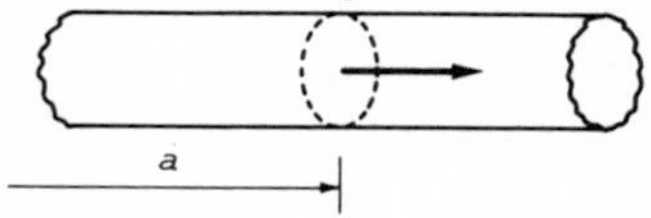	Uniformly Distributed Force p_{x1} 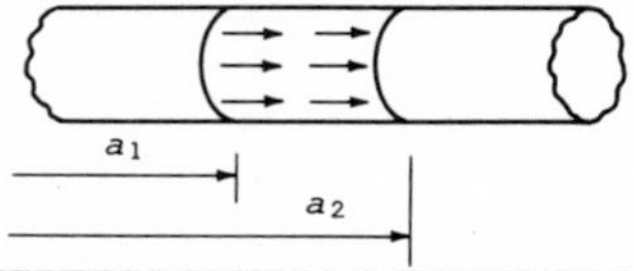	Temperature Change ΔT
$F_u(x)$	$-\dfrac{P_1\langle x-a\rangle}{EA}$	$-\dfrac{p_{x1}}{2EA}(\langle x-a_1\rangle^2 - \langle x-a_2\rangle^2)$	$\alpha\,\Delta T\,x$
$F_P(x)$	$-P_1\langle x-a\rangle^0$	$-p_{x1}(\langle x-a_1\rangle - \langle x-a_2\rangle)$	0

TABLE 4-2 Initial Parameters for Bar of Constant Cross Section (Eqs. 4.6)

Right End → / Left End ↓	1. Fixed 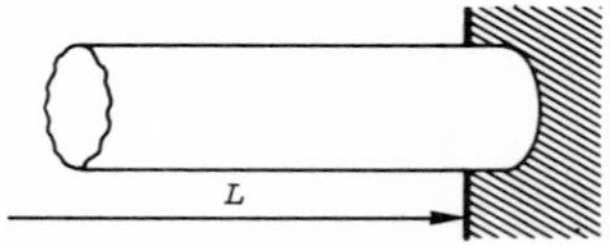	2. Free 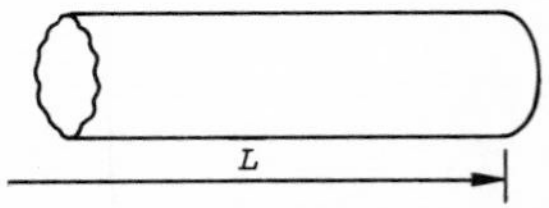
1. Fixed 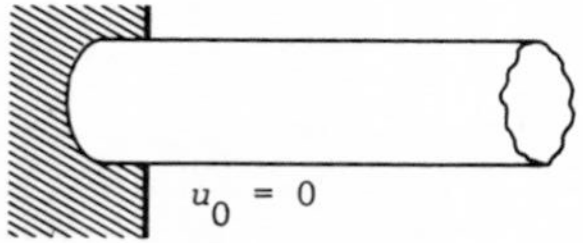 $u_0 = 0$	$P_0 = -\dfrac{EA}{L}(F_u)_{x=L}$	$P_0 = -(F_P)_{x=L}$
2. Free 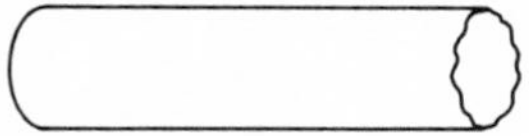 $P_0 = 0$	$u_0 = -(F_u)_{x=L}$	Kinematically Unstable*

EXAMPLE 4.2 Thermal Stresses Consider the effect on a prismatic bar of a uniform increase in temperature ΔT. Substitution in Eqs. (4.6) of the loading functions F_u, F_P from Table 4-1 gives

$$\begin{aligned} u &= u_0 + P_0 \frac{x}{EA} + \alpha\, \Delta T\, x \\ P &= P_0 \end{aligned} \tag{1}$$

If the bar is fixed at both ends, then, from Table 4-2, the initial parameters are

$$\begin{aligned} u_0 &= 0 \\ P_0 &= -\frac{EA}{L} F_u \big|_{x=L} = -EA\alpha\, \Delta T \end{aligned} \tag{2}$$

From (1), (2), and Eq. (4.1)

$$\begin{aligned} u &= -\alpha\, \Delta T\, x + \alpha\, \Delta T\, x = 0 \\ P &= -EA\alpha\, \Delta T \\ \sigma &= P/A = -E\alpha\, \Delta T \end{aligned}$$

Thus, for the completely restrained bar, the displacement is zero everywhere, while the stress is constant.

For a bar free at both ends, the initial parameter (Table 4-2) P_0 is zero, while u_0 is arbitrary since the bar is free to translate. Set $u_0 = 0$. Then, from (1) and Eq. (4.1)

$$\begin{aligned} u &= \alpha\, \Delta T\, x \\ P &= \sigma = 0 \end{aligned}$$

Thus, in contrast to a fixed bar, the stress is zero everywhere in a free bar while the extension is nonzero.

4.4 Natural Frequencies

The natural frequencies ω_n, $n = 1, 2, \ldots,$ for the extension of uniform bars are given by Eq. (3.8) with G/ρ^* replaced by E/ρ^* or EA/ρ.

B. COMPLEX BARS

4.5 Notation for Complex Bars

M_i	Concentrated mass (mass)
$\dfrac{\Delta p_x}{\Delta \ell}$	Gradient of distributed axial force, linearly varying in x direction (force/length^2)
k_x	Elastic foundation modulus (force/length^2)
ℓ	Length of segment, span of transfer matrix
t	Time
c_x	External or viscous damping coefficient (mass/(length · time), force · time/length^2)

c_ρ Proportional viscous damping coefficient (1/time). If c_x is chosen to be proportional to the mass, that is, $c_x = c_\rho \rho$, then c_ρ is the constant of proportionality.

$\mathbf{U}_i$ Field matrix of the ith segment

$\overline{\mathbf{U}}_i$ Point matrix at $x = a_i$

The notation for transfer matrices is:

$$\mathbf{U}_i = \begin{bmatrix} U_{uu} & U_{uP} & F_u \\ U_{Pu} & U_{PP} & F_P \\ 0 & 0 & 1 \end{bmatrix}$$

4.6 Differential Equations

The fundamental equations of motion in first-order form for the axial extension of bars and the motion of spring-mass systems are

$$\begin{aligned} \frac{\partial u}{\partial x} &= \frac{P}{AE} + \alpha\, \Delta T \\ \frac{\partial P}{\partial x} &= k_x u + \rho \frac{\partial^2 u}{\partial t^2} - p_x(x, t) \end{aligned} \tag{4.7}$$

In higher-order form these equations become

$$\begin{aligned} p_x &= -\frac{\partial}{\partial x}\left(EA \frac{\partial u}{\partial x}\right) + k_x u + \rho \frac{\partial^2 u}{\partial t^2} + \frac{\partial}{\partial x}(AE\alpha\, \Delta T) \\ P &= AE \frac{\partial u}{\partial x} - \alpha AE\, \Delta T \end{aligned} \tag{4.8}$$

4.7 Transfer Matrices for Uniform Segments

The transfer matrices for a shaft segment of length ℓ can be taken from Tables 3-4 to 3-7 by replacing GJ by EA, ϕ by u, T by P, k_t by k_x, ρr_p^2 by ρ, and m_x by p_x.

The transfer matrix for a simple bar, including thermal loading, is repeated in Table 4-3.

TABLE 4-3 Massless Bar with Uniform Temperature Change ΔT

$$\mathbf{U}_i = \begin{bmatrix} 1 & \frac{\ell}{AE} & F_u \\ 0 & 1 & F_P \\ 0 & 0 & 1 \end{bmatrix}$$

$$F_u = -p_{x1} \frac{\ell^2}{2AE} - \frac{\Delta p_x}{\Delta \ell} \frac{\ell^3}{6AE} + \alpha \ell\, \Delta T$$

$$F_P = -p_{x1} \ell - \frac{\Delta p_x}{\Delta \ell} \frac{\ell^2}{2}$$

4.8 Point Matrices

The transfer matrices that take into account concentrated (point) occurrences are given in Table 4-4.

TABLE 4-4 Point Matrix at $x = a_i$

Definitions: Care should be taken in using this table in that case (b) can not occur at the same axial location as cases (c) and (d). That is, based on physical reasoning either $1/k = 0$ or $k_b - M_i\omega^2 = 0$ at a particular location. Multiply point matrices according to the sequence of occurrences. See Table 4-6 for some commonly occurring combinations.

$$\bar{\mathbf{U}}_i = \begin{bmatrix} 1 & \frac{1}{k} & 0 \\ k_b - M_i\omega^2 & 1 & -P_i \\ 0 & 0 & 1 \end{bmatrix}$$

(a) Concentrated Applied Force

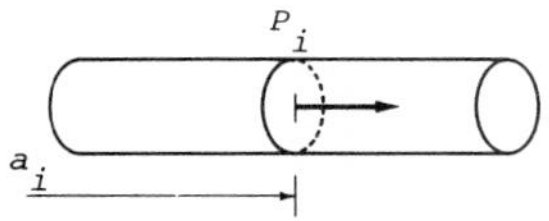

(b) Extension Spring or Bar Section

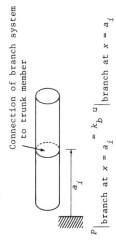

The spring constant k is equivalent to $\frac{EA}{\ell}$ of Table 4-4 for a bar being treated as an extension spring.

For springs in parallel

$$k = k_1 + k_2$$

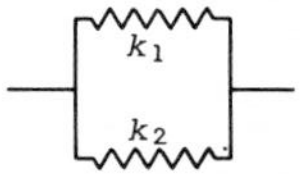

For a coil spring

$$k = \frac{Ed^4}{8ND^3}$$

E = Modulus of elasticity

d = Spring wire diameter

D = Mean coil diameter

N = Number of coils

(c) Branch System

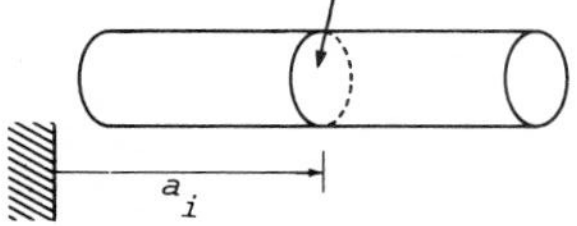

$$P\big|_{\text{branch at } x = a_i} = k_b\, u\big|_{\text{branch at } x = a_i}$$

(d) Concentrated Mass

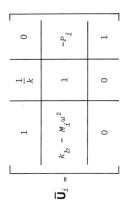

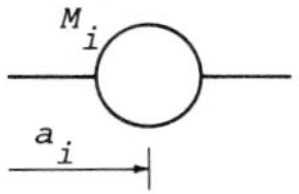

4.9 Other Information

Extension problems can be solved with the information in Chap. 3 for torsion if the substitutions indicated in Table 4-5 are made. All formulas, tables, and examples for loadings, static solutions, free dynamics, and forced dynamics can be converted in this fashion. Also, if temperature loadings are desired for the extension case, simply add the term $\int_0^L P_n(x)\alpha\,\Delta T\,dx$ to the resulting extension form of the loading function, $f_n(t)$.

TABLE 4-5 Equivalence of Extension and Torsion

Extension	Torsion
u	ϕ
P	T
E	G
A	J
ρ	ρr_p^2
p_x	m_x
k_x	k_t
c_x	c_t
M_i	I_{pi}

4.10 Applications

Static and dynamic problems of extension of bars are solved much as similar problems for the bending of beams and the twisting of shafts are solved. The examples of previous chapters should suffice as references to assist in setting up extension problems. The following examples deal with spring-mass systems.

EXAMPLE 4.3 Natural Frequencies of Spring-Mass System Find the natural frequencies of the spring-mass system of Fig. 4-3*a*. The global transfer matrix is formed as Eq. (3.15):

$$\begin{bmatrix} u \\ P \end{bmatrix}_{a_2} = \overline{\mathbf{U}}_4\overline{\mathbf{U}}_3\overline{\mathbf{U}}_2\overline{\mathbf{U}}_1\begin{bmatrix} u \\ P \end{bmatrix}_0 = \mathbf{U}\begin{bmatrix} u \\ P \end{bmatrix}_0 \tag{1}$$

where, from Table 4-4,

$$\overline{\mathbf{U}}_1 = \begin{bmatrix} 1 & 1/k_1 \\ 0 & 1 \end{bmatrix} \qquad \overline{\mathbf{U}}_3 = \begin{bmatrix} 1 & 1/k_2 \\ 0 & 1 \end{bmatrix}$$

$$\overline{\mathbf{U}}_2 = \begin{bmatrix} 1 & 0 \\ -M_1\omega^2 & 1 \end{bmatrix} \qquad \overline{\mathbf{U}}_4 = \begin{bmatrix} 1 & 0 \\ -M_2\omega^2 & 1 \end{bmatrix}$$

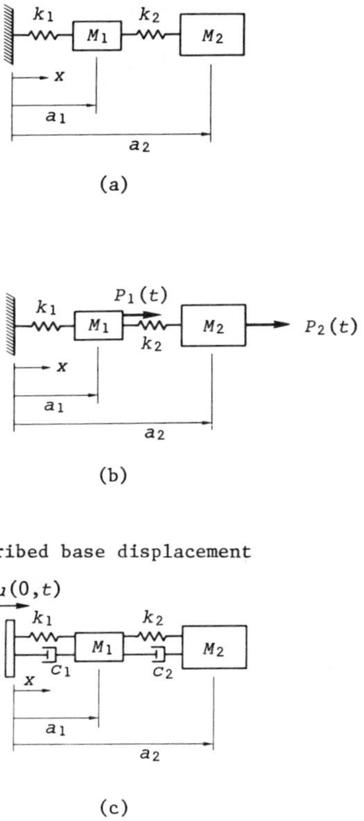

Fig. 4-3

The subscripts of the point matrices indicate the occurrence from left to right. This contrasts with the usual definition of a point matrix wherein the subscript designates location. Note that the third column and third row of the point matrix of Table 4-4 are dropped, since they are not needed. The frequencies are found by setting ∇ of Table 3-10 equal to zero. For fixed-free end conditions

$$\nabla = (U_{PP})_{a_2} = 1 - \frac{\omega^2 M_2}{k_1} - \frac{\omega^2 M_2}{k_2} - \frac{\omega^2 M_1}{k_1} + \omega^4 \frac{M_1 M_2}{k_1 k_2} = 0$$

or

$$\omega^4 - \left(\frac{k_2}{M_2} + \frac{k_1 + k_2}{M_1} \right) \omega^2 + \frac{k_1 k_2}{M_1 M_2} = 0$$

Two frequencies, corresponding to the two possible modes of motion of the system, can be extracted from this relation. For example, if $k_1 = k_2 = k$, $M_1 = M_2$,

$$\omega_1^2 = \frac{k}{M_1} \frac{3 - \sqrt{5}}{2} \qquad \omega_2^2 = \frac{k}{M_1} \frac{3 + \sqrt{5}}{2}$$

and if $k_2 = 2k_1$, $M_2 = 2M_1$,

$$\omega_1^2 = \frac{k_1}{M_1}(2 - \sqrt{3}) \qquad \omega_2^2 = \frac{k_1}{M_1}(2 + \sqrt{3})$$

EXAMPLE 4.4 Transient Response of Spring-Mass System Find the response of the spring-mass system of Fig. 4-3*a* if time-dependent forces are placed on the masses as shown in Fig. 4-3*b*.

The forced dynamic response is provided by the equations of Section 3.13. These equations require the mode shapes of Eq. (3.14). The initial parameters are given by Eqs. (3.17) for this fixed-free member as

$$u_0 = 0 \qquad \text{and} \qquad P_0 = P_0$$

and the mode shapes become [from Eq. (3.14) and the previous problem]

$$\begin{aligned} \mathbf{s}_{a_1} &= \overline{\mathbf{U}}_2\overline{\mathbf{U}}_1\mathbf{s}_0 \qquad \text{or} \qquad u_n(a_1) = P_0\frac{1}{k_1} \qquad n = 1, 2 \\ \mathbf{s}_{a_2} &= \overline{\mathbf{U}}_4\overline{\mathbf{U}}_3\overline{\mathbf{U}}_2\overline{\mathbf{U}}_1\mathbf{s}_0 \qquad \text{or} \qquad u_n(a_2) = \frac{P_0}{k_1k_2}(k_1 + k_2 - \omega_n^2 M_1) \qquad n = 1, 2 \end{aligned} \tag{1}$$

From Eq. (3.19*a*) (displacement method),

$$u(x, t) = A_1(t)u_1(x) + A_2(t)u_2(x)$$

The coordinate x does not have the usual meaning since the springs are of indeterminate length. The motion of the masses is given by

$$\begin{aligned} u(a_1, t) &= A_1(t)u_1(a_1) + A_2(t)u_2(a_1) \\ u(a_2, t) &= A_1(t)u_1(a_2) + A_2(t)u_2(a_2) \end{aligned} \tag{2}$$

where, according to Eqs. (3.20) to (3.24) with $\zeta_n = 0$ (no damping), and from Table 4-5,

$$A_n(t) = \frac{\displaystyle\int_0^t \int_0^L p(x, \tau)u_n\,dx\,\frac{\sin\omega_n(t-\tau)}{\omega_n}\,d\tau}{\displaystyle\int_0^L \rho u_n u_n\,dx}$$

for a system which is initially at rest $[\eta(0) = \partial\eta(0)/\partial t = 0]$.

Since

$$p(x, t) = P_1(t)\langle x - a_1\rangle^{-1} + P_2(t)\langle x - a_2\rangle^{-1}$$

$$\rho = M_1\langle x - a_1\rangle^{-1} + M_2\langle x - a_2\rangle^{-1}$$

it follows that

$$A_n(t) = \frac{\displaystyle\int_0^t [P_1(\tau)u_n(a_1) + P_2(\tau)u_n(a_2)]\,\frac{\sin\omega_n(t-\tau)}{\omega_n}\,d\tau}{M_1u_n^2(a_1) + M_2u_n^2(a_2)} \tag{3}$$

Equations (1), (2), and (3) express the complete response of the system.

EXAMPLE 4.5 Damped Spring-Mass system with Base Displacement Find the response of the two-mass system of Fig. 4-3*c* which is subjected to a prescribed base displacement $u(0, t)$. The dashpot constants c_1, c_2 are not proportional to the masses or the spring constants.

The solution procedure is similar to that of the two previous problems. For free motion, the base displacement $u(0, t)$ is set equal to zero. Since nonproportional viscous damping is present, the frequency will now be a complex number. The transfer matrices for the global transfer matrix

$$\mathbf{s}_{a_2} = \overline{\mathbf{U}}_4\overline{\mathbf{U}}_3\overline{\mathbf{U}}_2\overline{\mathbf{U}}_1\mathbf{s}_0 = \mathbf{U}\mathbf{s}_0 \tag{1}$$

are

$$\overline{\mathbf{U}}_1 = \begin{bmatrix} 1 & \dfrac{1}{sc_1 + k_1} \\ 0 & 1 \end{bmatrix} \qquad \overline{\mathbf{U}}_2 = \begin{bmatrix} 1 & 0 \\ s^2M_1 & 1 \end{bmatrix}$$
$$\overline{\mathbf{U}}_3 = \begin{bmatrix} 1 & \dfrac{1}{sc_2 + k_2} \\ 0 & 1 \end{bmatrix} \qquad \overline{\mathbf{U}}_4 = \begin{bmatrix} 1 & 0 \\ s^2M_2 & 1 \end{bmatrix} \tag{2}$$

where s is the complex frequency. The boundary conditions are fixed-free. Thus, from Table 3-10, the damped frequencies s_1, s_2 are found by setting $\nabla = (U_{PP})_{a_2} = 0$, where $(U_{PP})_{a_2}$ is an element of $\mathbf{U}$ of (1). This gives

$$s_n^4 + s_n^3\left(\frac{c_2}{M_1} + \frac{c_1}{M_1} + \frac{c_2}{M_2}\right) + s_n^2\left(\frac{k_2}{M_1} + \frac{c_1c_2}{M_1M_2} + \frac{k_1}{M_1} + \frac{k_2}{M_2}\right)$$
$$+ s_n\left(\frac{c_2k_1}{M_1M_2} + \frac{k_2c_1}{M_1M_2}\right) + \frac{k_1k_2}{M_1M_2} = 0 \tag{3}$$

where s_n is the complex frequency. Equation (3) may be solved by any suitable numerical polynomial-root finder when values of c_i, k_i, and M_i are given. In general, there will be four complex roots which occur in complex conjugate pairs. Thus only two damped free-motion frequencies will result.

Following Example 4.4, with $u_0 = 0$ and P_0 selected to be 1, we find the mode shapes

$$\begin{bmatrix} u_n(a_1) \\ P_n(a_1) \end{bmatrix} = \overline{\mathbf{U}}_2\overline{\mathbf{U}}_1 \begin{bmatrix} 0 \\ 1 \end{bmatrix} = \begin{bmatrix} \dfrac{1}{s_nc_1 + k_1} \\ 1 + \dfrac{s_n^2M_1}{s_nc_1 + k_1} \end{bmatrix} \tag{4}$$

$$\begin{bmatrix} u_n(a_2) \\ P_n(a_2) \end{bmatrix} = \overline{\mathbf{U}}_4\overline{\mathbf{U}}_3\overline{\mathbf{U}}_2\overline{\mathbf{U}}_1 \begin{bmatrix} 0 \\ 1 \end{bmatrix}$$
$$= \begin{bmatrix} \dfrac{1}{s_nc_1 + k_1} + \dfrac{1}{s_nc_2 + k_2} + \dfrac{M_1s_n^2}{(s_nc_1 + k_1)(s_nc_2 + k_2)} \\ 0 \end{bmatrix} \tag{5}$$

The dynamic motion of the masses is given by

$$\begin{aligned} u(a_1, t) &= A_1(t)u_1(a_1) + A_2(t)u_2(a_1) \\ u(a_2, t) &= A_1(t)u_1(a_2) + A_2(t)u_2(a_2) \end{aligned} \tag{6}$$

Similar expressions are written for the force $P(a_i, t)$. It follows from Eqs. (3.20) and (3.35), with $f_n(t)$ selected for applied base motion, that

$$A_n(t) = \frac{\int_0^t e^{s_n(t-\tau)} \left\{ u_n(a_1) \left[k_1 u(0, \tau) + c_1 \frac{\partial u(0, \tau)}{\partial \tau} \right] \right\} d\tau}{N_n} \tag{7}$$

where the system is assumed to be initially at rest, that is, $\eta_n(0) = 0$. Finally, from Eq. (3.34),

$$N_n = 2s_n\left[u_n^2(a_1)M_1 + u_n^2(a_2)M_2\right] + u_n^2(a_1)(c_1 + c_2) - 2c_2u_n(a_2)u_n(a_1) + c_2u_n^2(a_2)$$

C. COMPUTER PROGRAM AND EXAMPLES

4.11 Benchmark Examples

Complicated extension problems should be solved with the assistance of a computer program. The following examples are provided as benchmark examples against which a reader's own program can be checked. The computer program EXTENSION was used for these examples. For static and steady-state axial loads, EXTENSION finds the axial displacement and force. It also computes the natural frequencies and mode shapes of longitudinal vibration. The extension system can be a sequence of springs and masses, or a bar of uniform segments with arbitrary loading, foundations, and boundary conditions.

EXAMPLE 4.6 Static Response of Cantilevered Bar Find the axial displacement and force along the cantilevered bar of Fig. 4-4. Here $A_1 = 0.09$ m^2, $A_2 = 0.06$ m^2, $E_1 = E_2 = 200$ GN/m^2 = 200(10^9)N/m^2.

Partial results are provided in Fig. 4-5.

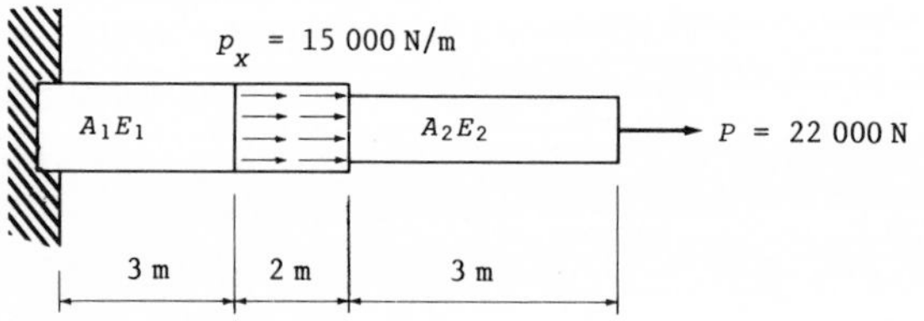

Fig. 4-4 Example 4.6.

AXIAL LOCATION	EXTENSION	AXIAL FORCE
0.	0.	5.20000E+04
3.00000E+00	8.66667E-06	5.20000E+04
5.00000E+00	1.27778E-05	2.20000E+04
8.00000E+00	1.82778E-05	2.20000E+04

Fig. 4-5 Partial output for Example 4.6.

EXAMPLE 4.7 Frequencies of a Spring-Mass System Find the two lower natural frequencies of the spring-mass system of Fig. 4-6. Let $k = 100$ lb/in, $M = 1$ lb · sec^2/in.

The frequencies are found to be 0.87 and 1.81 Hz.

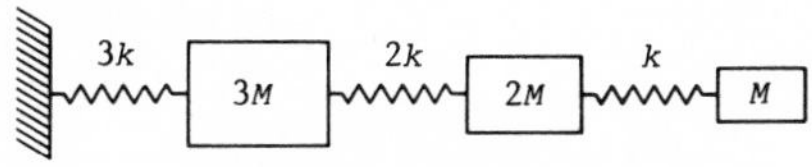

Fig. 4-6 Example 4.7.

FIVE

Torsion of Thin-Walled Beams

This chapter treats the static twisting motion, axial instability, and torsional vibration of thin-walled beams. Response formulas for simple thin-walled beams and fundamental stress formulas are provided in Part A. The formulas for more complex beams are given in Part B. Included are beams of variable cross section. Part C contains some examples of problems solved by computer programs. These examples include the computation of angle of twist, bimoment, torsional moment, and warping torque as well as the torsional constant, warping constant, and all stresses for a cross section of any shape.

The bending of a thin-walled beam can be treated with the formulas of Chap. 2. The formulas of Chap. 3 suffice to handle the torsion of most solid bars. However, twisted thin-walled beams frequently experience restrained warping, and the formulas of this chapter, instead of those of Chap. 3, must be employed.

Restrained warping occurs in thin-walled beams subjected to the twisting moments and/or boundary conditions in such a way that the internal twisting moment varies from section to section along the beam axis. This may be found in thin-walled beams with end or in-span conditions which do not permit the cross sections to warp freely.

The material here applies to beams for which bending and twisting effects can be uncoupled. This is the case if (1) there is no axial force or (2) the axial force passes through the shear center.

A. STRESS FORMULAS AND SIMPLE BEAMS

5.1 Notation and Conventions

ϕ	Angle of twist, rotation (radians)
Ψ	Rate of angle of twist (radians/length)
B	Bimoment (force · length^2)
T_ω	Warping torque (force · length)
T	Twisting moment, torque (force · length)
E	Modulus of elasticity (force/length^2)
G	Shear modulus of elasticity (force/length^2)
J	Torsional constant (length^4). In the case of a circular cross section, J is the polar moment of inertia (I_x) of the cross-sectional area with respect to the centroidal axis of the bar. Table 3-1 provides values of J for several shapes.
GJ	Torsional rigidity
ν	Poisson's ratio
Γ	Warping constant (length^6). Values of Γ for several shapes are given in Table 3-1. A computer program such as BEAMSTRESS (Chap. 14) can be used to calculate Γ for any shape.
$E\Gamma$	Warping rigidity
Γ_1, Γ_2	Primary, secondary warping constants. These constants are defined by Eqs. (5.7).
C	$\sqrt{GJ/E\Gamma}$
A	Cross-sectional area (length^2)
L	Length of beam
m_x	Distributed torque, twisting moment intensity (force · length/length)
m_{x1}	Magnitude of distributed torque that is uniform in the x direction (force · length/length)
$\dfrac{\Delta m_x}{\Delta \ell}$	Gradient of distributed torque, linearly varying in the x direction (force · length/length^2)
σ	Normal stress due to warping (force/length^2)
τ	Shear stress due to warping (force/length^2)
ω	Principal sectorial coordinate (length^4) with respect to the shear center, or warping of the cross section with respect to the plane of average warping
$\omega_S, \omega_C, \omega_P$	Sectorial coordinates with respect to shear center S, centroid C, and any point P, respectively
Q_ω	First sectorial moment (length^4)
S	Designation of shear center. The location of the shear center for several shapes is given in Table 3-1. The computer program BEAMSTRESS will calculate this location for a section of any shape.
t	Thickness of thin wall (length)
s	Coordinate along the center line of the wall profile, positive in counter-clockwise direction with respect to the pole P
r_S	The perpendicular distance from the shear center to the tangent of the center line of the wall profile
P	Compressive axial force that passes through shear center. Replace P by $-P$ for tensile axial forces.

e — Coordinate locating shear center (Table 3-1). Values for shapes not listed in Table 3-1 can be found in such references as 5.1 and 5.3 or by using a sectional analysis computer program.

y, z — Centroidal coordinates or principal centroidal coordinates if the axial forces are present

e_y, e_z — Distance in y, z directions between centroid and shear center

$$C_P = \begin{cases} (I_y + I_z)/A & \text{if shear center and centroid coincide} \\ (I_y + I_z)/A + \dfrac{e_z}{I_y}\displaystyle\int_A z(y^2+z^2)\,dA + \dfrac{e_y}{I_z}\displaystyle\int_A y(y^2+z^2)\,dA - (e_y^2 + e_z^2) & \\ \text{if the shear center and centroid do not coincide and the axial force passes through the shear center} & \end{cases}$$

$I_y = I$ — Moment of inertia of the cross section about the y axis

I_z — Moment of inertia of the cross section about the z axis

Positive angle of twist ϕ, twisting moment T, and bimoment B are shown in Fig. 5-1.

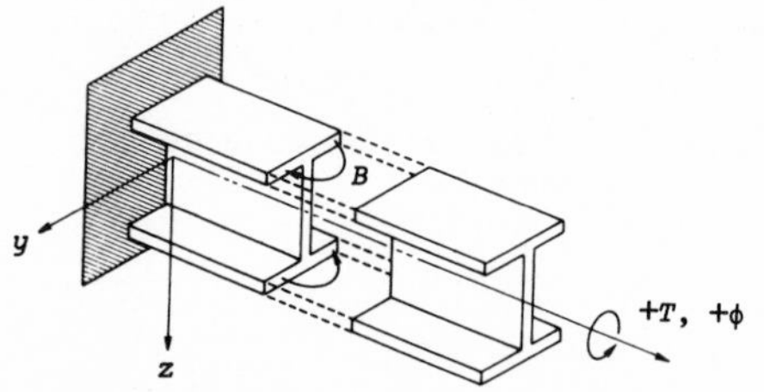

Fig. 5-1 Bimoment on the positive face of the cross section due to positive twisting moment. Also shown is positive angle of twist.

5.2 Warping Stresses

The tables of this chapter provide the internal bimoment and warping torque at any point along a beam. Once the bimoment and warping torque are known, the stresses due to warping can be calculated from the stress formulas in this section. For many members with thin-walled open sections, the normal and shear stresses due to restrained warping or nonuniform torsion are of primary concern in structural design, since they are frequently higher than the nonwarping stresses. See Chap. 14 for examples illustrating the calculation of warping sectional properties and stresses.

Normal Stress The normal stresses due to warping act perpendicular to the surface of the cross section. They are assumed to be constant across the thickness of these thin-walled sections. These stresses are given by

$$\sigma_x = \sigma = -\frac{B\omega}{\Gamma} \tag{5.1}$$

where ω is a principal sectorial coordinate defined as $\omega = \omega_0 - \omega_S$ with

$$\omega_0 = \frac{1}{A}\int_A \omega_S \, dA = \frac{1}{A}\int_A \omega_S t \, ds \tag{5.2}$$

$$\omega_S = \begin{cases} \displaystyle\int_0^s r_S \, ds & \text{for open cross sections} \\ \displaystyle\int_0^s r_S \, ds - \frac{\oint r_S \, ds}{\oint \frac{1}{t} \, ds} \int_0^s \frac{1}{t} \, ds & \text{for closed cross sections} \end{cases} \tag{5.3}$$

The integration in Eq. (5.3) is taken from any one of the free edges to the point at which the stress σ_x is desired. $\oint$ indicates an integral taken completely around the closed section.

The ω_0 defined by Eq. (5.2) is sometimes referred to as a sectorial centroid because of its resemblance to the usual definition of a centroid. Also, ω_0 forms an average warping plane which is parallel to the cross section of interest. The ω_S is a relative warping between any two points a and b on the walled profile ($\omega_S = \int_a^b r_S \, ds$ = double the sectorial area swept by r_S from a to b) and ω is the warping of the cross section with respect to the average warping plane. It should be emphasized that the shear center is taken as a pole. It is interesting to note that ω_0 can be eliminated from the calculations by choosing a point S_c for the s origin in the integration where the warping of the cross section with respect to the average warping plane is zero. This simplifies the calculation of the sectorial coordinate in some thin-walled sections in which the locations of S_c are known; for example, for symmetrical sections, S_c is on the axis of symmetry. If the shear stress is sought, it is more advantageous to select a free edge of the open section for the s origin.

Values of ω along some sections formed of straight thin elements are provided in Table 3-1. Also shown are patterns of the distribution of ω on the cross sections, along with peak values.

Shear Stress The shear stresses due to warping act parallel to the center line of the wall profile. These stresses are taken to be constant across the thickness of these thin-walled sections. The formula for these shear stresses is

$$\tau = \begin{cases} \dfrac{T_\omega Q_\omega}{t\Gamma} & \text{for open cross sections} \\ \dfrac{T_\omega}{t\Gamma}\left[Q_\omega - \dfrac{\oint \frac{Q_\omega}{t} \, ds}{\oint \frac{1}{t} \, ds} \right] & \text{for closed cross sections} \end{cases} \tag{5.4}$$

where Q_ω is the first sectorial moment defined by

$$Q_\omega = \int_{A_0} \omega \, dA = \int_0^{s_0} \omega t \, ds \tag{5.5}$$

The integration is taken over the area A_0 that lies between the position at which the stress is desired (s_0) and the outer fiber of the cross section. For a closed section the integration required for Q_ω is performed as though the section were open at an arbitrary point. Table 3-1 provides the peak values of Q_ω on the cross section for some shapes, along with illustrations of how the shear stress is distributed over the cross section.

Warping Constant The warping constant of a thin-walled section is calculated as

$$\Gamma = \begin{cases} \displaystyle\int_A \omega^2 \, dA & \text{for open cross sections} \\[2ex] \displaystyle\int_A \omega^2 \, dA - \frac{\oint r_S \, ds}{\oint \frac{1}{t} \, ds} \oint \frac{Q_\omega}{t} \, ds & \text{for closed cross sections} \end{cases} \tag{5.6}$$

Formulas for the warping constant of sections consisting of straight elements are given in Table 3-1.

In the theory of thin-walled open sections subjected to twisting moments, the warping across the thickness of the wall is assumed to be constant. Under this assumption, the warping constant Γ can then be calculated from the concept of the sectorial coordinate. In reality, the warping is not constant across the thickness; hence, the normal stresses vary across the wall thickness and this variation in turn leads to a shear stress across the wall thickness. If the wall thickness is not small, these shear stresses cannot be neglected. In general the warping constant can be expressed as

$$\Gamma = \Gamma_1 + \Gamma_2 \tag{5.7a}$$

where Γ_1 is due to the primary warping (warping along the wall profile) and Γ_2 is due to the secondary warping (warping across the wall thickness).

For sections with more than one intersection, such as the wide-flange, channel and zee sections, $\Gamma_1 \gg \Gamma_2$, and Γ_2 can usually be neglected. For sections such as the tee, angle, and cross, $\Gamma_1 = 0$ and therefore $\Gamma = \Gamma_2$.

The secondary warping constants for thin-walled open sections can be calculated as

$$\Gamma_2 = \frac{1}{12} \int_0^{S_A} t^3 r_n^2 \, ds \tag{5.7b}$$

where S_A is the total length of the wall profile, r_n is the perpendicular distance from the shear center to a normal to the center line of the wall.

EXAMPLE 5.1 **Stress Calculations** The beam shown in Fig. 5-2 is subjected to a concentrated load of 40,000 lb, which is eccentric by 1 in from the vertical plane of the web. The flanges are 10 in by 0.5 in and the web is 12 in by 0.5 in.

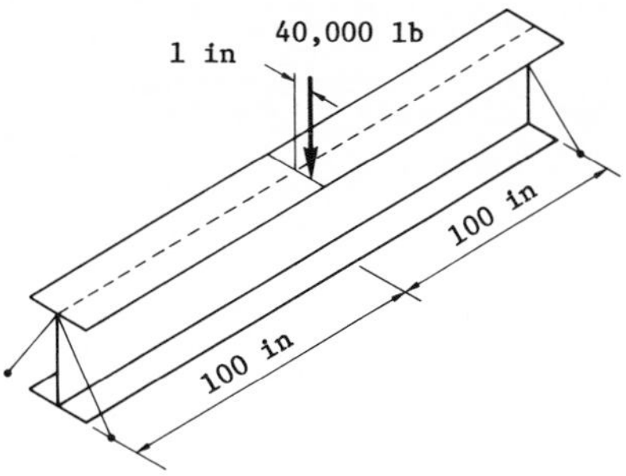

Fig. 5-2 Thin-walled beam.

The ends experience no rotation, but are free to warp. Determine the maximum normal and shear stresses due to warping.

From Tables 3-1, 2-1 we have $\Gamma = 3000\ \text{in}^6$, $I_y = 432\ \text{in}^4$. The formulas given later in this chapter will provide a bimoment at midspan of $B = 1{,}321{,}200\ \text{lb}\cdot\text{in}^2$ and a warping torque $T_\omega = 20{,}000\ \text{lb}\cdot\text{in}$.

The normal stress due to warping is given by Eq. (5.1). At the right edge of the top flange, the sectorial coordinate $\omega = 30\ \text{in}^2$. See Table 3-1. Then

$$\sigma = -\frac{B\omega}{\Gamma} = -\frac{1{,}321{,}200(30)}{3000} = -13{,}212\ \text{lb/in}^2$$

The usual bending stress of Eq. (2.1) is found from a bending moment at midspan of $M = PL/4 = 2{,}000{,}000\ \text{lb}\cdot\text{in}$.

If the two stresses are superimposed for the left and right edges of the top flange,

$$\sigma = \frac{2{,}000{,}000(6)}{432} \mp 13{,}212 = 40{,}990\ \text{lb/in}^2 \text{ or } 14{,}566\ \text{lb/in}^2$$

The shear stress of Eq. (5.4) requires Q_ω. At the intersection of the upper flange and the web,

$$Q_\omega = \int_0^{s_0} \omega t\, ds = \int_0^{b/2} \frac{h}{2}\left(s - \frac{b}{2}\right) t\, ds = \frac{ht}{4} s(s-b)\Big|_0^{b/2} = -\frac{ht}{16} b^2$$

where b is the width of the flange and h the height of the web. Then

$$\tau = \frac{T_\omega Q_\omega}{t\Gamma} = -\frac{T_\omega h b^2}{16\Gamma} = -\frac{20{,}000(12)10^2}{16(3000)} = -500\ \text{lb/in}^2$$

The usual transverse shear is given by Eq. (2.5) with $Q = tbh/2$. Then

$$\tau = \frac{V}{tI_y}\frac{tbh}{2} + 500 = \frac{20{,}000(10)12}{432(2)} + 500 = 3280\ \text{lb/in}^2$$

5.3 ANGLE OF TWIST

The fundamental equations of motion for the twisting of a thin-walled beam are

$$E\Gamma \frac{d^4\phi}{dx^4} - GJ \frac{d^2\phi}{dx^2} = m_x$$

$$E\Gamma \frac{d^3\phi}{dx^3} - GJ \frac{d\phi}{dx} = -T$$

$$E\Gamma \frac{d^3\phi}{dx^3} = -T_\omega \qquad (5.8)$$

$$E\Gamma \frac{d^2\phi}{dx^2} = -B$$

$$\frac{d\phi}{dx} = -\Psi$$

These relations are integrated to provide the angle of twist, rate of angle of twist, bimoment, twisting moment, and warping torque.

Angle of twist:

$$\phi = \phi_0 - \Psi_0 \frac{\sinh Cx}{C} + B_0 \frac{1 - \cosh Cx}{GJ} + T_0 \frac{Cx - \sinh Cx}{CGJ} + F_\phi(x) \qquad (5.9a)$$

Rate of angle of twist:

$$\Psi = \Psi_0 \cosh Cx + B_0 \frac{\sinh Cx}{CE\Gamma} - T_0 \frac{1 - \cosh Cx}{GJ} + F_\Psi(x) \qquad (5.9b)$$

Bimoment:

$$B = \Psi_0 CE\Gamma \sinh Cx + B_0 \cosh Cx + T_0 \frac{\sinh Cx}{C} + F_B(x) \qquad (5.9c)$$

Twisting moment:

$$T = T_0 + F_T(x) \qquad (5\text{-}9d)$$

Warping torque:

$$T_\omega = \Psi_0 GJ \cosh Cx + B_0 C \sinh Cx + T_0 \cosh Cx + F_T(x) + GJF_\Psi(x) \qquad (5.9e)$$

In these equations $C^2 = GJ/E\Gamma$. The F_ϕ, F_Ψ, F_B, F_T are loading functions given in Table 5-1. If there is more than one load on a beam, the F_ϕ, F_Ψ, F_B, F_T functions are formed by adding the terms given in Table 5-1 for each load. The initial parameters ϕ_0, Ψ_0, B_0, T_0 (values of ϕ, Ψ, B, T at the left end of the bar, $x = 0$) are provided in Table 5-2. These formulas are applicable to statically determinate and indeterminate beams.

The simply supported or pinned end conditions occur when the cross section at the end of the beam is prevented from rotating, but is allowed to warp freely.

TABLE 5-1 Loading Functions $F_\phi(x)$, $F_\Psi(x)$, $F_B(x)$, $F_T(x)$ for a Beam of Constant Cross Section (Eqs. 5.9)

	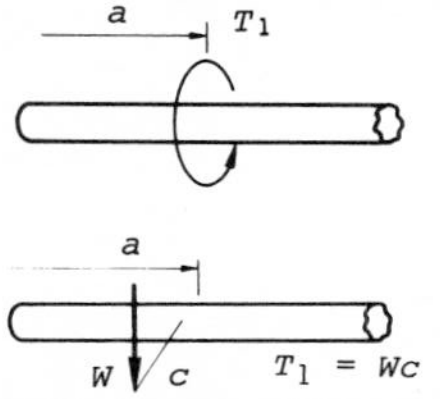	Uniformly Distributed Torque 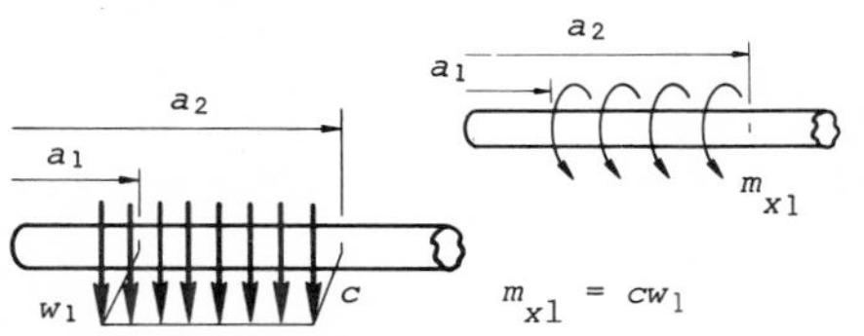	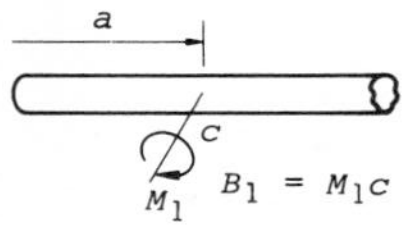	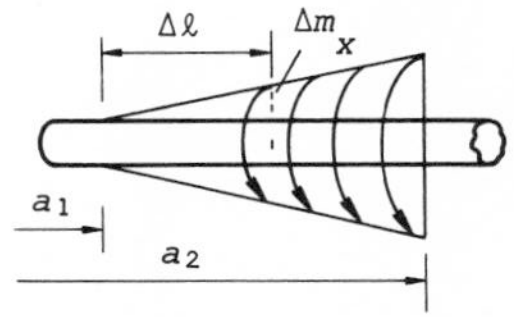
$F_\phi(x)$	$\frac{T_1}{CGJ}(-C\langle x-a\rangle + \sinh C\langle x-a\rangle)$	$\frac{m_{x1}}{E\Gamma}(\cosh C\langle x-a_1\rangle - \langle x-a_1\rangle^0 - C^2\frac{\langle x-a_1\rangle^2}{2} - \cosh C\langle x-a_2\rangle + \langle x-a_2\rangle^0 + C^2\frac{\langle x-a_2\rangle^2}{2})$	$\frac{B_1}{GJ}(-\langle x-a\rangle^0 + \cosh C\langle x-a\rangle)$	$\frac{\Delta m_x}{\Delta\ell\ E\Gamma}(\frac{1}{C}\sinh C\langle x-a_1\rangle - \langle x-a_1\rangle - C^2\frac{\langle x-a_1\rangle^2}{6} - \frac{1}{C}\sinh C\langle x-a_2\rangle + \langle x-a_2\rangle + C^2\frac{\langle x-a_2\rangle^2}{6})$
$F_\psi(x)$	$\frac{T_1}{GJ}(\langle x-a\rangle^0 - \cosh C\langle x-a\rangle)$	$\frac{m_{x1}}{CGJ}(C\langle x-a_1\rangle - \sinh C\langle x-a_1\rangle - C\langle x-a_2\rangle + \sinh C\langle x-a_2\rangle)$	$-\frac{B_1}{CE\Gamma}\sinh C\langle x-a\rangle$	$-\frac{\Delta m_x}{\Delta\ell CGJ}(\frac{1}{C}\cosh C\langle x-a_1\rangle - \frac{\langle x-a_1\rangle^0}{C} - \frac{C}{2}\langle x-a_1\rangle^2 - \frac{1}{C}\cosh C\langle x-a_2\rangle + \frac{\langle x-a_2\rangle^0}{C} + \frac{C}{2}\langle x-a_2\rangle^2)$
$F_B(x)$	$-\frac{T_1}{C}\sinh C\langle x-a\rangle$	$-\frac{m_{x1}}{C^2}(\cosh C\langle x-a_1\rangle - \langle x-a_1\rangle^0 - \cosh C\langle x-a_2\rangle + \langle x-a_2\rangle^0)$	$-B_1\cosh C\langle x-a\rangle$	$\frac{\Delta m_x}{\Delta\ell}(\langle x-a_1\rangle - \frac{1}{C}\sinh C\langle x-a_1\rangle - \langle x-a_2\rangle + \frac{1}{C}\sinh C\langle x-a_2\rangle)$
$F_T(x)$	$-T_1\langle x-a\rangle^0$	$-m_{x1}(\langle x-a_1\rangle - \langle x-a_2\rangle)$	0	$-\frac{\Delta m_x}{\Delta\ell\ 2}(\langle x-a_1\rangle^2 - \langle x-a_2\rangle^2)$

Definition: $C = \sqrt{E\Gamma/GJ}$

TABLE 5-2 Initial Parameters for a Thin-Walled Beam of Constant Cross Section (Eqs. 5.9)

Left End ↓ / Right End →	1. Simply Supported 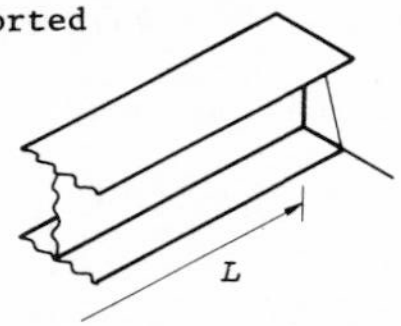	2. Fixed 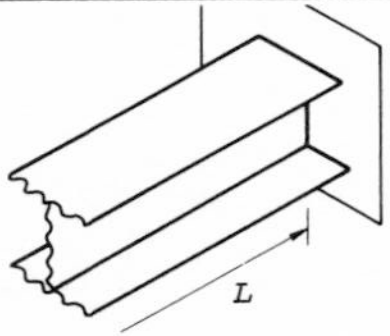	3. Free 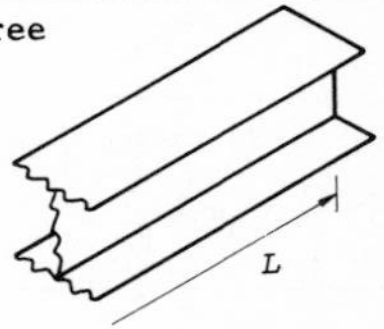
1. Simply Supported 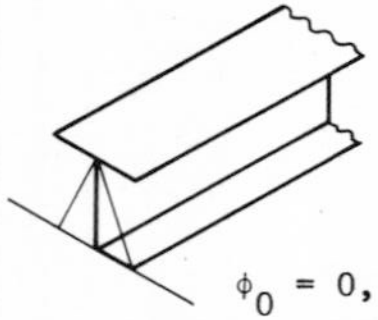 $\phi_0 = 0,\ B_0 = 0$	$\Psi_0 = \frac{F_B}{GJL}\left(\frac{CL}{\sinh CL} - 1\right) + \frac{F_\phi}{L}$ $T_0 = -\frac{1}{L}(GJF_\phi + F_B)$	$\Psi_0 = \frac{1}{\nabla}[(CL - \sinh CL)F_\Psi$ $-C(\cosh CL - 1)F_\phi]$ $T_0 = \frac{GJ}{\nabla}(F_\phi C \cosh CL + F_\Psi \sinh CL)$ $\nabla = \sinh CL - CL \cosh CL$	$\Psi_0 = \frac{1}{GJ}\left(F_T - \frac{CF_B}{\sinh CL}\right)$ $T_0 = -F_T$
2. Fixed 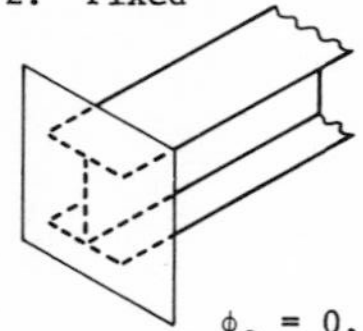 $\phi_0 = 0,\ \Psi_0 = 0$	$B_0 = \frac{1}{\nabla}[(CL - \sinh CL)F_B - GJF_\phi \sinh CL]$ $T_0 = F_\phi E\Gamma C^3 \cosh CL - C(1 - \cosh CL)F_B$ $\nabla = \sinh CL - CL \cosh CL$	$B_0 = \frac{1}{\nabla}[E\Gamma C(CL - \sinh CL)F_\Psi$ $- GJ(\cosh CL - 1)F_\phi]$ $T_0 = \frac{1}{\nabla}[GJCF_\phi \sinh CL - GJ(1 - \cosh CL)F_\Psi]$ $\nabla = \cosh CL$	$B_0 = \frac{F_T}{C}\tanh CL - \frac{F_B}{\cosh CL}$ $T_0 = -F_T$
3. Free 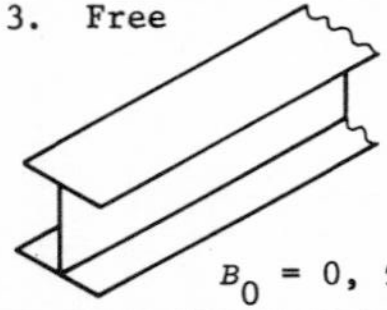 $B_0 = 0,\ T_0 = 0$	$\phi_0 = -\frac{F_B}{GJ} - F_\phi$ $\Psi_0 = -\frac{F_B}{E\Gamma C \sinh CL}$	$\phi_0 = -\frac{F_\Psi}{C}\tanh CL - F_\phi$ $\Psi_0 = -\frac{F_\Psi}{\cosh CL}$	Kinematically Unstable

Definitions: $C = \sqrt{E\Gamma/GJ}$, $F_\phi = F_\phi\big|_{x=L}$, $F_\Psi = F_\Psi\big|_{x=L}$, $F_B = F_B\big|_{x=L}$, $F_T = F_T\big|_{x=L}$

Both the angle of twist and bimoment must be zero at a pinned end. A fixed end occurs when the rotation and warping of the cross section at the end of the beam are prevented. That is, the angle of twist and the rate of angle of twist are zero for a fixed end. At a free end condition, both rotation and warping of the cross section are unrestrained. However, the bimoment and twisting moment are zero at a free end.

EXAMPLE 5.2 Deformation and Stresses in a Cantilevered Bar Find the angles of twist, moments, and stresses in the cantilevered bar of Fig. 5-3.

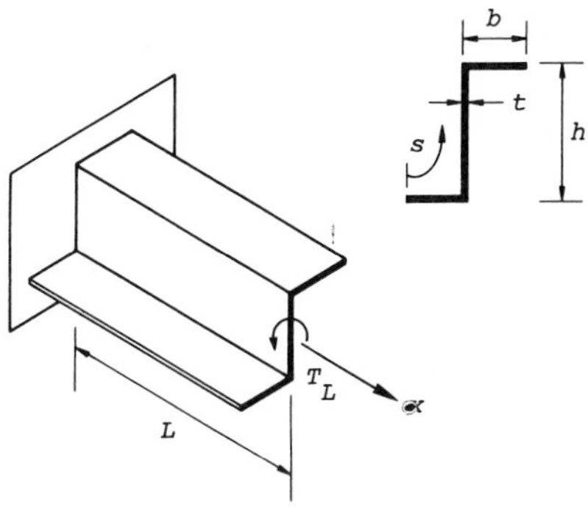

Fig. 5-3

The geometric quantities are taken from Tables 3-1 as

$$J = \frac{t^3}{3}(2b + h), \qquad \Gamma = \frac{tb^3h^2(b + 2h)}{12(2b + h)} \tag{1}$$

If $b = h/2$,

$$J = \frac{4}{3}t^3b, \qquad \Gamma = \frac{5}{12}tb^5 \tag{2}$$

The angles of twist and moments along the bar are given by Eqs. (5.9), with the loading functions F_ϕ, F_Ψ, F_B, F_T taken from Table 5-1. We find

$$F_\phi = -T_L \frac{C\langle x - L\rangle - \sinh C\langle x - L\rangle}{CGJ} \qquad F_\Psi = T_L \frac{\langle x - L\rangle^0 - \cosh C\langle x - L\rangle}{GJ}$$

$$F_B = -T_L \frac{\sinh C\langle x - L\rangle}{C} \qquad F_T = -T_L\langle x - L\rangle^0 \tag{3}$$

The four initial parameters are found from the boundary conditions. In this case, the conditions $\phi_{x=0} = 0$, $\Psi_{x=0} = 0$, $B_{x=L} = 0$, $T_{x=L} = 0$ applied to Eqs. (5.9) give ϕ_0, Ψ_0, B_0, T_0. Alternatively, the formulas of Table 5-2 can be used. Note that the torque applied at the end of the bar is taken into account by the loading functions (3) rather than through the boundary conditions. From Table

5-2 for fixed-free ends, with the loading functions of (3),

$$\phi_0 = 0 \qquad \Psi_0 = 0 \qquad B_0 = -\frac{T_L}{C}\tanh CL \qquad T_0 = T_L \tag{4}$$

In evaluating the loading functions at $x = L$, the equalities $\langle L - L\rangle^0 = 1$, $\langle L - L\rangle = 0$ should be employed.

Substitution of (3) and (4) into Eqs. (5.9) provides the desired expressions for the angles of twist and moments. The stresses are computed using Eqs. (5.1), and (5.4). The normal stress is

$$\sigma_x = -\frac{B\omega}{\Gamma} = -\frac{T_L}{C}(\sinh Cx - \tanh CL \cosh Cx - \sinh C\langle x - L\rangle)\frac{\omega}{\Gamma} \tag{5}$$

where peak values of ω can be taken for this **Z** section from Table 3-1. This normal stress should be added to normal stresses due to other loadings, for example, axial force or bending moments.

The shear stress is

$$\tau = \frac{T_\omega Q_\omega}{t\Gamma} = T_L(-\tanh CL \sinh Cx + \cosh Cx - \cosh C\langle x - L\rangle)\frac{Q_\omega}{t\Gamma} \tag{7}$$

with Table 3-1 providing critical values of Q_ω. Shear stresses such as (7) are due to nonuniform torques or from restrained warping. The total shear stress is found by superimposing this stress and the torsional stress as calculated from the formulas of Chap. 3.

If this same beam were loaded with a concentrated vertical force W situated on the outer edge of a flange, then the statically equivalent problem of a beam with an end twisting moment of Wb plus a force W acting through the shear center, that is, vertical along the web, can be treated. The bending effects due to W are found from the formulas of Chap. 2, while the torsion of this thin-walled section is handled by the formulas of this example problem.

EXAMPLE 5.3 Twisting of a Bar with Concentrated Bimoment Find the angle of twist and warping torque along the beam of Fig. 5-4 subjected to an eccentric bending moment.

This loading is equivalent to the application of a bimoment $B_1 = M_1 c$ to the beam. With loading functions taken from Table 5-1, Eqs. 5.9 for the two desired variables become

$$\phi = \phi_0 - \Psi_0\frac{\sinh Cx}{C} + B_0\frac{1 - \cosh Cx}{GJ} + T_0\frac{Cx - \sinh Cx}{CGJ}$$
$$+ \frac{M_1 c}{GJ}(-\langle x - a\rangle^0 + \cosh C\langle x - a\rangle) \tag{1}$$

$$T_\omega = \Psi_0 GJ\cosh Cx + B_0 C\sinh Cx + T_0\cosh Cx - \frac{M_1 cGJ}{CE\Gamma}\sinh C\langle x - a\rangle$$

The initial parameters are taken from Table 5-2. These formulas require $F_\phi|_{x=L}$, $F_B|_{x=L}$. From Table 5-1 we find

$$F_\phi|_{x=L} = \frac{M_1 c}{GJ}[-1+\cosh C(L-a)],\ F_B|_{x=L} = -M_1 c \cosh C(L-a) \tag{2}$$

Table 5-2 then provides the initial parameters

$$\phi_0 = 0, \qquad \Psi_0 = 0, \qquad \nabla = \sinh CL - CL \cosh CL$$

$$B_0 = -\frac{M_1 c}{\nabla}[CL \cosh C(L-a) - \sinh CL] \tag{3}$$

$$T_0 = -\frac{M_1 c}{\nabla} C[\cosh CL - \cosh C(L-a)]$$

Substitution of (3) into (1) completes the expressions for ϕ and T_ω.

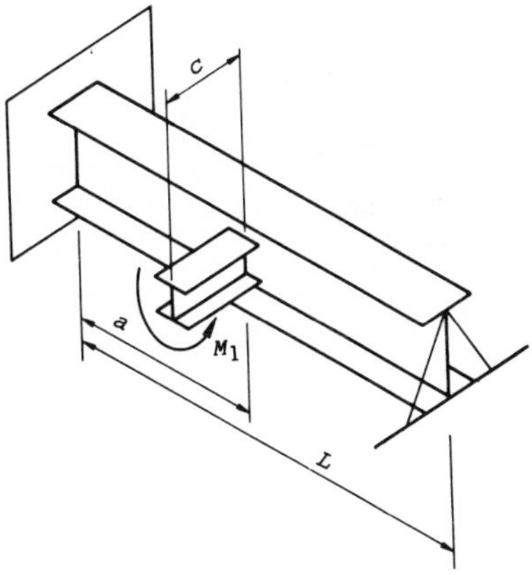

Fig. 5-4

EXAMPLE 5.4 Beam with Multiple Loadings To illustrate the procedure for finding the response of a beam with multiple loadings, consider the configuration of Fig. 5-5 with two concentrated forces.

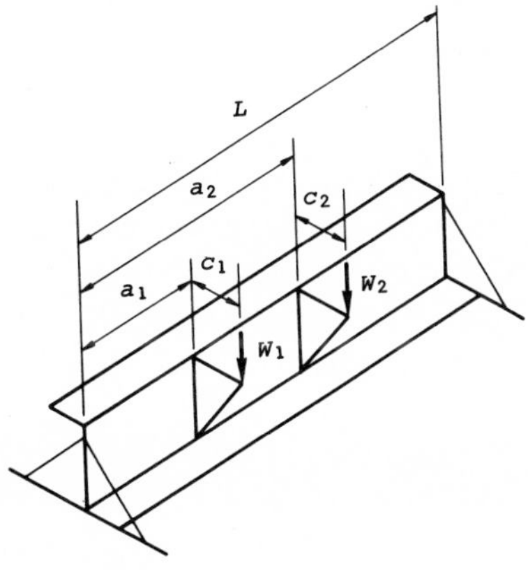

Fig. 5-5

The loading functions for the two forces are the sums of the loading functions for each force. Thus, from Table 5-1

$$
\begin{aligned}
F_\phi(x) &= \frac{W_1 c_1}{CGJ}\left(-C\langle x-a_1\rangle + \sinh C\langle x-a_1\rangle\right) \\
&\quad + \frac{W_2 c_2}{CGJ}\left(-C\langle x-a_2\rangle + \sinh C\langle x-a_2\rangle\right) \\
F_\Psi(x) &= \frac{W_1 c_1}{GJ}\left(\langle x-a_1\rangle^0 - \cosh C\langle x-a_1\rangle\right) \\
&\quad + \frac{W_2 c_2}{GJ}\left(\langle x-a_2\rangle^0 - \cosh C\langle x-a_2\rangle\right) \\
F_B(x) &= -\frac{W_1 c_1}{C}\sinh C\langle x-a_1\rangle - \frac{W_2 c_2}{C}\sinh C\langle x-a_2\rangle \\
F_T(x) &= -W_1 c_1\langle x-a_1\rangle^0 - W_2 c_2\langle x-a_2\rangle^0
\end{aligned}
\tag{1}
$$

These are ready to be placed in Eqs. (5.9). The initial parameters are taken from Table 5-2, in which the loading functions F_ϕ and F_B evaluated at $x = L$ are required. From (1)

$$
\begin{aligned}
F_\phi\big|_{x=L} &= \frac{W_1 c_1}{CGJ}\left[-C(L-a_1) + \sinh C(L-a_1)\right] + \frac{W_2 c_2}{CGJ}\left[-C(L-a_2)\right. \\
&\quad \left. + \sinh C(L-a_2)\right] \\
F_B\big|_{x=L} &= -\frac{W_1 c_1}{C}\sinh C(L-a_1) - \frac{W_2 c_2}{C}\sinh C(L-a_2)
\end{aligned}
\tag{2}
$$

Equations (5.9) supplemented with (1) and (2) now provide the response variables along the beam.

5.4 Stability

For thin-walled beams subjected to axial forces, buckling can occur as the result of bending instability as described in Chap. 2, or because of torsional instability. The latter is found with the formulas of this chapter. The torsional buckling loads P_{cr} for uniform shafts with axial forces passing through the shear center are given by

$$
P_{cr} = \frac{1}{C_P}\left(\frac{C_1\pi^2}{L^2}E\Gamma + GJ\right) \tag{5.10}
$$

$$\text{where} \quad C_1 = \begin{cases} \frac{1}{4} & \text{for fixed-free ends} \\ 1 & \text{for simply supported–simply supported ends} \\ 2.045 & \text{for fixed simply supported ends} \\ 4 & \text{for fixed-fixed ends} \end{cases}$$

B. COMPLEX BEAMS

5.5 Notation for Complex Beams

ρ Mass per unit length (mass/length)

ω Natural frequency (radian/second)

t Time

k_t Elastic foundation modulus (force · length/length)

r_p Polar radius of gyration; r_p is the radius of gyration of the cross-sectional area about the longitudinal (x) axis of the beam (Table 2-1).

I_{pi} Polar mass moment of inertia of concentrated mass at station i (mass · length^2). This can be calculated as $I_{pi} = \Delta a\ \rho r_p^2$ where Δa is the length of beam lumped at station i.

c_t' External or viscous damping coefficient (mass · length/time, force · length · time/length)

c_ρ Proportional viscous damping coefficient (1/time). If c_t' is chosen to be proportional to the mass—that is, $c_t' = c_\rho \rho r_p^2$—then c_ρ is a constant of proportionality.

$\mathbf{U}_i$ Field matrix of the ith segment

$\overline{\mathbf{U}}_i$ Point matrix at $x = a_i$

The notation for transfer matrices is:

$$\mathbf{U}_i = \begin{bmatrix} U_{\phi\phi} & U_{\phi\Psi} & U_{\phi B} & U_{\phi T} & F_{\phi} \\ U_{\Psi\phi} & U_{\Psi\Psi} & U_{\Psi B} & U_{\Psi T} & F_{\Psi} \\ U_{B\phi} & U_{B\Psi} & U_{BB} & U_{BT} & F_{B} \\ U_{T\phi} & U_{T\Psi} & U_{TB} & U_{TT} & F_{T} \\ 0 & 0 & 0 & 0 & 1 \end{bmatrix} \tag{5.11}$$

5.6 Differential Equations

The fundamental equations of motion in first-order form for the twisting of a thin-walled beam are

$$\frac{\partial \phi}{\partial x} = -\Psi \tag{5.12a}$$

$$\frac{\partial \Psi}{\partial x} = \frac{B}{E\Gamma} \tag{5.12b}$$

$$\frac{\partial B}{\partial x} = T + (GJ - C_P P)\Psi = T_\omega \tag{5.12c}$$

$$\frac{\partial T}{\partial x} = k_t \phi + \rho r_p^2 \frac{\partial^2 \phi}{\partial t^2} - m_x(x, t) \tag{5.12d}$$

TABLE 5-3 Massless Beam

Definitions:

(a) With No Axial Force

$$C^2 = GJ/E\Gamma, \quad d_1 = C^2$$

$$s = \frac{\sinh C\ell}{C}, \quad c = \cosh C\ell$$

(b) With Axial Force Through Shear Center

P is a compressive axial force.

For tensile axial forces replace P by $-P$

$$\alpha = \frac{C_P P - GJ}{E\Gamma}$$

If $\alpha<0$:

$$C^2 = \frac{GJ - C_P P}{E\Gamma}, \quad d_1 = C^2$$

$$s = \frac{\sinh C\ell}{C}, \quad c = \cosh C\ell$$

If $\alpha>0$:

$$C^2 = \frac{C_P P - GJ}{E\Gamma}, \quad d_1 = -C^2$$

$$s = \frac{\sin C\ell}{C}, \quad c = \cos C\ell$$

$$\mathbf{U}_i = \begin{bmatrix} 1 & -s & \dfrac{1-c}{d_1 E\Gamma} & \dfrac{\ell - s}{d_1 E\Gamma} & F_\phi \\ 0 & c & \dfrac{s}{E\Gamma} & \dfrac{-1+c}{d_1 E\Gamma} & F_\Psi \\ 0 & E\Gamma s\, d_1 & c & s & F_B \\ 0 & 0 & 0 & 1 & F_T \\ 0 & 0 & 0 & 0 & 1 \end{bmatrix}$$

$$F_\phi = \frac{m_{x1}}{d_1 E\Gamma}\left(\frac{c-1}{d_1} - \frac{\ell^2}{2}\right) + \frac{\Delta m_x}{\Delta\ell}\,\frac{1}{d_1 E\Gamma}\left(\frac{s}{d_1} - \frac{\ell^3}{6} - \frac{\ell}{d_1}\right)$$

$$F_\Psi = m_{x1}\,\frac{\ell - s}{d_1 E\Gamma} + \frac{\Delta m_x}{\Delta\ell}\,\frac{1}{d_1 E\Gamma}\left(\frac{\ell^2}{2} + \frac{1-c}{d_1}\right)$$

$$F_B = m_{x1}\,\frac{1-c}{d_1} + \frac{\Delta m_x}{\Delta\ell}\,\frac{\ell - s}{d_1}$$

$$F_T = -m_{x1}\ell + \frac{\Delta m_x}{\Delta\ell}\,\frac{\ell^2}{2}$$

TABLE 5-4 Beam with Mass

Definitions:

$$\zeta = (C_p P - GJ)/E\Gamma$$

$$\lambda = -\frac{\rho r_p^2 \omega^2}{E\Gamma}$$

$$a^2 = \sqrt{-\lambda + \zeta^2/4} - \zeta/2$$

$$b^2 = \sqrt{-\lambda + \zeta^2/4} + \zeta/2$$

$$g^2 = a^2 + b^2$$

$$A = \cosh a\ell, \quad B = \cos b\ell$$

$$C = \sinh a\ell, \quad D = \sin a\ell$$

$$e_0 = \frac{1}{g}(a^3 C - b^3 D)$$

$$e_1 = \frac{1}{g}(a^2 A + b^2 B)$$

$$e_2 = \frac{1}{g}(aC + bD)$$

$$e_3 = \frac{1}{g}(A - B)$$

$$e_4 = \frac{1}{g}\left(\frac{C}{a} - \frac{D}{b}\right)$$

$$e_5 = \frac{1}{g}\left(\frac{A}{a^2} + \frac{B}{b^2}\right) - \frac{1}{a^2 b^2}$$

$$e_6 = \frac{1}{g}\left(\frac{C}{a^3} + \frac{D}{b^3}\right) - \frac{\ell}{a^2 b^2}$$

$\mathbf{U}_i =$

$e_1 + \zeta e_3$	$-e_2$	$-e_3/E\Gamma$	$-e_4/E\Gamma$	F_ϕ
λe_4	e_1	$e_2/E\Gamma$	$e_3/E\Gamma$	F_ψ
$\lambda E\Gamma e_3$	$E\Gamma e_0$	e_1	e_2	F_B
$\lambda E\Gamma(e_2 + \zeta e_4)$	$-\lambda E\Gamma e_3$	$-\lambda e_4$	$e_1 + \zeta e_3$	F_T
0	0	0	0	1

$$F_\phi = \left(m_{x1} e_5 + \frac{\Delta m_x}{\Delta\ell} e_6\right)/E\Gamma$$

$$F_\psi = -m_{x1}\frac{e_4}{E\Gamma} - \frac{\Delta m_x}{\Delta\ell}\frac{e_5}{E\Gamma}$$

$$F_B = -m_{x1} e_3 - \frac{\Delta m_x}{\Delta\ell} e_4$$

$$F_T = -m_{x1}(e_2 + \zeta e_4) - \frac{\Delta m_x}{\Delta\ell}(e_3 + \zeta e_5)$$

Equations (5.12) are appropriate for a beam with a tensile axial force if P is replaced by $-P$. In the present form Eqs. (5.12) apply to beams with a compressive axial force P that passes through the shear center of the section. In higher-order form the equations become

$$
\begin{aligned}
m_x &= \frac{\partial^2}{\partial x^2} E\Gamma \frac{\partial^2 \phi}{\partial x^2} - \frac{\partial}{\partial x}(GJ - C_P P)\frac{\partial \phi}{\partial x} + k_t \phi + \rho r_p^2 \frac{\partial^2 \phi}{\partial t^2} \\
T &= -\frac{\partial}{\partial x} E\Gamma \frac{\partial^2 \phi}{\partial x^2} + (GJ - C_P P)\frac{\partial \phi}{\partial x} \\
B &= -E\Gamma \frac{\partial^2 \phi}{\partial x^2} \\
\Psi &= -\frac{\partial \phi}{\partial x}
\end{aligned}
\qquad (5.13)
$$

5.7 Transfer Matrices for Uniform Segments

The transfer matrices for a thin-walled beam segment of length ℓ are provided in Tables 5-3 and 5-4. In these tables, uniformly distributed torques m_{x1} and linearly varying torques of gradient $\Delta m_x / \Delta \ell$ are taken into account.

TABLE 5-5 Equivalence of the Bending of Beams and the Twisting of Thin-Walled Beams in Use of Table 2-13

Bending of Beams	Twisting of Thin-Walled Beams
v	ϕ
θ	Ψ
M	B
V	T
w_1	m_{x1}
$\frac{\Delta w}{\Delta \ell}$	$\frac{\Delta m_x}{\Delta \ell}$
$\frac{1}{GA_s}$	0
$\lambda = (k - \rho\omega^2)/EI$	$\lambda = (k_t - \rho r_p^2 \omega^2)/E\Gamma$
$\eta = (k - \rho\omega^2)/GA_s$	$\eta = 0$
$\zeta = (P + \rho r^2 \omega^2 - k^*)/EI$	$\zeta = (C_p P - GJ)/E\Gamma$

For transfer matrices for thin-walled beams on elastic foundations or for segments of beams with a combination of axial force, foundation, and mass, employ the general transfer matrix of Table 2-13. In so doing, utilize the definitions of Table 5-5.

5.8 Point Matrices

The transfer matrices that take into account concentrated point occurrences are listed in Table 5-6. With appropriate substitutions as indicated in Table 5-5, the point matrices of Chap. 2 apply for thin-walled beams. Particularly useful are the point matrices for in-span indeterminates, for example, in-span rigid support, of Table 2-15.

5.9 Loading Functions

The loading functions are provided in the previous transfer matrix tables for most common loadings. These functions can be calculated for other loadings from the formulas

$$\begin{aligned}
F_\phi &= -\int_0^\ell m_x(x)U_{\phi T}(\ell - x)\,dx = -\int_0^\ell m_x(\ell - x)U_{\phi T}(x)\,dx \\
F_\Psi &= -\int_0^\ell m_x(x)U_{\Psi T}(\ell - x)\,dx = -\int_0^\ell m_x(\ell - x)U_{\Psi T}(x)\,dx \\
F_B &= -\int_0^\ell m_x(x)U_{BT}(\ell - x)\,dx = -\int_0^\ell m_x(\ell - x)U_{BT}(x)\,dx \\
F_T &= -\int_0^\ell m_x(x)U_{TT}(\ell - x)\,dx = -\int_0^\ell m_x(\ell - x)U_{TT}(x)\,dx
\end{aligned} \tag{5.14}$$

where m_x is an arbitrary torque intensity. The notation $U_{ij}(\ell - x)$ refers to the U_{ij} given in the transfer matrix tables with ℓ replaced by $\ell - x$. A similar definition applies for $m_x(\ell - x)$.

5.10 Static Response

The procedure for calculating the angle of twist, rate of angle of twist, internal bimoment, and internal twisting moment due to static loading follows the procedure given in Section 2.16 for the static response of beams. This discussion and the examples in Chap. 2 apply to the twisting of thin-walled beams if v, θ, M, V are replaced by ϕ, Ψ, B, T, respectively.

Values of the Initial Parameters Use Table 2-19 or Eqs. (2.31), (2.32), again substituting ϕ, Ψ, B, T for v, θ, M, V.

Warping Torque The computations as set forth here provide ϕ, Ψ, B, and T. However, stress calculations using Eq. (5.4) or a computer program require knowledge of the warping torque T_ω. This response is readily found from Ψ and T with Eq. (5.12):

$$T_\omega = T + (GJ - C_P P)\Psi \tag{5.15}$$

TABLE 5-6 Point Matrices at $x = a_i$

Definitions:

(a) Concentrated Applied Torque

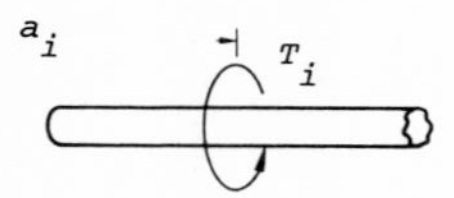

(b) Concentrated Applied Moments Equivalent to Bimoment B_i

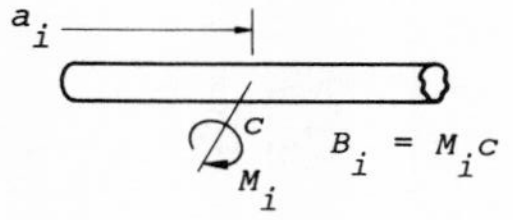

(c) Concentrated Mass

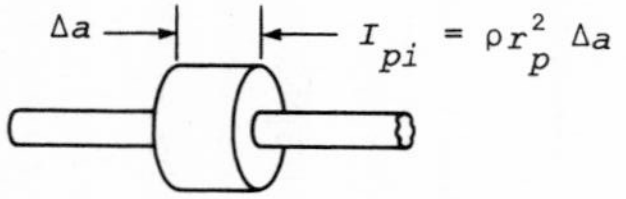

$$\bar{\mathbf{U}}_i = \begin{bmatrix} 1 & 0 & 0 & 0 & 0 \\ 0 & 1 & 0 & 0 & 0 \\ 0 & C_p P_{Li} & 1 & 0 & -B_i \\ k_{t1} - I_{pi}\omega^2 & 0 & 0 & 1 & -T_i \\ 0 & 0 & 0 & 0 & 1 \end{bmatrix}$$

(d) Torsional Spring

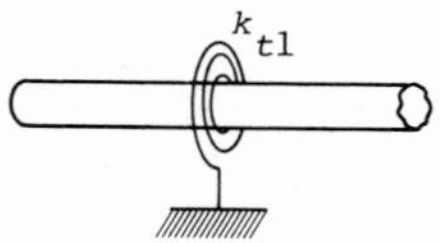

(e) Lumped Axial Force

$$P_{Li} = \Delta a \, P$$

Substitute $-P$ for P if force is tensile. For a distributed axial force p (force/length), P is found using the condition of equilibrium $\sum F_x = 0$.

5.11 Stability

The axial force leading to torsional instability can be ascertained with the equations of this chapter. The uncoupled theory of this chapter applies as long as the axial force passes through the shear center. Otherwise, a theory coupling bending and twisting effects must be considered.

The lower of the critical bending and torsional axial forces is of primary interest in practical applications. Both critical loads are found in the same fashion. The critical axial load (the buckling or unstable load) is the lowest value of the compressive axial force P for which ∇ of Table 2-19 or Eqs. (2.31) or (2.32), as appropriate, is zero. In using the table or equations for torsional instability, replace v, θ, M, V by ϕ, Ψ, B, T.

EXAMPLE 5.5 Critical Loads of Uniform Beams Derive expressions for the critical axial loads P_{cr} for thin-walled beams that are simply supported or fixed on both ends.

The critical load is found as the lowest value of P that satisfies $\nabla = 0$, where ∇ is the characteristic relationship resulting from application of the boundary conditions. From Table 2-19 for two simply supported ends,

$$\nabla = (U_{\phi\Psi}U_{BT} - U_{B\Psi}U_{\phi T})_{x=L} \tag{1}$$

Insertion of the elements of the transfer matrix of Table 5-3 with $C^2 = (C_P P - GJ)/E\Gamma$, in (1), leads to

$$\begin{aligned}\nabla &= -\frac{\sin CL}{C}\frac{\sin CL}{C} - (-E\Gamma C\sin CL)\left(-\frac{CL - \sin CL}{E\Gamma C^3}\right)\\ &= -\frac{L}{C}\sin CL\end{aligned}$$

The critical axial load P_{cr} then is found from $\sin CL = 0$. This gives $CL = n\pi$, $n = 1, 2, \ldots$. Since the critical load occurs for $n = 1$, we find

$$P_{\text{cr}} = \frac{1}{C_P}\left(E\Gamma\frac{\pi^2}{L^2} + GJ\right)$$

In the case of a fixed-fixed beam

$$\nabla = (U_{\phi B}U_{\Psi T} - U_{\Psi B}U_{\phi T})_{x=L}$$

From Table 5-3,

$$\begin{aligned}\nabla &= \left(-\frac{1-\cos CL}{E\Gamma C^2}\right)\left(-\frac{-1+\cos CL}{E\Gamma C^2}\right) - \left(\frac{\sin CL}{E\Gamma C}\right)\left(-\frac{CL - \sin CL}{E\Gamma C^3}\right)\\ &= \frac{-2 + 2\cos CL + CL\sin CL}{(E\Gamma C^2)^2} = \frac{4\sin\dfrac{CL}{2}\left(\dfrac{CL}{2}\cos\dfrac{CL}{2} - \sin\dfrac{CL}{2}\right)}{(E\Gamma C^2)^2}\end{aligned}$$

Setting $\nabla = 0$ leads to $\sin CL/2 = 0$ and $(CL/2)\cos CL/2 - \sin CL/2 = 0$. The critical load is taken as the smallest root of these relations. It will be found from $\sin CL/2 = 0$, and the root is $CL = 2n\pi$, $n = 1, 2, \ldots,$. Finally

$$P_{cr} = \frac{1}{C_P}\left(E\Gamma\frac{4\pi^2}{L^2} + GJ\right)$$

Note that the two critical loads derived above are identical to those given in Eq. (5.10) for these boundary conditions.

5.12 Free Dynamic Response—Natural Frequencies

The natural frequencies ω_n, $n = 1, 2, \ldots,$ are roots of the characteristic equation $\nabla = 0$ resulting from the evaluation of the initial parameters. Follow the procedure described in Section 2.18 for beams with ϕ, Ψ, B, T substituted for v, θ, M, V.

EXAMPLE 5.6 Natural Frequencies of a Pinned-Pinned Beam Find the natural torsional frequencies of a thin-walled beam with pinned ends.

The natural frequencies are those values of ω_n that make the determinant ∇ of Table 2-19 zero. For pinned-pinned ends

$$\nabla = (U_{\phi\Psi}U_{BT} - U_{\phi T}U_{B\Psi})_{x=L} \tag{1}$$

Substitution of the transfer matrix elements of Table 5-4 into (1) leads to

$$\nabla = [(-e_2)e_2 - (-e_4/E\Gamma)(E\Gamma e_0)]_{x=L} = \left(2ab + \frac{a^3}{b} + \frac{b^3}{a}\right)\sinh aL \sin bL = 0$$

This implies that $\sin bL = 0$ or

$$\omega_n = \frac{n\pi}{L^2}\sqrt{\frac{E\Gamma(n^2\pi^2 - \zeta L^2)}{\rho r_P^2}} \qquad n = 1, 2, \ldots$$

where $\zeta = (C_P P - GJ)/E\Gamma$.

5.13 Forced Harmonic Motion

If the forcing functions vary as $\sin \Omega t$ or $\cos \Omega t$, where Ω is the frequency of the loading, the state variables ϕ, Ψ, B, and T also vary as $\sin \Omega t$ or $\cos \Omega t$. The spatial distributlons of ϕ, Ψ, B, and T are found by a static solution using transfer matrices containing ρ. In these transfer matrices ω should be replaced by Ω. The solution procedure is exactly the same as for the static response, since time need not enter the analysis. The initial parameters ϕ_0, Ψ_0, B_0, T_0 are given by Table 2-19 [Eqs. (2.31) and Eqs. (2.32)] with ϕ, Ψ, B, T replacing v, θ, M, V.

5.14 Dynamic Response Due to Arbitrary Loading

In terms of a modal solution, the time-dependent state variables resulting from arbitrary dynamic loading are expressed by

Displacement method:

$$\phi(x, t) = \sum_n A_n(t)\phi_n(x)$$

$$\Psi(x, t) = \sum_n A_n(t)\Psi_n(x)$$

$$B(x, t) = \sum_n A_n(t)B_n(x) \tag{5.16a}$$

$$T(x, t) = \sum_n A_n(t)T_n(x)$$

$$T_\omega(x, t) = \sum_n A_n(t)T_{\omega n}(x)$$

Acceleration method:

$$\phi(x, t) = \phi_s(x, t) + \sum_n B_n'(t)\phi_n(x)$$

$$\Psi(x, t) = \Psi_s(x, t) + \sum_n B_n'(t)\Psi_n(x)$$

$$B(x, t) = B_s(x, t) + \sum_n B_n'(t)B_n(x) \tag{5.16b}$$

$$T(x, t) = T_s(x, t) + \sum_n B_n'(t)T_n(x)$$

$$T_\omega(x, t) = T_{\omega s}(x, t) + \sum_n B_n'(t)T_{\omega n}(x)$$

where

$$A_n(t) = \frac{\eta_n(t)}{N_n} \qquad B_n'(t) = \frac{\xi_n(t)}{N_n} \tag{5.17}$$

These relations provide the complete dynamic response. That is, these are the solutions to Eqs. (5.12).

No Damping or Proportional Damping For viscous damping, the term $c_t' \, \partial\phi/\partial t$ should be added to the right-hand side of Eq. (5.12*d*). If the damping is "proportional," then $c_t' = c_\rho \rho r_p^2$. Set $c_t' = 0$ for no viscous damping.

In Eqs. (5.16*b*) the terms with the subscript s are static solutions, as given in Section 5.10, that are determined for the applied loading at each point in time. The terms ϕ_n, Ψ_n, B_n, T_n, $T_{\omega n}$ are the mode shapes, that is, ϕ, Ψ, B, T, T_ω with $\omega = \omega_n$, $n = 1, 2, \ldots,$ as explained in Section 5.12. The quantities in Eq. (5.17) are defined as:

$$N_n = \int_0^L \rho r_p^2 \phi_n^2 \, dx \tag{5.18}$$

$$\eta_n(t) = e^{-\zeta_n \omega_n t}\left(\cos \alpha_n t + \frac{\zeta_n \omega_n}{\alpha_n} \sin \alpha_n t\right)\eta_n(0) + e^{-\zeta_n \omega_n t} \frac{\sin \alpha_n t}{\alpha_n} \frac{\partial \eta}{\partial t}(0)$$

$$+ \int_0^t f_n(\tau) e^{-\zeta_n \omega_n (t-\tau)} \frac{\sin \alpha_n (t-\tau)}{\alpha_n} \, d\tau \tag{5.19}$$

where

$$\alpha_n = \omega_n \sqrt{1 - \zeta_n^2} \qquad \zeta_n = c_\rho / 2\omega_n$$

$\xi_n(t)$ is taken from $\eta_n(t)$ by replacing

$$\eta_n(0) \quad \text{by} \quad \eta_n(0) - \frac{f_n(0)}{\omega_n^2}$$

$$\frac{\partial \eta_n(0)}{\partial t} \quad \text{by} \quad \frac{\partial \eta_n(0)}{\partial t} - \frac{1}{\omega_n^2} \frac{\partial f_n(0)}{\partial t}$$

$$f_n(\tau) \quad \text{by} \quad -\frac{1}{\omega_n^2}\left(\frac{\partial^2}{\partial \tau^2} + c_\rho \frac{\partial}{\partial \tau}\right) f_n(\tau)$$

If $\zeta_n > 1$, replace sin by sinh, cos by cosh, and $\sqrt{1 - \zeta_n^2}$ by $\sqrt{\zeta_n^2 - 1}$. For zero viscous damping, $\zeta_n = 0$, and

$$\eta_n(0) = \int_0^L \rho r_p^2 \phi(x, 0) \phi_n \, dx \tag{5.20}$$

$$f_n(t) = \int_0^L m_x(x, t) \phi_n \, dx + h_n(a_k, t) \tag{5.21}$$

The function $m_x(x, t)$ is the applied torque intensity. The function $h_n(a_k, t)$ accounts for nonhomogeneous angles of twist $\phi(a_k, t)$ at $x = a_k$, such as supports or prescribed time-dependent angles of twist located in-span or on the boundary:

$$h_n(a_k, t) = -\phi(a_k, t) \Delta T_n(a_k)$$

$$\Delta T_n(a_k) = T_n(a_k^-) - T_n(a_k^+) \qquad +(-) \text{ means just to the right (left) of } x = a_k \tag{5.22}$$

If the right end has a nonzero (for example, time-dependent) angle of twist, then Eq. (5.22) reduces for $a_k = L$ to

$$h_n(a_k, t) = -\phi(L, t) T_n(L) \tag{5.23}$$

If the left end has a nonhomogeneous displacement, then Eq. (5.22) reduces for

$a_k = 0$ to

$$h_n(a_k, t) = +\phi(0, t)T_n(0) \tag{5.24}$$

In the case of the acceleration method, $f_n(t)$ can alternatively be expressed as

$$f_n(t) = \omega_n^2 \int_0^L \rho r_p^2 \phi_s(x, t)\phi_n \, dx \tag{5.25}$$

Angular ground motion at $x = 0$ may be included by adding $k_t\phi_0(t)\phi_n(0) + c_t'\phi_{0,t}(t)\phi_n(0)$ to $f_n(t)$. For ground motion at $x = L$, add $k_t\phi_L(t)\phi_n(a_L) + c_t'\phi_{L,t}\phi_n(a_L)$ to $f_n(t)$, where $\phi_{a_k}(t)$ and $\phi_{a_k,t}(t)$ are the applied angular displacement and velocity, respectively, at $x = a_k$.

Nonproportional Damping The formulas of this chapter need considerable adjustment in order to make them applicable to materials with nonproportional viscous damping. As noted above, Eq. (5.12*d*) should be supplemented by the term $c_t'\partial\phi/\partial t$ on the right-hand side. The frequency s_n is now a complex number. Also, the mode shapes and some of the transfer matrices are complex functions.

Equations (5.16) and (5.17) remain valid but now the static solutions refer to

$$\phi_s(x, t) = -\sum_n \frac{\phi_n(x)f_n(t)}{s_n N_n} \tag{5.26}$$

where $\phi_n(x)$ and N_n are the mode shape and norm associated with the complex eigenvalue s_n. Similar expansions hold for the other state variables.

The definitions of the previous subsection must be adjusted so that

$$N_n = 2s_n \int_0^L \rho r_p^2 \phi_n^2 \, dx + \int_0^L c_t' \phi_n^2 \, dx \tag{5.27}$$

$$\eta_n(t) = e^{s_n t}\eta_n(0) + \int_0^t e^{s_n(t-\tau)} f_n(\tau) \, d\tau \tag{5.28}$$

$$\xi_n(t) = e^{s_n t}\left[\frac{1}{s_n^2}\frac{\partial f_n(0)}{\partial t} + \frac{1}{s_n} f_n(0) + \eta_n(0)\right]$$
$$- \frac{1}{s_n^2}\frac{\partial f_n(t)}{\partial t} + \frac{1}{s_n^2}\int_0^t e^{s_n(t-\tau)} \frac{\partial^2 f_n}{\partial \tau^2}(\tau) \, d\tau \tag{5.29}$$

$$\eta_n(0) = \int_0^L \left[s_n \rho r_p^2 \phi(x, 0)\phi_n + c_t'\phi(x, 0)\phi_n(x) + \rho r_p^2 \frac{\partial\phi}{\partial t}(x, 0)\phi_n(x)\right] dx \tag{5.30}$$

In the case of the acceleration method, $f_n(t)$ can alternatively be given by

$$f_n(t) = -s_n \int_0^L c_t'\phi_s(x, t)\phi_n(x) \, dx$$
$$- s_n^2 \int_0^L \rho r_p^2 \phi_s(x, t)\phi_n(x) \, dx \tag{5.31}$$

For discrete damping with dashpot constants c_i (force · length · time) at $x = a_i$, use the point matrix of Table 5-6 with k_{t1} replaced by $k_{t1} + sc_i$ and $-I_{pi}\omega^2$ by $I_{pi}s^2$. Also, in N_n set

$$\int_0^L c_t' \phi_n^2 \, dx = \sum_i c_i \phi_n^2(a_i) \tag{5.32}$$

and in $\eta_n(0)$ set

$$\int_0^L c_t' \phi(x, 0)\phi_n \, dx = \sum_i c_i \phi(a_i, 0)\phi_n(a_i) \tag{5.33}$$

Voigt-Kelvin Material The formulas of this chapter are applicable to a beam of Voigt-Kelvin material. In the equations of motion, Eqs. (5.12), replace E by $E(1 + \epsilon\, \partial/\partial t)$, where ϵ is the Voigt-Kelvin damping coefficient, and where similar behavior in dilatation and shear is assumed. Also, add a term $c_\rho \rho r_p^2\, \partial\phi/\partial t$ to the right-hand side of Eq. (5.12*d*) if the beam is externally damped. The response formulas of this section apply if c_ρ is replaced by $c_\rho + \epsilon\omega_n^2$ and ζ_n is redefined as $\zeta_n = c_\rho/(2\omega_n) + \epsilon\omega_n/2$. Continue to use the same undamped mode shapes employed above, where ω_n is the corresponding nth undamped frequency.

C. COMPUTER PROGRAMS AND EXAMPLES

5.15 Warping Stresses and Constants

Restrained warping stresses are difficult to compute for many cross-sectional shapes. Computers can be used to great advantage in these calculations. For a bar of arbitrary cross section, the computer program BEAMSTRESS can be employed for such calculations as well as for the evaluation of the warping constant. Chapter 14 contains analytical and computational examples of the calculation of restrained warping stresses and constants.

5.16 Benchmark Examples for Twisting Response

Complicated thin-walled beam response problems should be solved with the assistance of a computer program. The following examples are provided as benchmark examples against which a reader's own program can be checked. The computer program THINBEAM was used for these examples. For static and steady-state torsional loads, THINBEAM calculates the angle of twist, the rate of twist, bimoment, twisting moment, and warping torque. It also computes the critical axial force and mode shapes for torsional instability, and the natural frequencies and mode shapes for torsional vibrations. The beam can be formed of uniform segments with any loading, in-span supports, foundations, and boundary conditions.

EXAMPLE 5.7 Twisting of Beam with Offset Force Find the angles of twist and internal moments at midspan of the thin-walled beam of Fig. 5-2. The flanges are 10 in by 0.5 in, the web 12 in by 0.5 in. The cross-sectional area is 16 in^2, the modulus of elasticity is $3(10^7)$lb/in^2, Poisson's ratio is 0.3, the moment of inertia about horizontal and vertical axes through the centroid are 432 in^4 and 83.33 in^4, the torsional constant is 1.3333 in^4, the warping constant is 3000 in^6. The ends experience no rotation, but are free to warp; that is, the ends are simply supported.

The following results are computed: angle of twist = 0.04 rad, rate of twist = 0, bimoment = $1.32(10^6)$ lb · in^2, torque = warping torque = −20,000 lb · in.

EXAMPLE 5.8 Buckling of a Thin-Walled Beam Calculate the magnitude of the axial force that leads to instability of the thin-walled beam of Fig. 5-2 and Example 5.7.

For torsional instability the critical axial load is computed to be $1.167(10^6)$ lb. The critical load for bending is

$$P_{cr} = \frac{EI\pi^2}{200^2} = \frac{3(10^7)83.333\pi^2}{200^2} = 6.168(10^5)\text{ lb}$$

Thus, the beam will fail in the bending mode.

EXAMPLE 5.9 Twisting of a Wide-Flange Beam Compute the angles of twist and internal moments of the wide-flanged (W10 × 21) beam of Fig. 5-6. The modulus of elasticity is 200 GN/m^2, Poisson's ratio is 0.27, the torsional constant is 8.7409 cm^4, the warping constant is 0.06606 dm^6.

Some output is given in Fig. 5-7.

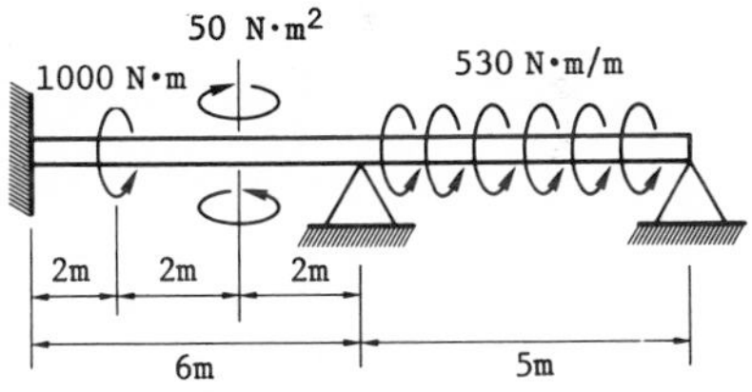

Fig. 5-6 Example 5.9.

AXIAL LOCATION	ANGLE OF TWIST	RATE OF TWIST	BIMOMENT	TORQUE	WARPING TORQUE
0.00	0.	0.	-.5615E+03	.6241E+03	.6241E+03
1.00	.1410E-01	-.2161E-01	-.3445E+02	.6241E+03	.7728E+03
2.00	.3092E-01	-.5676E-02	.4739E+03	-.3759E+03	-.3369E+03
3.00	.2325E-01	.1697E-01	.1507E+03	-.3759E+03	-.4927E+03
4.00	.3684E-02	.1915E-01	-.4062E+02	-.3759E+03	-.5077E+03
5.00	-.1070E-01	.6158E-02	-.3176E+03	-.3759E+03	-.4183E+03
6.00	.4974E-13	-.3318E-01	-.7672E+03	.1478E+04	.1707E+04
8.50	.1023E+00	-.8681E-02	.5686E+03	.1534E+03	.2132E+03
11.00	-.3437E-12	.7350E-01	.5588E-08	-.1172E+04	-.1677E+04

Fig. 5-7 Output for Example 5.9.

REFERENCES

5.1 Hopkins, R. B.: *Design Analysis of Shafts and Beams*, McGraw-Hill, New York, 1970.
5.2 Oden, J. T.: *Mechanics of Elastic Structures*, McGraw-Hill, New York, 1967.
5.3 Reissner, E., and Tsai, W. T.: "On the Determination of the Centers of Twist and Shear of Cylindrical Shell Beams," ASME Paper No. 72-APM-XX, 1972.
5.4 Timoshenko, S. P., and Gere, J. M.: *Mechanics of Materials*, Van Nostrand, New York, 1972.
5.5 Column Research Committee of Japan: *Handbook of Structural Stability*, Corona Publishing Company, Tokyo, 1971.
5.6 Galambos, T. V.: *Structural Members and Frames*, Prentice-Hall, Englewood Cliffs, N. J., 1968.

SIX

Rotating Shafts

In this chapter the critical speeds of a rotating shaft and the response of a shaft to unbalanced forces are treated. The transient response of the shaft to loadings on the shaft or through the bearing systems is also found. The formulas required for unbalanced response, critical speeds, and transient response are provided in Part A. Part B contains some examples of problems solved with a computer program.

The theory presented is for general shafts on anisotropic bearings. Sometimes this is reduced to the special case of isotropic bearings.

A. FORMULAS

6.1 Notation and Conventions

The notation and sign conventions differ from those used in the rest of this book. Rather, they more nearly conform to those normally employed in practice by engineers dealing with rotating shafts.

z, y	Stationary coordinates
v_z, v_y	Deflection in the z and y directions (length)
θ_y, θ_z	Slope components of the deflection curves about the y and z axes (radians)
M_y, M_z	Bending moment components about the y and z axes (force · length)

V_z, V_y	Shear force components in the z and y directions (force)
w_z, w_y	Applied loading intensities in the z and y directions (force/length)
$\bar{v}, \bar{v}_n$	Displacement and displacement mode shape of the pedestal mass in the bearing system
E	Modulus of elasticity of the material (force/length^2)
I	Area moment of inertia about the transverse neutral axis (length^4)
L	Length of shaft
$A_s = Ak_s$	Equivalent shear area, A = cross-sectional area, k_s = shear form factor; see Table 2-1 or use the computer program BEAMSTRESS for k_s.
G	Shear modulus of elasticity (force/length^2)
P	Axial force; plus for compression, minus for tension
k_z, k_y	Winkler (elastic) foundation moduli for the z and y directions (force/length^2)
k_z^*, k_y^*	Rotary foundation moduli (force · length/length)
$k_{zi}, k_{zyi}, k_{yi}, k_{yzi}, i = 3, 4$	Extension stiffness coefficients for bearing system (force/length) (Fig. 6-1)
$c_{zi}, c_{zyi}, c_{yi}, c_{yzi}, i = 3, 4$	Extension damping coefficients for bearing system (force · time/length) (Fig. 6-1)
$k_{zi}^*, k_{zyi}^*, k_{yi}^*, k_{yzi}^*, i = 3, 4$	Rotary stiffness coefficients for bearing system (force · length/radian)
$c_{zi}^*, c_{zyi}^*, c_{yi}^*, c_{yzi}^*, i = 3, 4$	Rotary damping coefficients for bearing system (force · time · length/radian)
$k_{z3}, c_{z3}, k_{y3}, c_{y3}$	Constants for bearing system (Fig. 6-1)
$k_{z4}, c_{z4}, k_{y4}, c_{y4}$	Constants for pedestal of bearing system (Fig. 6-1)
t	time
ℓ	Length of segment, span of transfer matrix
ω	Angular speed of rotation (radians/second)
Ω	Whirl frequency (radians/time)
s_n	Complex nth critical speed
ρ	Mass per unit length (mass/length), (force · time^2/length^2)
r	Radius of gyration of cross sectional area about y or z axis (Table 2-1) (length)
r_p	Polar radius of gyration (Table 2-1) (length)
r_{outer}	Outer radius of a hollow circular section (length)
r_{inner}	Inner radius of a hollow circular section (length)
$I_T = \rho r^2$	Rotary inertia, transverse or diametrical mass moment of inertia per unit length (mass · length). For a hollow circular section $I_T = \rho(r_{\text{outer}}^2 + r_{\text{inner}}^2)/4$.
$I_p = \rho r_p^2$	Polar mass moment of inertia per unit length (mass · length). For a hollow circular section $I_p = \rho(r_{\text{outer}}^2 + r_{\text{inner}}^2)/2$.
I_{Ti}	Rotary inertia, transverse or diametrical mass moment of inertia of concentrated mass at station i (mass · length^2). This can be calculated as $I_{Ti} = \Delta a\, \rho r^2$, where Δa is the length of shaft lumped at station i. For a hollow cylinder of length Δa and mass M_i, $I_{Ti} = M_i(r_{\text{outer}}^2 + r_{\text{inner}}^2)/4 +$

	$M_i(\Delta a)^2/12$. For a disk of mass M_i, $I_{Ti} = M_i(r_{\text{outer}}^2 + r_{\text{inner}}^2)/4$.
I_{pi}	Polar mass moment of inertia of concentrated mass at station i (mass · length^2). This can be calculated as $I_{pi} = \Delta a\ \rho r_p^2$ where Δa is the length of shaft lumped at station i. For a disk of mass M_i, $I_{pi} = M_i(r_{\text{outer}}^2 + r_{\text{inner}}^2)/2$.
M_i	Concentrated mass (mass)
e	Eccentricity arm for offset mass (length)
δ_z, δ_y	Static deflection in the z, y direction of the geometric center of the shaft from the line of bearing centers
Q_n	Quality factor for damping
$\zeta = c/c_c$	Damping ratio for single degree of freedom system. c_c is the critical damping.
λ_n	Damping exponent of the nth mode
Ω_n	Damped natural frequency of the nth mode
δ_s	Loss factor for consideration of structural damping
U	Unbalance moment (force-length)
g, $\dot{g}$	Prescribed displacement, velocity applied at base of bearing system (length, length/time)
$\mathbf{U}_i$	Field matrix of the ith segment
$\overline{\mathbf{U}}_i$	Point matrix at $x = a_i$

Positive deflections, slopes, moments, and shear forces are indicated in Fig. 6-1.

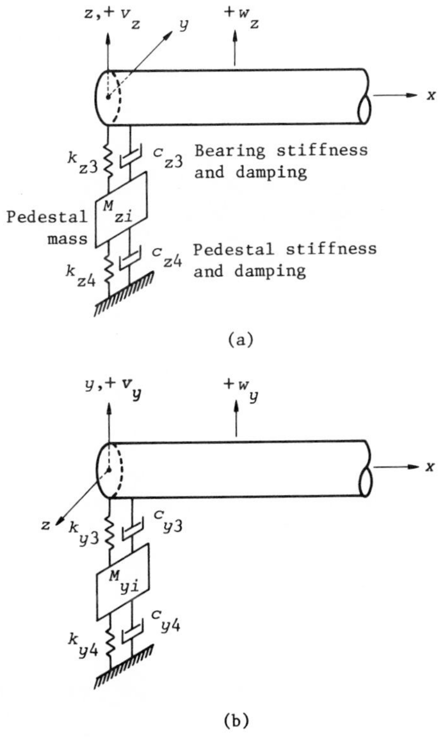

Fig. 6-1 (*a*) x, z plane. (*b*) x, y plane. (*c*) x-section view, x axis increases into the page. (*d*) x-section view if cross-coupling bearing coefficients are included. The x axis increases in the direction into the page. (*e*) Positive forces, moments, and slopes.

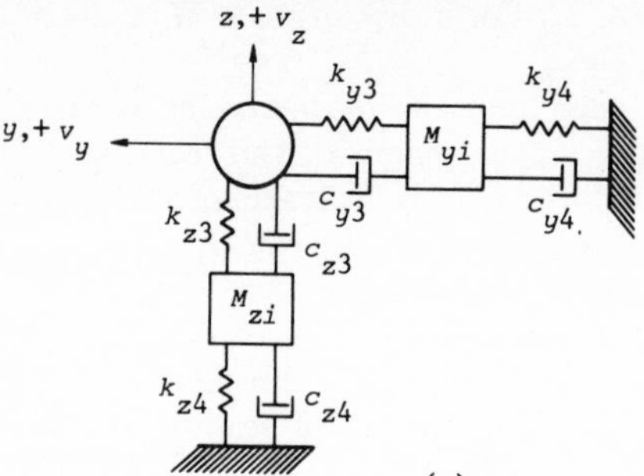

(c)

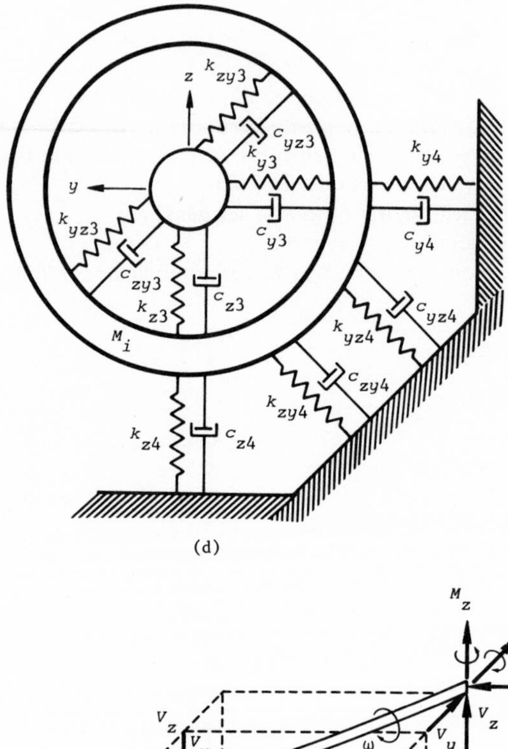

(d)

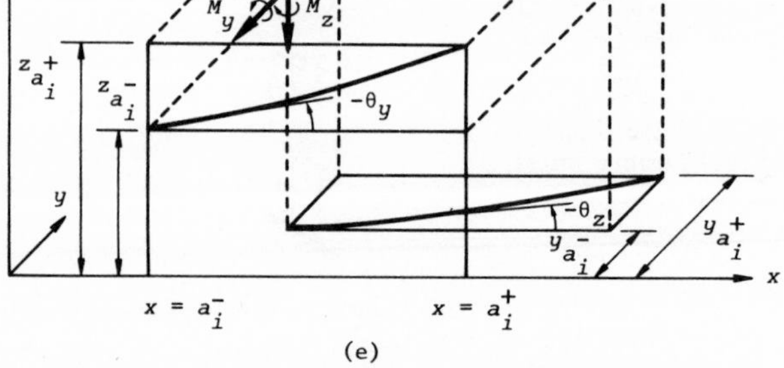

(e)

Fig. 6-1 (*Continued*)

6.2 Differential Equations

The fundamental equations of motion in first-order form for the shaft are similar to those for a beam:

z direction:

$$\frac{\partial v_z}{\partial x} = -\theta_y + \frac{V_z}{GA_s}$$

$$\frac{\partial \theta_y}{\partial x} = \frac{M_y}{EI}$$

$$\frac{\partial M_y}{\partial x} = V_z + (k_y^* - P)\theta_y + I_T \frac{\partial^2 \theta_y}{\partial t^2} + \omega I_p \frac{\partial \theta_z}{\partial t} \qquad (6.1a)$$

$$\frac{\partial V_z}{\partial x} = k_z v_z + \rho \frac{\partial^2 v_z}{\partial t^2} - w_z$$

y direction:

$$\frac{\partial v_y}{\partial x} = -\theta_z + \frac{V_y}{GA_s}$$

$$\frac{\partial \theta_z}{\partial x} = \frac{M_z}{EI}$$

$$\frac{\partial M_z}{\partial x} = V_y + (k_z^* - P)\theta_z + I_T \frac{\partial^2 \theta_z}{\partial t^2} - \omega I_p \frac{\partial \theta_y}{\partial t} \qquad (6.1b)$$

$$\frac{\partial V_y}{\partial x} = k_y v_y + \rho \frac{\partial^2 v_y}{\partial t^2} - w_y$$

These equations can be adjusted to account for a static bow in the shaft by replacing v_z with $v_z - \delta_z$ and v_y with $v_y - \delta_y$, in which the new v_z, v_y are total deflections and δ_z, δ_y are measures of the static bow of the geometric shaft center relative to the line of centers of the bearing system.

For the steady-state motion of a rotating shaft, the motion can be expressed as the real part of the response variables, that is, $v_z(x, t) = \text{Re}[v_z^c(x) + iv_z^s(x)]\, e^{i\Omega t}$, etc.:

$$\begin{aligned} v_z &= v_z^c \cos \Omega t - v_z^s \sin \Omega t \\ \theta_y &= \theta_y^c \cos \Omega t - \theta_y^s \sin \Omega t \\ M_y &= M_y^c \cos \Omega t - M_y^s \sin \Omega t \\ V_z &= V_z^c \cos \Omega t - V_z^s \sin \Omega t \end{aligned} \qquad (6.2a)$$

$$\begin{aligned} v_y &= v_y^c \cos \Omega t - v_y^s \sin \Omega t \\ \theta_z &= \theta_z^c \cos \Omega t - \theta_z^s \sin \Omega t \\ M_z &= M_z^c \cos \Omega t - M_z^s \sin \Omega t \\ V_y &= V_y^c \cos \Omega t - V_y^s \sin \Omega t \end{aligned} \qquad (6.2b)$$

where Ω is the whirl frequency. For synchronous precession $\omega = \Omega$, where ω is the spin frequency. For nonsynchronous precession $\omega \neq \Omega$. Substitution of Eq. (6.2) and similar expressions for applied loadings in Eq. (6.1) leads to two sets of similar but independent equations. The equations associated with the cosine and sine terms appear as:

Cosine terms

z direction:

$$\begin{aligned}\frac{dv_z^c}{dx} &= -\theta_y^c + \frac{V_z^c}{GA_s}\\ \frac{d\theta_y^c}{dx} &= \frac{M_y^c}{EI}\\ \frac{dM_y^c}{dx} &= V_z^c + (k_y^* - P)\theta_y^c - \Omega^2 I_T \theta_y^c - \omega\Omega I_p \theta_z^s\\ \frac{dV_z^c}{dx} &= k_z v_z^c - \Omega^2 \rho v_z^c - w_z^c\end{aligned} \tag{6.3a}$$

y direction:

$$\begin{aligned}\frac{dv_y^c}{dx} &= -\theta_z^c + \frac{V_y^c}{GA_s}\\ \frac{d\theta_z^c}{dx} &= \frac{M_z^c}{EI}\\ \frac{dM_z^c}{dx} &= V_y^c + (k_z^* - P)\theta_z^c - \Omega^2 I_T \theta_z^c + \omega\Omega I_p \theta_y^s\\ \frac{dV_y^c}{dx} &= k_y v_y^c - \Omega^2 \rho v_y^c - w_y^c\end{aligned} \tag{6.3b}$$

Sine terms

z direction:

$$\begin{aligned}\frac{dv_z^s}{dx} &= -\theta_y^s + \frac{V_z^s}{GA_s}\\ \frac{d\theta_y^s}{dx} &= \frac{M_y^s}{EI}\\ \frac{dM_y^s}{dx} &= V_z^s + (k_y^* - P)\theta_y^s - \Omega^2 I_T \theta_y^s + \omega\Omega I_p \theta_z^c\\ \frac{dV_z^s}{dx} &= k_z v_z^s - \Omega^2 \rho v_z^s - w_z^s\end{aligned} \tag{6.3c}$$

y direction:

$$\begin{aligned}\frac{dv_y^s}{dx} &= -\theta_z^s + \frac{V_y^s}{GA_s}\\ \frac{d\theta_z^s}{dx} &= \frac{M_z^s}{EI}\\ \frac{dM_z^s}{dx} &= v_y^s + (k_z^* - P)\theta_z^s - \Omega^2 I_T \theta_z^s - \omega\Omega I_p \theta_y^c\\ \frac{dV_y^s}{dx} &= k_y v_y^s - \Omega^2 \rho v_y^s - w_y^s\end{aligned} \tag{6.3d}$$

Complex Notation With the complex quantities

$$\begin{aligned} v_z &= v_z^c + iv_z^s \\ \theta_z &= \theta_z^c + i\theta_z^s \\ M_z &= M_z^c + iM_z^s \\ V_z &= V_z^c + iV_z^s \\ w_z &= w_z^c + iw_z^s \end{aligned} \tag{6.4a}$$

$$\begin{aligned} v_y &= v_y^c + iv_y^s \\ \theta_y &= \theta_y^c + i\theta_y^s \\ M_y &= M_y^c + iM_y^s \\ V_y &= V_y^c + iV_y^s \\ w_y &= w_y^c + iw_y^s \end{aligned} \tag{6.4b}$$

Eqs. (6.3) can be combined into two, rather than four, sets of first order equations. This is accomplished by multiplying Eqs. (6.3*c*) and (6.3*d*) by i and adding to Eqs. (6.3*a*) and (6.3*b*), respectively. The resulting equations appear the same as Eqs. (6.3*a*) and Eqs. (6.3*b*) if the c and s superscripts are removed and the quantity $\omega\Omega I_p$ is replaced by $-i\omega\Omega I_p$. Much of this chapter will be based on Eqs. (6.3) rather than on the combined complex equations of motion.

Uncoupled Equations Of particular practical interest are problems for which the equations of motion can be uncoupled. As indicated subsequently this occurs if the cross-coupling coefficients in the bearings are negligible. If, in addition, the bearings are the same in the y and z directions, that is, the bearings are *isotropic*, the number of equations of motion can be reduced.

Define

$$v = v_z + iv_y \qquad \theta = \theta_y + i\theta_z \qquad M = M_y + iM_z$$
$$V = V_z + iV_y \qquad w = w_z + iw_y \tag{6.5}$$

The equations of motion (6.1) then become

$$\frac{\partial v}{\partial x} = -\theta + \frac{V}{GA_s} \tag{6.6a}$$

$$\frac{\partial \theta}{\partial x} = \frac{M}{EI} \tag{6.6b}$$

$$\frac{\partial M}{\partial x} = V + (k^* - P)\theta + I_T\frac{\partial^2\theta}{\partial t^2} - i\omega I_p\frac{\partial \theta}{\partial t} \tag{6.6c}$$

$$\frac{\partial V}{\partial x} = kv + \rho\frac{\partial^2 v}{\partial t^2} - w \tag{6.6d}$$

6.3 Transfer Matrix for Uniform Segments

The transfer matrix for a shaft segment of length ℓ is given in Table 6-1. The same matrix, with static deformation taken into account, is provided in Table

6-2. The state vector is composed of the four vectors

$$\mathbf{s}_z^j = \begin{bmatrix} v_z^j \\ \theta_y^j \\ M_y^j \\ V_z^j \end{bmatrix} \qquad \mathbf{s}_y^j = \begin{bmatrix} v_y^j \\ \theta_z^j \\ M_z^j \\ V_y^j \end{bmatrix} j = c,\ s \tag{6.7}$$

The complete state vector $\mathbf{s}$ is

$$\mathbf{s} = \begin{bmatrix} \mathbf{s}_z^s \\ \mathbf{s}_z^c \\ \mathbf{s}_y^s \\ \mathbf{s}_y^c \\ 1 \end{bmatrix} \tag{6.8}$$

TABLE 6-1 Transfer Matrix for Uniform Shaft Section

$$\mathbf{s} = \begin{bmatrix} \mathbf{s}_z^s \\ \mathbf{s}_z^c \\ \mathbf{s}_y^s \\ \mathbf{s}_y^c \\ 1 \end{bmatrix} \qquad \mathbf{U}_i = \begin{bmatrix} \mathbf{T} & & & & \\ & \mathbf{T} & & & \\ & & \mathbf{T} & & \\ & & & \mathbf{T} & \\ & & & & 1 \end{bmatrix}$$

All blank spaces indicate zeros

The 4 x 4 matrix $\mathbf{T}$ is taken from the general beam solution of Table 2-13. In this table

$$\mathbf{T} = \begin{bmatrix} U_{vv} & U_{v\theta} & U_{vM} & U_{vV} \\ U_{\theta v} & U_{\theta\theta} & U_{\theta M} & U_{\theta V} \\ U_{Mv} & U_{M\theta} & U_{MM} & U_{MV} \\ U_{Vv} & U_{V\theta} & U_{VM} & U_{VV} \end{bmatrix}$$

Definitions: See Eqs. (6.7) for definitions of $\mathbf{s}_z^j$, $\mathbf{s}_y^j$, $j=c,s$.

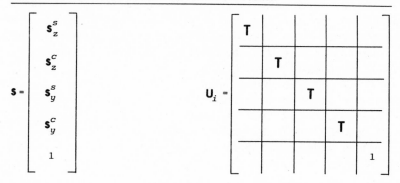

TABLE 6-2 Transfer Matrix for a Uniform Shaft Section with Static Bow

Definitions:

(a) **S** remains the same as for Table 6-1.

(b) $D = (\delta_y)_i - (\delta_y)_{i-1}$

$E = (\delta_z)_i - (\delta_z)_{i-1}$

δ_i is the static bow at the right end of the segment.

δ_{i-1} is the static bow at the left end of the segment.

$$
\mathbf{U}_i = \left[\begin{array}{c|c|c|c|c}
\mathbf{T} & & & & \begin{array}{c} -D \\ - \\ - \\ - \end{array} \\ \hline
 & \mathbf{T} & & & \begin{array}{c} E \\ - \\ - \\ - \end{array} \\ \hline
 & & \mathbf{T} & & \begin{array}{c} -E \\ - \\ - \\ - \end{array} \\ \hline
 & & & \mathbf{T} & \begin{array}{c} D \\ - \\ - \\ - \end{array} \\ \hline
 & & & & 1
\end{array}\right]
$$

All blank spaces indicate zeros.

TABLE 6-3 Point Matrix at $x = a_i$ for Anisotropic Bearing Systems

$$
\mathbf{s} = \begin{bmatrix} V_z^s \\ \Theta_y^s \\ M_y^s \\ V_z^s \\ \hline V_z^c \\ \Theta_y^c \\ M_y^c \\ V_z^c \\ \hline V_y^s \\ \Theta_z^s \\ M_z^s \\ V_y^s \\ \hline V_y^c \\ \Theta_z^c \\ M_z^c \\ V_y^c \\ \hline 1 \end{bmatrix}
\qquad
\bar{\mathbf{U}}_i = \left[\begin{array}{cccc|cccc|cccc|cccc|c}
1 & & & & & & & & & & & & & & & & \\
 & 1 & & & & & & & & & & & & & & & \\
 & k^*_{z1} & 1 & & & -\Omega c^*_{z1} & & & & k^*_{zy1} & & & & -\Omega c^*_{zy1} & & & \\
k_{z1} & & & 1 & -\Omega c_{z1} & & & & k_{zy1} & & & & -\Omega c_{zy1} & & & & \\
\hline
 & & & & 1 & & & & & & & & & & & & \\
 & & & & & 1 & & & & & & & & & & & \\
 & -\Omega c^*_{z1} & & & & k^*_{z1} & 1 & & & \Omega c^*_{zy1} & & & & k^*_{zy1} & & & \\
\Omega c_{z1} & & & & k_{z1} & & & 1 & \Omega c_{zy1} & & & & k_{zy1} & & & & \\
\hline
 & & & & & & & & 1 & & & & & & & & \\
 & & & & & & & & & 1 & & & & & & & \\
 & k^*_{yz1} & & & & -\Omega c^*_{yz1} & & & & k^*_{y1} & 1 & & & -\Omega c^*_{y1} & & & \\
k_{yz1} & & & & -\Omega c_{yz1} & & & & k_{y1} & & & 1 & -\Omega c_{y1} & & & & \\
\hline
 & & & & & & & & & & & & 1 & & & & \\
 & & & & & & & & & & & & & 1 & & & \\
 & \Omega c^*_{yz1} & & & & k^*_{yz1} & & & & \Omega c^*_{y1} & & & & k^*_{y1} & 1 & & \\
\Omega c_{yz1} & & & & k_{yz1} & & & & \Omega c_{y1} & & & & k_{y1} & & & 1 & \\
\hline
 & & & & & & & & & & & & & & & & 1
\end{array}\right]
$$

$$d_1 = k_{z3}(k_{z4} - M_i\Omega^2)(k_{z3} + k_{z4} - M_i\Omega^2) + \Omega^2(k_{z3}c_{z4}^2 + k_{z4}c_{z3}^2) - \Omega^2 c_{z3}^2 M_i$$

$$d_2 = (k_{z3} + k_{z4} - M_i\Omega^2)^2 + \Omega^2(c_{z4} + c_{z3})^2$$

$$d_3 = c_{z3}(k_{z4} - M_i\Omega^2)^2 + \Omega^2(c_{z3}^2 c_{z4} + c_{z3}c_{z4}^2) + \Omega c_{z4}k_{z3}^2$$

$$d_4 = k_{z3}(k_{z4}e_4 - k_{zy4}e_3) + k_{yz3}(k_{zy4}e_1 - k_{z4}e_2)$$

$$d_5 = k_{yz3}(k_{z4}e_4 - k_{zy4}e_3) + k_{y3}(k_{zy4}e_1 - k_{z4}e_2)$$

$$d_6 = k_{z3}(k_{yz4}e_4 - k_{y4}e_3) + k_{yz3}(k_{y4}e_1 - k_{yz4}e_2)$$

$$d_7 = k_{zy3}(k_{yz4}e_4 - k_{y4}e_3) + k_{y3}(k_{y4}e_1 - k_{yz4}e_2)$$

$$d = e_1e_4 - e_3e_2$$

$$e_1 = k_{z3} + k_{z4} - M_i\Omega^2$$

$$e_2 = k_{zy3} + k_{zy4}$$

$$e_3 = k_{yz3} + k_{yz4}$$

$$e_4 = k_{y3} + k_{y4} - M_i\Omega^2$$

The constants k_{z1}, k_{y1}, c_{z1}, c_{y1}, k_{zy1}, k_{yz1}, c_{zy1}, c_{yz1} for the more complicated bearing system of Fig. 6-1d are computed as described in Section 6.4.

The values of rotary spring constants k^*_{z1}, k^*_{y1}, c^*_{z1}, c^*_{y1}, k^*_{zy1}, k^*_{yz1}, c^*_{zy1}, c^*_{yz1} for a rotary bearing system are obtained from this table (and Section 6.4 in the case of Fig. 6-1d) by replacing the extension constants k_{z1}, c_{z1}, k_{z3}, k_{z4}, c_{z3}, c_{z4}, etc. by the rotary constants k^*_{z1}, c^*_{z1}, k^*_{z3}, k^*_{z4}, c^*_{z3}, c^*_{z4}, etc.

The spring constants for the bearing system in the y direction that are not given here are found from this table by changing the subscripts z to y.

TABLE 6-3 Point Matrix at $x = a_i$ for Anisotropic Bearing Systems (continued)

→ Shaft Center	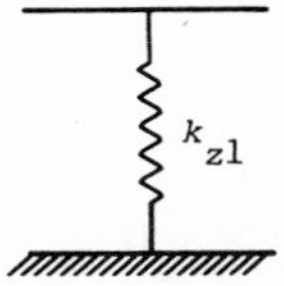	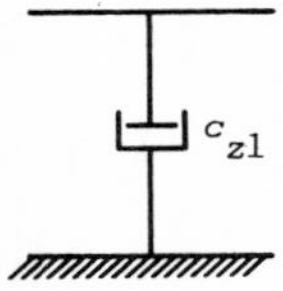	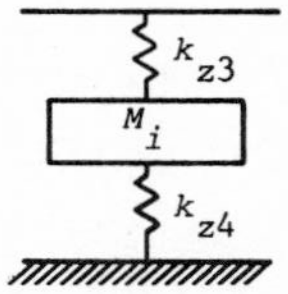	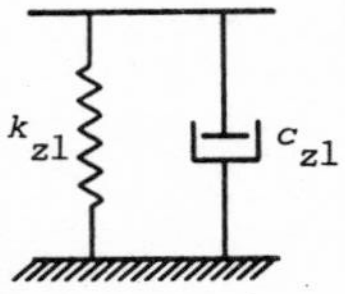	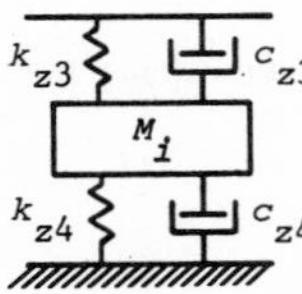	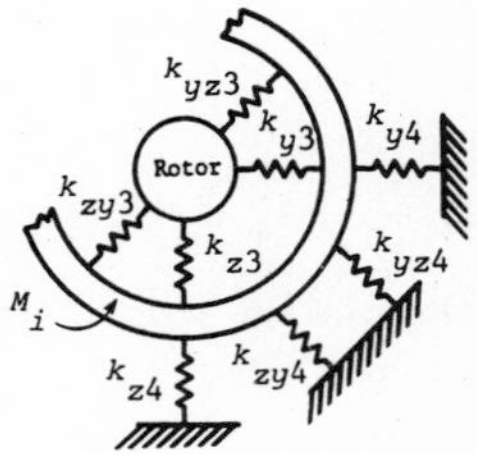
k_{z1}	k_{z1}	0	d_0	k_{z1}	d_1/d_2	d_4/d
c_{z1}	0	c_{z1}	0	c_{z1}	$d_3/(d_2\Omega)$	0
k_{zy1}	0	0	0	0	0	d_5/d
k_{yz1}	0	0	0	0	0	d_6/d
k_{y1}	0	0	0	0	0	d_7/d

TABLE 6-4 Point Matrix at $x = a_i$ for Concentrated Mass with Eccentricity e

$$\mathbf{s} = \begin{bmatrix} v_z^s & \theta_y^s & M_y^s & V_z^s & \bar{c} & v_z^c & \theta_y^c & M_y^c & V_z^c & \bar{s} & v_y^s & \theta_z^s & M_z^s & V_y^s & \bar{c} & v_y^c & \theta_z^c & M_z^c & V_y^c & \bar{y} & 1 \end{bmatrix}^T$$

$\bar{\mathbf{U}}_i =$

1																
	1															
	A_1	1											A_2			
D			1													B
				1												
					1											
					A_1	1			$-A_2$							
				D			1									C
								1								
									1							
					$-A_2$				A_1	1						
								D			1					$-C$
												1				
													1			
	A_2												A_1	1		
												D			1	$-B$
																1

All blank spaces indicate zeros.

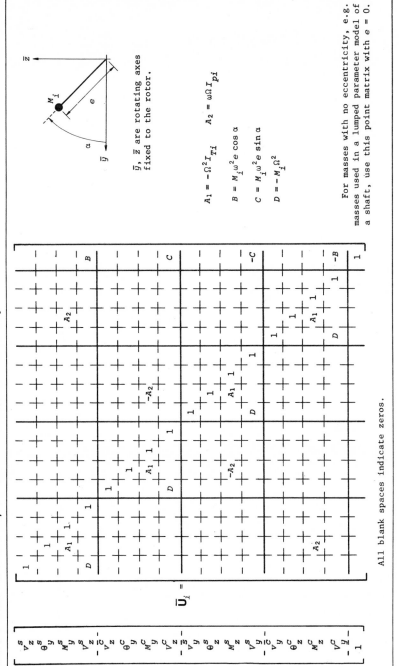

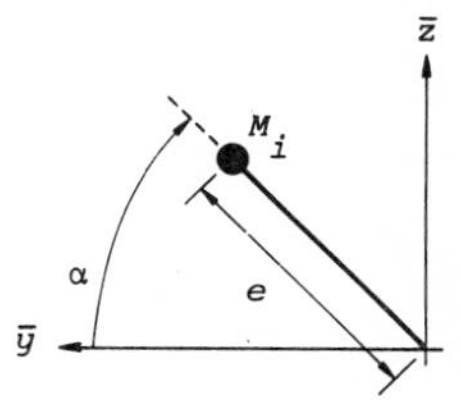

$\bar{y}$, $\bar{z}$ are rotating axes fixed to the rotor.

$A_1 = -\Omega^2 I_{Ti}$ $\quad A_2 = \omega\Omega I_{pi}$

$B = M_i \omega^2 e \cos\alpha$

$C = M_i \omega^2 e \sin\alpha$

$D = -M_i \Omega^2$

For masses with no eccentricity, e.g. masses used in a lumped parameter model of a shaft, use this point matrix with $e = 0$.

TABLE 6-5 Point Matrix at $x = a_i$ for Isotropic Hinges

Definitions:

(a) **S** remains the same as for Tables 6-1, 6-2, 6-3, 6-4.

(b) Linear Hinge (shear release with spring and dashpot)

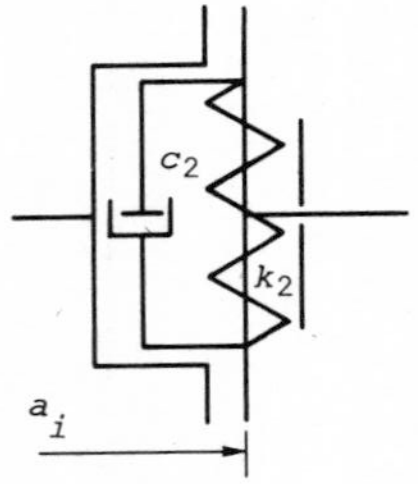

(c) Rotary Hinge (moment release with spring and dashpot)

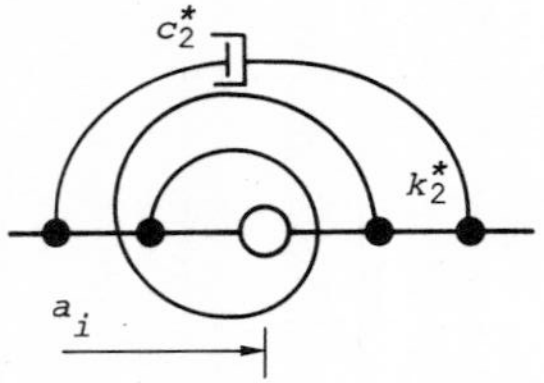

$$\bar{\mathbf{U}}_i = \begin{bmatrix} z_1 & z_2 & & & \\ z_2 & z_1 & & & \\ & & z_1 & z_2 & \\ & & z_2 & z_1 & \\ & & & & 1 \end{bmatrix}$$

$$z_1 = \begin{bmatrix} 1 & 0 & 0 & k_2/(k_2^2 + c_2^2\Omega^2) \\ 0 & 1 & k_2^*/(k_2^{*2} + c_2^{*2}\Omega^2) & 0 \\ 0 & 0 & 1 & 0 \\ 0 & 0 & 0 & 1 \end{bmatrix}$$

$$z_2 = \begin{bmatrix} 0 & 0 & 0 & -c_2\Omega/(k_2^2 + c_2^2\Omega^2) \\ 0 & 0 & -c_2^*\Omega/(k_2^{*2} + c_2^{*2}\Omega^2) & 0 \\ 0 & 0 & 0 & 0 \\ 0 & 0 & 0 & 0 \end{bmatrix}$$

All blank spaces indicate zeros.

TABLE 6-6 Point Matrix at $x = a_i$ for Isotropic Bearing System

Definitions:

(a) M_i is the mass of the bearing pedestal (Fig. 6-1).

(b) See Eqs. (6.5) for the definitions of v, θ, M, V.

(c) $s = \lambda_n + i\Omega_n$ for a stability analysis or a damped free vibration.

$s = i\Omega$ for an unbalanced response.

$s = i\Omega$ for an undamped critical speed.

$$\mathbf{S} = \begin{bmatrix} v \\ \theta \\ M \\ V \\ 1 \end{bmatrix} \qquad \bar{\mathbf{U}}_i = \begin{bmatrix} 1 & 0 & 0 & 0 & 0 \\ 0 & 1 & 0 & 0 & 0 \\ 0 & 0 & 1 & 0 & 0 \\ -z_i & 0 & 0 & 1 & 0 \\ 0 & 0 & 0 & 0 & 1 \end{bmatrix}$$

$$z_i = \frac{(k_3 + sc_3)\ (k_4 + sc_4 + s^2 M_i)}{k_3 + sc_3 + k_4 + sc_4 + s^2 M_i}$$

6.4 Point Matrices

The transfer matrices that take into account concentrated (point) occurrences are listed in Tables 6-3 through 6-6. In the case of Table 6-3 for the bearing systems, the reaction in the z direction between the bearing system at $x = a_i$ and the shaft is of the form

$$k_{z1}v_z\big|_{x=a_i} + k_{zy1}v_y\big|_{x=a_i} + c_{z1}\frac{\partial v_z}{\partial t}\bigg|_{x=a_i} + c_{zy1}\frac{\partial v_y}{\partial t}\bigg|_{x=a_i} \tag{6.9}$$

for extension bearing systems, and of the form

$$k_{z1}^*\theta_y\big|_{x=a_i} + k_{zy1}^*\theta_z\big|_{x=a_i} + c_{z1}^*\frac{\partial \theta_y}{\partial t}\bigg|_{x=a_i} + c_{zy1}^*\frac{\partial \theta_z}{\partial t}\bigg|_{x=a_i} \tag{6.10}$$

for rotary bearing systems. Similar expressions apply for the bearing system in the y direction. Formulas for most of the spring and damping constants, for example, k_{z1}, c_{z1}, k_{z1}^*, c_{z1}^*, are provided in Table 6-3.

The cross-coupling coefficients, for example, k_{zy1}, are sometimes determined empirically. The constant k_{zy1} is the force in the z direction resulting from a unit displacement of the bearing system in the y direction. The constants for the general bearing system of Fig. 6-1d can be determined from

$$\mathbf{K} = \mathbf{A}_1(\mathbf{A}_2 + \mathbf{A}_1)^{-1}\mathbf{A}_2 \tag{6.11}$$

where

$$\mathbf{K} = \begin{bmatrix} k_{z1} & -\Omega c_{z1} & k_{zy1} & -\Omega c_{zy1} \\ \Omega c_{z1} & k_{z1} & \Omega c_{zy1} & k_{zy1} \\ k_{yz1} & -\Omega c_{yz1} & k_{y1} & -\Omega c_{y1} \\ \Omega c_{yz1} & k_{yz1} & \Omega c_{y1} & k_{y1} \end{bmatrix}$$

$$\mathbf{A}_1 = \begin{bmatrix} k_{z4} - M_i\Omega^2 & -\Omega c_{z4} & k_{zy4} & -\Omega c_{zy4} \\ \Omega c_{z4} & k_{z4} - M_i\Omega^2 & \Omega c_{zy4} & k_{zy4} \\ k_{zy4} & -\Omega c_{yz4} & k_{y4} - M_i\Omega^2 & -\Omega c_{y4} \\ \Omega c_{yz4} & k_{yz4} & \Omega c_{y4} & k_{y4} - M_i\Omega^2 \end{bmatrix}$$

$$\mathbf{A}_2 = \begin{bmatrix} k_{z3} & -\Omega c_{z3} & k_{zy3} & -\Omega c_{zy3} \\ \Omega c_{z3} & k_{z3} & \Omega c_{zy3} & k_{zy3} \\ k_{yz3} & -\Omega c_{yz3} & k_{y3} & -\Omega c_{y3} \\ \Omega c_{yz3} & k_{yz3} & \Omega c_{y3} & k_{y3} \end{bmatrix}$$

6.5 Unbalanced Response

The procedure for calculating the unbalanced response of a rotating shaft is essentially the same technique described in Section 2.16 for the static response of beams.

First the shaft must be modeled in terms of continuous, lumped segments, bearing systems, and unbalance forces. The unbalance forces in Table 6-4 are introduced in terms of an offset mass. As shown in Fig. 6-2 the lumped mass M_i is identified as being offset by the arm length e at phase angle α. The usual masses along the shaft axis of a lumped-mass model can also be accounted for with Table 6-4 with the eccentricity (e) set equal to zero.

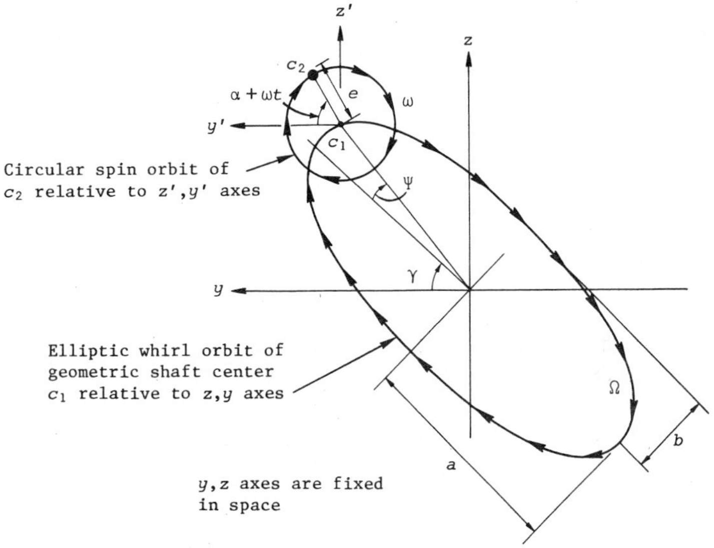

Fig. 6-2 Spinning and whirl of the center of mass c_2 of an unbalanced lumped mass.

In many instances the unbalance of a rotating shaft may be designated in terms other than the e and α used in Table 6-4. If the moment unbalance U and phase angle α are given, then set $e = U/(M_i g)$ in Table 6-4, where g is the gravitational constant.

Sometimes the rotor unbalance is defined by its components U_z, U_y. The unbalance forces $w_z = \omega^2 U_z$, $w_y = \omega^2 U_y$ can be written from

$$U_z = U \sin(\alpha + \omega t) = U(\sin\alpha \cos\omega t + \cos\alpha \sin\omega t) = U_z^c \cos\omega t + U_z^s \sin\omega t$$
$$U_y = U \cos(\alpha + \omega t) = U(\cos\alpha \cos\omega t - \sin\alpha \sin\omega t) = U_y^c \cos\omega t - U_y^s \sin\omega t \tag{6.12}$$

where

$$U_z^c = U \sin\alpha \qquad U_z^s = U\cos\alpha$$
$$U_y^c = U\cos\alpha \qquad U_y^s = U\sin\alpha$$

Then

$$\alpha = \tan^{-1}\left(\frac{U_z^c}{U_z^s}\right) = \tan^{-1}\left(\frac{U_y^s}{U_y^c}\right)$$
$$e = \frac{1}{gM_i}\sqrt{(U_z^c)^2 + (U_z^s)^2} = \frac{1}{gM_i}\sqrt{(U_y^c)^2 + (U_y^s)^2} \tag{6.13}$$

Similarly, the unbalance excitation may be caused by the static bow of the shaft described by the static displacements, δ_y and δ_z. Then $\delta_z^c = -\delta_y^s = \delta_z$, $-\delta_z^s = \delta_y^c = \delta_y$.

The state vector for the general rotor

$$\mathbf{s} = \begin{bmatrix} \mathbf{s}_z^s \\ \mathbf{s}_z^c \\ \mathbf{s}_y^s \\ \mathbf{s}_y^c \\ 1 \end{bmatrix} \quad \text{where} \quad \mathbf{s}_z^j = \begin{bmatrix} v_z^j \\ \theta_y^j \\ M_y^j \\ V_z^j \end{bmatrix} \quad \mathbf{s}_y^j = \begin{bmatrix} v_y^j \\ \theta_z^j \\ M_z^j \\ V_y^j \end{bmatrix} \quad j = c, s \tag{6.14}$$

and for the isotropic case

$$\mathbf{s} = \begin{bmatrix} v \\ \theta \\ M \\ V \\ 1 \end{bmatrix} \tag{6.15}$$

at any point along the shaft is found by progressive multiplication of the transfer matrices for all occurrences from left to right up to that point. That is, the state variables at any point j are given by

$$\mathbf{s}_j = \mathbf{U}_j \mathbf{U}_{j-1} \cdots \overline{\mathbf{U}}_k \cdots \mathbf{U}_2 \mathbf{U}_1 \mathbf{s}_0 \tag{6.16}$$

where $\mathbf{s}_0 = \mathbf{s}_{x=0}$. If the right end of the shaft occurs at station n, the state variables at the right end of the shaft become

$$\mathbf{s}_{x=L} = \mathbf{s}_n = \mathbf{U}_n \mathbf{U}_{n-1} \cdots \overline{\mathbf{U}}_k \cdots \mathbf{U}_2 \mathbf{U}_1 \mathbf{s}_0 = \mathbf{U}\mathbf{s}_0 \tag{6.17}$$

The matrix $\mathbf{U}$ is the overall or global transfer matrix. The initial parameters composing $\mathbf{s}_0$ are found from the boundary conditions that occur at each end of the shaft. The boundary conditions for the z direction (y direction) apply to both $\mathbf{s}_z^c(\mathbf{s}_y^c)$ and $\mathbf{s}_z^s(\mathbf{s}_y^s)$. This gives eight conditions at each of the two ends for the 16 unknowns.

Most rotating shafts are treated as having free ends. Then the 16 conditions are

$$\begin{aligned} & M_y^j|_{x=0} = V_z^j|_{x=0} = M_z^j|_{x=0} = V_y^j|_{x=0} = 0 \\ \text{and} \quad & M_y^j|_{x=L} = V_z^j|_{x=L} = M_z^j|_{x=L} = V_y^j|_{x=L} = 0 \end{aligned} \tag{6.18}$$

for $j = s, c$.

After the state variables are calculated at points of interest from Eq. (6.16), the deflection, slope, bending moment, and shear force components can be found at various times with Eqs. (6.2). In general, the motion in the y, z plane will describe an elliptic orbit (Fig. 6-2) and the state variables are functions of time. The displacement of the rotor's elastic center in the radial direction will be

$$\sqrt{v_z^2 + v_y^2} = \sqrt{(v_z^c \cos \Omega t - v_z^s \sin \Omega t)^2 + (v_y^c \cos \Omega t - v_y^s \sin \Omega t)^2} \tag{6.19}$$

If the major and minor semiaxes are denoted by a and b, then

$$a = \left\{ \tfrac{1}{2}\left[(v_y^c)^2 + (v_y^s)^2 + (v_z^c)^2 + (v_z^s)^2\right] + \sqrt{\tfrac{1}{4}\left[(v_y^c)^2 + (v_y^s)^2 - (v_z^c)^2 - (v_z^s)^2\right]^2 + \left[(v_y^c)(v_z^c) + (v_z^s)(v_y^s)\right]^2} \right\}^{\frac{1}{2}} \tag{6.20}$$

$$b = \frac{(v_y^s)(v_z^c) - (v_y^c)(v_z^s)}{a}$$

These correspond to the maximum and minimum displacement amplitudes, respectively. The angle between the major semiaxis and the y axis is

$$\gamma = \tfrac{1}{2} tan^{-1} \left\{ \frac{2\left[(v_y^c)(v_z^c) + (v_y^s)(v_z^s)\right]}{(v_y^c)^2 + (v_y^s)^2 - (v_z^c)^2 - (v_z^s)^2} \right\} \tag{6.21}$$

Equation (6.21) gives the "time angle" at which the maximum displacement occurs.The phase angle is given as

$$\Psi = \tfrac{1}{2} tan^{-1} \left\{ \frac{2\left[(v_y^c)(v_z^c) + (v_z^s)(v_y^s)\right]}{(v_y^c)^2 - (v_y^s)^2 + (v_z^c)^2 - (v_z^s)^2} \right\} \tag{6.22}$$

The bearing reactions can be found in several ways. They can be calculated directly from the forces and moments crossing a bearing. For example, the reaction force component in the z direction at a bearing at $x = a_i$ is given as

$$V_z\big|_{x=a_i^+} - V_z\big|_{x=a_i^-}$$

where a_i^+ refers to a point just to the right of $x = a_i$ and a_i^- indicates a point just to the left of $x = a_i$. In a similar fashion the other bearing reaction forces and moments are given by

$$V_y\big|_{x=a_i^+} - V_y\big|_{x=a_i^-} \qquad M_y\big|_{x=a_i^+} - M_y\big|_{x=a_i^-} \qquad M_z\big|_{x=a_i^+} - M_z\big|_{x=a_i^-}$$

In terms of shaft displacements at the bearings, the reactions can be calculated as

$$\begin{bmatrix} \text{Force in } z \text{ direction} \\ \text{Force in } y \text{ direction} \end{bmatrix} = \begin{bmatrix} k_{z1} & k_{zy1} \\ k_{yz1} & k_{y1} \end{bmatrix} \begin{bmatrix} v_z^c \\ v_y^c \end{bmatrix} \cos\Omega t - \begin{bmatrix} k_{z1} & k_{zy1} \\ k_{yz1} & k_{y1} \end{bmatrix} \begin{bmatrix} v_z^s \\ v_y^s \end{bmatrix} \sin\Omega t$$
$$- \Omega \begin{bmatrix} c_{z1} & c_{zy1} \\ c_{yz1} & c_{y1} \end{bmatrix} \begin{bmatrix} v_z^s \\ v_y^s \end{bmatrix} \cos\Omega t - \Omega \begin{bmatrix} c_{z1} & c_{zy1} \\ c_{yz1} & c_{y1} \end{bmatrix} \begin{bmatrix} v_z^c \\ v_y^c \end{bmatrix} \sin\Omega t \tag{6.23}$$

$$\begin{bmatrix} \text{Moment about } y \text{ axis} \\ \text{Moment about } z \text{ axis} \end{bmatrix} = \begin{bmatrix} k_{z1}^* & k_{zy1}^* \\ k_{yz1}^* & k_{y1}^* \end{bmatrix} \begin{bmatrix} \theta_z^c \\ \theta_y^c \end{bmatrix} \cos\Omega t - \begin{bmatrix} k_{z1}^* & k_{zy1}^* \\ k_{yz1}^* & k_{y1}^* \end{bmatrix} \begin{bmatrix} \theta_z^s \\ \theta_y^s \end{bmatrix} \sin\Omega t$$
$$- \Omega \begin{bmatrix} c_{z1}^* & c_{zy1}^* \\ c_{yz1}^* & c_{y1}^* \end{bmatrix} \begin{bmatrix} \theta_z^s \\ \theta_y^s \end{bmatrix} \cos\Omega t - \Omega \begin{bmatrix} c_{z1}^* & c_{zy1}^* \\ c_{yz1}^* & c_{y1}^* \end{bmatrix} \begin{bmatrix} \theta_z^c \\ \theta_y^c \end{bmatrix} \sin\Omega t \tag{6.24}$$

where the constants can be taken from Table 6-3.

Uncoupled Problems In many instances the cross-coupling bearing coefficients k_{zy1}, k_{yz1}, c_{zy1}, c_{yz1}, k^*_{zy1}, k^*_{yz1}, c^*_{zy1}, c^*_{yz1} can be treated as being negligible. This permits the problem to be uncoupled in that the z and y directions can be treated separately. Thus, eight variables v^j_z, θ^j_y, M^j_y, V^j_z, $j = s, c$ for this "orthotropic" situation can be found from 9×9 transfer matrices rather than the more general 17×17 matrices of Tables 6-1 to 6-5. In finding the variables use the bearing coefficients k_{z1}, k_{z3}, k_{z4}, c_{z1}, and so forth that correspond to the motion in the x, z plane. The same reasoning applies to the calculation of the eight variables v^j_y, θ^j_z, M^j_z, V^j_y, $j = c, s$ in the y direction. In this case, the bearing coefficients k_{y1}, k_{y3}, k_{y4}, c_{y1}, and so forth are employed. The reduction of the transfer matrices of Tables 6-1 through 6-5 to 9×9 is straightforward.

Since the bearings are different in the y and z directions, the motion in the y, z plane will be an elliptic orbit as shown in Fig. 6-2. The displacements, angles, and bearing reactions are found from Eqs. (6.19) through (6.24).

Isotropic Bearings If in addition to the cross-coupling bearing coefficients being regarded as negligible, the bearings are taken as being the same in the y and z directions, then it is necessary only to solve a problem characterized by five state variables. These are isotropic bearing systems for which the point matrix of Table 6-6 with $s = i\Omega$ applies. The state variables shown in Table 6-6 —that is v, θ, M, V—are defined by Eqs. (6.5).

For the case of isotropic bearings, the loadings in the z and y directions are the same magnitude but differ in phase angle by 90°. Thus if the response in the z direction is found—that is, if

$$\begin{bmatrix} v_z \\ \theta_y \\ M_y \\ V_z \end{bmatrix} = -\begin{bmatrix} v^s_z \\ \theta^s_y \\ M^s_y \\ V^s_z \end{bmatrix} \sin \Omega t + \begin{bmatrix} v^c_z \\ \theta^c_y \\ M^c_y \\ V^c_z \end{bmatrix} \cos \Omega t \tag{6.25}$$

is known—then the response in the y direction can be calculated from

$$\begin{bmatrix} v_y \\ \theta_z \\ M_z \\ V_y \end{bmatrix} = -\begin{bmatrix} v^s_z \\ \theta^s_y \\ M^s_y \\ V^s_z \end{bmatrix} \sin (\Omega t + \pi/2) + \begin{bmatrix} v^c_z \\ \theta^c_y \\ M^c_y \\ V^c_z \end{bmatrix} \cos (\Omega t + \pi/2) \tag{6.26}$$

Since the bearings are isotropic, the orbit of the shaft is a circle. The radius of the circle, that is, the peak magnitude of the displacement, is

$$\begin{aligned} &\sqrt{(v^c_z \cos \Omega t - v^s_z \sin \Omega t)^2 + [v^c_z \cos (\pi/2 + \Omega t) - v^s_z \sin (\pi/2 + \Omega t)]^2} \\ &\qquad = \sqrt{(v^c_z)^2 + (v^s_z)^2} \end{aligned} \tag{6.27}$$

which is independent of time. The phase angle is given by

$$\psi = \tan^{-1} (v^s_z / v^c_z) \tag{6.28}$$

Thus, both the magnitude and the phase angle can be calculated for each location of interest along the shaft.

Uncoupled and Undamped Problems For the case of an uncoupled bearing system with no damping, the rotating-shaft unbalanced-response problem is reduced to a beam problem with unbalance loads. These loads can be taken from Table 6-4 and incorporated in a beam analysis.

6.6 Critical Speeds

The critical speeds can be calculated in the same fashion as natural frequencies. These speeds, Ω_n, $n = 1, 2, \ldots,$ are the roots of the characteristic equation resulting from the evaluation of the initial parameters for an undamped model. If this characteristic equation is represented by ∇, then the critical speeds Ω_n are those values of Ω in ∇ that make ∇ equal to zero.

If desired, the damped critical speeds can be found for a rotor model with dashpots in the bearings. They can also be found for a shaft with structural damping by replacing the usual bending modulus by the complex term $E(1 - i\delta_s)$. The critical speed search is conducted in the complex plane with Ω replaced by the complex critical speed s_n, which appears as

$$s_n = \lambda_n + i\Omega_n \tag{6.29}$$

where λ_n is the damping exponent of the nth mode and Ω_n is the damped natural frequency for the nth mode. Frequently the quality factor Q_n of the system is computed as $Q_n = -\Omega_n/2\lambda_n$. Also, Q_n can be expressed as $1/2\zeta$ where $\zeta = c/c_c$ is the damping ratio for a single-degree-of-freedom system. This permits Q_n to be viewed as a modal damping factor so that in the vicinity of s_n the flexible rotor may be represented by a single-degree-of-freedom model ($\ddot{x} + (\Omega_n/Q_n)\dot{x} + \Omega_n^2 x = 0$ in the standard elementary vibration theory notation).

Frequently, lumped-mass models of rotors are employed. As explained in Sec. 2.18, in such cases the critical speed search can be set up to lead to a frequency equation which is a polynomial in the complex s_n. The use of a polynomial can avoid the need for repetitive shaft analyses that are usually required for critical speed and instability studies (Section 6.7).

The simplifications presented in Section 6.5 for the unbalanced response of an uncoupled rotor apply also for the critical speed problem. That is, if there are no cross-coupling coefficients, the z and y directions can be treated separately, with only half the state variables required for the coupled problem. If the bearings are modeled with dampers, then an 8×8 transfer matrix is required (the loading vector of the 9×9 transfer matrix for unbalanced response can be ignored if desired). For bearing systems with no damping, the problem can be solved by a beam analysis with 4×4 transfer matrices. For the uncoupled, orthotropic bearing problems, different critical speeds are calculated for the motion in the two directions. For isotropic uncoupled bearings, motion in only one of the directions need be treated, since the same critical speeds would be found for motion in the z and y directions.

See the natural-frequency calculations of the other chapters for details and examples for the manipulations required to find the critical speeds and corresponding mode shapes. As with other structural members, the mass can be modeled as being continuously distributed or it can be lumped at particular locations along the shaft.

6.7 Threshold Speed of Instability

The amplitude of a rotor response can be written as

$$v_z = \mathrm{Re}\{[v_z^c + iv_z^s]e^{s_n t}\} \qquad \text{etc.} \tag{6.30}$$

where s_n is given in Eq. (6.29). Normally λ_n is negative, and according to Eq. (6.30) the vibration dies out exponentially with time. However, for positive λ_n the amplitude will increase inordinately with time and the rotor becomes unstable. The speed Ω_n at which this first occurs corresponds to the threshold of instability. Thus, a damped-critical-speed analysis can provide both the damped natural frequency and the threshold speed of instability.

6.8 Transient Response

In terms of a modal solution, the time-dependent state variables for a rotating shaft with isotropic bearings and negligible cross-coupling bearing coefficients are expressed by

Displacement method:

$$\begin{aligned} v(x,t) &= \Sigma A_n(t) v_n(x) \\ \theta(x,t) &= \Sigma A_n(t)\theta_n(x) \\ M(x,t) &= \Sigma A_n(t) M_n(x) \\ V(x,t) &= \Sigma A_n(t) V_n(x) \\ \bar{v}(x,t) &= \Sigma A_n(t)\bar{v}_n(x) \end{aligned} \tag{6.31a}$$

Acceleration method:

$$\begin{aligned} v(x,t) &= v_s(x,t) + \Sigma B_n(t) v_n(x) \\ \theta(x,t) &= \theta_s(x,t) + \Sigma B_n(t)\theta_n(x) \\ M(x,t) &= M_s(x,t) + \Sigma B_n(t) M_n(x) \\ V(x,t) &= V_s(x,t) + \Sigma B_n(t) V_n(x) \\ \bar{v}(x,t) &= \bar{v}_s(x,t) + \Sigma B_n(t)\bar{v}_n(x) \end{aligned} \tag{6.31b}$$

where

$$A_n(t) = \frac{\eta_n(t)}{N_n} \qquad B_n(t) = \frac{\xi_n(t)}{N_n} \tag{6.32}$$

These relations provide the complete dynamic response in terms of the complex isotropic state variables. That is, these are the solutions to Eqs. (6.6), where the state variables v, θ, M, V represent the motion in both planes as defined in Eqs. (6.5). Similarly, v_n, θ_n, M_n, V_n represent complex mode shapes for the motion in both planes, again using notation of the sort in Eqs. (6.5).

In Eqs. (6.31) the terms with the subscript s are steady-state solutions that are determined for the applied loading at each point in time. These are of the form

$$v_s(x, t) = -\sum_n \frac{v_n(x) f_n(t)}{s_n N_n} \tag{6.33}$$

where $v_n(x)$ and N_n are the shaft deflection mode shape and norm associated with the complex eigenvalue s_n. Similar expressions apply for the other four state variables.

The remaining quantities in Eqs. (6.31) to (6.33) are defined as

$$\eta_n(t) = e^{s_n t}\eta_n(0) + \int_0^t e^{s_n(t-\tau)} f_n(\tau)\, d\tau \tag{6.34}$$

$$\begin{aligned}\eta_n(0) = \int_0^L \Big\{ & s_n\big[v_n \rho v(x, 0) + \theta_n I_T \theta(x, 0)\big] + v_n \rho \frac{\partial v}{\partial t}(x, 0) \\ & + \theta_n I_T \frac{\partial \theta}{\partial t}(x, 0) - i\omega I_P \theta_n \theta(x, 0)\Big\}\, dx \\ & + \sum_{i=1}^{N} \Big[s_n \bar{v}_n(a_i) M_i \bar{v}(a_i, 0) + \bar{v}_n(a_i) M_i \frac{\partial \bar{v}}{\partial t}(a_i, 0) \\ & + v_n(a_i) c_3 \big[v(a_i, 0) - \bar{v}(a_i, 0)\big] \\ & + \bar{v}_n(a_i) c_3 \big[\bar{v}(a_i, 0) - v(a_i, 0)\big] + \bar{v}_n(a_i) c_4 \bar{v}(a_i, 0)\Big]\end{aligned} \tag{6.35}$$

$$\begin{aligned}\xi_n(t) = e^{s_n t}\left[\frac{1}{s_n^2}\frac{\partial f_n}{\partial t}(0) + \frac{1}{s_n} f_n(0) + \eta_n(0)\right] - \frac{1}{s_n^2}\frac{\partial f_n}{\partial t}(t) \\ + \frac{1}{s_n^2}\int_0^t e^{s_n(t-\tau)} \frac{\partial^2 f_n}{\partial \tau^2}(\tau)\, d\tau\end{aligned} \tag{6.36}$$

$$\begin{aligned}N_n(t) = 2s_n \int_0^L (\rho v_n^2 + I_T \theta_n^2)\, dx - i\omega \int_0^L I_P \theta_n^2\, dx + \sum_{i=1}^{N} \Big(2s_n M_i \{\bar{v}_n(a_i)\}^2 \\ + c_3\{v_n(a_i)[v_n(a_i) - \bar{v}_n(a_i)] + \bar{v}_n(a_i)[\bar{v}_n(a_i) - v_n(a_i)]\} + c_4[\bar{v}_n(a_i)]^2\Big)\end{aligned} \tag{6.37}$$

$$f_n(t) = \int_0^L v_n w\, dx + \sum_{i=1}^{N} \{[k_4 g(a_i, t) + c_4 \dot{g}(a_i, t)]\bar{v}_n(a_i)\} \tag{6.38}$$

where $w = w_z + iw_y$ and $g(a_i, t)$ and $\dot{g}(a_i, t)$ are the complex applied displacement and velocity of the bearing support at $x = a_i$. Set $g = \dot{g} = 0$ for no bearing base motion. The quantities $\bar{v}$ and $\bar{v}_n$ are the displacement and displacement mode shape of the pedestal mass in the bearing system. They can be calculated from

$$\begin{matrix} \bar{v}(a_i, 0) \\ \text{or} \\ \bar{v}_n(a_i) \end{matrix} = \left(\frac{k_3 + s_n c_3}{k_3 + s_n c_3 + k_4 + s_n c_4 + s_n^2 M_i} \right) \begin{bmatrix} v(a_i, 0) \\ \text{or} \\ v_n(a_i) \end{bmatrix} \tag{6.39}$$

The terms $v(a_i, 0)$ and $\partial v(a_i, 0)/\partial t$, permit a prescribed shaft displacement and velocity at the location of the bearing system to be introduced.

B. COMPUTER PROGRAM AND EXAMPLES

6.9 Benchmark Examples

Most rotating-shaft problems should be solved with the assistance of a computer program. The following examples are provided as benchmark examples against which a reader's own program can be checked. The computer program SHAFT was used for these examples. SHAFT calculates the unbalanced response, critical speeds, and transient response of a rotating shaft. For the unbalanced problem, it calculates the component and resultant deflection, slope, bending moment, and shear force along the shaft. In the case of the critical speed option, damped and undamped critical speeds and mode shapes are computed. The transient response due to prescribed initial conditions and to applied force and displacement loadings on the bearing system can also be found.

The shaft can be formed of lumped or continuous-mass segments with foundations, any boundary conditions, and any distribution of unbalanced masses. The user can include any or all of bending, shear deformation, and rotary inertia effects. The bearing systems can include springs, dampers, and a pedestal mass.

EXAMPLE 6.1 Critical Speeds of an Undamped Shaft Find the first three critical speeds of an undamped uniform shaft. The ends are free. The shaft is 70 in long. The elastic modulus is 10^7 lb/in^2 and Poisson's ratio is 0.3. The mass per unit length is 0.000813 lb · sec^2/in^2. The shaft is solid, of radius 1 in. Consider shear deformation and rotary inertia effects.

These three critical speeds are calculated to be 71.3, 195.8, 381.7 Hz.

EXAMPLE 6.2 Unbalanced Response of Simple Lumped-Mass Model Find the unbalanced response of the shaft of Fig. 6-3 that is rotating at 100 Hz. The shaft is modeled as a massless shaft with a single mass at the center that is offset at 0.01 in. at a phase angle of 45°. The moment of inertia of the shaft is 1000 in^4 and the elastic modulus is $3(10^7)$ lb/in^2.

Some results are given in Fig. 6-4.

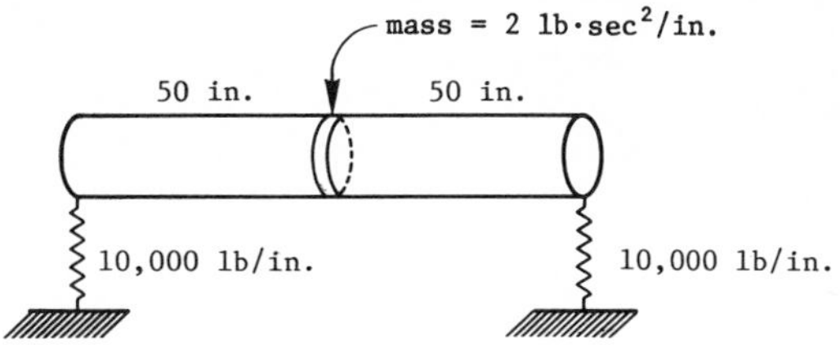

Fig. 6-3 Example 6.2.

AXIAL LOCATION	DEFLECTION	SLOPE	MOMENT	SHEAR	ANGLE
0.	9.9704E-03	-8.4632E-06	0.	9.9704E+01	4.4999E+01
50.00	1.0253E-02	-5.2266E-17	4.9852E+03	-9.9704E+01	4.4999E+01
100.00	9.9704E-03	8.4632E-06	-6.0581E-08	-9.9704E+01	4.4999E+01

BEARING REACTIONS

AXIAL LOCATION	FORCE	FORCE ANGLE	MOMENT	MOMENT ANGLE
0.	9.9704E+01	44.999	0.	0.
100.000	9.9704E+01	44.999	0.	0.

Fig. 6-4 Partial output for Example 6.2.

EXAMPLE 6.3 Unbalanced Response of Shaft with Pedestal Mass in the Bearing System

Calculate the unbalanced response of the rotor system of Fig. 6-5 that is rotating at 50 Hz. The elastic modulus is $3(10^7)$ lb/in^2. The moment of inertia of the massless sections is 12.566 in^4. Each shaft mass is 50 lb · sec^2/in. The

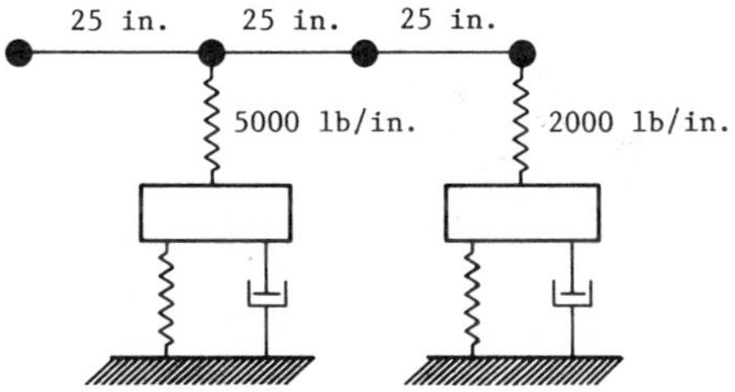

Fig. 6-5 Example 6.3.

AXIAL LOCATION	DEFLECTION	SLOPE	MOMENT	SHEAR	ANGLE
0.00	-.1014E-05	-.2049E-03	0.	.5003E+01	.2611E+03
25.00	.5116E-02	-.2042E-03	.1251E+03	.2824E+02	.1171E+01
50.00	.1001E-01	-.1802E-03	.7572E+03	-.3029E+02	.1790E-01
75.00	.1412E-01	-.1575E-03	-.1617E-05	-.3029E+02	.1680E+01

BEARING REACTIONS

LOCATION	FORCE	FORCE ANGLE	MOMENT	MOMENT ANGLE
25.000	.2698E+02	.645	0.	0.000
75.000	.2884E+02	1.884	0.	0.000

Fig. 6-6 Partial results for Example 6.3.

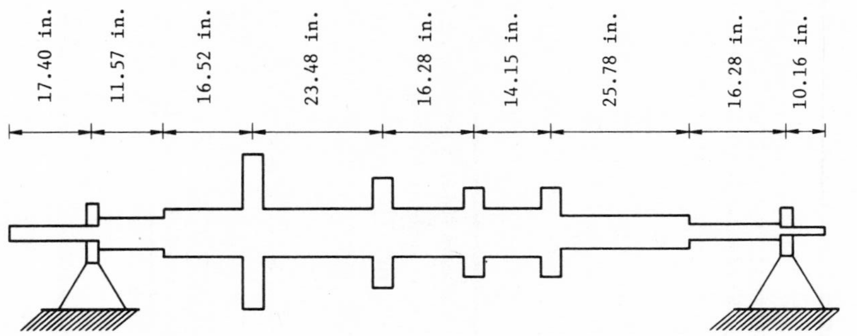

Data:

Lumped masses:	Station No. 4	5	6	7
Mass $(lb\text{-}sec^2/in)$	7.472	1.7	1.304	1.315
Eccentricity (in)	0	0.1	0	0

Bearings:	Station No. 2	9
Bearing Stiffness (lb/in.)	250,000	250,000

Sections:	Section No. 1	2	3	4	5	6	7	8	9
$I(in^4)$	155.355	211.359	1018.4	1018.4	1018.4	1018.4	665.71	117.87	80.456
$\rho(lb\text{-}sec^2/in^2)$	0.0331	0.0425	0.08488	0.08488	0.08488	0.08488	0.06811	0.02888	0.0212
Area (in^2)	44.18	51.53	113.11	113.11	113.11	113.11	91.463	38.486	31.796

Fig. 6-7 Turbine, Example 6.4.

pedestal masses are each 1 lb · sec²/in. The pedestal spring and damping constants are 5000 lb/in and 50 lb · sec/in, respectively. The third shaft mass from the left is offset at an eccentricity of 0.01 in.

The computer printout for several locations along the shaft is shown in Fig. 6-6.

EXAMPLE 6.4 Unbalanced Response of Turbine System Find the unbalanced response of the turbine of Fig. 6.7, which rotates at 60 Hz. The bearing systems are modeled as simple springs with spring constants of magnitude 250,000 lb/in. The elastic modulus is $3(10^7)$ lb/in². The second mass from the left end is offset with an eccentricity of 0.1 in. Take into account shear deformation effects.

Some results are shown in Fig. 6-8.

AXIAL LOCATION	DEFLECTION	SLOPE	MOMENT	SHEAR	ANGLE
0.	-1.1542E-03	-1.5713E-04	0.	0.	0.
17.40	1.5775E-03	-1.5671E-04	1.7310E+02	3.7699E+02	0.
28.79	3.3473E-03	-1.5302E-04	3.6165E+03	2.0734E+02	0.
28.97	3.3748E-03	-1.5300E-04	3.6535E+03	2.0369E+02	0.
45.49	5.8857E-03	-1.5130E-04	8.1103E+01	-6.9698E+03	0.
68.97	9.9630E-03	-2.1995E-04	-1.8725E+05	1.2577E+04	0.
85.25	1.4073E-02	-2.6830E-04	-5.0389E+02	7.6215E+03	0.
99.40	1.7763E-02	-2.4635E-04	8.8830E+04	1.5803E+03	0.
125.20	2.2581E-02	-1.3191E-04	6.6383E+04	-3.5205E+03	0.
141.50	2.3034E-02	2.2648E-05	-3.5197E+03	6.9793E+02	0.
151.60	2.2843E-02	1.7740E-05	-1.4901E-08	0.	0.

BEARING REACTIONS

AXIAL LOCATION	FORCE	FORCE ANGLE	MOMENT	MOMENT ANGLE
17.400	3.9437E+02	0.	0.	0.
141.500	5.7585E+03	0.	0.	0.

Fig. 6-8 Partial output for Example 6.4.

EXAMPLE 6.5 Damped Critical Speeds Find the damped critical speeds of a uniform shaft 70 in long. The two ends of the shaft rest on dashpots with constants 40 lb · sec/in. The mass per length is $8.13(10^{-4})$ lb · sec²/in², $E = 10(10^6)$ lb/in², $G = 3.846(10^6)$ lb/in² and the cross-sectional area is 3.142 in². Include shear deformation effects using a shear form factor of 0.75.

Some of the results beyond the rigid body mode are:

Mode	Damped natural frequency, Hz	Damped exponent, 1/sec
1	31.47	−22.77
2	127.11	−90.82
3	288.17	−200.41

EXAMPLE 6.6 Asynchronous Whirl The shaft of Fig. 6-9 is rotating at 1000 rpm. It is formed of two solid circular shafts, each 50 in long. Let $E = 10(10^6)$

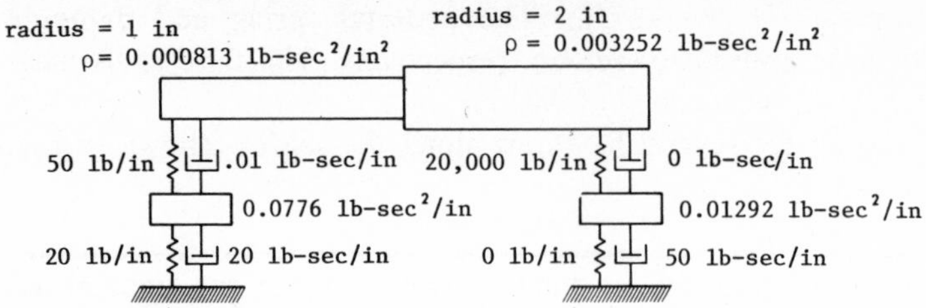

Fig. 6-9 Example 6.6.

lb/in^2 and $G = 3.846(10^6)$ lb/in^2. Find the damped whirl frequencies. Include shear deformation effects with $k_s = 0.8864$.

The lower frequencies for this asynchronous whirl problem are:

Mode	Damped natural frequency, Hz	Damped exponent, 1/sec
1	5.51	−1.62
2	30.27	−8.73
3	148.94	−41.98
4	275.62	−25.46

EXAMPLE 6.7 Transient Response Due to Initial Conditions Suppose the right end of the rotor of Fig. 6-10 is displaced 1 in upward at time zero and released. Find the transient deflection at $x = 80$ in for $t > 0$. Let the bases of the bearing system remain stationary.

A plot of the results is given in Fig. 6-11.

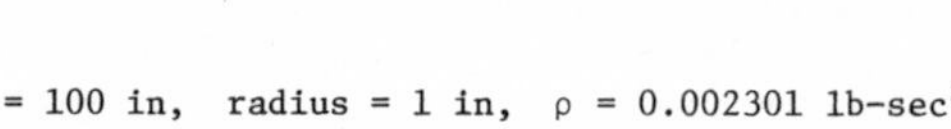

Rotor: $L = 100$ in, radius = 1 in, $\rho = 0.002301$ lb-sec^2/in^2,
$E = 30(10^6)$ lb/in^2, $G = 11.5(10^6)$ psi, $\Omega = 10{,}000$ rpm.

Bearings: $k_1 = 50{,}000$ lb/in, $c_1 = 20$ lb-sec/in, $M_1 = 0.005176$ lb-sec^2/in,
$k_2 = 20{,}000$ lb/in, $c_2 = 50$ lb-sec/in.

Fig. 6-10 Example 6.7.

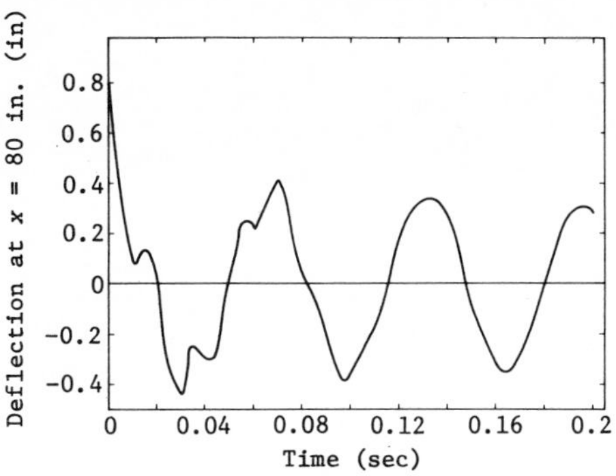

Fig. 6-11 Example 6.7, z component of shaft deflection.

SEVEN

Grillages

This chapter deals with the static, stability, and dynamic analysis of grillages. A *grillage* or *gridwork* consists of two sets of mutually perpendicular beams rigidly connected at the intersections (Fig. 7-1). The loading is transverse to the plane of the sets of beams. The beams in one direction are called *girders* and those in the other direction are called *stiffeners*. In theory, there is no basis for deciding which of the two sets of beams is to be labeled girders. In practice, the heavier and wider-spaced beams are usually designated as the girders while the closer-

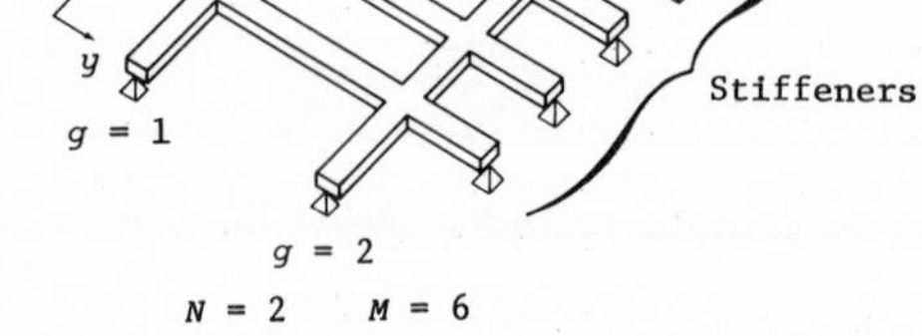

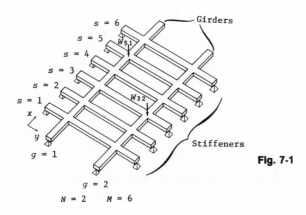

Fig. 7-1

spaced, lighter beams are the stiffeners. Many of the formulas of this chapter apply to *uniform* grillages. A uniform grillage has girders that are identical in size, end conditions, and spacing. The same holds for the stiffeners, although the stiffeners and girders may differ from each other.

The beams may be of open or closed cross section, although for closed sections there may be an error of up to 5 percent because the torsional rigidity of the beams is not taken into account.

A. SIMPLE GRILLAGES

7.1 Notation and Conventions

Girders	Beams parallel to the x axis
Stiffeners	Beams parallel to the y axis
N, M	Number of girders and stiffeners, respectively
g, s	Index for girders and stiffeners, respectively
v_g, θ_g, M_g, V_g	Deflection, slope, bending moment, and shear force of the gth girder
E	Modulus of elasticity of the material (force/length^2)
I_g, I_s	Moments of inertia of girders and stiffeners, respectively (length^4)
L_g, L_s	Length of girders and stiffeners, respectively
S_g, S_s	Spacing of girders and stiffeners, respectively (length). For evenly spaced stiffeners $S_g = L_s/N$ and for evenly spaced girders $S_s = L_g/M$.
M_{sg}	Lumped mass at the intersection x_s, y_g (mass)
P_g, P_s	Axial forces of girders and stiffeners, respectively (force)
w_s	Loading intensity along the sth stiffener (force/length)
W_{sg}	Concentrated force at the intersection x_s, y_g (force)
ω	Natural frequency (radians/second)
ρ_g, ρ_s	Mass per unit length of girders and stiffeners (mass/length), (force · time^2/length^2)

The sign convention for displacements and forces for the beams of Chap. 2 apply to grillage beams.

7.2 Static Loading

If the ends of the girders and stiffeners of a uniform grillage are simply supported, the deflection, slope, bending moment, and shear force of the gth girder are given by

Deflection:
$$v_g = \sin \frac{\pi g}{N+1} \sum_{j=1}^{\infty} K_j \sin \frac{j\pi x}{L_g} \tag{7.1a}$$

Slope:
$$\theta_g = -\sin \frac{\pi g}{N+1} \sum_{j=1}^{\infty} K_j \frac{j\pi}{L_g} \cos \frac{j\pi x}{L_g} \tag{7.1b}$$

Bending moment:
$$M_g = EI_g \sin \frac{\pi g}{N+1} \sum_{j=1}^{\infty} K_j \left(\frac{j\pi}{L_g} \right)^2 \sin \frac{j\pi x}{L_g} \tag{7.1c}$$

$$\text{Shear force:}\quad V_g = EI_g \sin\frac{\pi g}{N+1} \sum_{j=1}^{\infty} K_j \left[\left(\frac{j\pi}{L_g}\right)^3 \cos\frac{j\pi x}{L_g} + \frac{\pi^4 I_s}{(N+1)L_s^3 I_g} \times \sum_{s=1}^{M} \langle x - x_s \rangle^0 \sin\frac{j\pi x_s}{L_g} \right] \tag{7.1d}$$

The parameters K_j are given in Table 7-1 for various loadings. It is usually sufficiently accurate to include M terms in Eqs. (7.1), that is,

$$\sum_{j=1}^{\infty} \approx \sum_{j=1}^{M}$$

TABLE 7-1 K_j **for Eqs. (7.1)**

Loading	K_j
1. For concentrated loads W_{sg} at x_s, y_g	$\dfrac{\dfrac{2L_s^3}{EI_s\pi^4}\dfrac{P_e}{P_e - P_s}\sum\limits_{s=1}^{M}\sum\limits_{g=1}^{N} W_{sg}\sin\dfrac{\pi g}{N+1}\sin\dfrac{j\pi s}{M+1}}{\dfrac{N+1}{2}j^4\left(\dfrac{L_s}{L_g}\right)^3\left(\dfrac{I_g}{I_s}\right)\left(1-\dfrac{P_g}{jP_c}\right)+\dfrac{M+1}{2}}$
2. For a uniform force w_s along the sth stiffener	$\dfrac{\dfrac{4L_s^4}{EI\pi^5}\dfrac{P_e}{P_e - P_s}\sum\limits_{s=1}^{M} w_s\sin\dfrac{j\pi s}{M+1}}{\dfrac{N+1}{2}j^4\left(\dfrac{L_s}{L_g}\right)^3\left(\dfrac{I_g}{I_s}\right)\left(1-\dfrac{P_g}{jP_c}\right)+\dfrac{M+1}{2}}$
3. If the uniform force w_s is the same for all stiffeners	Only the first term ($j = 1$) in Eqs. (7.1) is required $K_1 = \dfrac{\dfrac{4L_s^4}{EI_s\pi^5}\dfrac{P_e}{P_e - P_s}\sum\limits_{s=1}^{M} w_s\sin\dfrac{\pi s}{M+1}}{\dfrac{N+1}{2}\left(\dfrac{L_s}{L_g}\right)^3\left(\dfrac{I_g}{I_s}\right)\left(1-\dfrac{P_g}{P_c}\right)+\dfrac{M+1}{2}}$

$$P_e = \frac{\pi^2 EI_s}{L_s^2}, \quad P_c = \frac{\pi^2 EI_g}{L_g^2}$$

EXAMPLE 7.1 Gridwork with Uniform Force Find the deflection at the intersections of the grillage of Fig. 7-2. The grillage is loaded with a uniform force of 10 lb/in^2.

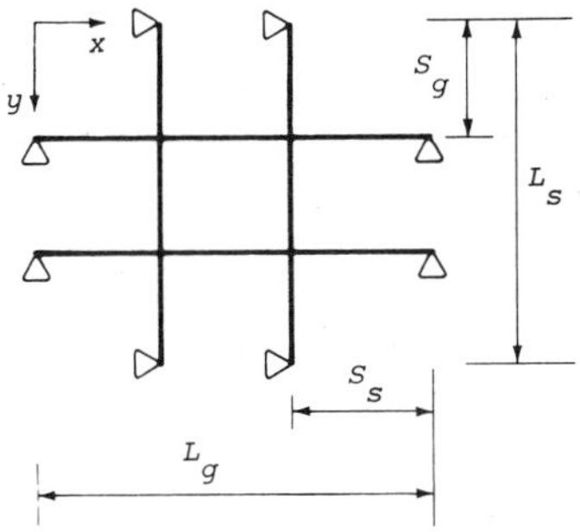

Fig. 7-2 $I_s = I_g = 100 \text{ in}^4$, $E = 3(10^7)$ psi
$L_s = L_g = 100$ in, $P_s = P_g = 0$
$S_g = S_s = 100/3$ in

The loading intensity along either of the stiffeners will be $w_s = 10 \text{ lb/in}^2 \times L_g/(M+1) = 10(100/3) = 333.33$ lb/in. From Eq. (7.1a)

$$v_g = \sin\frac{\pi g}{N+1} K_1 \sin\frac{\pi x}{L_g} = K_1 \sin\frac{\pi g}{3} \sin\frac{\pi x}{100}$$

where, from case 3 of Table 7-1,

$$K_1 = \frac{\dfrac{4L_s^4}{EI_s\pi^5}\displaystyle\sum_{s=1}^{2} w_s \sin\frac{\pi s}{3}}{\dfrac{3}{2}+\dfrac{3}{2}} = \frac{\dfrac{4L_s^4 w_s}{EI_s\pi^5}\left(\dfrac{\sqrt{3}}{2}+\dfrac{\sqrt{3}}{2}\right)}{\dfrac{3}{2}+\dfrac{3}{2}}$$

Then

$$v_1\big|_{x=L_g/3} = v_2\big|_{x=L_g/3} = v_1\big|_{x=2L_g/3} = v_2\big|_{x=2L_g/3} = \frac{4L_s^4 w_s}{EI_s\pi^5}\frac{\sqrt{3}}{3}\sin\frac{\pi}{3}\sin\frac{\pi}{3}$$

$$= 0.062886 \text{ in}$$

The computer program GRILLAGE gives 0.06287 in for this problem.

EXAMPLE 7.2 Gridwork with Concentrated Force Suppose the grillage of Fig. 7-2 is loaded with concentrated loads of 10,000 lb acting at each intersection.

Since all of the concentrated forces are equal, one term of Eqs. (7.1) will probably be sufficiently accurate. Thus

$$v_g = K_1 \sin\frac{\pi g}{3} \sin\frac{\pi x}{100}$$

TABLE 7-2 Critical Axial Loads in Girders

End Conditions of Girders		D_1	P_{cr}
Simply-Supported	1.	≤ 1	$(1 + D_1)P_e$
	2.	>1	$D_2 P_e$
Fixed	3.	≤ 1	$(4 + D_1)P_e$
	4.	>1	$(3 + D_2)P_e$

$$D_1 = 0.0866 \frac{L_g^2}{\sqrt{C_1 S_s L_s^3 I_g / I_s}}$$

$$D_2 = 0.202 \frac{L_g^2}{\sqrt{C_1 S_s L_s^3 I_g / I_s}}$$

$$P_e = \frac{\pi^2 E I_g}{L_g^2}$$

Take C_1 from Table 7-3.

with (from case 1 of Table 7-1)

$$K_1 = \frac{\dfrac{2L_s^3}{EI_s\pi^4} 10{,}000 \sum_{s=1}^{2} \sum_{g=1}^{2} \sin \dfrac{\pi g}{3} \sin \dfrac{\pi s}{3}}{\dfrac{3}{2} + \dfrac{3}{2}} = \frac{2L_s^3}{EI_s\pi^4} 10{,}000$$

We find

$$v_1|_{x=L_g/3} = v_2|_{x=L_g/3} = v_2|_{x=2L_g/3} = v_2|_{x=2L_g/3} = 0.0514 \text{ in}$$

This same result is computed with the program GRILLAGE. See Example 7.5.

If the concentrated forces differ for the different intersections, then more terms of the series of Eqs. (7.1) must be employed. For such problems, it may well be easier to use a computer program.

7.3 Stability

The critical axial loads $P_g = P_{cr}$ in the girders of uniform grillages with more than five stiffeners are given in Table 7-2. In some instances, these formulas will be sufficiently accurate for as few as three stiffeners (see Example 7.8).

EXAMPLE 7.3 Buckling of Simply Supported Grillage Compute the critical axial forces in the girders of the grillage of Fig. 7-3. The stiffeners are simply

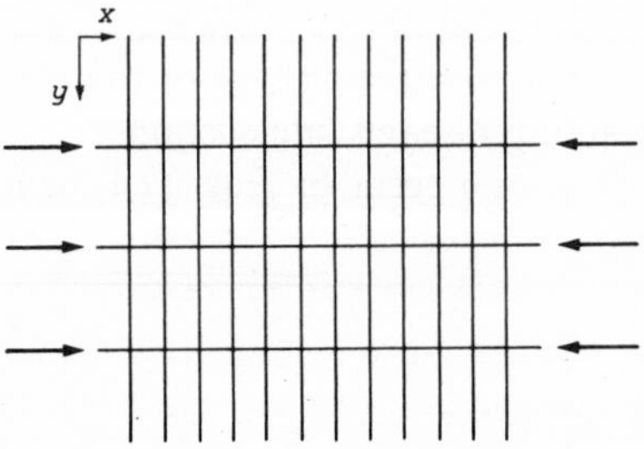

Fig. 7-3

TABLE 7-3 Values of C_1 for Stability

Number of Girders N	C_1 End Conditions of Stiffeners	
	Simply-Supported	Fixed
1	0.020833	0.0052083
2	0.030864	0.0061728
3	0.041089	0.0080419
4	0.051342	0.010009
5	0.061603	0.011997
6	0.071866	0.013990
7	0.082131	0.015986
8	0.092396	0.017982
9	0.10266	0.019979
10	0.11293	0.021976

For simply-supported stiffener ends the formula

$$C_1 = \frac{N+1}{\pi^4}\left[1 + \sum_{j=1}^{\infty} \{[2j(N+1)+1]^{-4} - [2j(N+1)-1]^{-4}\}\right]$$

applies for any N

supported. Here $I_g = I_s$, $L_g = L_s = L$, $N = 3$, $M = 12$, $S_g = L/4$, $S_s = L/13$.

To use Table 7-2 we first calculate D_1. The constant C_1 is given as 0.041089 in Table 7-3 for $N = 3$ and simply supported stiffeners.Then

$$D_1 = 0.0866 \frac{L_g^2}{\sqrt{C_1 S_s L_s^3 I_g / I_s}} = \frac{0.0866 L^2}{\sqrt{C_1 (L/13) L^3}} = 0.0866\sqrt{\frac{13}{C_1}} = 1.54$$

Since $D_1 > 1$, the formulas for cases 2 and 4 in Table 7-2 are used. We find $P_{cr} = D_2 P_e = 3.5930 P_e$ if the girders' ends are simply supported and $P_{cr} = 6.5930 P_e$ for fixed girder ends.

7.4 Free Dynamic Response—Natural Frequencies

The natural frequencies ω_n, $n = 1, 2, 3, \ldots$, can be more meaningfully represented if a second subscript is added, ω_{mn}. The subscript m indicates the number of mode-shape half waves in the y (stiffener) direction, and n indicates the number of half waves in the x (girder) direction. Some example mode shapes associated with ω_{mn} are shown in Fig. 7-4.

The lower natural frequencies of a uniform grillage with simply supported stiffeners are given by

$$\omega_{mn} = \frac{EI_sL_g\left(\frac{\pi m}{L_s}\right)^4 + EI_g\frac{N+1}{C_nL_g^3} - P_s\left(\frac{m\pi}{L_s}\right)^2}{\rho_sL_s + \rho_gL_g} \tag{7.2}$$

where N is the number of girders. The parameter C_n should be taken from Table 7-4 for the girders with simply supported or built-in ends. For other end conditions use the method explained in Section 7.7. As mentioned previously, either set of grillage beams can be designated as the girders.

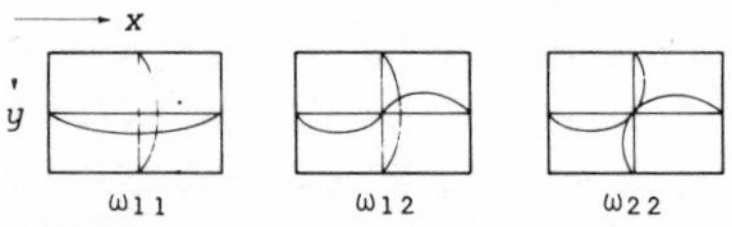

Fig. 7-4

If all of the girders are subjected to an axial force P_g, Eq. (7.2) applies if C_n is replaced by

$$C_n\frac{P_e}{P_e - P_g} \tag{7.3}$$

where $P_e = \pi^2EI_g/L_g^2$.

EXAMPLE 7.4 Natural Frequencies of Simply Supported Grillage Calculate the natural frequencies of a grillage with $N = M = 3$, $I_g = I_s = I = 100$ in^4, $\rho_g = \rho_s = \rho = 1$ lb · sec^2/in, $L_g = L_s = L = 100$ in, $S_g = S_s = S$, $P_g = P_s = 0$, $E = 3(10^7)$ lb/in^2. All ends are simply supported. From Eq. (7.2),

$$\omega_{mn}^2 = \frac{\frac{3(10^7)100}{100^3}\left(m^4\pi^4 + \frac{3+1}{C_n}\right)}{2(100)} = 15\left(m^4\pi^4 + \frac{4}{C_n}\right) \tag{1}$$

From Table 7-4 with $N = 3$, $C_1 = 0.041089$, $C_2 = 0.0026042$, $C_3 = 0.00057767$. From (1),

$$\omega_{11}^2 = 15\left(\pi^4 + \frac{4}{C_1}\right) = 2921.69 \qquad \text{or} \qquad \omega_{11} = 54 \text{ rad/sec}$$

$$\omega_{21}^2 = 15\left(16\pi^4 + \frac{4}{C_1}\right) = 24{,}838.347 \qquad \text{or} \qquad \omega_{21} = 157.6 \text{ rad/sec}$$

$$\omega_{12}^2 = 15\left(\pi^4 + \frac{4}{C_2}\right) = 24{,}500.831 \qquad \text{or} \qquad \omega_{12} = 156.5 \text{ rad/sec}$$

$$\omega_{22}^2 = 15\left(16\pi^4 + \frac{4}{C_2}\right) = 46{,}417.80 \qquad \text{or} \qquad \omega_{22} = 215 \text{ rad/sec}$$

Other frequencies can be obtained in a similar fashion.

TABLE 7-4 Value of C_n for Eqs. (7.2) and (7.6)

Simply Supported Girders

Number of Stiffeners M	C_1	C_2	C_3	C_4	C_5	C_6	C_7	C_8	C_9	C_{10}
1	0.020833									
2	0.030864	0.0020576								
3	0.041089	0.0026042	0.00057767							
4	0.051342	0.0032240	0.00065790	0.0002462						
5	0.061603	0.0038580	0.00077160	0.00025720	0.00012564					
6	0.071866	0.0044962	0.00089329	0.00028895	0.00012688	0.000073890				
7	0.082131	0.0051361	0.0010177	0.00032552	0.00013769	0.000072209	0.000047321			
8	0.092396	0.0057767	0.0011431	0.00036387	0.00015157	0.000076208	0.000045226	0.000032215		
9	0.10266	0.0064178	0.0012691	0.00040301	0.00016667	0.000082237	0.000046681	0.000030328	0.000022963	
10	0.11293	0.0070590	0.0013954	0.00044252	0.00018233	0.000089133	0.000049521	0.000030753	0.000021400	0.000016967
Any M	$C_n = \frac{M+1}{\pi^4}\left[\frac{1}{n^4} + \sum_{j=1}^{\infty}\{[2j(M+1)+n]^{-4} - [2j(M+1)-n]^{-4}\}\right]$									

Girders with Fixed Ends

Number of Stiffeners M	C_1	C_2	C_3	C_4	C_5	C_6	C_7	C_8	C_9	C_{10}
1	0.0052083									
2	0.0061728	0.0011431								
3	0.0080419	0.0011393	0.00042165							
4	0.010009	0.0013459	0.00039075	0.00020078						
5	0.011997	0.0015917	0.00043081	0.00018009	0.00011111					
6	0.013990	0.0018480	0.00048904	0.00018923	0.000098217	0.000067910				
7	0.015986	0.0021078	0.00055303	0.00020779	0.000099794	0.000059682	0.000044545			
8	0.017982	0.0023691	0.00061925	0.00022977	0.00010668	0.000059226	0.000039097	0.000030804		
9	0.019970	0.0026311	0.00068645	0.00025320	0.00011572	0.000061961	0.000038155	0.000027067	0.000022193	
10	0.021976	0.0028934	0.00075415	0.00027732	0.00012573	0.000066109	0.000039232	0.000026101	0.000019547	0.000016522

In order to demonstrate the accuracy of Eq. (7.2), the quantity $\rho\omega_{mn}^2 S^4/EI$ calculated by this equation and by other methods is listed below.

mn	*Eq. (7.2)*	*Ref. 7.4, finite differences*	*Ref. 7.5, finite differences*	*Orthotropic plate model using eqs. of Ref. 7.6*	*Ref. 7.7, finite elements*
11	0.7853	0.7853	0.7854	0.7854	0.7854
12	1.3364	1.336	1.339	1.3406	1.3393
13	1.9244	1.924	1.963	1.9878	1.5708
22	1.5679	1.565	1.571	1.5708	1.9629
23	2.0767	2.061	2.057	2.0734	2.0569
33	2.3192	2.281	2.356	2.3562	2.3562

B. COMPUTER PROGRAMS AND EXAMPLES

7.5 Static Analysis

Complicated gridwork problems should be solved with the assistance of a computer program. The following examples are provided as benchmark examples for static problems against which a reader's own program can be checked. The computer program GRILLAGE was used for these examples. This program accepts a variety of geometric, loading, and boundary conditions which are somewhat more general than the formulas of this chapter.

EXAMPLE 7.5 Concentrated Loads Consider the 2×2 grillage with the concentrated loads of Example 7.2.

Partial results are provided in Fig. 7-5. X stations refer to points along the girders and the ends. The Y stations are points along the stiffeners.

```
DEFLECTION

 Y STATIONS       1          2
X STATION

 1        0.                0.
 2   5.1440E-02   5.1440E-02
 3   5.1440E-02   5.1440E-02
 4   3.8332E-20  -3.8332E-20
```

```
BENDING MOMENTS

 Y STATIONS       1          2
X STATION

 1        0.                0.
 2   1.6667E+05   1.6667E+05
 3   1.6667E+05   1.6667E+05
 4  -5.2684E-09  -5.2684E-09
```

```
SLOPE

 Y STATIONS       1          2
X STATION

 1  -1.8519E-03  -1.8518E-03
 2  -9.2593E-04  -9.2592E-04
 3   9.2593E-04   9.2592E-04
 4   1.8519E-03   1.8518E-03
```

```
SHEAR FORCE

 Y STATIONS       1          2
X STATION

 1   5.0000E+03   5.0000E+03
 2  -1.2348E-10  -1.2347E-10
 3  -5.0000E+03  -5.0000E+03
 4  -5.0000E+03  -5.0000E+03
```

Fig. 7-5 Displacements and forces of the girders of Example 7.5.

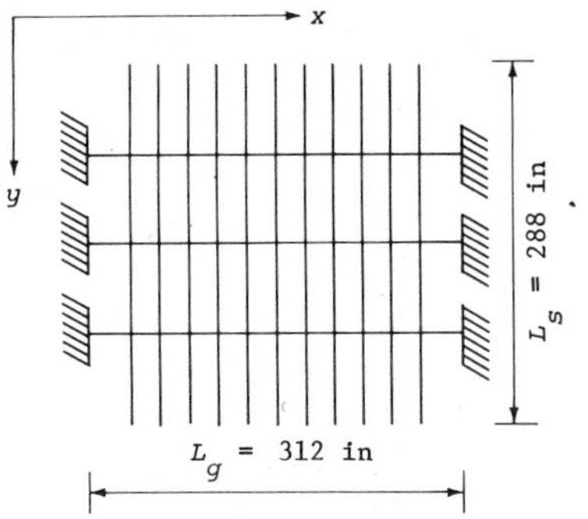

Fig. 7-6 Example 7.6.

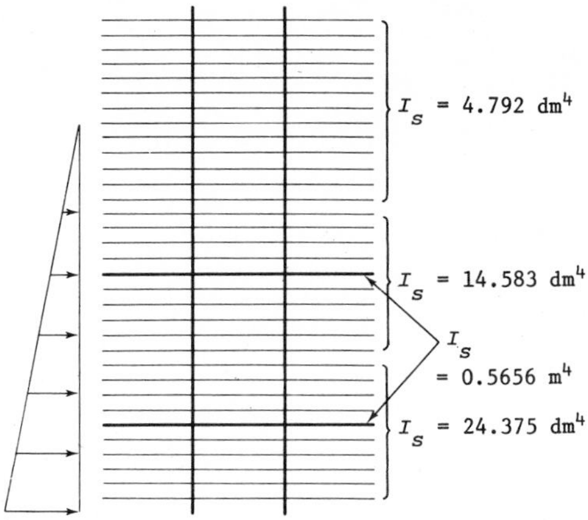

Fig. 7-7 Ship bulkhead modeled as a grillage, Example 7.7.

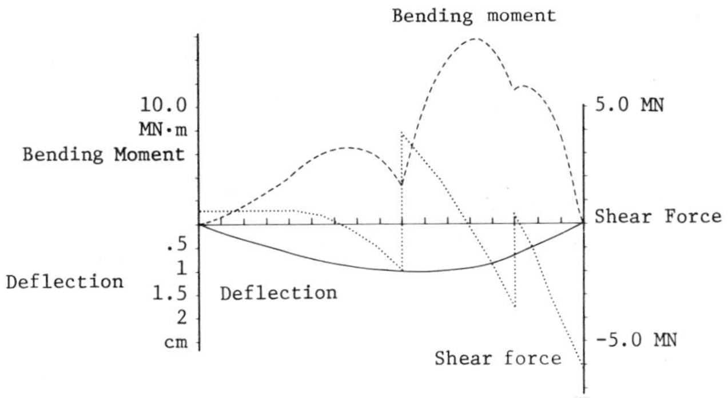

Fig. 7-8 The responses along the girders of Example 7.7.

EXAMPLE 7.6 Uniform Load The 3×11 grillage of Fig. 7-6 is subjected to a uniform load of 15 psi. The ends of the girders are fixed. The stiffeners are simply supported. $E = 3(10^7)$ lb/in^2, $I_g = 19{,}250$ in^4, $I_s = 6100$ in^4.

At the center of the middle girder the deflection is found to be 0.047 in.

EXAMPLE 7.7 Ship Bulkhead Consider the problem of calculating the deflections, bending moments, and shear forces of an oil-tight bulkhead of a 380,000-ton tanker. Since the bulkhead stiffeners and girders are much stiffer than the small longitudinal frames, they can be treated as simply supported at both ends, This bulkhead (Fig. 7-7) is subjected to a 22.3 m water head. The hydrostatic load begins at 6.67 m from the top and continues to the bottom of the bulkhead. The intensity of the load is 10 051 N/m^3. Use $E = 200$ GN/m^2, $I_g = 0.44795$ m^4, $L_g = 29.07$ m, $L_s = 18.036$ m.

The results are shown in Fig. 7-8.

7.6 Stability Analysis

The critical axial load in the girders of a grillage can be found by a beam-response program. The properties of a girder are entered in the beam program as though the girder were a beam with extension springs at the intersections with the stiffeners. For a stability analysis it is not necessary to include transverse loads. Only the length of the girder, end conditions, modulus of elasticity, moment of inertia, and the equivalent spring constants need be entered in the beam program. A general program such as BEAMRESPONSE can account for such effects as elastic foundations, shear deformation, changes in modulus of elasticity, changes in moment of inertia, and in-span supports for the girders. All girders must be physically identical and must be equally spaced. However, the stiffener spacing and moments of inertia can vary. The spring constants for the extension springs that should be introduced at the location of stiffener intersections are given by

$$\frac{EI_s}{C_1 L_s^3} \tag{7.4}$$

where C_1 is given in Table 7-3. Since the moments of inertia of the stiffeners can vary, the spring constants may differ along the girder.

EXAMPLE 7.8 Buckling Load Consider a grillage with three girders of 40 in length and three stiffeners of 40 in length. The modulus of elasticity is $1(10)^7$ lb/in^2. Also, $I_s = I_g = 128$ in^4. All ends are fixed.

From Table 7-3 for $N = 3$ and fixed stiffeners, we find $C_1 = 0.00804$. Then

$$k = \frac{EI_s}{C_1 L_s^3} = 2.487(10^6)$$

If the girder is now entered into a beam-response program as a beam with springs with constant $k = 2.487(10^6)$ at $x = 10$ in, 20 in, and 30 in, it is found that the critical axial load is $6(10^7)$ lb. Using the approximate formula of Table 7.2, we find $P_{cr} = 5.9256(10^7)$ lb.

7.7 Natural Frequencies

The natural frequencies of a grillage can be computed with a beam-response program by employing an equivalent beam formed from a girder. The mass and stiffness of the stiffeners are accounted for with concentrated masses and extension springs, respectively, located at the stiffener intersections. The length, end conditions, modulus of elasticity, moment of inertia, and mass density of the girder are used in the beam program. For a general beam program such as BEAMRESPONSE, the effects of elastic foundations, shear deformation, rotary inertia, in-span supports, changes in moment of inertia, and changes in modulus of elasticity can also be taken into account as long as all girders remain identical. The girders must have uniform spacing, but the stiffener spacing and moments of inertia can vary. The concentrated mass at the intersections resulting from the stiffeners is of magnitude $\rho_s S_g$. The spring constants to be introduced at each stiffener intersection are of magnitude

$$k = \frac{EI_s}{C_m L_s^3} \tag{7.5}$$

where C_m is given in Table 7-5. For a particular calculation using the beam program, the C_m in Eq. (7.5) is the same for each spring constant along the beam.

Since the moments of inertia can vary from stiffener to stiffener, the spring constants may differ along the girder. The index m indicates the number of half waves in the stiffener direction. There are an infinite number of frequencies associated with each C_m, since the girders are considered to have continuously distributed mass. There are as many C_m as there are girders. Thus, for a grillage with three girders, only C_1, C_2 and C_3 (and three different values of spring constants) are used. If k using C_1—that is, $k = EI_s/(C_1 L_s^3)$—is used along the girder in the beam program, then the first frequency calculated will be ω_{11}, the second ω_{12}, and so forth. If k is formed of C_2, the first frequency will be ω_{21}, the second ω_{22}, the third ω_{23}, and so forth. These relationships are illustrated in Fig. 7-9.

With the above formulation it is possible to obtain the number of mode shapes in the y direction up to the number of C_m or the number of girders. But since the girders can be treated as beams with distributed mass or lumped masses, an infinite number of frequencies—for example, ω_{m10}, ω_{m100}, $m = 1, 2, 3$—can occur. More modes can be calculated if the roles of the stiffeners and girders are interchanged. Then, the effect of the girders would be taken into account with equivalent spring constants. The stiffeners would be entered as beams into the beam program with springs

$$k = \frac{EI_g}{C_n L_g^3} \tag{7.6}$$

where the C_n are taken from Table 7-4.

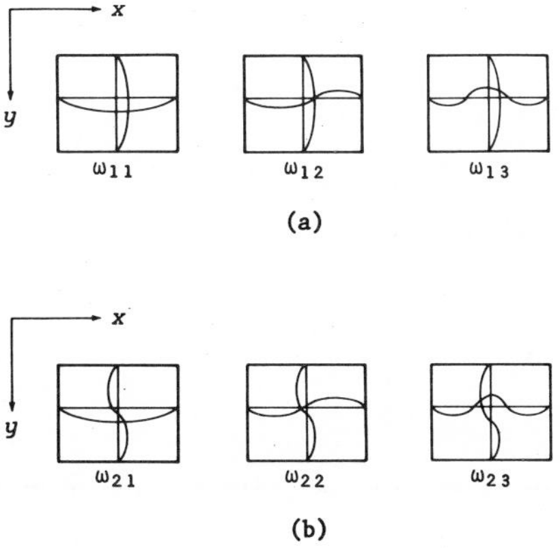

Fig. 7-9 Mode shapes for various values of C_m. (*a*) k using C_1. (*b*) k using C_2.

TABLE 7-5 Values of C_m for use with Eqs. (7.5) and a Beam Computer Program

Stiffeners with Simply-Supported Ends

Number of Girders N	C_1	C_2	C_3	C_4	C_5	C_6	C_7	C_8	C_9	C_{10}
1	0.020833									
2	0.030864	0.0020576								
3	0.041089	0.0026042	0.00057767							
4	0.051342	0.0032240	0.00065790	0.0002462						
5	0.061603	0.0038580	0.00077160	0.00025720	0.00012564					
6	0.071866	0.0044962	0.00089329	0.00028895	0.00012688	0.000073890				
7	0.082131	0.0051361	0.0010177	0.00032552	0.00013769	0.000072209	0.000047321			
8	0.092396	0.0057767	0.0011431	0.00036387	0.00015157	0.000076208	0.000045226	0.000032215		
9	0.10266	0.0064178	0.0012691	0.00040301	0.00016667	0.000082237	0.000046681	0.000030328	0.000022963	
10	0.11293	0.0070590	0.0013954	0.00044252	0.00018233	0.000089133	0.000049521	0.000030753	0.000021400	0.000016967
Any N	$C_m = \frac{N+1}{\pi^4}\left[\frac{1}{m^4} + \sum_{j=1}^{\infty}\{[2j(N+1)+m]^{-1/4} - [2j(N+1)-m]^{-1/4}\}\right]$									

Stiffeners with Fixed Ends

Number of Girders N	C_1	C_2	C_3	C_4	C_5	C_6	C_7	C_8	C_9	C_{10}
1	0.0052083									
2	0.0061728	0.0011431								
3	0.0080419	0.0011393	0.00042165							
4	0.010009	0.0013459	0.00039075	0.00020078						
5	0.011997	0.0015917	0.00043081	0.00018009	0.00011111					
6	0.013990	0.0018480	0.00048904	0.00018923	0.000098217	0.000067910				
7	0.015986	0.0021078	0.00055303	0.00020779	0.000099794	0.000059682	0.000044545			
8	0.017982	0.0023691	0.00061925	0.00022977	0.00010668	0.000059226	0.000039097	0.000030804		
9	0.019979	0.0026311	0.00068645	0.00025320	0.00011572	0.000061961	0.000038155	0.000027067	0.000022193	
10	0.021976	0.0028934	0.00075415	0.00027732	0.00012573	0.000066109	0.000039232	0.000026101	0.000019547	0.000016522

EXAMPLE 7.9   Interchange Girders and Stiffeners   For a demonstration of the interchanging of girders and stiffeners, consider the 5×3 grillage of Fig. 7-10a. The usual model is shown in Fig. 7-10b. The option in which the girders and stiffeners are interchanged is displayed in Fig. 7-10c. Note the differences in the frequencies calculated. The frequency ω_{11} should be the same for both options. In fact, the results for the first few modes should be very close for both options. The differences will be greater for the higher modes. In general, the second option (Fig. 7-10c) will lead to more accurate frequencies because the redistribution of mass during the modeling is less, since $S_g > S_s$.

Note that in the first option, the frequencies ω_{11}, ω_{12}, ω_{13} are obtained from the beam program in one run when three frequencies are requested. However, when frequencies are determined for the model of the second option (Fig. 7-10c), the beam program must be run three times for three different sets of $k(C_n, n = 1, 2, 3)$.

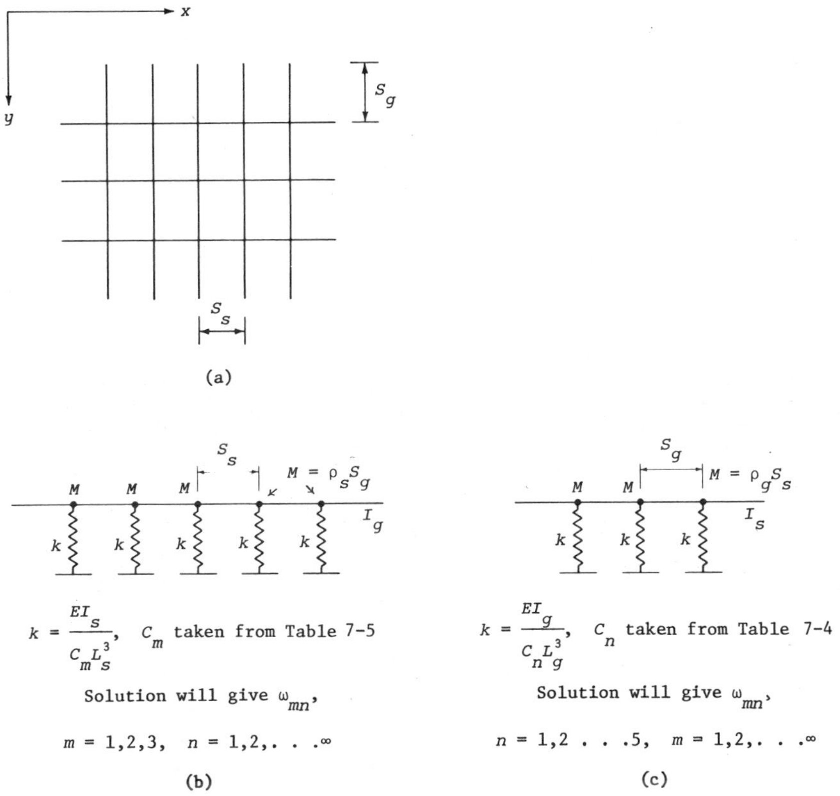

(a)

$k = \dfrac{EI_s}{C_m L_s^3}$, C_m taken from Table 7–5

Solution will give ω_{mn},

$m = 1,2,3$, $n = 1,2, \ldots \infty$

(b)

$k = \dfrac{EI_g}{C_n L_g^3}$, C_n taken from Table 7–4

Solution will give ω_{mn},

$n = 1,2 \ldots 5$, $m = 1,2, \ldots \infty$

(c)

Fig. 7-10   (a) A 5×3 grillage. (b) Calculation of natural frequencies with a beam program using the girders modeled as beams. (c) Calculation of the natural frequencies with a beam program using the stiffeners modeled as beams.

EXAMPLE 7.10 Frequency Calculation Find the natural frequencies of a 3×3 grillage with $L_s = L_g = 40$ in, $I_s = I_g = 100$ in^4, $E = 10(10^6)$ lb/in^2, $\rho_s = \rho_g = 1$ mass per inch. All ends are simply supported.

Since the girders are spaced 10 in apart, $S_g = 10$ in and the stiffener mass M_{sg} to be lumped at the intersections is $\rho_s S_g = 10$. In addition to these concentrated masses, the girder beam model will have springs of constants given by Eq. (7.5) at the stiffener intersection locations (10 in, 20 in, and 30 in). If, for example, for $N = 3$ in Table 7-5, we choose $C_1 = 0.041089$ associated with the ω_{1n} mode of the grillage, then $k = EI_s/C_1 L_s^3 = 3.8(10^5)$. The beam program finds the first three frequencies in hertz to be $\omega_{11} = 31.2$, $\omega_{12} = 90.4$, $\omega_{13} = 195.2$.

For ω_{11}, ω_{12}, ω_{13} we must use C_1 associated with the first mode of the stiffeners. For ω_{21}, ω_{22}, ω_{23} use C_2.

For this problem the two modeling options of Fig. 7-10*b* and *c* lead to the same results for modes (1, 1), (2, 2), and (3, 3), but give slightly different results for (2, 1), (2, 3), (1, 2), and (3, 2).

The frequencies for this grillage can also be found from the formula of Eq. (7.2):

$$\omega_{11} = \left[\frac{10^7(100)40\left(\dfrac{\pi}{40}\right)^4 + \dfrac{4(10^7)100}{(0.04109)40^3}}{40 + 40}\right]^{\frac{1}{2}} = 195.03 \text{ rad/sec}$$

$$\omega_{12} = \left[\frac{10^7(100)40\left(\dfrac{2\pi}{40}\right)^4 + \dfrac{4(10^7)100}{(0.04109)40^3}}{40 + 40}\right]^{\frac{1}{2}} = 568.69 \text{ rad/sec}$$

$$\omega_{13} = \left[\frac{10^7(100)40\left(\dfrac{3\pi}{40}\right) + \dfrac{4(10^7)100}{(0.04109)40^3}}{40 + 40}\right]^{\frac{1}{2}} = 1249.0 \text{ rad/sec}$$

$$\omega_{21} = \left[\frac{10^7(100)40\left(\dfrac{\pi}{40}\right)^4 + \dfrac{4(10^7)100}{(0.002604)40^3}}{40 + 40}\right]^{\frac{1}{2}} = 564.84 \text{ rad/sec}$$

or $\omega_{11} = 31.04$, $\omega_{12} = 90.15$, $\omega_{13} = 198.78$, $\omega_{21} = 89.9$ in hertz.

REFERENCES

7.1 Chang, P. Y.: "A Simple Method for Elastic Analysis of Grillages," *J. Ship. Res.*, vol. 12, no. 2, June 1968.

7.2 Chang, P. Y., and Michelsen, F. C.: "On the Stability of Grillage Beams," *J. Ship Res.*, vol. 13, no. 1, March 1969.

7.3 Chang, P. Y., and Michelsen, F. C.: "A Vibration Analysis of Grillage Beams," *J. Ship Res.*, vol. 13, no. 1, March 1969.

7.4 Ellington, J. P., and McCallin, H.: "The Free Vibration of Grillages," *J. Appl. Mech.*, vol. 26, 1959, pp. 603–607.

7.5 Wah, T.: "Analysis of Laterally Loaded Gridworks," *Eng. Mech. Div., Proc. Am. Soc. Civil Eng.*, Vol. 90, 1964, pp. 83–106.

7.6 Frederick, D., and Falgout, T. E.: "On the Dynamic Response of Stiffened Rectangular Plates Subjected to Time Dependent Edge Conditions," David Taylor Model Basin Report (181) 56489A(x), 1964.

7.7 Cheng, F. Y.: "Dynamics of Prismatic and Open Section Member Grids," *Applications of Finite Element Methods in Civil Engineering, Proc. Am. Soc. Civ. Eng.*, 1969, pp. 339–373.

EIGHT

Disks

This chapter treats the static radial motion and radial vibration of disks. Displacement and stress formulas for simple disks are handled in Part A. The formulas for more complex disks are provided in Part B. Part C contains some examples of problems solved by a computer program.

The formulas of this chapter apply to plane stress models for which the axial normal stress and several shear stresses are zero. All loadings, displacements, and internal forces are axially symmetric.

In the case of disks formed of anisotropic material, use the formulas for cylinders of Chap. 9.

A. STRESS FORMULAS AND SIMPLE DISKS

8.1 Notation and Conventions

u	Radial displacement (length)
$P = \sigma_r b$	Internal radial force per unit circumferential length (normal force per unit length on r face), (force/length)
σ_r	Radial stress (force/length^2)
b	Thickness of disk (length)
$P_\phi = \sigma_\phi b$	Tangential force per unit radial length (normal force per unit length on ϕ face), (force/length)
σ_ϕ	Tangential stress (force/length^2)
p	Applied pressure, radial force per unit circumferential length (force/length). An applied pressure in the positive radial (increasing r) direction is taken to be positive.

q_r	Change in pressure in the radial direction, radial loading intensity (force/length^2); q_r is positive if its direction is along the positive r direction.
E	Modulus of elasticity (force/length^2)
ν	Poisson's ratio
α	Coefficient of thermal expansion (length/length · degree)
ΔT	Change in temperature (degree), that is, the temperature rise with respect to the reference temperature
ΔT_1	Magnitude of temperature change, uniform in r direction
ρ	Mass per unit volume (mass/length^3), (force · time^2/length^4)
Ω	Angular velocity of rotating disk (radians/time)
r, ϕ	Radial, circumferential coordinates
a_0	Radius of inner surface of disk (length)
a_L	Radius of outer surface of disk (length)

Positive displacement u and forces P, P_ϕ are shown in Fig. 8-1.

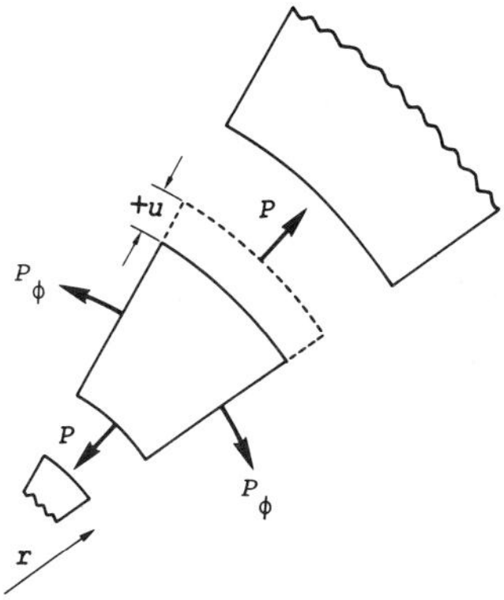

Fig. 8-1 Positive radial displacement u and forces per unit length, P, P_ϕ [or stresses (force/length^2)].

8.2 Tangential Stress

The tables of this chapter provide the radial stress (σ_r or P) and displacement (u) throughout the disk. The tangential stress can then be derived from the relation

$$\sigma_\phi = \frac{E}{r} u + \nu\sigma_r - E\alpha\,\Delta T \tag{8.1}$$

or for the tangential force per unit length

$$P_\phi = \frac{Eb}{r} u + \nu P - Eb\alpha\,\Delta T \tag{8.2}$$

8.3 Radial Displacement and Stress

The differential equations relating the radial displacement and stresses to the applied loadings for a disk with constant thickness and material properties are

$$\frac{d^2u}{dr^2} + \frac{1}{r}\frac{du}{dr} - \frac{1}{r^2}u = -\frac{(1-\nu^2)}{Eb} q_r + (1+\nu)\alpha\frac{d\Delta T}{dr}$$

$$\frac{Eb}{1-\nu^2}\frac{du}{dr} + \frac{\nu Eb}{1-\nu^2}\frac{u}{r} - \frac{Eb\alpha}{1-\nu}\Delta T = P \tag{8.3}$$

These relations are integrated to provide the radial displacement and stress (or force/circumferential length) for a disk with its inner boundary at $r = a_0$.

Radial displacement:
$$u = u_0 \frac{r}{a_0}\left[1 - \frac{1+\nu}{2}\frac{(r^2 - a_0^2)}{r^2}\right] + P_0 \frac{1-\nu^2}{2Eb}\frac{(r^2 - a_0^2)}{r} + F_u \tag{8.4a}$$

Radial force:
$$P = u_0 \frac{Eb}{2a_0 r^2}(r^2 - a_0^2) + P_0\left[1 - \frac{1-\nu}{2}\frac{(r^2 - a_0^2)}{r^2}\right] + F_P \tag{8.4b}$$

The F_u, F_P are loading functions given in Table 8-1. If there is more than one load on the disk, the F_u, F_P functions are formed by adding the terms given in Table 8-1 for each load. The initial parameters u_0, P_0 (values of u, P at the inner circumference of the disk, $r = a_0$) are provided in Table 8-2.

EXAMPLE 8.1 Rotating Disk with Internal Pressure Find the displacement and internal forces in a disk of constant thickness, rotating at angular velocity Ω, and subject to an internal pressure p (force/length) on the inner periphery at $r = a_0$.

The radial displacements and forces are given by Eqs. (8.4), with the assistance of Table 8-1 and 8-2. From Table 8-1, the loading functions for the centrifugal loading and internal pressure are given by

$$F_u(r) = -\frac{p}{Eb}\frac{1-\nu^2}{2}\frac{(r^2 - a_0^2)}{r} - \frac{(r^2 - a_0^2)^2}{r}\frac{(1-\nu^2)}{8}\frac{\rho\Omega^2}{E}$$

$$F_P(r) = -p\left[1 - \frac{1-\nu}{2}\frac{(r^2 - a_0^2)}{r^2}\right] - (r^2 - a_0^2)\frac{\rho\Omega^2 b}{4} \times\left[(1+\nu) + \frac{(1-\nu)}{2}\frac{(r^2 + a_0^2)}{r^2}\right] \tag{1}$$

The initial parameters in Eqs. (8.4) are provided by Table 8-2. The boundaries are taken to be free-free. Even with the internal pressure, the inner boundary is treated as being free since the pressure is accounted for as a loading and not as a boundary condition. From Table 8-2,

$$P_0 = 0$$

$$u_0 = -\frac{2F_P|_{r=a_L} a_0 a_L^2}{Eb(a_L^2 - a_0^2)} = \frac{p2a_0a_L^2\left[1 - (1-\nu)(a_L^2 - a_0^2)/2a_L^2\right]}{Eb(a_L^2 - a_0^2)} + \frac{a_0 a_L^2}{E}\frac{\rho\Omega^2}{2}\left[(1+\nu) + \frac{(1-\nu)}{2}\frac{(a_L^2 + a_0^2)}{a_L^2}\right] \tag{2}$$

TABLE 8-1 Loading Functions F_u, F_P for a Disk of Constant Thickness and Material Properties (Eqs. 8.4)

	Radial Force per Unit Circumferential Length (Pressure)	Constant Temperature Change ΔT_1 (Independent of r)	Centrifugal Loading Due to Rotation of Disk at the Angular Velocity Ω	Arbitrary Temperature Change $\Delta T(r)$
	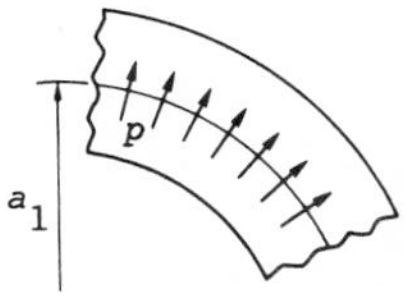			
$F_u(r)$	$-\dfrac{p}{Eb}\,\dfrac{1-\nu^2}{2}\,\dfrac{(r^2-a_1^2)}{r}\langle r-a_1\rangle^0$	$\dfrac{(r^2-a_0^2)}{2r}(1+\nu)\alpha\,\Delta T_1$	$-\dfrac{(r^2-a_0^2)^2}{r}\,\dfrac{(1-\nu^2)}{8}\,\dfrac{\rho\Omega^2}{E}$	$\dfrac{1+\nu}{r}\,\alpha\int_{a_0}^{r}\xi\,\Delta T(\xi)\,d\xi$
$F_p(r)$	$-p\left[1-\dfrac{1-\nu}{2}\,\dfrac{(r^2-a_1^2)}{r^2}\right]\langle r-a_1\rangle^0$	$-b\dfrac{(r^2-a_0^2)}{2r^2}\,E\alpha\,\Delta T_1$	$-(r^2-a_0^2)\,\dfrac{\rho\Omega^2 b}{4}\left[(1+\nu)+\dfrac{(1-\nu)}{2}\,\dfrac{(r^2+a_0^2)}{r^2}\right]$	$-\dfrac{bE\alpha}{r^2}\int_{a_0}^{r}\xi\,\Delta T(\xi)\,d\xi$

TABLE 8-2 Initial Parameters for Disk of Constant Thickness and Material Properties (Eqs. 8.4)

Outer Edge → / Inner Edge ↓	1. Fixed 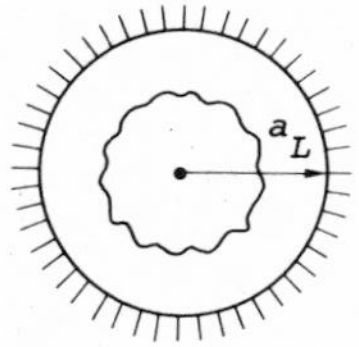	2. Free 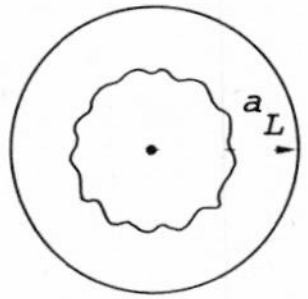
1. Fixed 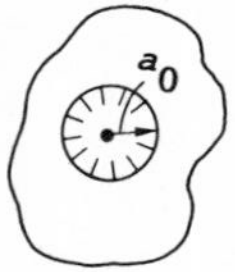 $u_0 = 0$	$P_0 = -F_u \Big/ \left[\frac{1-\nu^2}{2Eb}\frac{(a_L^2 - a_0^2)}{a_L}\right]$	$P_0 = -F_P \Big/ \left[1 - \frac{1-\nu}{2}\frac{(a_L^2 - a_0^2)}{a_L^2}\right]$
2. Free 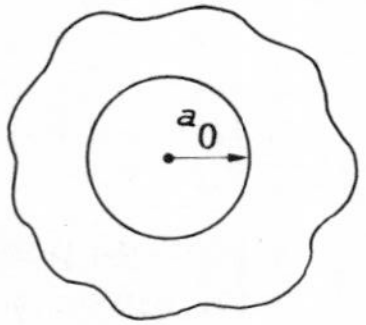 $P_0 = 0$	$u_0 = -F_u \Big/ \left[\frac{a_L}{a_0}\left(1 - \frac{1+\nu}{2}\frac{(a_L^2 - a_0^2)}{a_L^2}\right)\right]$	$u_0 = -F_P \Big/ \left[\frac{Eb}{2a_0 a_L^2}(a_L^2 - a_0^2)\right]$

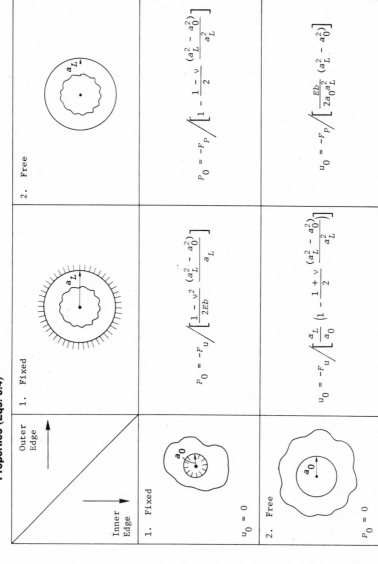

Definitions: $F_u = F_u\big|_{r=a_L}$, $F_P = F_P\big|_{r=a_L}$

where, from (1), $F_P = F_P|_{r=a_L}$ was taken to be

$$F_P = F_P|_{r=a_L} = F_P(a_L) = -p\left[1 - \frac{1-\nu}{2}\frac{(a_L^2 - a_0^2)}{a_L^2}\right]$$

$$-(a_L^2 - a_0^2)\frac{\rho\Omega^2 b}{4}\left[(1+\nu) + \frac{1-\nu}{2}\frac{(a_L^2 + a_0^2)}{a_L^2}\right]$$

With (1) and (2), Eqs. (8.4) now provide the radial displacement and force throughout the disk. The tangential force is given by Eq. (8.2) using the u and P found as just described.

B. COMPLEX DISKS

8.4 Notation for Complex Disks

t Time
ω Natural frequency (radian/second)
c' External or viscous damping coefficient (mass/(time · length^3), force · time/length^4)
c_ρ Proportional viscous damping coefficient (1/time). If c' is chosen to be proportional to the mass, that is, $c' = c_\rho \rho$, then c_ρ is the constant of proportionality.
M_i Concentrated ring mass (mass/length)
Ω^* Frequency of steady-state forces and responses (radians/time)
$\mathbf{U}_i$ Field matrix of the ith segment
$\bar{\mathbf{U}}_i$ Point matrix at $r = a_i$

The notation for transfer matrices is:

$$\mathbf{U}_i = \begin{bmatrix} U_{uu} & U_{uP} & F_u \\ U_{Pu} & U_{PP} & F_P \\ 0 & 0 & 1 \end{bmatrix} \tag{8.5}$$

8.5 Differential Equations

The fundamental equations in first-order form for the radial motion of a disk are

$$\frac{\partial u}{\partial r} = -\nu\frac{u}{r} + \frac{1-\nu^2}{Eb}P + (1+\nu)\alpha\,\Delta T \tag{8.6a}$$

$$\frac{\partial P}{\partial r} = \frac{Ebu}{r^2} + \frac{\nu-1}{r}P + \rho b\frac{\partial^2 u}{\partial t^2} - \frac{bE\alpha\,\Delta T}{r} - q_r(r, t) \tag{8.6b}$$

These relations can be reorganized to form higher-order equations similar to Eqs. (8.3).

8.6 Anisotropic Disks

The formulas of Chap. 9 for an anisotropic cylinder apply as well to the plane-stress transversely isotropic disk if the material constants c_{11}, c_{12}, c_{22}, c_{13},

TABLE 8-3 Massless Ring

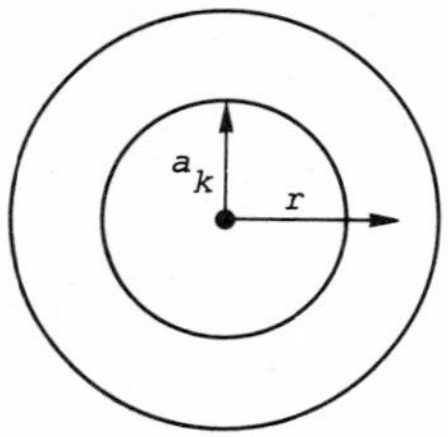

Definitions:

(a) ΔT_1 = a constant temperature change that is independent of r. Set $\Delta T(\xi) = 0$ if only a constant temperature change is present.

(b) $\Delta T(r)$ = arbitrary temperature change.

(c) Ω = angular velocity of rotation of disk that leads to centrifugal loading force.

(d) $q_r(r)$ = arbitrary loading intensity in r direction.

(e) a_k is the radial coordinate of the inner surface of the ith section. That is, the ith section begins at a_k and continues to r.

$$\mathbf{U}_i = \begin{bmatrix} \dfrac{r}{a_k}\left[1 - \dfrac{1+\nu}{2}\dfrac{(r^2 - a_k^2)}{r^2}\right] & \dfrac{1}{Eb}\,\dfrac{1-\nu^2}{2}\,\dfrac{(r^2 - a_k^2)}{r} & F_u \\ \dfrac{Eb}{2a_k r^2}(r^2 - a_k^2) & 1 - \dfrac{1-\nu}{2}\,\dfrac{(r^2 - a_k^2)}{r^2} & F_P \\ 0 & 0 & 1 \end{bmatrix}$$

$$F_u = \frac{(r^2 - a_k^2)}{2r}(1+\nu)\alpha\,\Delta T_1 + \frac{(1+\nu)}{r}\,\alpha \int_{a_k}^{r} \xi\,\Delta T(\xi)\,d\xi - \frac{(r^2 - a_k^2)^2}{r}\,\frac{(1-\nu^2)}{8}\,\frac{\rho\Omega^2}{E} - \frac{(1-\nu^2)}{Eb}\int_{a_k}^{r}\left[\eta\int_{a_k}^{\eta} q_r(\xi)\,d\xi\right]d\eta$$

$$F_P = -\frac{b(r^2 - a_k^2)}{2r^2}E\alpha\,\Delta T_1 - \frac{bE\alpha}{r^2}\int_{a_k}^{r}\xi\,\Delta T(\xi)\,d\xi - (r^2 - a_k^2)\frac{\rho\Omega^2 b}{4}\left[(1+\nu) + \frac{(1-\nu)}{2}\,\frac{(r^2 + a_k^2)}{r^2}\right] + \frac{(1-\nu)}{r^2}\int_{a_k}^{r}\eta\left[\int_{a_k}^{\eta} q_r(\xi)\,d\xi\right]d\eta - \int_{a_k}^{r} q_r(\xi)\,d\xi$$

c_{23}, λ_1, λ_2, λ_3, are defined as

$$\begin{aligned} c_{11} = c_{22} &= \frac{a_{11}}{a_{11}^2 - a_{12}^2} \qquad c_{12} = -\frac{a_{12}}{a_{11}^2 - a_{12}^2} \qquad c_{13} = c_{23} = 0 \\ \lambda_1 &= \frac{\alpha_1 a_{11} - \alpha_2 a_{12}}{a_{11}^2 - a_{12}^2} \qquad \lambda_2 = \frac{\alpha_2 a_{11} - \alpha_1 a_{12}}{a_{11}^2 - a_{12}^2} \qquad \lambda_3 = 0 \end{aligned} \tag{8.7}$$

where a_{11}, a_{12}, α_1, α_2 are material constants defined by the stress-strain relations (Ref. 8.1)

$$\begin{aligned} \epsilon_r &= \frac{\partial u}{\partial r} = a_{11}\sigma_r + a_{12}\sigma_\phi + \alpha_1\,\Delta T \\ \epsilon_\phi &= \frac{u}{r} = a_{12}\sigma_r + a_{11}\sigma_\phi + \alpha_2\,\Delta T \end{aligned} \tag{8.8}$$

8.7 Transfer Matrices for Uniform Segments

The transfer matrices for ring segments are provided in Tables 8-3 to 8-6. If desired, the transfer matrices for segments extending from radial coordinate a_k to r can be employed for a center (holeless) segment by considering a small hole to be at the center. The boundary condition for this imaginary hole should be fixed.

8.8 Point Matrices

The transfer matrices that take into account concentrated (point) occurrences are listed in Table 8-7.

8.9 Transfer Matrix for a Disk of Variable Width

A transfer matrix for a ring segment of variable width is listed in Table 8-8.

TABLE 8-4 Massless Disk Segment without Center Hole (Center Segment)

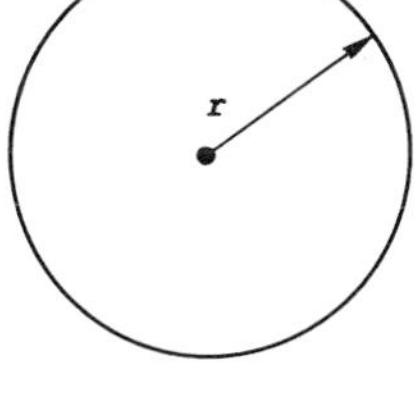

$$\mathbf{U}_i = \begin{bmatrix} 0 & \frac{r}{Eb}(1-\nu) & F_u \\ 0 & 1 & F_P \\ 0 & 0 & 0 \end{bmatrix}$$

Definitions:

The definitions of Table 8-3 still apply for ΔT_1, $\Delta T(r)$, Ω, q_r.

$$F_u = r\alpha\,\Delta T_1 + (1+\nu)\frac{\alpha}{r}\int_0^r \xi\,\Delta T(\xi)\,d\xi + r\alpha\,\frac{(1-\nu)}{2}\,\Delta T\Big|_{r=0} - (1-\nu^2)\frac{\rho\Omega^2 r^3}{8E} - \frac{(1-\nu^2)}{Ebr}\int \left[r\int q_r(r)\,dr\right] dr$$

$$F_P = -\frac{bE\alpha}{r^2}\int_0^r \xi\,\Delta T(\xi)\,d\xi + \frac{b\alpha}{2}E\,\Delta T\Big|_{r=0} - (3+\nu)\,\frac{\rho b\Omega^2 r^2}{8} + \frac{(1-\nu)}{r^2}\int r\left[\int q_r(r)\,dr\right]dr - \int q_r(r)\,dr$$

TABLE 8-5 Ring with Mass

$$
\mathbf{U}_i = \begin{bmatrix}
\frac{1}{e_1}\left[e_3(a_k)J_1(\beta r) - e_2(a_k)Y_1(\beta r)\right] & \frac{1}{e_1}\left[J_1(\beta a_k)Y_1(\beta r) - Y_1(\beta a_k)J_1(\beta r)\right] & F_u \\
\frac{1}{e_1}\left[e_3(a_k)e_2(r) - e_2(a_k)e_3(r)\right] & \frac{1}{e_1}\left[J_1(\beta a_k)e_3(r) - Y_1(\beta a_k)e_2(r)\right] & F_P \\
0 & 0 & 1
\end{bmatrix}
$$

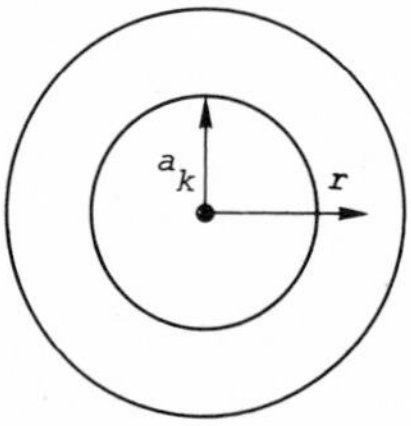

Definitions:

a_k is the radial coordinate of the inner surface of the ith section. That is, the ith section begins at a_k and continues to r.

$$\beta = \sqrt{\frac{\omega^2 \rho (1 - \nu^2)}{E}}$$

$$e_1 = e_3(a_k)J_1(\beta a_k) - e_2(a_k)Y_1(\beta a_k)$$

$$e_2(r) = \frac{1}{r}\frac{Eb}{1-\nu^2}[(1+\nu)J_1(\beta r) - \beta r J_2(\beta r)]$$

$$e_3(r) = \frac{1}{r}\frac{Eb}{1-\nu^2}[(1+\nu)Y_1(\beta r) - \beta r Y_2(\beta r)]$$

$J_1(\beta r)$, $J_2(\beta r)$ are Bessel functions of the first kind of orders one and two, respectively.

$Y_1(\beta r)$, $Y_2(\beta r)$ are Bessel functions of the second kind of orders one and two, respectively.

TABLE 8-6 Disk Segment without Center Hole (Center Segment), including Mass

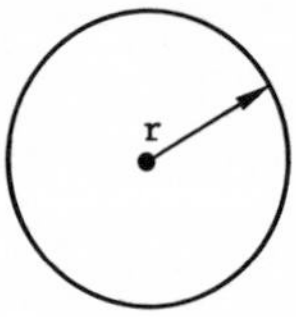

Definitions:

$$\beta = \sqrt{\frac{\omega^2 \rho (1 - \nu^2)}{E}}$$

$J_1(\beta r)$, $J_2(\beta r)$ are Bessel functions of the first kind of orders one and two, respectively.

$$
\mathbf{U}_i = \begin{bmatrix}
0 & \frac{2J_1(\beta r)}{\beta E b}(1 - \nu) & F_u \\
0 & \frac{2}{\beta(1 + \nu) r}[(1 + \nu)J_1(\beta r) - \beta r J_2(\beta r)] & F_P \\
0 & 0 & 1
\end{bmatrix}
$$

TABLE 8-7 Point Matrices at $r = a_i$

Definitions:

(a) Pressure (Radial Force Per Unit Circumferential Length)

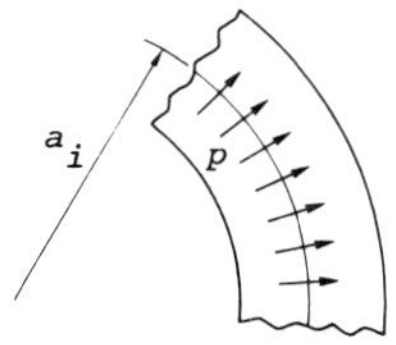

$$\bar{\mathbf{U}}_i = \begin{bmatrix} 1 & 0 & 0 \\ k - M_i\omega^2 & 1 & -p \\ 0 & 0 & 1 \end{bmatrix}$$

(b) Elastic Support

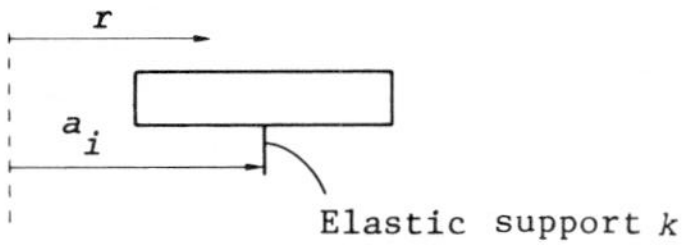

(c) Concentrated Mass

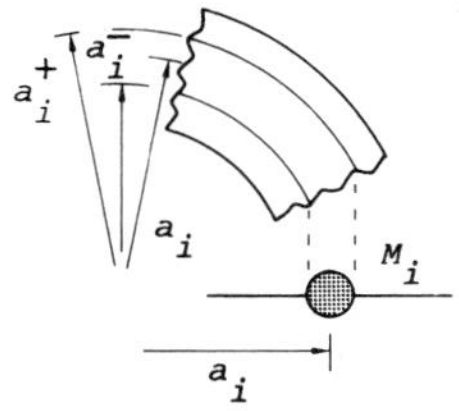

M_i is the mass per unit circumferential length.

M_i can be calculated as

$$M_i = \frac{(a_i^+)^2 - (a_i^-)^2}{2a_i} b\rho = \Delta a \, b\rho$$

where ρ is the mass per unit volume

$$\Delta a = a_i^+ - a_i^-$$

TABLE 8-8 Ring of Hyperbolic Profile

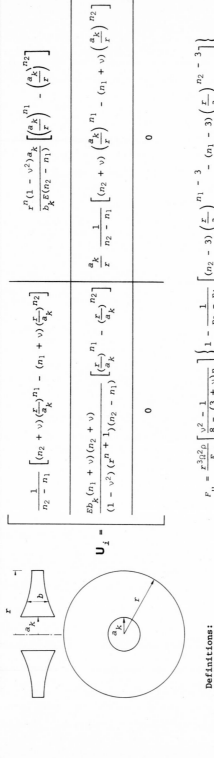

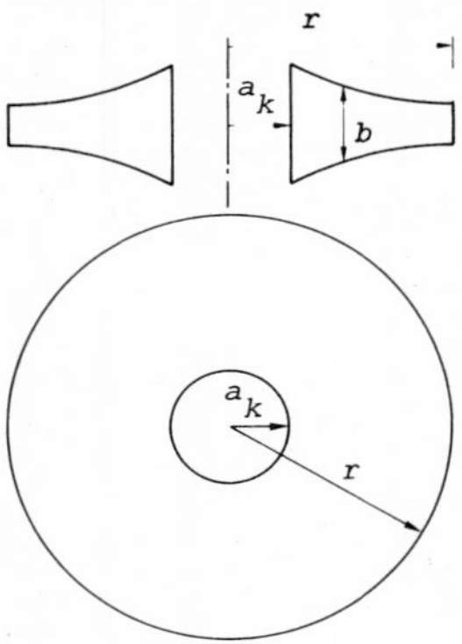

$$\mathbf{U}_i = \begin{bmatrix} \frac{1}{n_2 - n_1}\left[(n_2 + \nu)\left(\frac{r}{a_k}\right)^{n_1} - (n_1 + \nu)\left(\frac{r}{a_k}\right)^{n_2}\right] & \frac{r^n(1 - \nu^2)a_k}{b_k E(n_2 - n_1)}\left[\left(\frac{a_k}{r}\right)^{n_1} - \left(\frac{a_k}{r}\right)^{n_2}\right] & F_u \\ \frac{Eb_k(n_1 + \nu)(n_2 + \nu)}{(1 - \nu^2)(r^{n+1})(n_2 - n_1)}\left[\left(\frac{r}{a_k}\right)^{n_1} - \left(\frac{r}{a_k}\right)^{n_2}\right] & \frac{a_k}{r}\,\frac{1}{n_2 - n_1}\left[(n_2 + \nu)\left(\frac{a_k}{r}\right)^{n_1} - (n_1 + \nu)\left(\frac{a_k}{r}\right)^{n_2}\right] & F_P \\ 0 & 0 & 1 \end{bmatrix}$$

Definitions:

$b(\text{width}) = b_k/r^n$

b_k **is a reference width**

$$n_1 = \frac{n}{2} + \sqrt{\frac{n^2}{4} + \nu n + 1}$$

$$n_2 = \frac{n}{2} - \sqrt{\frac{n^2}{4} + \nu n + 1}$$

$$F_u = \frac{r^3\Omega^2\rho}{E}\left[\frac{\nu^2 - 1}{8 - (3 + \nu)n}\right]\left\{1 - \frac{1}{n_2 - n_1}\left[(n_2 - 3)\left(\frac{r}{a_k}\right)^{n_1 - 3} - (n_1 - 3)\left(\frac{r}{a_k}\right)^{n_2 - 3}\right]\right\}$$

$$F_P = -\frac{\rho\Omega^2 b_k r^{2 - n}}{8 - (3 + \nu)n}\left\{(3 + \nu) - \frac{1}{n_2 - n_1}\left[(n_1 + \nu)(n_2 - 3)\left(\frac{r}{a_k}\right)^{n_1 - 3} - (n_2 + \nu)(n_1 - 3)\left(\frac{r}{a_k}\right)^{n_2 - 3}\right]\right\}$$

Ω **= angular velocity of rotation of disk that leads to centrifugal loading force.**

a_k is the radial coordinate of the inner surface of the *i*th section. That is, the *i*th section begins at a_k and continues to r.

8.10 Loading Functions

The loading functions are provided in the transfer matrix table for most common loadings. These functions can be calculated for other loadings from the formulas:

$$\begin{aligned} F_u &= -\int_{a_i}^{r} q_r(\xi, a_i) U_{uP}(r, \xi)\, d\xi = -\int_{a_i}^{r} q_r(r, \xi) U_{uP}(\xi, a_i)\, d\xi \\ F_P &= -\int_{a_i}^{r} q_r(\xi, a_i) U_{PP}(r, \xi)\, d\xi = -\int_{a_i}^{r} q_r(r, \xi) U_{PP}(\xi, a_i)\, d\xi \end{aligned} \tag{8.9}$$

Here the notation $q_r(r, a_i)$ is employed to indicate a loading intensity (force/length^2) that begins at radius a_i and extends to radius r. The notation $U_{kj}(r, a_i)$ has a similar meaning. $U_{kj}(r, a_i)$ denotes a transfer matrix element for a ring section that begins at radius a_i and extends to radius r. Do not use U_{kj} from transfer matrices for disk segments with no center hole; use U_{kj} from transfer matrices for ring segments.

8.11 Static Response

The procedure for calculating the radial displacement and internal force due to static loading follows the technique given in Section 2.16 for beams and Section 3.10 for bars in torsion. In brief, the state vector

$$\mathbf{s} = \begin{bmatrix} u \\ P \\ 1 \end{bmatrix} \tag{8.10}$$

at any point along the radial direction is found by progressive multiplication of the transfer matrices for all occurrences from the inner radius outward up to that point. That is, the state variables at any point j are given by

$$\mathbf{s}_j = \mathbf{U}_j \mathbf{U}_{j-1} \cdots \overline{\mathbf{U}}_k \cdots \mathbf{U}_2 \mathbf{U}_1 \mathbf{s}_0 \tag{8.11}$$

where $\mathbf{s}_0 = \mathbf{s}_{r=0}$ or $\mathbf{s}_{r=a_0}$. If the outer surface of the disk occurs at station n, the state variables at the outer surface become

$$\mathbf{s}_{r=a_L} = \mathbf{s}_n = \mathbf{U}_n \mathbf{U}_{n-1} \cdots \overline{\mathbf{U}}_k \cdots \mathbf{U}_2 \mathbf{U}_1 \mathbf{s}_0 = \mathbf{U}\mathbf{s}_0 \tag{8.12}$$

The matrix $\mathbf{U}$ is the overall or global transfer matrix. The initial parameters u_0, P_0 composing $\mathbf{s}_0$ are found from the boundary conditions that occur at the inner and outer surfaces of the disk.

For a disk with no inner surface, that is, a disk without a center hole, the inner boundary condition is $u_{r=0} = u_0 = 0$. Since pressures on the inner and outer surfaces are introduced as applied loads by point matrices and not as boundary conditions, the boundary conditions for surfaces with applied pressures are taken as being free; that is, $P = 0$.

Formulas for the Initial Parameters The application of the boundary conditions to Eq. (8.12) suffices to determine the initial parameters u_0, P_0 for all disks. Formulas to accomplish the same thing are listed in Table 8-9. In this table values of u_0, P_0 are listed according to boundary conditions. The U_{kj} and F_k in

TABLE 8-9 Initial Parameters

Outer Edge → Inner Edge ↓	1. Fixed 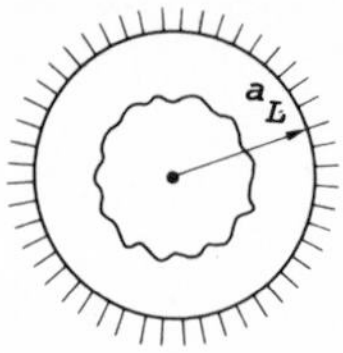	2. Free 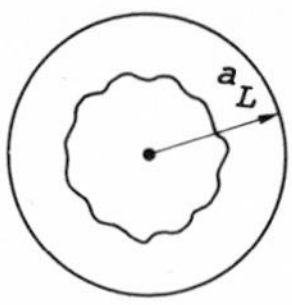
1. Fixed or Holeless Disk ($a_0 = 0$) 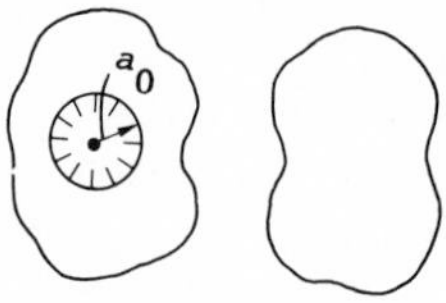 $u_0 = 0$	$P_0 = -F_u/\nabla$ $\nabla = U_{uP}$	$P_0 = -F_P/\nabla$ $\nabla = U_{PP}$
2. Free 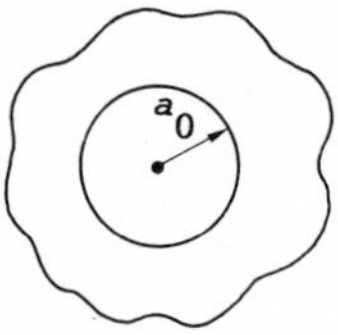 $P_0 = 0$	$u_0 = -F_u/\nabla$ $\nabla = U_{uu}$	$u_0 = -F_P/\nabla$ $\nabla = U_{Pu}$

Definitions: $F_u = F_u\big|_{r=a_L}$, $F_P = F_P\big|_{r=a_L}$, $U_{kj} = U_{kj}\big|_{r=a_L}$, $k, j = u, P$

the table are elements of the overall or global transfer matrix extending from $r = 0$ or a_0 to $r = a_L$, that is, the elements of **U** of Eq. (8.12).

Table 8-9 can be condensed into a single set of formulas. Let s_1, s_2 be the state variables (u, P but not necessarily in this order) with initial values s_{10}, s_{20}. One of the initial parameters is known by observation to be zero at the inner surface of the disk. Let this be s_{10}. In the case of a disk without a center hole, the displacement u is zero at $r = 0$ and $s_{10} = u_0$.

If s_k, $k = 1$ or 2, is the state variable that is zero at the outer surface, that is, $(s_k)_{r=a_L} = 0$, then the initial parameters are given by

$$\begin{aligned} s_{10} &= 0 \\ s_{20} &= -F_{s_k}/\nabla \\ \nabla &= U_{s_k s_2} \end{aligned} \tag{8.13}$$

where $U_{s_k s_2}$ and F_{s_k} are elements of the transfer matrix extending from $r = 0$ or a_0 to $r = a_L$, that is, the elements of **U** of Eq. (8.12).

EXAMPLE 8.2 Solid Rotating Disk with External Pressure Find the radial displacement and internal forces in a disk of constant width b with no center hole.

The disk rotates at angular velocity Ω and is subject to an external pressure p on the periphery at radius a_L.

The desired radial displacement and force are given in Table 8-4. The external pressure would normally be introduced with the assistance of the point matrix of Table 8-7. It can also be calculated using from loading functions of Table 8-4 with

$$q_r(r) = -p\langle r - a_L\rangle^{-1} \tag{1}$$

The minus sign is required since applied pressures are positive in the increasing r direction and the pressure in this case is on the outer surface pressing inward. Then, from Table 8-4,

$$\begin{aligned} u &= u_0 \cdot 0 + P_0 \frac{r}{Eb}(1-\nu) + F_u \\ P &= u_0 \cdot 0 + P_0 \cdot 1 + F_P \end{aligned} \tag{2}$$

with

$$\begin{aligned} F_u &= -(1-\nu^2)\frac{\rho\Omega^2 r^3}{8E} + \frac{(1-\nu^2)}{Ebr}\int\left[r\int p\langle r-a_L\rangle^{-1}\,dr\right]dr \\ &= -(1-\nu^2)\frac{\rho\Omega^2 r^3}{8E} + \frac{p}{Ebr}\frac{(1-\nu^2)}{2}\langle r^2 - a_L^2\rangle \end{aligned} \tag{3}$$

$$\begin{aligned} F_P &= -(3+\nu)\frac{\rho b\Omega^2 r^2}{8} - \frac{(1-\nu)}{r^2}\int r\left[\int p\langle r-a_L\rangle^{-1}\,dr\right]dr + \int p\langle r-a_L\rangle^{-1}\,dr \\ &= -(3+\nu)\frac{\rho b\Omega^2 r^2}{8} - p\frac{(1-\nu)}{2}\frac{\langle r^2-a_L^2\rangle}{r^2} + p\langle r-a_L\rangle^0 \end{aligned}$$

Note that the terms in (3) accounting for the pressure p could also have been taken from Table 8-1. Equations (3) can be reduced somewhat by recognizing that there is no meaningful case wherein r is larger than a_L. Then (2) and (3) become

$$\begin{aligned} u &= P_0 \frac{r}{Eb}(1-\nu) - (1-\nu^2)\frac{\rho\Omega^2 r^3}{8E} \\ P &= P_0 - (3+\nu)\frac{\rho b\Omega^2 r^2}{8} + p\langle r-a_L\rangle^0 \end{aligned} \tag{4}$$

According to Table 8-9, the initial parameters for this disk take the form

$$u_0 = 0 \qquad P_0 = -F_P/\nabla = -F_P|_{r=a_L}/U_{PP}|_{r=a_L} = \left[(3+\nu)\frac{\rho b\Omega^2 a_L^2}{8} - p\right]\Big/1 \tag{5}$$

Note that the outer surface is considered to be free even though a pressure is applied on the surface. The applied pressure has already been taken into account through the loading functions F_u and F_P, and not through the boundary conditions.

Substitution of (5) into (4) yields the final expressions for u and P:

$$u = -p\,\frac{r(1-\nu)}{Eb} + \frac{\rho\Omega^2 r(1-\nu)}{8E}\left[(3+\nu)a_L^2 - (1+\nu)r^2\right]$$

$$P = -p + (3+\nu)\,\frac{\rho b\Omega^2}{8}\,(a_L^2 - r^2) + p\langle r - a_L\rangle^0$$

From Eq. (8.2) the tangential force per unit length is

$$P_\phi = -p + \frac{(3+\nu)}{8}\,\rho\Omega^2 b\left[a_L^2 - \frac{1+3\nu}{3+\nu}\,r^2\right] + p\nu\langle r - a_L\rangle^0$$

EXAMPLE 8.3 Disk of Hyperbolic Profile A steel disk of hyperbolic profile is loosely attached to a rigid post as shown in Fig. 8-2. The configuration rotates at frequency Ω. Find the radial stress distribution in the disk.

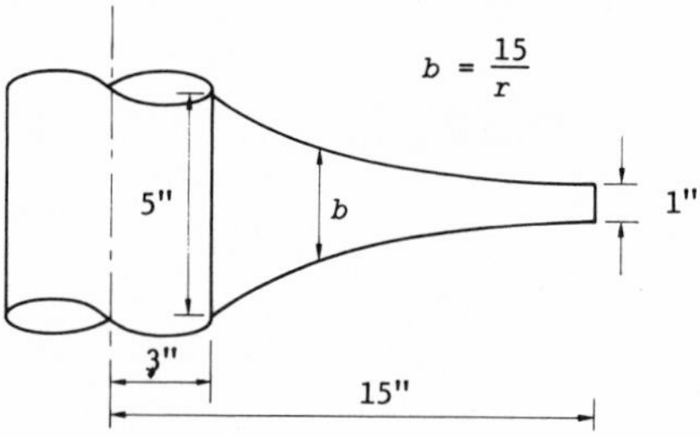

Fig. 8-2

The transfer matrix of Table 8-8 applies to this problem. For this disk $n = 1$, $b_k = 15$, $E = 3(10)^7$ lb/in^2, $\nu = 0.3$, $n_1 = 1.75$, $n_2 = -0.75$. The stress is given by

$$\sigma = \frac{P}{b} = u_0 U_{Pu}/b + P_0 U_{PP}/b + F_P/b \qquad (1)$$

where, from Table 8-8 with $a_k = 3$ in,

$$U_{Pu} = 182.5(10)^6\,\frac{1}{r^2}\left[\left(\frac{r}{3}\right)^{1.75} - \left(\frac{r}{3}\right)^{-0.75}\right]$$

$$F_P = -3.19\rho\Omega^2 r\left[3.3 - 3.075\left(\frac{r}{3}\right)^{-1.25} - 0.225\left(\frac{r}{3}\right)^{-3.75}\right] \qquad (2)$$

It is not necessary to look up U_{PP} since P_0 of (1) is zero as a result of the loose fit at the inner boundary. Also, at the outer circumference, where $r = a_L = 15$ in, the force P is zero. From Table 8-9 for free-free edge conditions

$$P_0 = 0,\ u_0 = -F_P/\nabla\,|_{r=a_L} \qquad \nabla = U_{Pu}|_{r=a_L} \qquad (3)$$

Substitution of $r = a_L = 15$ in into (2) gives

$$U_{Pu}|_{r=a_L} = 13.32(10)^6 \qquad F_P|_{r=a_L} = -138.2\ \rho\Omega^2 \qquad (4)$$

Equations (3) and (4) lead to $u_0 = 1.04(10)^{-5}\rho\Omega^2$. If this value of u_0 is placed in

(1), we find

$$\sigma = \left(21.05r^{0.75} - \frac{286}{r^{1.75}} - 0.70r^2\right)\rho\Omega^2 \tag{5}$$

The radial stress in the disk is completely defined by (5).

EXAMPLE 8.4 Shrink-Fit Disk-Shaft System A disk-shaft system (Fig. 8-3) can be used to illustrate both the solution to a complicated disk problem and the treatment of shrink-fit.

Suppose the shaft can be modeled as a solid (holeless) disk of thickness equal to that (b_1) of the outer disk where it is connected to the shaft (Fig. 8-4*a*). The outer disk, which possesses a variable width, will be handled as a succession of disks, each with a constant width (Fig. 8-4*b*). Assume the disk to be shrink-fitted to the shaft.

Expressions for the displacements and forces in the shaft can be treated separately and then the results joined to account for shrink-fit effects. In the case of the shaft, the responses are taken from Table 8-4 as

$$u_{r=a_0} = P_0 U_{uP}\big|_{r=a_0} + F_u\big|_{r=a_0} = P_0 \frac{a_0}{Eb_1}(1-\nu) - (1-\nu^2)\frac{\rho\Omega^2 a_0^3}{8E} \tag{1a}$$

$$P_{r=a_0} = P_0 U_{PP}\big|_{r=a_0} + F_P\big|_{r=a_0} = P_0 - (3+\nu)\frac{\rho b_1 \Omega^2 a_0^2}{8} \tag{1b}$$

For the disk of variable cross section, the overall transfer matrix **U** of Eq. (8.12) must be developed. The transfer matrix in Eq. (8.12) for each ring of the disk is, from Table 8-3 with $r = a_i$,

$$\mathbf{U}_i = \begin{bmatrix} \dfrac{a_i}{a_{i-1}}\left[1 - \dfrac{1+\nu}{2} \times \dfrac{(a_i^2 - a_{i-1}^2)}{a_i^2}\right] & \dfrac{1-\nu^2}{2Eb_i}\dfrac{(a_i^2 - a_{i-1}^2)}{a_i} & -\dfrac{(a_i^2 - a_{i-1}^2)}{a_i} \times \dfrac{(1-\nu^2)}{8}\dfrac{\rho\Omega^2}{E} \\ \dfrac{Eb_i}{2a_{i-1}a_i^2} \times (a_i^2 - a_{i-1}^2) & 1 - \dfrac{1-\nu}{2}\dfrac{(a_i^2 - a_{i-1}^2)}{a_i^2} & -\dfrac{(a_i^2 - a_{i-1}^2)\rho\Omega^2 b_i}{4} \times \left[(1+\nu) + \dfrac{(1-\nu)}{2}\dfrac{(a_i^2 + a_{i-1}^2)}{a_i^2}\right] \\ 0 & 0 & 1 \end{bmatrix}$$

where $i = 1, \ldots, n$. (2)

In nonmatrix form the displacement and force at $r = a_L$ are

$$u_{r=a_L} = u_{a_0} U_{uu}(a_L, a_0) + P_{a_0} U_{uP}(a_L, a_0) + F_u(a_L, a_0) \tag{3a}$$

$$P_{r=a_L} = u_{a_0} U_{Pu}(a_L, a_0) + P_{a_0} U_{PP}(a_L, a_0) + F_P(a_L, a_0) \tag{3b}$$

where $U_{jk}(a_L, a_0)$ are components of the overall transfer matrix from $r = a_0$ to $r = a_L$.

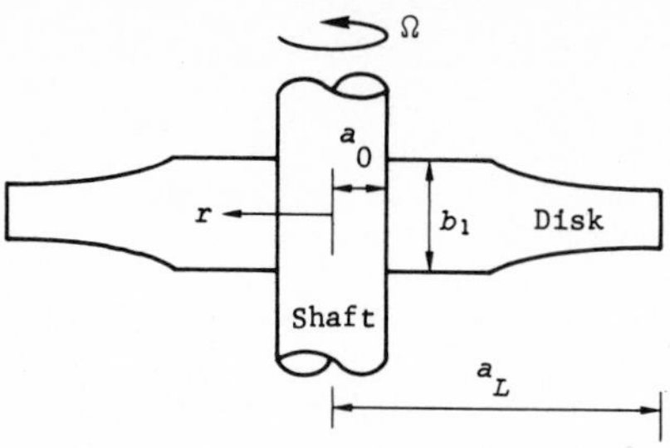

Fig. 8-3

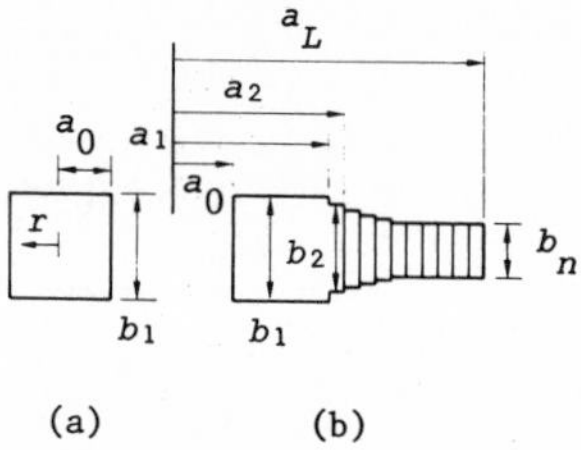

Fig. 8-4 (*a*) Shaft. (*b*) Disk.

Generation of a solution requires knowledge of the initial parameters u_0, P_0 of (1) and u_{a_0}, P_{a_0} of (3). For a member with no shrink-fit, these can be looked up as the initial parameters in Table 8-9. In the case of a shrink-fit problem, the interaction between the shaft and disk provides the condition necessary to evaluate the initial parameters. Suppose that r_s is the outer radius of the shaft before shrinking, r_D is the inner radius of the disk before shrinking, and a_0 is the inner radius of the disk and the outer radius of the shaft after shrinking. Also, let $u_D|_{r=a_0}$ be the displacement of the inner radius of the disk upon shrink fitting and $-u_s|_{r=a_0}$ the displacement of the outer radius of the shaft after shrinking. The shrink-fit deformation, or shrinkage, is

$$\Delta_{sf} = r_s - r_D = u_D|_{r=a_0} - u_s|_{r=a_0} \tag{4}$$

The shaft displacement $u_s|_{r=a_0}$ is found in terms of $P_{r=a_0} = P_{a_0}$ by eliminating P_0 from (1*a*) and (1*b*).

$$u_s|_{r=a_0} = \frac{P_{a_0} - F_P|_{r=a_0}}{U_{PP}|_{r=a_0}} U_{uP}|_{r=a_0} + F_u|_{r=a_0} \tag{5}$$

The disk displacement $u_D|_{r=a_0}$ is given by (3*b*) as a function of P_{a_0}

$$u_D|_{r=a_0} = u_{a_0} = \frac{P_{r=a_L} - F_P(a_L, a_0)}{U_{Pu}(a_L, a_0)} - P_{a_0}\frac{U_{PP}(a_L, a_0)}{U_{Pu}(a_L, a_0)} \tag{6}$$

Substitution of (5) and (6) into (4) provides a relationship sufficient to solve a variety of shrink-fit problems. For example, for a prescribed shrinkage Δ_{sf} the pressure P_{a_0} between the shaft and disk can be computed from (4). Or, the shrinkage Δ_{sf} necessary to achieve a given interaction pressure P_{a_0} can be calculated. In each case, a specified external pressure on the disk can be included. In problems involving cylinders (Chap. 9), it is common to include the effect of an internal pressure on an inner cylinder. With either Δ_{sf} or P_{a_0} given, it is possible to calculate all displacements and stresses in the shaft and disk.

EXAMPLE 8.5 Shrink-Fit Analysis This example is from Ref. 8.2. For the disk shaft system with a disk as shown in Fig. 8-5, the pressure between the shaft and disk is required to be $P_{a_0} = -5580$ lb/in. An external pressure of 3360 lb/in is exerted on the outer rim of the disk. Find the displacements on the inner and outer rims if the disk spins at 3000 rpm, $E = 30(10)^6$ lb/in^2, $\nu = 0.3$. The weight

$W_i/2\pi$ of each ring mass in pounds per radian (calculated for a specific weight of 0.284 lb/in^3) is shown in Fig. 8-5.

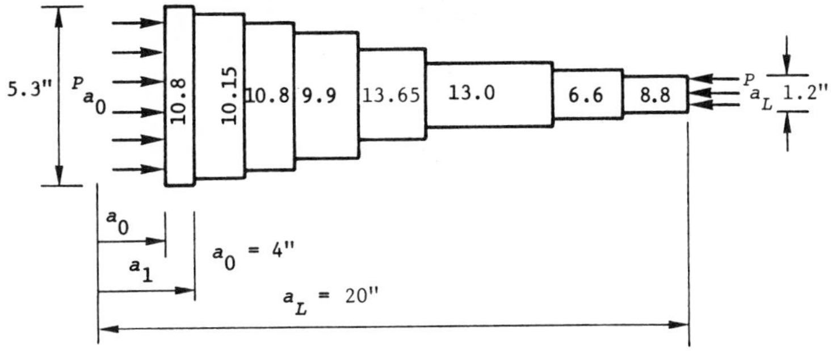

Fig. 8-5 Example 8.5.

With this information, Δ_{sf} is calculated from (4) of the previous example. The stresses and displacements are given by (3*a*), (3*b*), with the transfer matrices of (2). The external pressure is incorporated by employing a point matrix of Table 8-7 with $p = -3360$ lb/in as the final term in the progressive matrix multiplication. In (3), P_{a_0} is given in the problem statement and u_{a_0} is calculated from (6). The displacement of the inner rim of the disk is found to be 0.002 in and of the outer rim, 0.004 in.

8.12 Free Dynamic Response—Natural Frequencies

The natural frequencies ω_n, $n = 1, 2, \ldots,$ are roots of the characteristic equation resulting from the evaluation of the initial parameters. If this characteristic equation is represented by ∇, then the natural frequencies ω_n are those values of ω in ∇ that make ∇ equal to zero.

The procedure for calculating frequencies and mode shapes involves many of the same steps used in a static analysis. The model employed for free dynamic analyses requires the inclusion of mass. As with other structural members, the mass can be modeled as being continuously distributed or it can be lumped at particular locations along the radius. See Sections 2.18 or 3.11 for details and examples on computing the natural frequencies and mode shapes.

Formulas for the Initial Parameters The natural frequencies ω_n, $n = 1, 2, \ldots,$ of a disk are those values of ω that make ∇ of Table 8-9 or Eqs. (8.13) equal to zero. This ∇ is formed from the elements of the global transfer matrix **U** of Eq. (8.12). For complex disks it is necessary to evaluate the frequencies from $\nabla = 0$ with a computational root-finding technique.

The mode shapes for the radial displacement and internal force are given by Eq. (8.11) using the initial parameters

$$\left.\begin{aligned} u_0 &= 0 \\ P_0 &= P_0 \end{aligned}\right\} \quad \text{for fixed inner surfaces or holeless disks}$$
$$\left.\begin{aligned} u_0 &= u_0 \\ P_0 &= 0 \end{aligned}\right\} \quad \text{for free inner surfaces} \tag{8.14}$$

Note that in each case one initial parameter remains unspecified. This is to be expected since mode shapes describe only a shape of a state variable and not a precise amplitude. Frequently this unspecified initial parameter is set equal to unity. The third-column terms in the transfer matrices, that is, the loading functions F_u, F_P, can be ignored during the calculation of natural frequencies and mode shapes.

In terms of the definitions of Eq. (8.13), the initial parameters of the mode shapes are defined by

$$\begin{aligned} s_{10} &= 0 \\ s_{20} &= s_{20} \end{aligned} \tag{8.15}$$

where s_{20} is assigned to be the initial parameter that can be given any value (usually 1).

8.13 Forced Harmonic Motion

If the forcing functions vary as $\sin \Omega^* t$ or $\cos \Omega^* t$, where Ω^* is the frequency of the loading, the state variables u and P also vary as $\sin \Omega^* t$ or $\cos \Omega^* t$. The spatial distribution of u and P are found by replacing ω in the appropriate transfer matrices by Ω^*. The solution procedure is the same as for the static response since time need not enter the analysis. The initial parameters are given by Table 8-9 or Eqs. (8.13).

8.14 Dynamic Response Due to Arbitrary Loading

In terms of a modal solution, the time-dependent state variables resulting from arbitrary dynamic loading are expressed by

Displacement method:

$$\begin{aligned} u(r,t) &= \sum_n A_n(t) u_n(r) \\ P(r,t) &= \sum_n A_n(t) P_n(r) \end{aligned} \tag{8.16a}$$

Acceleration method:

$$\begin{aligned} u(r,t) &= u_s(r,t) + \sum_n B_n(t) u_n(r) \\ P(r,t) &= P_s(r,t) + \sum_n B_n(t) P_n(r) \end{aligned} \tag{8.16b}$$

where

$$A_n(t) = \frac{\eta_n(t)}{N_n} \qquad B_n(t) = \frac{\xi_n(t)}{N_n} \tag{8.17}$$

These relations provide the complete dynamic response. That is, these are the solutions to Eqs. (8.6).

No Damping or Proportional Damping For viscous damping the term $c'b\, \partial u/\partial t$ should be added to the right-hand side of Eq. (8.6*b*). If the damping is "proportional," then $c' = c_p \rho$. Set $c' = 0$ for no viscous radial damping.

In Eqs. (8.16) the terms with the subscript s are static solutions, as given in Section 8.11, that are determined for the applied loading at each point in time. The terms u_n, P_n are the mode shapes, that is, u, P with $\omega = \omega_n$, $n = 1, 2, \ldots,$ as explained in Section 8.12. The quantities in Eq. (8.17) are defined as

$$N_n = \int_{a_0}^{a_L} \rho b r u_n^2 \, dr$$

$$\eta_n(t) = e^{-\zeta_n \omega_n t}\left[\cos \alpha_n t + \frac{\zeta_n \omega_n}{\alpha_n} \sin \alpha_n t\right]\eta_n(0) + e^{-\zeta_n \omega_n t}\,\frac{\sin \alpha_n t}{\alpha_n}\,\frac{\partial \eta_n}{\partial t}(0)$$

$$+ \int_0^t f_n(\tau) e^{-\zeta_n \omega_n (t-\tau)} \frac{\sin \alpha_n (t-\tau)}{\alpha_n}\, d\tau \tag{8.18}$$

where

$$\alpha_n = \omega_n \sqrt{1 - \zeta_n^2} \qquad \zeta_n = c_\rho / 2\omega_n$$

$\xi_n(t)$ is taken from $\eta_n(t)$ by replacing

$$\eta_n(0) \quad \text{by} \quad \eta_n(0) - \frac{f_n(0)}{\omega_n^2}$$

$$\frac{\partial \eta_n}{\partial t}(0) \quad \text{by} \quad \frac{\partial \eta_n}{\partial t}(0) - \frac{1}{\omega_n^2}\frac{\partial f_n}{\partial t}(0)$$

$$f_n(\tau) \quad \text{by} \quad -\frac{1}{\omega_n^2}\left(\frac{\partial^2}{\partial \tau^2} + c_\rho \frac{\partial}{\partial \tau}\right) f_n(\tau)$$

If $\zeta_n > 1$, replace sin by sinh, cos by cosh, and $\sqrt{1-\zeta_n^2}$ by $\sqrt{\zeta_n^2 - 1}$. For zero viscous damping, $\zeta_n = 0$, and

$$\eta_n(0) = \int_{a_0}^{a_L} \rho b r u(r, 0) u_n \, dr \tag{8.19}$$

$$f_n(t) = \int_{a_0}^{a_L} \{q_r(r,t) r u_n + \alpha\, \Delta T(r,t)[E b u_n + r(1+\nu)P_n]\}\, dr + a_0 u(a_0, t) P_n(a_0)$$

$$- a_L u(a_L, t) P_n(a_L) - a_0 p(a_0, t) u_n(a_0) + a_L p(a_L, t) u_n(a_L) \tag{8.20}$$

where $q_r(r, t)$ is the applied pressure intensity; $u(a_0, t)$ and $u(a_L, t)$ are the applied time-dependent radial displacements at the inner and outer surfaces, respectively; $p(a_0, t)$ and $p(a_L, t)$ are the applied time-dependent radial pressures on the inner and outer surfaces, respectively.

Replace a_0 by 0 for a disk with no center hole.

In the case of the acceleration method, $f_n(t)$ can alternatively be expressed as

$$f_n(t) = \omega_n^2 \int_{a_0}^{a_L} \rho b r u_s(r, t) u_n \, dr \tag{8.21}$$

Nonproportional Damping The formulas of this chapter need considerable adjustment in order to make them applicable to materials with nonproportional viscous damping. As noted above, Eqs. (8.6*b*) should be supplemented by the term $c'b\,\partial u/\partial t$ on the right-hand side. The frequency s_n is now a complex number. The real part of s_n is commonly referred to as the damping exponent whereas the imaginary part of s_n is called the frequency of the damped free vibration. Also, the mode shapes and some of the transfer matrices are complex functions.

Equations (8.16) and (8.17) remain valid but now the static solutions refer to

$$u_s(r,t) = -\sum_n \frac{u_n(r) f_n(t)}{s_n N_n} \qquad P_s(r,t) = -\sum_n \frac{P_n(r) f_n(t)}{s_n N_n} \tag{8.22}$$

where $u_n(r)$, $P_n(r)$, and N_n are the mode shapes and norm associated with the complex eigenvalue s_n.

The definitions of this section must be adjusted so that

$$N_n = 2s_n \int_{a_0}^{a_L} \rho b r u_n^2 \, dr + \int_{a_0}^{a_L} c' b r u_n^2 \, dr \tag{8.23}$$

$$\eta_n(t) = e^{s_n t}\eta_n(0) + \int_0^t e^{s_n(t-\tau)} f_n(\tau)\, d\tau \tag{8.24}$$

$$\xi_n(t) = e^{s_n t}\left[\frac{1}{s_n^2}\frac{\partial f_n}{\partial t}(0) + \frac{1}{s_n} f_n(0) + \eta_n(0)\right] - \frac{1}{s_n^2}\frac{\partial f_n}{\partial t}(t) + \frac{1}{s_n^2}\int_0^t e^{s_n(t-\tau)} \frac{\partial^2 f_n}{\partial \tau^2}(\tau)\, d\tau \tag{8.25}$$

$$\eta_n(0) = \int_{a_0}^{a_L}\left[\rho b s_n u(r,0) u_n + c' b u(r,0) u_n(r) + \rho b \frac{\partial u}{\partial t}(r,0) u_n(r)\right] r\, dr \tag{8.26}$$

In the case of the acceleration method, $f_n(t)$ can alternatively be expressed as

$$f_n(t) = -s_n \int_{a_0}^{a_L} c' b u_s(r,t) u_n(r) r\, dr - s_n^2 \int_{a_0}^{a_L} \rho b u_s(r,t) u_n(r) r\, dr \tag{8.27}$$

For discrete damping with dashpot constants c_i at $r = a_i$, use the point matrix of Table 8-7 for $r = a_i$ with $1/k$ replaced by $1/sc_i$ and $-M_i\omega^2$ by $M_i s^2$. Also, in N_n set

$$\int_{a_0}^{a_L} c' b r u_n^2 \, dr = b \sum_i c_i a_i u_n^2(a_i) \tag{8.28}$$

and in $\eta_n(0)$ set

$$\int_{a_0}^{a_L} c' b u(r,0) u_n(r) r\, dr = b \sum_i c_i a_i u(a_i, 0) u_n(a_i) \tag{8.29}$$

Voigt-Kelvin Material The formulas of this chapter are applicable to a disk of Voigt-Kelvin material. In the equations of motion, Eqs. (8.6), replace E by $E(1 + \epsilon\, \partial/\partial t)$ where ϵ is the Voigt-Kelvin damping coefficient, and where similar behavior in dilatation and shear is assumed. Also, add a term $(c_\rho \rho b\, \partial u/\partial t)$ to the right-hand side of Eq. (8.6*b*) if the disk is externally damped in the radial direction. The formulas of this section apply if c_ρ is replaced by $c_\rho + \epsilon\omega_n^2$ and ζ_n is redefined as $\zeta_n = c_\rho/(2\omega_n) + \epsilon\omega_n/2$. Continue to use the same undamped mode shapes employed above, where ω_n is the corresponding nth undamped frequency.

C. COMPUTER PROGRAMS AND EXAMPLES

8.15 Benchmark Examples

Complicated disk problems should be solved with the assistance of a computer program. The following examples are provided as benchmark examples against which a reader's own computer program can be checked. The computer program DISK was used for these examples. DISK can be used to compute the radial displacement, stresses, and natural frequencies in a disk of any complexity. The disk can be formed of layers of different materials with arbitrary mechanical or thermal loading and boundary conditions.

EXAMPLE 8.6 Thermally Loaded Disk Find the displacements and stresses in a disk subjected to piecewise constant temperature changes as shown in Fig. 8-6.

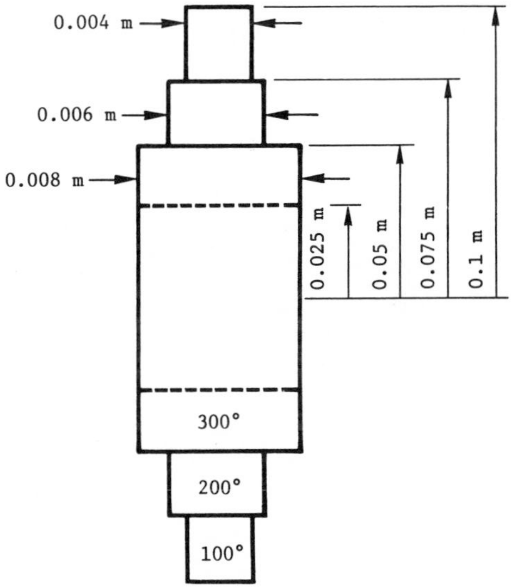

Fig. 8-6 Example 8.6.

RADIAL LOCATION	DISPLACEMENT	RADIAL STRESS	TANGENTIAL STRESS
.25000E-01	.94569E-04	0.	-.32344E+00
.50000E-01	.22856E-03	-.16172E+00	.14572E+00
.75000E-01	.30121E-03	-.11448E+00	.40887E+00
.10000E+00	.32719E-03	-.34694E-14	.29439E+00

Fig. 8-7 Results for Example 8.6.

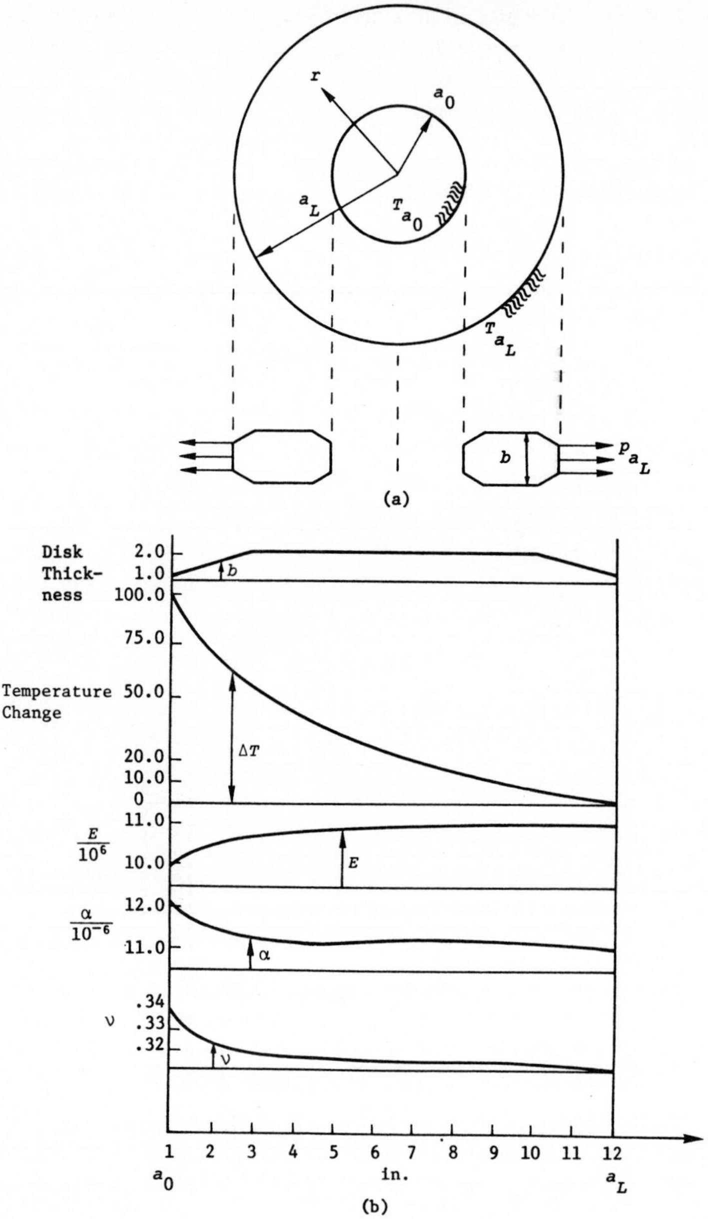

Fig. 8-8 Example 8.7.

The inner radius is 0.025 m and the outer radius is 0.1 m. Let $E = 200\ \text{GN/m}^2$, $\nu = 0.3$, $\alpha = 1.8(10^{-5})$/degree. The inner and outer boundaries are free to expand.

The results are given in Fig. 8-7.

EXAMPLE 8.7 Disk with Temperature-Dependent Material Properties Find the peak stresses and displacements in the disk of Fig. 8-8a. The temperatures of the inner and outer surfaces are T_{a_0}, T_{a_L} relative to a reference temperature. Let $T_{a_L} - T_{a_0} = 100°$, $a_0 = 1.0$ in, $a_L = 12.0$ in, $p_{a_L} = 1000$ lb/in. The material properties vary with temperature as

$$E = [11 - \Delta T(r)/100]10^6 \text{ psi} \qquad \nu = 0.32 + 0.00025\,\Delta T(r)$$

$$\alpha = \left(11 + \frac{\Delta T}{100}\right) \text{in/in} \cdot \text{degree}$$

The width b of the disk varies linearly from 1 in at the inner radius of 1.0 in to 2 in at $r = 3$ in. Then, b remains constant from $r = 3$ in to $r = 10$ in, where it decreases linearly to 1 in at $r = a_L = 12$ in. The steady-state temperature distribution in the disk is given by the following equation from Ref. 8.3:

$$\Delta T(r) = (T_{a_L} - T_{a_0}) \ln (a_L/r)/\ln (a_L/a_0)$$

The resulting variation of the essential parameters is shown in Fig. 8-8*b*.

The disk can be replaced by a model of piecewise constant segments. An average constant ΔT_1 can be chosen for each segment. Similar constant properties are chosen for b, E, ν, and α. The transfer matrix of Table 8-3, with a loading column of

$$F_u = \frac{r^2 - a_k^2}{2r}(1 + \nu_k)\alpha_k(\Delta T_1)_k \qquad F_P = -\frac{b_k(r^2 - a_k^2)}{2r^2} E_k\alpha_k(\Delta T_1)_k$$

is employed for each segment. The external pressure is taken into account with the point matrix of Table 8-7.

The peak values are computed as approximately $u = 0.003$ in on the outer rim, $P = -4000$ lb/in at $r = 2.5$ in, and $P_\phi = 5000$ lb/in at $r = 10$ in.

REFERENCES

8.1 Lekhnitskii, S.: *Theory of Elasticy of an Anisotropic Elastic Body*, Holden-Day, San Francisco, 1963.

8.2 Pestel, E., and Leckie, F.: *Matrix Methods in Elastomechanics*, McGraw-Hill, New York, 1963.

8.3 Zudans, Z., Yen, T. C., and Steigelmann, W. H.: *Thermal Stress Techniques*, American Elsevier, New York, 1965.

NINE

Thick Cylinders

This chapter treats the static and dynamic radial motion of long thick cylinders. Displacement and stress formulas for simple cylinders are handled in Part A. The formulas for more complex cylinders are provided in Part B. Part C contains some examples of problems solved by a computer program.

The formulas of this chapter apply to plane strain models for which the axial strain in the cylinder is zero. All loadings, displacements, and stresses are axially symmetric.

A. STRESS FORMULAS AND SIMPLE CYLINDERS

9.1 Notation and Conventions

u	Radial displacement (length)
$\sigma = \sigma_r$	Radial stress (force/length^2)
σ_ϕ	Tangential stress (force/length^2)
σ_z	Axial stress, positive in tension (force/length^2)
p	Applied radial pressure (force/length^2). An applied pressure in the positive radial (increasing r) direction is taken to be positive.
q_r	Change in pressure in the radial direction, radial loading

	intensity (force/length^3); q_r is positive if its direction is along the positive r direction.
E	Modulus of elasticity (force/length^2)
ν	Poisson's ratio
$G = E/2(1+\nu)$	Shear modulus, Lamé coefficient (force/length^2)
$\lambda = E\nu/[(1+\nu)(1-2\nu)]$	Lamé coefficient (force/length^2)
α	Coefficient of thermal expansion (length/length · degree)
ΔT	Change in temperature (degrees), that is, the temperature rise with respect to the reference temperature
ρ	Mass per unit volume (mass/length^3) (force · time^2/length^4)
Ω	Angular velocity of rotating disk (radians/time)
r, ϕ	Radial, circumferential coordinates
a_0	Radius of inner surface of cylinder (length)
a_L	Radius of outer surface of cylinder (length)
ΔT_1	Magnitude of temperature change, uniform in r direction

Positive displacement u and stresses σ, σ_ϕ are shown in Fig. 9-1.

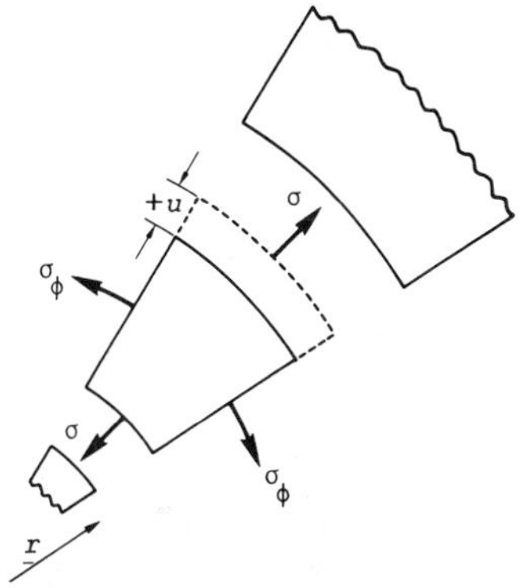

Fig. 9-1 Positive radial displacement u and stresses σ, σ_ϕ.

c_{11}, c_{12}, c_{13}, c_{22}, c_{23}, λ_1, λ_2, λ_3, are elastic constants for anisotropic material (usually called transversely isotropic material) defined according to the stress displacement relations (Ref. 9.1):

$$\sigma_r = c_{11}\frac{\partial u}{\partial r} + c_{12}\frac{u}{r} - \lambda_1\,\Delta T \qquad \sigma_\phi = c_{12}\frac{\partial u}{\partial r} + c_{22}\frac{u}{r} - \lambda_2\,\Delta T$$
$$\sigma_z = c_{13}\frac{\partial u}{\partial r} + c_{23}\frac{u}{r} - \lambda_3\,\Delta T \tag{9.1}$$

In terms of the stress-strain relations (Ref. 9.2),

$$\epsilon_r = \frac{\partial u}{\partial r} = a_{11}\sigma_r + a_{12}\sigma_\phi + a_{13}\sigma_z + \alpha_1\,\Delta T$$
$$\epsilon_\phi = \frac{u}{r} = a_{12}\sigma_r + a_{22}\sigma_\phi + a_{13}\sigma_z + \alpha_2\,\Delta T \qquad (a_{11} = a_{22}) \tag{9.2}$$
$$\epsilon_z = a_{13}(\sigma_r + \sigma_\phi) + a_{33}\sigma_z + \alpha_3\,\Delta T$$

The constants of Eqs. (9.1) and (9.2) are interrelated by

$$c_{11} = c_{22} = \frac{\left(a_{11} - a_{13}^2/a_{33}\right)^2}{\left(a_{11} - a_{13}^2/a_{33}\right)^2 - \left(a_{12} - a_{13}^2/a_{33}\right)^2}$$

$$c_{12} = \frac{\left(a_{12} - a_{13}^2/a_{33}\right)^2}{\left(a_{11} - a_{13}^2/a_{33}\right)^2 - \left(a_{12} - a_{13}^2/a_{33}\right)^2}$$

$$c_{13} = c_{23} = -\frac{a_{13}(c_{11} + c_{12})}{a_{33}}$$

$$\lambda_1 = \left[\alpha_1 a_{11} - \alpha_2 a_{12} - (\alpha_1 - \alpha_2)\frac{a_{13}^2}{a_{33}} - \alpha_3(a_{11} - a_{12})\frac{a_{13}}{a_{33}}\right] \Big/ A_1$$

$$\lambda_2 = \left[\alpha_2 a_{11} - \alpha_1 a_{12} - (\alpha_2 - \alpha_1)\frac{a_{13}^2}{a_{33}} - \alpha_3(a_{11} - a_{12})\frac{a_{13}}{a_{33}}\right] \Big/ A_1$$

$$A_1 = (a_{11}^2 - a_{12}^2) - 2(a_{11} - a_{12})\frac{a_{13}^2}{a_{33}} \qquad \lambda_3 = \frac{a_{13}}{a_{33}}(\lambda_1 + \lambda_2) - \frac{\alpha_3}{a_{33}}$$

9.2 Tangential and Axial Stress

The tables of this chapter provide the radial stress ($\sigma_r = \sigma$) and displacement (u) throughout the cylinder. The tangential and axial stresses can then be derived from the relations:

$$\sigma_\phi = \frac{4G(\lambda + G)}{\lambda + 2G}\frac{u}{r} + \frac{\lambda}{(\lambda + 2G)}\sigma - \frac{E\alpha\,\Delta T}{1 - \nu} \tag{9.3}$$

Isotropic material:

$$\sigma_z = \frac{\lambda 2G}{(\lambda + 2G)}\frac{u}{r} + \frac{\lambda}{(\lambda + 2G)}\sigma - \frac{E\alpha\,\Delta T}{1 - \nu} \tag{9.4}$$

$$\sigma_\phi = \left(c_{22} - \frac{c_{12}^2}{c_{11}}\right)\frac{u}{r} + \frac{c_{12}}{c_{11}}\sigma + \left(\frac{c_{12}\lambda_1}{c_{11}} - \lambda_2\right)\Delta T \tag{9.5}$$

Transverse curvilinear isotropic material:

$$\sigma_z = \left(c_{23} - \frac{c_{12}c_{13}}{c_{11}}\right)\frac{u}{r} + \frac{c_{13}}{c_{11}}\sigma + \left(\frac{c_{13}\lambda_1}{c_{11}} - \lambda_3\right)\Delta T \tag{9.6}$$

9.3 Radial Displacement and Stress

The differential equations relating the radial displacement and stresses to the applied loadings on a cylinder with constant material properties are

$$\begin{gathered}\frac{d^2u}{dr^2} + \frac{1}{r}\frac{du}{dr} - \frac{u}{r^2} = -\frac{1}{\lambda + 2G}q_r + \frac{3\lambda + 2G}{\lambda + 2G}\alpha\frac{d\,\Delta T}{dr} \\ (\lambda + 2G)\frac{du}{dr} + \frac{\lambda}{r}u - (3\lambda + 2G)\alpha\,\Delta T = \sigma\end{gathered} \tag{9.7}$$

These relations are integrated to provide the radial displacement and stress for a cylinder with its inner boundary at $r = a_0$.

Radial displacement:

$$u = u_0 \frac{1}{(1-\nu)}\left(\frac{G\nu}{\lambda}\frac{r}{a_0} + \frac{a_0}{2r}\right) + \sigma_0 \frac{\nu}{2(1-\nu)\lambda}\left(\frac{r^2 - a_0^2}{r}\right) + F_u \qquad (9.8a)$$

Radial stress:

$$\sigma = u_0 \frac{G}{(1-\nu)}\left(\frac{r^2 - a_0^2}{a_0 r^2}\right) + \sigma_0 \frac{1}{(1-\nu)}\left[\frac{1}{2} + \frac{G\nu}{\lambda}\left(\frac{a_0}{r}\right)^2\right] + F_\sigma \qquad (9.8b)$$

The F_u, F_σ are loading functions given in Table 9-1. If there is more than one load on the cylinder, the F_u, F_σ functions are formed by adding the terms given in Table 9-1 for each load. The initial parameters u_0, σ_0 (values of u, σ at the inner surface of the cylinder, $r = a_0$) are provided in Table 9-2.

EXAMPLE 9.1 Thermally Loaded Cylinder Find the stresses and displacements in a thermally loaded cylinder with zero tractions on the inner and outer boundaries.

For isotropic material, the radial stress and displacement are given by Eqs. (9.8) with the thermal loading taken from Table 9-1 and the initial parameters taken from Table 9-2 for free-free boundaries. From Table 9-1 for an arbitrary temperature change $\Delta T(r)$,

$$F_u(r) = \frac{1+\nu}{1-\nu}\frac{\alpha}{r}\int_{a_0}^{r} \Delta T(\xi)\xi\, d\xi \qquad F_\sigma(r) = -\frac{E}{1-\nu}\frac{\alpha}{r^2}\int_{a_0}^{r} \Delta T(\xi)\xi\, d\xi \qquad (1)$$

From Table 9-2, the initial parameters are $\sigma_0 = 0$ and

$$u_0 = -\frac{F_\sigma|_{r=a_L}}{\dfrac{G}{(1-\nu)}\left(\dfrac{a_L^2 - a_0^2}{a_0 a_L^2}\right)} = \frac{\dfrac{E}{1-\nu}\dfrac{\alpha}{a_L^2}\displaystyle\int_{a_0}^{a_L} \Delta T(\xi)\xi\, d\xi}{\dfrac{G}{(1-\nu)}\left(\dfrac{a_L^2 - a_0^2}{a_0 a_L^2}\right)} = \frac{a_0 E\alpha \displaystyle\int_{a_0}^{a_L} \Delta T(\xi)\xi\, d\xi}{G(a_L^2 - a_0^2)} \qquad (2)$$

With (1) and (2), Eqs. (9.8) give the radial variables of interest. The tangential and axial stress are then provided by Eqs. (9.3) and (9.4).

In the case of a cylinder with a temperature change of T_{a_0} on the inner surface and T_{a_L} on the outer surface, both relative to a reference temperature, the axially symmetric steady-state temperature distribution is given by this formula (from Ref. 9.3):

$$\Delta T(r) = \frac{T_{a_0} \ln \dfrac{a_L}{r} - T_{a_L} \ln \dfrac{a_0}{r}}{\ln \dfrac{a_L}{a_0}} \qquad (3)$$

TABLE 9-1 Loading Functions F_u, F_σ for a Cylinder of Constant Material Properties (Eqs. 9.8)

	Radial Pressure (force/length^2) 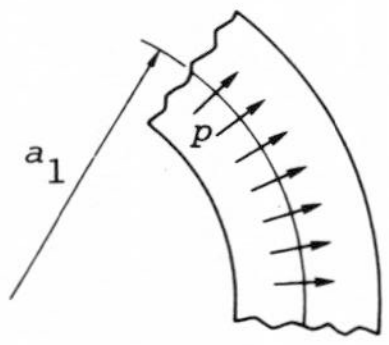	Constant Temperature Change ΔT_1 (Independent of r)	Centrifugal Loading Due to Rotation of Cylinder at the Angular Velocity Ω	Arbitrary Temperature Change $\Delta T(r)$
$F_u(r)$	$-\dfrac{p\nu}{2E(1-\nu)\lambda}\dfrac{(r^2-a_1^2)}{r}\langle r-a_1\rangle^0$	$\dfrac{r^2-a_0^2}{2r}\dfrac{(1+\nu)}{(1-\nu)}\alpha\,\Delta T_1$	$-\dfrac{\rho\Omega^2}{8}\dfrac{(r^2-a_0^2)^2}{r}\dfrac{(1+\nu)(1-2\nu)}{E(1-\nu)}$	$\dfrac{1+\nu}{1-\nu}\dfrac{\alpha}{r}\int_{a_0}^{r}\Delta T(\xi)\xi\,d\xi$
$F_\sigma(r)$	$-\dfrac{p}{E(1-\nu)}\left[\dfrac{1}{2}+\dfrac{G\nu}{2}\left(\dfrac{a_1}{r}\right)^2\right]\langle r-a_1\rangle^0$	$-\dfrac{r^2-a_0^2}{2r^2}\dfrac{\alpha E}{1-\nu}\Delta T_1$	$-(r^2-a_0^2)\dfrac{\rho\Omega^2}{4}\left[2-\dfrac{1-2\nu}{1-\nu}\dfrac{(r^2-a_0^2)}{2r^2}\right]$	$-\dfrac{E}{1-\nu}\dfrac{\alpha}{r^2}\int_{a_0}^{r}\Delta T(\xi)\xi\,d\xi$

TABLE 9-2 Initial Parameters for Sphere of Constant Material Properties (Eqs. 9.8)

Inner Edge ↓ / Outer Edge →	1. Fixed 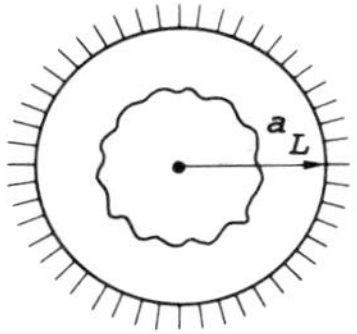	2. Free 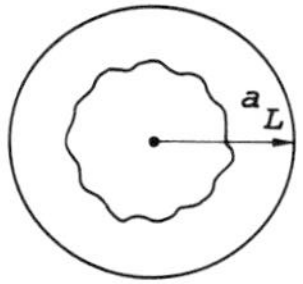
1. Fixed 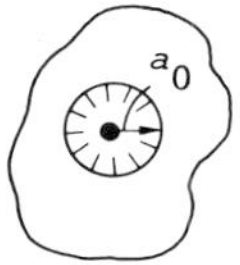 $u_0 = 0$	$\sigma_0 = -F_u \Big/ \left[\frac{\nu}{2(1-\nu)\lambda} \left(\frac{a_L^2 - a_0^2}{a_L} \right) \right]$	$\sigma_0 = -F_\sigma \Big/ \left\{ \frac{1}{(1-\nu)} \left[\frac{1}{2} + \frac{G\nu}{\lambda} \left(\frac{a_0}{a_L} \right)^2 \right] \right\}$
2. Free 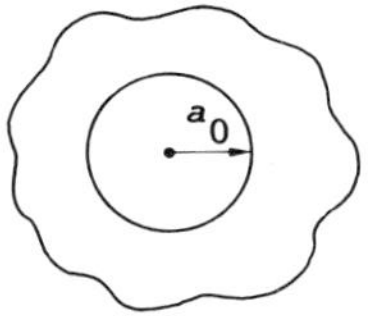 $\sigma_0 = 0$	$u_0 = -F_u \Big/ \left[\frac{1}{(1-\nu)} \left(\frac{G\nu}{\lambda} \frac{a_L}{a_0} + \frac{a_0}{2a_L} \right) \right]$	$u_0 = -F_\sigma \Big/ \left[\frac{G}{(1-\nu)} \left(\frac{a_L^2 - a_0^2}{a_0 a_L^2} \right) \right]$

Definitions: $F_u = F_u \big|_{r=a_L}$, $F_\sigma = F_\sigma \big|_{r=a_L}$

The above solution for stresses and displacements requires $\int_{a_0}^{r} \Delta T(\xi)\xi \, d\xi$ to be calculated. We find

$$\int_{a_0}^{r} \Delta T(\xi)\xi \, d\xi = \frac{\int_{a_0}^{r} T_{a_0} \ln \frac{a_L}{\xi} - T_{a_L} \ln \frac{a_0}{\xi}}{\ln (a_L/a_0)} \xi \, d\xi = \frac{r^2}{4} \frac{T_{a_0} - T_{a_L}}{\ln (a_L/a_0)} + \frac{r^2}{2} \Delta T(r) \tag{4}$$

From (4),

$$\int_{a_0}^{a_L} \Delta T(\xi)\xi \, d\xi = \frac{a_L^2}{4} \frac{T_{a_0} - T_{a_L}}{\ln (a_L/a_0)} + \frac{a_L^2}{2} \frac{T_{a_L}}{\ln (a_L/a_0)} \tag{5}$$

Equations (4) and (5) placed in (1) and (2) complete the solution.

B. COMPLEX CYLINDERS

9.4 Notation for Complex Cylinders

t	Time
ω	Natural frequency (radian/second)
c'	External or viscous radial damping coefficient (mass/time · length³), (force · time/length⁴)
c_ρ	Proportional viscous damping coefficient (1/time). If c' is chosen to be proportional to the mass, that is, $c' = c_\rho \rho$, then c_ρ is the constant of proportionality.
M_i	Concentrated cylindrical mass (mass/length²), that is, the mass is lumped as a thin cylindrical shell
$\mathbf{U}_i$	Field matrix of the ith segment
$\overline{\mathbf{U}}_i$	Point matrix at $r = a_i$

The notation for transfer matrices is:

$$\mathbf{U}_i = \begin{bmatrix} U_{uu} & U_{u\sigma} & F_u \\ U_{\sigma u} & U_{\sigma\sigma} & F_\sigma \\ 0 & 0 & 1 \end{bmatrix} \tag{9.9}$$

9.5 Differential Equations

The fundamental equations in first-order form for the radial motion of a cylinder are

$$\frac{\partial u}{\partial r} = -\frac{\lambda}{r(\lambda + 2G)} u + \frac{1}{\lambda + 2G} \sigma + \frac{3\lambda + 2G}{\lambda + 2G} \alpha \, \Delta T \tag{9.10a}$$

$$\frac{\partial \sigma}{\partial r} = \frac{4G}{r^2} \frac{(G + \lambda)}{(\lambda + 2G)} u - \frac{2G}{\lambda + 2G} \frac{1}{r} \sigma + \rho \frac{\partial^2 u}{\partial t^2} - \frac{2G(3\lambda + 2G)}{\lambda + 2G} \frac{1}{r} \alpha \, \Delta T - q_r(r, t) \tag{9.10b}$$

These relations can be reorganized to form higher-order equations similar to Eqs. (9.7).

TABLE 9-3 Massless Cylinder Segment

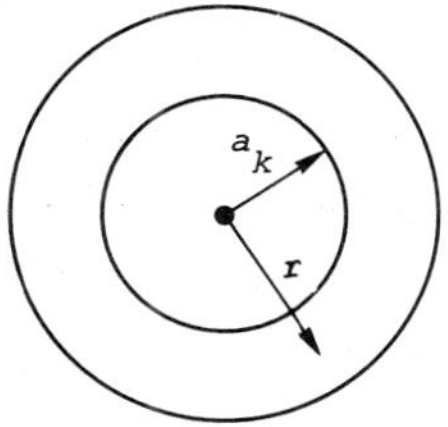

$$\mathbf{U}_i = \begin{bmatrix} \frac{1}{(1-\nu)}\left(\frac{G\nu}{\lambda}\frac{r}{a_k} + \frac{a_k}{2r}\right) & \frac{\nu}{2(1-\nu)\lambda}\frac{r^2 - a_k^2}{r} & F_u \\ \frac{G}{(1-\nu)}\frac{r^2 - a_k^2}{a_k r^2} & \frac{1}{(1-\nu)}\left[\frac{1}{2} + \frac{G\nu}{\lambda}\left(\frac{a_k}{r}\right)^2\right] & F_\sigma \\ 0 & 0 & 1 \end{bmatrix}$$

Definitions:

(a) ΔT_1 = a constant temperature change that is independent of r. Set $\Delta T(\xi) = 0$ if only a constant temperature change is present.

(b) $\Delta T(r)$ = arbitrary temperature change.

(c) Ω = angular velocity of rotation of cylinder that leads to centrifugal loading force.

(d) $q_r(r)$ = arbitrary loading intensity in r direction.

(e) a_k = radial coordinate of the inner surface of the ith section. That is, the ith section begins at a_k and continues to r.

$$F_u = \frac{(r^2 - a_k^2)}{2r}\frac{(1+\nu)}{(1-\nu)}\alpha\,\Delta T_1 + \frac{1+\nu}{1-\nu}\frac{\alpha}{r}\int_{a_k}^{r}\xi\,\Delta T(\xi)\,d\xi - \frac{\rho\Omega^2}{8}\frac{(r^2 - a_k^2)^2}{r}\frac{(1+\nu)(1-2\nu)}{E(1-\nu)} - \frac{(1+\nu)(1-2\nu)}{E(1-\nu)}\frac{1}{r}\int_{a_k}^{r}\eta\left[\int_{a_k}^{\eta}q_r(\xi)\,d\xi\right]d\eta$$

$$F_\sigma = -\frac{(r^2 - a_k^2)}{2r^2}\frac{\alpha E}{(1-\nu)}\Delta T_1 - \frac{E}{1-\nu}\frac{\alpha}{r^2}\int_{a_k}^{r}\xi\,\Delta T(\xi)\,d\xi - (r^2 - a_k^2)\frac{\rho\Omega^2}{4}\left[2 - \frac{1-2\nu}{1-\nu}\frac{(r^2 - a_k^2)}{2r^2}\right] + \frac{1-2\nu}{1-\nu}\frac{1}{r^2}\int_{a_k}^{r}\eta\left[\int_{a_k}^{\eta}q_r(\xi)\,d\xi\right]d\eta - \int_{a_k}^{r}q_r(\xi)\,d\xi$$

TABLE 9-4 Massless Cylinder without Center Hole (Center Segment)

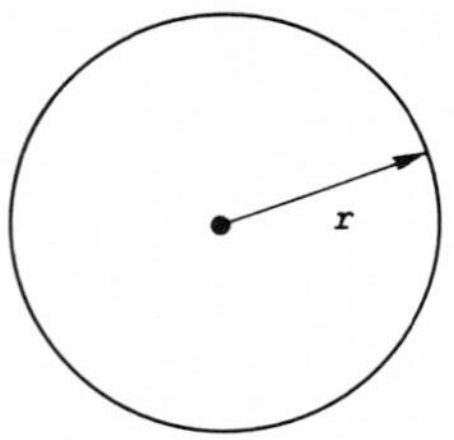

$$\mathbf{U}_i = \begin{bmatrix} 0 & \frac{\nu r}{\lambda} & F_u \\ 0 & 1 & F_\sigma \\ 0 & 0 & 1 \end{bmatrix}$$

Definitions:

The definitions of Table 9-3
still apply for ΔT_1, $\Delta T(r)$, Ω, q_r.

$$F_u = (1+\nu) r\alpha\,\Delta T_1 + \frac{1+\nu}{1-\nu}\frac{\alpha}{r}\int_0^r \xi\,\Delta T(\xi)\,d\xi + \frac{\alpha r E\nu\,\Delta T\big|_{r=0}}{2\lambda(1-\nu)} - \frac{\rho\Omega^2 r^3}{8}\frac{(1+\nu)(1-2\nu)}{E(1-\nu)} - \frac{(1+\nu)(1-2\nu)}{E(1-\nu)r}\int r\left[\int q_r(r)\,dr\right]dr$$

$$F_\sigma = -\frac{E}{1-\nu}\frac{\alpha}{r^2}\int_0^r \xi\,\Delta T(\xi)\,d\xi + \frac{\alpha E\,\Delta T\big|_{r=0}}{2(1-\nu)} - \frac{\rho\Omega^2 r^2}{8}\frac{(3-2\nu)}{1-\nu} + \frac{1-2\nu}{1-\nu}\frac{1}{r^2}\int r\left[\int q_r(r)\,dr\right]dr - \int q_r(r)\,dr$$

TABLE 9-5 Massless Anisotropic Cylinder Segment

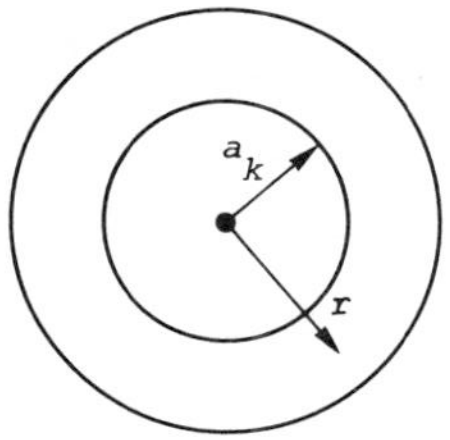

$$\mathbf{U}_i = \left[\begin{array}{cc|c} \dfrac{1}{2\beta c_{11}}\left[\beta_1\left(\dfrac{r}{a_k}\right)^{\beta} + \beta_2\left(\dfrac{a_k}{r}\right)^{\beta}\right] & \dfrac{a_k}{2\beta c_{11}}\left[\left(\dfrac{r}{a_k}\right)^{\beta} - \left(\dfrac{a_k}{r}\right)^{\beta}\right] & F_u \\ \dfrac{\beta_1\beta_2}{2\beta c_{11} a_k}\left[\left(\dfrac{r}{a_k}\right)^{\beta-1} - \left(\dfrac{a_k}{r}\right)^{\beta+1}\right] & \dfrac{1}{2\beta c_{11}}\left[\beta_1\left(\dfrac{r}{a_k}\right)^{\beta-1} + \beta_2\left(\dfrac{a_k}{r}\right)^{\beta+1}\right] & F_\sigma \\ 0 & 0 & 1 \end{array}\right]$$

Definitions:

(a) $\beta^2 = c_{22}/c_{11}$, $\beta_1 = \beta c_{11} - c_{12}$, $\beta_2 = \beta c_{11} + c_{12}$

(b) ΔT = arbitrary temperature variation.

(c) $R(r) = \dfrac{\lambda_1}{c_{11}}\dfrac{d\,\Delta T}{dr} + \dfrac{\lambda_1 - \lambda_2}{c_{11}}\dfrac{\Delta T}{r}$

(d) a_k = the radial coordinate of the inner surface of the ith section. That is, the ith section begins at a_k and continues to r.

$$F_u = \frac{1}{2\beta}\left[r^{\beta}\int_{a_k}^{r}\xi^{-\beta+1}R(\xi)\,d\xi - r^{-\beta}\int_{a_k}^{r}\xi^{\beta+1}R(\xi)\,d\xi\right] + \frac{\lambda_1 a_k}{2\beta c_{11}}\left[\left(\frac{r}{a_k}\right)^{\beta} - \left(\frac{a_k}{r}\right)^{\beta}\right]\Delta T\Bigg|_{r=a_k}$$

$$F_\sigma = \frac{1}{2\beta}\left[\beta_2 r^{\beta-1}\int_{a_k}^{r}\xi^{-\beta+1}R(\xi)\,d\xi + \beta_1 r^{-\beta-1}\int_{a_k}^{r}\xi^{\beta+1}R(\xi)\,d\xi\right]$$

$$-\lambda_1\,\Delta T + \frac{\lambda_1}{2\beta c_{11}}\left[\beta_2\left(\frac{r}{a_k}\right)^{\beta-1} + \beta_1\left(\frac{a_k}{r}\right)^{\beta+1}\right]\Delta T\Bigg|_{r=a_k}$$

TABLE 9-6 Massless Anisotropic Segment without Center Hole (Center Segment)

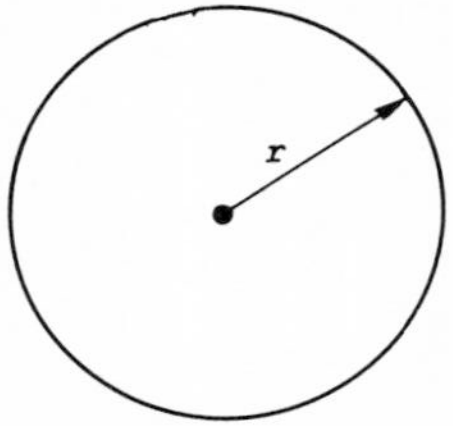

$$\mathbf{U}_i = \left[\begin{array}{c|c|c} 0 & r^{\beta} & F_u \\ \hline 0 & (c_{11}\beta + c_{12})r^{\beta-1} & F_\sigma \\ \hline 0 & 0 & 1 \end{array}\right]$$

Definitions:

(a) $\beta^2 = c_{22}/c_{11} \geq 1$

$$F_u = \frac{1}{2\beta}\left[r^{\beta}\int^{r} r^{-\beta+1}R(r)\,dr - r^{-\beta}\int^{r} r^{\beta+1}R(r)\,dr\right]$$

(b) ΔT = arbitrary temperature variation.

(c) $R(r) = \frac{\lambda_1}{c_{11}}\frac{d(\Delta T)}{dr} + \frac{\lambda_1 - \lambda_2}{c_{11}}\frac{\Delta T}{r}$

$$F_\sigma = \frac{1}{2\beta}\left[(c_{11}\beta + c_{12})r^{\beta-1}\int^{r} r^{-\beta+1}R(r)\,dr + (c_{11}\beta - c_{12})r^{-\beta-1}\int^{r} r^{\beta+1}R(r)\,dr\right] - \lambda_1\Delta T$$

This $\mathbf{U}_i$ is not truly a transfer matrix, as the vector at radius $r = 0$ contains arbitrary constants rather than state variables. However, the vector at other radii is the state vector. All operations remain the same as those for the usual transfer matrices.

TABLE 9-7 Cylinder Segment with Mass

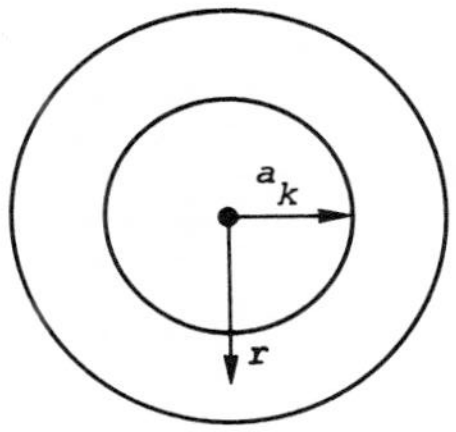

$$U_i = \begin{bmatrix} \frac{1}{e_1}\left[e_3(a_k)J_\gamma(\beta r) - e_2(a_k)Y_\gamma(\beta r)\right] & \frac{1}{e_1}\left[J_\gamma(\beta a_k)Y_\gamma(\beta r) - Y_\gamma(\beta a_k)J_\gamma(\beta r)\right] & F_u \\ \frac{1}{e_1}\left[e_3(a_k)e_2(r) - e_2(a_k)e_3(r)\right] & \frac{1}{e_1}\left[J_\gamma(\beta a_k)e_3(r) - Y_\gamma(\beta a_k)e_2(r)\right] & F_\sigma \\ 0 & 0 & 1 \end{bmatrix}$$

Definitions:

(a) Isotropic

$\gamma = 1, \; \beta^2 = \rho\omega^2/(\lambda + 2G)$

$$e_1 = e_3(a_k)J_1(\beta a_k) - e_2(a_k)Y_1(\beta a_k)$$

$$e_2(r) = \frac{\lambda + 2G}{r}\left[\frac{2(\lambda + G)}{\lambda + 2G} J_1(\beta r) - \beta r J_2(\beta r)\right]$$

$$e_3(r) = \frac{\lambda + 2G}{r}\left[\frac{2(\lambda + G)}{\lambda + 2G} Y_1(\beta r) - \beta r Y_2(\beta r)\right]$$

(b) Anisotropic (Transverse Curvilinear Isotropic Material)

$\gamma^2 = c_{22}/c_{11}, \quad \beta^2 = \omega^2\rho/c_{11}$

$$e_1 = e_3(a_k)J_\gamma(\beta a_k) - e_2(a_k)Y_\gamma(\beta a_k)$$

$$e_2(r) = \frac{c_{11}}{r}\left[(\gamma + \frac{c_{12}}{c_{11}})J_\gamma(\beta r) - \beta r J_{\gamma+1}(\beta r)\right]$$

$$e_3(r) = \frac{c_{11}}{r}\left[(\gamma + \frac{c_{12}}{c_{11}})Y_\gamma(\beta r) - \beta r Y_{\gamma+1}(\beta r)\right]$$

$J_\gamma(\beta r)$, $Y_\gamma(\beta r)$ are Bessel functions of order γ of the first and second kinds, respectively.

a_k is the radial coordinate of the inner surface of the ith section. That is, the ith section begins at a_k and continues to r.

In the case of transverse curvilinear isotropy, the differential equations are

$$\frac{\partial u}{\partial r} = -\frac{c_{12}}{c_{11}}\frac{u}{r} + \frac{1}{c_{11}}\sigma + \frac{\lambda_1}{c_{11}}\Delta T \tag{9.11a}$$

$$\frac{\partial \sigma}{\partial r} = \frac{1}{r^2}\left(c_{22} - \frac{c_{12}^2}{c_{11}}\right)u + \frac{1}{r}\left(\frac{c_{12}}{c_{11}} - 1\right)\sigma + \rho\frac{\partial^2 u}{\partial t^2} + \frac{1}{r}\left(\frac{c_{12}}{c_{11}}\lambda_1 - \lambda_2\right)\Delta T - q_r(r, t) \tag{9.11b}$$

9.6 Transfer Matrices

The transfer matrices for cylindrical segments are provided in Tables 9-3 to 9-9. If desired, the transfer matrices for segments extending from radial coordinate a_k to r can be employed for a center (holeless) segment by considering a small hole to be at the center. The boundary condition for this imaginary hole should be fixed.

9.7 Point Matrices

The transfer matrices that take into account concentrated (point) occurrences are listed in Table 9-10.

9.8 Loading Functions

The loading functions are provided in the transfer matrix tables for most common loadings. These functions can be calculated for other loadings from the formulas:

$$\begin{aligned} F_u &= -\int_{a_i}^{r} q_r(\xi, a_i)\, U_{u\sigma}(r, \xi)\, d\xi = -\int_{a_i}^{r} q_r(r, \xi)\, U_{u\sigma}(\xi, a_i)\, d\xi \\ F_\sigma &= -\int_{a_i}^{r} q_r(\xi, a_i)\, U_{\sigma\sigma}(r, \xi)\, d\xi = -\int_{a_i}^{r} q_r(r, \xi)\, U_{\sigma\sigma}(\xi, a_i)\, d\xi \end{aligned} \tag{9.12}$$

TABLE 9-8 Cylinder Segment without Center Hole (Center Segment), Including Mass

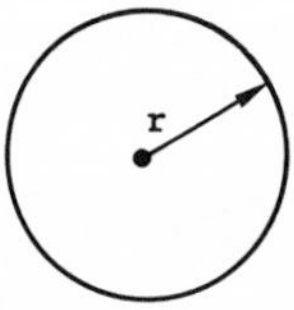

Definitions:

$$\beta = \sqrt{\frac{\omega^2\rho}{\lambda + 2G}}$$

$J_1(\beta r)$, $J_2(\beta r)$ are Bessel functions of the first kind of orders one and two, respectively.

$$\mathbf{U}_i = \begin{bmatrix} 0 & \dfrac{J_1(\beta r)}{(\lambda + G)\beta} & F_u \\ 0 & \dfrac{\lambda + 2G}{\beta(\lambda + G)r}\left[\dfrac{2(\lambda + G)}{(\lambda + 2G)} J_1(\beta r) - \beta r J_2(\beta r)\right] & F_\sigma \\ 0 & 0 & 1 \end{bmatrix}$$

TABLE 9-9 Anisotropic (Transverse Curvilinear Isotropic Material) Cylinder Segment without Center Hole (Center Segment), Including Mass

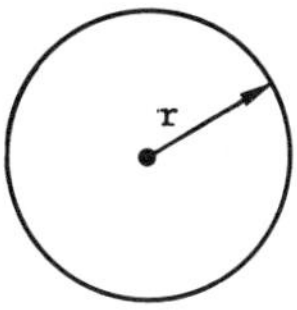

Definitions:

$\gamma^2 = (c_{22}/c_{11}) \geq 1$

$\beta^2 = \omega^2 \rho / c_{11}$

$J_\gamma(\beta r)$ are Bessel functions of order γ of the first kind.

This $\mathbf{U}_i$ is not truly a transfer matrix. See the note in Table 9-6.

$$\mathbf{U}_i = \begin{bmatrix} 0 & J_\gamma(\beta r) & F_u \\ 0 & \frac{c_{11}}{r}\left[\left(\gamma + \frac{c_{12}}{c_{11}}\right) J_\gamma(\beta r) - \beta r J_{\gamma+1}(\beta r)\right] & F_\sigma \\ 0 & 0 & 1 \end{bmatrix}$$

Here the notation $q_r(r, a_i)$ is employed to indicate a loading intensity (force/length^3) that begins at radius a_i and extends to radius r. The notation $U_{kj}(r, a_i)$ has a similar meaning. It denotes a transfer matrix element for a ring section that begins at radius a_i and extends to radius r. Do not use U_{kj} from transfer matrices for cylinder segments with no center hole.

9.9 Other Information

Cylinder problems can be solved with the procedures outlined in Chap. 8 if P used in Chap. 8 is replaced by σ for Chap. 9. This substitution applies to static problems as outlined in Section 8.11, free dynamics problems in Section 8.12, forced harmonic motion in Section 8.13, and the transient response of Section 8.14 if N_n of Eq. (8.18), $\eta_n(0)$ of Eq. (8.19), and $f_n(t)$ of Eq. (8.20) are replaced by

$$N_n = \int_{a_0}^{a_L} \rho r u_n^2 \, dr \tag{9.13}$$

$$\eta_n(0) = \int_{a_0}^{a_L} \rho r \, u(r, 0) u_n \, dr \tag{9.14}$$

$$f_n(t) = \int_{a_0}^{a_L} [q_r(r, t) r u_n + \alpha \, \Delta T(r, t) \, d_n] \, dr + a_0 u(a_0, t) \sigma_n(a_0)$$
$$- a_L u(a_L, t) \sigma_n(a_L) - a_0 p(a_0, t) u_n(a_0) + a_L p(a_L, t) u_n(a_L) \tag{9.15}$$

$$d_n = \begin{cases} \dfrac{1+\nu}{1-\nu}\left[\dfrac{E}{1+\nu} u_n + r\sigma_n\right] & \text{for isotropic material} \\ \left[\left(-\dfrac{c_{12}}{c_{11}}\lambda_1 + \lambda_2\right) u_n + \dfrac{\lambda_1}{c_{11}} r\sigma_n\right]\dfrac{1}{\alpha r} & \text{for anisotropic material} \end{cases} \tag{9.16}$$

TABLE 9-10 Point Matrices at $r = a_i$

Definitions:

(a) Radial Pressure (force/length^2)

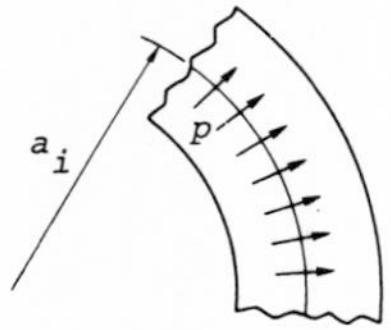

$$\bar{\mathbf{U}}_i = \begin{bmatrix} 1 & 0 & 0 \\ -M_i\omega^2 & 1 & -p \\ 0 & 0 & 1 \end{bmatrix}$$

(b) Concentrated Mass

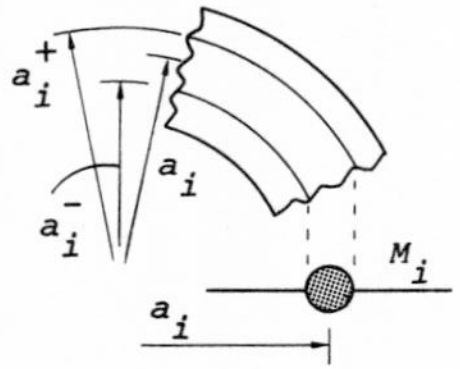

M_i is the mass per unit cylindrical surface area. M_i can be calculated as

$$M_i = \frac{(a_i^+)^2 - (a_i^-)^2}{2a_i}\rho = \Delta a\,\rho$$

where ρ is the mass per unit volume

$$\Delta a = a_i^+ - a_i^-$$

Note that in many instances the equations of Chap. 8 apply if b is dropped in the relations. For proportional damping $c' = c_\rho \rho$, so that a term $c_\rho \rho \, \partial u/\partial t$ should be added to the right-hand sides of Eqs. (9.10b) and (9.11b). Set $c' = 0$ for no damping. In the case of nonproportional damping for both isotropic and anisoptropic materials, add the term $c' \, \partial u/\partial t$ to the right-hand sides of Eqs. (9.10b) and (9.11b), and replace Eq. (8.23) by

$$N_n = 2s_n \int_{a_0}^{a_L} \rho r u_n^2 \, dr + \int_{a_0}^{a_L} c' r u_n^2 \, dr \tag{9.17}$$

and Eq. (8.26) by

$$\eta_n(0) = \int_{a_0}^{a_L} \left[\rho s_n u(r, 0) u_n(r) + c' u(r, 0) u_n(r) + \rho \frac{\partial u}{\partial t} (r, 0) u_n(r) \right] r \, dr \tag{9.18}$$

For discrete damping with dashpot constants c_i use

$$\int_{a_0}^{a_L} c' r u_n^2 \, dr = \sum_i c_i a_i u_n^2(a_i) \tag{9.19}$$

in N_n, and

$$\int_{a_0}^{a_L} c' u(r, 0) u_n(r) \, r \, dr = \sum_i c_i a_i u(a_i, 0) u_n(a_i) \tag{9.20}$$

in $\eta_n(0)$.

For isotropic Voigt-Kelvin material continue to apply the formulas of Chap. 8.

9.10 Applications

Since static and dynamic problems for the radial motion of cylinders are solved in much the same fashion as similar problems for other members, the example problems of previous chapters should suffice as references in setting up cylinder problems. A few additional examples follow.

EXAMPLE 9.2 Thermally Loaded Cylinder with No Center Hole Find the radial stress and displacement in a solid cylindrical shaft with thermal loading.

For isotropic material, the expressions for radial stress and displacement can be taken from Table 9-4.

$$\begin{aligned} u &= u_0 \cdot 0 + \sigma_0 \frac{\nu r}{\lambda} + \frac{1+\nu}{1-\nu} \frac{\alpha}{r} \int_0^r \xi \, \Delta T(\xi) \, d\xi + \frac{\alpha E \nu r \, \Delta T|_{r=0}}{2\lambda(1-\nu)} \\ \sigma &= u_0 \cdot 0 + \sigma_0 \cdot 1 - \frac{E}{1-\nu} \frac{\alpha}{r^2} \int_0^r \xi \, \Delta T(\xi) \, d\xi + \frac{\alpha E \, \Delta T|_{r=0}}{2(1-\nu)} \end{aligned} \tag{1}$$

The initial parameters are taken from Table 8-9 with $P = \sigma$ for a free outer surface:

$$u_0 = 0 \qquad \sigma_0 = -F_\sigma/\nabla = -F_\sigma|_{r=a_L} \Big/ U_{\sigma\sigma}|_{r=a_L} \tag{2}$$

Substitution of (1) in (2) gives

$$u_0 = 0 \qquad \sigma_0 = \frac{E}{1-\nu}\,\frac{\alpha}{a_L^2}\int_0^{a_L}\xi\,\Delta T(\xi)\,d\xi - \frac{\alpha E\,\Delta T|_{r=0}}{2(1-\nu)} \tag{3}$$

Finally, from (1) and (3), the radial displacement and stress become

$$u = \frac{\nu}{\lambda}\,\frac{E}{1-\nu}\,\frac{\alpha}{a_L^2}\,r\int_0^{a_L}\xi\,\Delta T(\xi)\,d\xi + \frac{1+\nu}{1-\nu}\,\frac{\alpha}{r}\int_0^{r}\xi\,\Delta T(\xi)\,d\xi$$

$$\sigma = \frac{E}{1-\nu}\,\frac{\alpha}{a_L^2}\int_0^{a_L}\xi\,\Delta T(\xi)\,d\xi - \frac{E}{1-\nu}\,\frac{\alpha}{r^2}\int_0^{r}\xi\,\Delta T(\xi)\,d\xi$$

EXAMPLE 9.3 Natural Frequencies of Anisotropic Cylinder Find the natural frequencies for a solid (holeless) anisotropic cylinder of radius a_L. The outer surface is free.

The frequencies are those values of ω that make ∇ of Table 8-9 equal to zero. This table with $P = \sigma$ shows, for a holeless inner surface and free outer surface,

$$\nabla = U_{\sigma\sigma}|_{r=a_L}$$

The transfer matrix component $U_{\sigma\sigma}$ is taken from Table 9-9 as

$$U_{\sigma\sigma}|_{r=a_L} = \frac{c_{11}}{a_L}\left[\left(\gamma + \frac{c_{12}}{c_{11}}\right)J_\gamma(\beta a_L) - \beta a_L J_{\gamma+1}(\beta a_L)\right]$$

or

$$\nabla = \eta J_\gamma(\beta a_L) - \beta a_L J_{\gamma+1}(\beta a_L) = 0 \tag{1}$$

where

$$\gamma^2 = c_{22}/c_{11} \qquad \beta^2 = \omega^2\rho/c_{11} \qquad \eta = \gamma + c_{12}/c_{11}$$

The natural frequencies ω, or equivalently, β, can be computed from (1) for specific values of γ and η. A plot of the first natural frequency as a function of γ and η is given in Fig. 9-2 (Ref. 9.4).

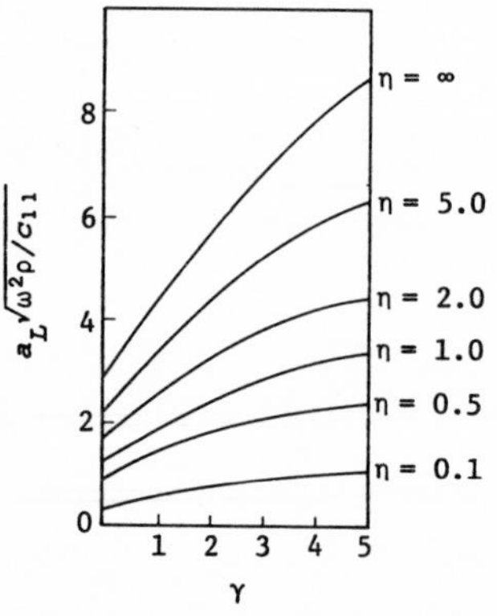

Fig. 9-2 Results for Example 9.3.

C. COMPUTER PROGRAMS AND EXAMPLES

9.11 Benchmark Examples

Complicated cylindrical problems should be solved with the assistance of a computer program. The following example is provided as a benchmark problem against which a reader's own computer program can be checked. The computer program CYLINDER was used for this example. CYLINDER can be used to compute the stresses, radial displacement, and natural frequencies in a cylinder of any complexity. The cylinder can be formed of layers of different materials with arbitrary mechanical or thermal loading and boundary conditions.

EXAMPLE 9.4 Rotating Cylinder Determine the displacements and stresses in a rotating cylinder with internal pressure (Fig. 9-3). The speed of rotation is 15.9155 Hz and the pressure is 3000 lb/in^2. The mass density is 0.000741 lb $\cdot$ sec^2/in^4, $E = 30(10^6)$ lb/in^2, $\nu = 0.3$.

The displacements and stresses are given in Fig. 9-4.

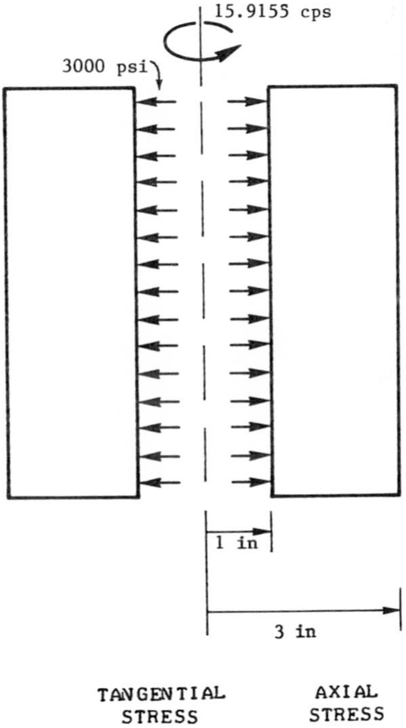

Fig. 9-3 Example 9.4.

RADIAL LOCATION	DISPLACEMENT	RADIAL STRESS	TANGENTIAL STRESS	AXIAL STRESS
.10000E+01	.15451E-03	-.30000E+04	.38081E+04	.24243E+03
.14000E+01	.11515E-03	-.13360E+04	.21390E+04	.24091E+03
.18000E+01	.94492E-04	-.65404E+03	.14503E+04	.23888E+03
.22000E+01	.82304E-04	-.31185E+03	.10997E+04	.23635E+03
.26000E+01	.74651E-04	-.11821E+03	.89589E+03	.23330E+03
.30000E+01	.69692E-04	-.14552E-10	.76585E+03	.22975E+03

Fig. 9-4 Results for Example 9.4.

REFERENCES

9.1 Eason, G.: "Thermal Stresses in Anisotropic Cylinders," *Proc. Edinburgh Math. Soc.*, vol 13 (series II), pt. 2, 1962, pp. 159–164.

9.2 Lekhnitskii, S.: *Theory of Elasticity of an Anisotropic Elastic Body*, Holden-Day, San Francisco, 1963.

9.3 Zudans, Z., Yen, T. C., and Steigelmann, W. H.: *Thermal Stress Techniques*, American Elsevier, New York, 1965.

9.4 Eason, G.: "On the Vibration of Anisotropic Cylinders and Spheres," *Appl. Sci. Res.*, A, vol. 12, 1962, pp. 81–85.

TEN

Thick Spherical Shells

This chapter treats the static and dynamic radial motion of thick spheres. Displacement and stress formulas for simple spheres are handled in Part A. The formulas for more complex spheres are provided in Part B. Part C contains some examples of problems solved by a computer program.

The formulas of this chapter apply to spheres with spherically symmetric loading. The displacements and stresses are also spherically symmetric.

A. STRESS FORMULAS AND SIMPLE SPHERES

10.1 Notation and Conventions

u	Radial displacement (length)
$\sigma = \sigma_r$	Radial stress (force/length^2)
σ_ϕ	Tangential stress (force/length^2)
p	Radial pressure (force/length^2). An applied pressure in the positive radial (increasing r) direction is taken to be positive.
q_r	Change in pressure in the radial direction, radial loading intensity (force/length^3); q_r is positive if its direction is along the positive r direction.
E	Modulus of elasticity (force/length^2)

ν	Poisson's ratio
$G = E/[2(1+\nu)]$	Shear modulus, Lamé coefficient (force/length^2)
$\lambda = E\nu/[(1+\nu)(1-2\nu)]$	Lamé coefficient (force/length^2)
α	Coefficient of thermal expansion (length/length · degree)
ΔT	Change in temperature (degrees), that is, the temperature rise with respect to the reference temperature
ΔT_1	Magnitude of temperature change, uniform in r direction
r, ϕ	Radial, circumferential coordinates
a_0	Radial coordinate of inner surface of sphere (length)
a_L	Radial coordinate of outer surface of sphere (length)

Positive displacement u and stresses σ, σ_ϕ are shown in Fig. 9-1.

c_{11}, c_{12}, c_{22}, c_{23}, λ_1, λ_2 are elastic constants for anisotropic material (usually called transversely isotropic material) defined according to the stress displacement relations (Ref. 10.1):

$$\sigma_r = c_{11}\frac{\partial u}{\partial r} + 2c_{12}\frac{u}{r} - \lambda_1\,\Delta T \qquad \sigma_\phi = c_{12}\frac{\partial u}{\partial r} + (c_{22}+c_{23})\frac{u}{r} - \lambda_2\,\Delta T \tag{10.1}$$

In terms of the stress-strain relations (Ref. 10.2),

$$\epsilon_r = \frac{\partial u}{\partial r} = a_{11}\sigma_r + a_{12}\sigma_\theta + a_{12}\sigma_\phi + \alpha_1\,\Delta T \qquad \theta \text{ is the "other" circumferential coordinate, orthogonal to } \phi.$$

$$\epsilon_\theta = \frac{u}{r} = a_{12}\sigma_r + a_{22}\sigma_\theta + a_{23}\sigma_\phi + \alpha_2\,\Delta T \qquad \sigma_\theta = \sigma_\theta,\quad \alpha_2 = \alpha_3 \tag{10.2}$$

$$\epsilon_\phi = \frac{u}{r} = a_{12}\sigma_r + a_{23}\sigma_\theta + a_{22}\sigma_\phi + \alpha_3\,\Delta T$$

The constants of Eqs. (10.1) and (10.2) are interrelated by

$$c_{11} = (a_{22}^2 - a_{23}^2)/|A| \qquad c_{12} = a_{12}(a_{23} - a_{22})/|A| \qquad c_{22} = (a_{11}a_{22} - a_{12}^2)/|A|$$

$$c_{23} = -(a_{11}a_{23} - a_{12}^2)/|A| \qquad \lambda_1 = \alpha_1 c_{11} + 2\alpha_2 c_{12} \qquad \lambda_2 = \alpha_1 c_{12} + \alpha_2(c_{22} + c_{23})$$

$$|A| = \begin{vmatrix} a_{11} & a_{12} & a_{12} \\ a_{12} & a_{22} & a_{23} \\ a_{12} & a_{23} & a_{22} \end{vmatrix}$$

10.2 Tangential Stress

The tables of this chapter provide the radial stress ($\sigma_r = \sigma$) and displacement (u) throughout the sphere. The tangential stress can then be derived from the relations:

Isotropic material:

$$\sigma_\phi = 2G\frac{(2G+3\lambda)}{2G+\lambda}\frac{u}{r} + \frac{\lambda}{2G+\lambda}\sigma - \frac{2G(3\lambda+2G)}{2G+\lambda}\alpha\,\Delta T \tag{10.3}$$

Transverse curvilinear isotropic material:

$$\sigma_\phi = \left(c_{22} + c_{23} - 2\frac{c_{12}^2}{c_{11}}\right)\frac{u}{r} + \frac{c_{12}}{c_{11}}\sigma + \left(\frac{c_{12}}{c_{11}}\lambda_1 - \lambda_2\right)\alpha\,\Delta T \tag{10.4}$$

TABLE 10-1 Loading Functions F_u, F_σ for a Sphere of Constant Material Properties (Eqs. 10.6)

	Radial Pressure (force/length^2) 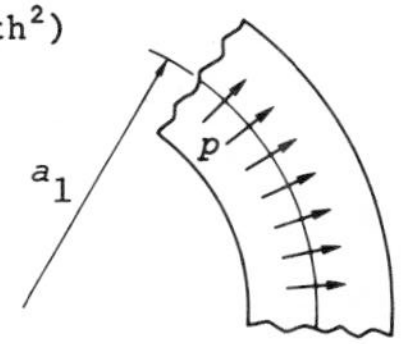	Constant Temperature Change ΔT_1 (Independent of r)	Arbitrary Temperature Change $\Delta T(r)$
$F_u(r)$	$-\dfrac{pa_1}{3(\lambda + 2G)}\left(\dfrac{r}{a_1} - \dfrac{a_1^2}{r^2}\right)\langle r - a_1\rangle^0$	$\dfrac{3\lambda + 2G}{3(\lambda + 2G)}\,\dfrac{(r^3 - a_0^3)}{r^2}\,\alpha\,\Delta T_1$	$\dfrac{3\lambda + 2G}{\lambda + 2G}\,\dfrac{\alpha}{r^2}\displaystyle\int_{a_0}^{r}\Delta T(\xi)\,\xi^2\,d\xi$
$F_\sigma(r)$	$-p\left(\dfrac{3\lambda + 2G + 4Ga_1^3/r^3}{3(\lambda + 2G)}\right)\langle r - a_1\rangle^0$	$\dfrac{-4G(3\lambda + 2G)}{3(\lambda + 2G)}\,\dfrac{(r^3 - a_0^3)}{r^3}\,\alpha\,\Delta T_1$	$\dfrac{-4G\alpha}{r^3}\,\dfrac{3\lambda + 2G}{\lambda + 2G}\displaystyle\int_{a_0}^{r}\Delta T(\xi)\,\xi^2\,d\xi$

TABLE 10-2 Initial Parameters for Sphere of Constant Material Properties (Eqs. 10.6)

Inner Edge ↓ \ Outer Edge →	1. Fixed 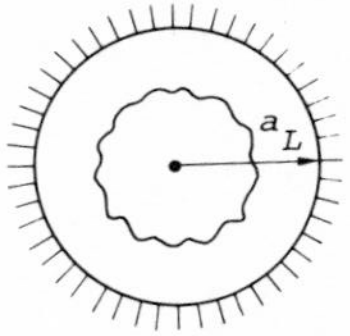	2. Free 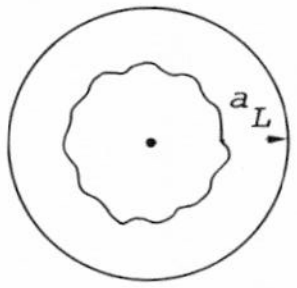
1. Fixed 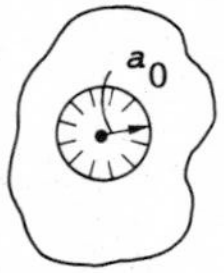 $u_0=0$	$\sigma_0 = -F_u \Big/ \left[\frac{a_0}{3(\lambda+2G)}\left(\frac{a_L}{a_0} - \frac{a_0^2}{a_L^2}\right)\right]$	$\sigma_0 = -F_\sigma \Big/ \left[\frac{3\lambda + 2G + 4Ga_0^3/a_L^3}{3(\lambda+2G)}\right]$
2. Free 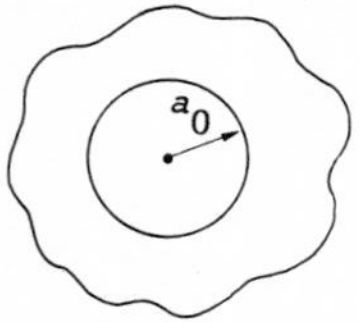 $\sigma_0=0$	$u_0 = -F_u \Big/ \left[\frac{4Ga_L^3 + (3\lambda+2G)a_0^3}{3(\lambda+2G)\,a_0 a_L^2}\right]$	$u_0 = -F_\sigma \Big/ \left[\frac{4G(3\lambda+2G)(a_L^3 - a_0^3)}{3(\lambda+2G)a_L^3\,a_0}\right]$

Definitions: $F_u = F_u\big|_{r=a_L}$, $F_\sigma = F_\sigma\big|_{r=a_L}$

10.3 Radial Displacement and Stress

The differential equations relating the radial displacement and stresses to the applied loadings for a sphere with constant material properties are

$$\frac{d^2u}{dr^2} + \frac{2}{r}\frac{du}{dr} - 2\frac{u}{r^2} = -\frac{1}{(\lambda + 2G)} q_r + \frac{3\lambda + 2G}{\lambda + 2G}\alpha\frac{d\,\Delta T}{dr}$$
$$(\lambda + 2G)\frac{du}{dr} + \frac{2\lambda}{r}u - (3\lambda + 2G)\alpha\,\Delta T = \sigma \tag{10.5}$$

These relations are integrated to provide the radial displacement and stress for a sphere with its inner boundary at $r = a_0$:

Radial displacement:

$$u = u_0 \frac{4Gr^3 + (3\lambda + 2G)a_0^3}{3(\lambda + 2G)a_0 r^2} + \sigma_0 \frac{a_0}{3(\lambda + 2G)}\left(\frac{r}{a_0} - \frac{a_0^2}{r^2}\right) + F_u \tag{10.6a}$$

Radial stress:

$$\sigma = u_0 \frac{4G(3\lambda + 2G)}{3(\lambda + 2G)a_0 r^3}(r^3 - a_0^3) + \sigma_0\left[\frac{3\lambda + 2G + 4Ga_0^3/r^3}{3(\lambda + 2G)}\right] + F_\sigma \tag{10.6b}$$

The F_u, F_σ are loading functions given in Table 10-1. If there is more than one load on the sphere, the F_u, F_σ functions are formed by adding the terms given in Table 10-1 for each load. The initial parameters u_0, σ_0 (values of u, σ at the inner surface of the sphere, $r = a_0$) are provided in Table 10-2.

EXAMPLE 10.1 Thermally Loaded Sphere An isotropic spherical thick shell is subject to a change in temperature of $\Delta T(r)$. Determine the variation of the stresses and displacements in the radial direction.

Equations (10.6) represent the radial stress and displacement. Once these are known, the tangential stress can be computed from Eq. (10.3). For pure thermal loading, the loading functions required for Eqs. (10.6) are given by Table 10-1 as

$$F_u(r) = \frac{3\lambda + 2G}{\lambda + 2G}\frac{\alpha}{r^2}\int_{a_0}^{r} \Delta T(\xi)\xi^2\,d\xi$$
$$F_\sigma(r) = -\frac{4G\alpha}{r^3}\frac{3\lambda + 2G}{\lambda + 2G}\int_{a_0}^{r} \Delta T(\xi)\xi^2\,d\xi \tag{1}$$

If there are neither mechanical loadings nor physical constraints on the surfaces ($r = a_0$, $r = a_L$), then the initial parameters for the free-free case of Table 10-2 should be used; that is,

$$\sigma_0 = 0,\ u_0 = -\frac{F_\sigma|_{r=a_L}}{\left[\dfrac{4G(3\lambda + 2G)(a_L^3 - a_0^3)}{3(\lambda + 2G)a_L^3 a_0}\right]} = \frac{3\alpha a_0 \int_{a_0}^{a_L} \Delta T(\xi)\,\xi^2\,d\xi}{(a_L^3 - a_0^3)} \tag{2}$$

Relations (1) and (2) placed in Eqs. (10.6) fully define the radial displacement and stress.

B. COMPLEX SPHERES

10.4 Notation for Complex Spheres

t Time

ω Natural frequency (radian/second)

c' External or viscous radial damping coefficient [mass/(time · length^3), force · time/length^4]

c_ρ Proportional viscous damping coefficient (1/time). If c' is chosen to be proportional to the mass, that is, $c' = c_\rho \rho$, then c_ρ is the constant of proportionality.

M_i Concentrated spherical mass (mass/length^2), that is, the mass is lumped as a thin spherical shell

$\mathbf{U}_i$ Field matrix of the ith segment

$\overline{\mathbf{U}}_i$ Point matrix at $r = a_i$

The notation for transfer matrices is:

$$\mathbf{U}_i = \begin{bmatrix} U_{uu} & U_{u\sigma} & F_u \\ U_{\sigma u} & U_{\sigma\sigma} & F_\sigma \\ 0 & 0 & 1 \end{bmatrix} \tag{10.7}$$

10.5 Differential Equations

The fundamental equations in first-order form for the radial motion of a sphere are

$$\frac{\partial u}{\partial r} = -\frac{2\lambda}{\lambda + 2G}\frac{u}{r} + \frac{1}{\lambda + 2G}\sigma + \frac{3\lambda + 2G}{\lambda + 2G}\alpha\,\Delta T \tag{10.8a}$$

$$\frac{\partial \sigma}{\partial r} = \frac{4G}{r^2}\left(\frac{3\lambda + 2G}{\lambda + 2G}\right)u - \frac{4G}{\lambda + 2G}\frac{1}{r}\sigma + \rho\frac{\partial^2 u}{\partial t^2}$$
$$-\frac{4G}{r}\frac{(3\lambda + 2G)}{(\lambda + 2G)}\alpha\,\Delta T - q_r(r, t) \tag{10.8b}$$

These relations can be reorganized to form higher-order equations similar to Eqs. (10.5).

In the case of transverse curvilinear isotropy, the differential equations are

$$\frac{\partial u}{\partial r} = -\frac{2c_{12}}{c_{11}}\frac{u}{r} + \frac{1}{c_{11}}\sigma + \frac{\lambda_1}{c_{11}}\Delta T \tag{10.9a}$$

$$\frac{\partial \sigma}{\partial r} = 2\left(c_{22} + c_{23} - \frac{2c_{12}^2}{c_{11}}\right)\frac{u}{r^2} - \frac{2}{r}\left(1 - \frac{c_{12}}{c_{11}}\right)\sigma$$
$$+\rho\frac{\partial^2 u}{\partial t^2} + \frac{2}{r}\left(\frac{c_{12}}{c_{11}}\lambda_1 - \lambda_2\right)\Delta T - q_r(r, t) \tag{10.9b}$$

TABLE 10-3 Massless Spherical Segment

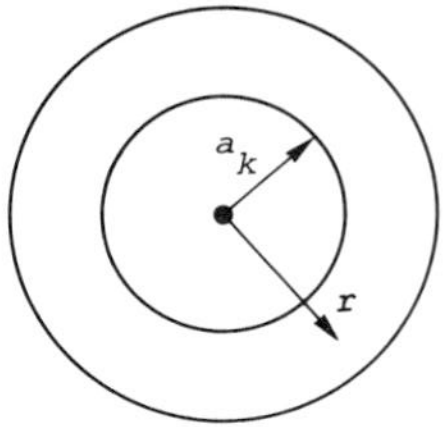

Definitions:

(a) ΔT_1 = a constant temperature change that is independent of r. Set $\Delta T(\xi) = 0$ if only a constant temperature change is present.

(b) $\Delta T(r)$ = arbitrary temperature change

(c) $q_r(r)$ = arbitrary loading intensity in r direction.

(d) a_k = the radial coordinate of the inner surface of the ith section. That is, the ith section begins at a_k and continues to r.

$$\mathbf{U}_i = \begin{bmatrix} \dfrac{4Gr^3 + (3\lambda + 2G)a_k^3}{3(\lambda + 2G)a_k r^2} & \dfrac{a_k}{3(\lambda + 2G)}\left(\dfrac{r}{a_k} - \dfrac{a_k^2}{r^2}\right) & F_u \\ \dfrac{4G(3\lambda + 2G)(r^3 - a_k^3)}{3a_k r^3(\lambda + 2G)} & \dfrac{3\lambda + 2G + 4Ga_k^3/r^3}{3(\lambda + 2G)} & F_\sigma \\ 0 & 0 & 1 \end{bmatrix}$$

$$F_u = \frac{3\lambda + 2G}{3(\lambda + 2G)}\frac{(r^3 - a_k^3)}{r^2}\alpha\,\Delta T_1 + \frac{3\lambda + 2G}{(\lambda + 2G)}\frac{\alpha}{r^2}\int_{a_k}^{r}\xi^2\,\Delta T(\xi)\,d\xi - \frac{1}{r^2(\lambda + 2G)}\int_{a_k}^{r}\eta^2\left[\int_{a_k}^{\eta} q_r(\xi)\,d\xi\right]d\eta$$

$$F_\sigma = -\frac{4G(3\lambda + 2G)}{3(\lambda + 2G)}\frac{(r^3 - a_k^3)}{r^3}\alpha\,\Delta T_1 - \frac{4G(3\lambda + 2G)}{\lambda + 2G}\frac{\alpha}{r^3}\int_{a_k}^{r}\xi^2\,\Delta T(\xi)\,d\xi + \frac{4G}{r^3(\lambda + 2G)}\int_{a_k}^{r}\eta^2\left[\int_{a_k}^{\eta} q_r(\xi)\,d\xi\right]d\eta - \int_{a_k}^{r} q_r(\xi)\,d\xi$$

TABLE 10-4 Massless Spherical Segment without Center Hole (Center Segment)

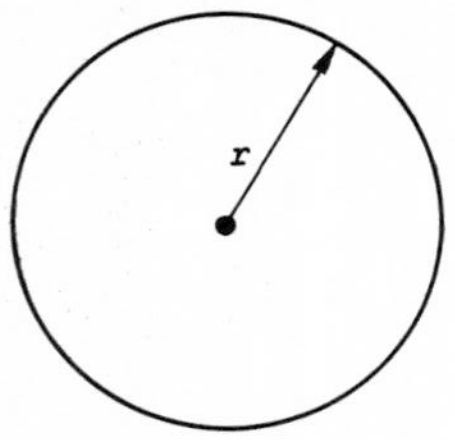

$$\mathbf{U}_i = \begin{bmatrix} 0 & \dfrac{r}{3\lambda + 2G} & F_u \\ 0 & 1 & F_\sigma \\ 0 & 0 & 1 \end{bmatrix}$$

Definitions:

The definitions of Table 10-3 still apply for ΔT_1, $\Delta T(r)$, $q_r(r)$.

$$F_u = r\alpha\,\Delta T_1 + \frac{3\lambda + 2G}{\lambda + 2G}\frac{\alpha}{r^2}\int_0^r \xi^2\,\Delta T(\xi)\,d\xi + \frac{4G\alpha r}{3(\lambda + 2G)}\,\Delta T\Big|_{r=0} + \frac{1}{3(\lambda + 2G)}\left(\frac{1}{r^2}\int_0^r q_r \xi^3\,d\xi - r\int_0^r q_r\,d\xi\right)$$

$$F_\sigma = -\frac{4G(3\lambda + 2G)}{\lambda + 2G}\frac{\alpha}{r^3}\int_0^r \xi^2\,\Delta T(\xi)\,d\xi + \frac{4G(3\lambda + 2G)\alpha}{3(\lambda + 2G)}\,\Delta T\Big|_{r=0} - \frac{1}{3(\lambda + 2G)}\left[\frac{4G}{r^3}\int_0^r q_r \xi^3\,d\xi + (3\lambda + 2G)\int_0^r q_r\,d\xi\right]$$

TABLE 10-5 Massless Anisotropic Spherical Segment

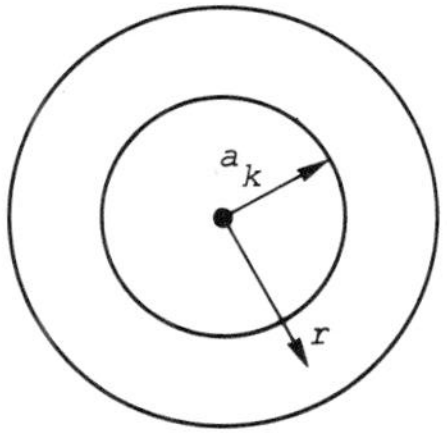

$$\mathbf{U}_i = \begin{bmatrix} \dfrac{1}{(2\beta+1)c_{11}}\left[\beta_1\left(\dfrac{r}{a_k}\right)^{\beta} + \beta_2\left(\dfrac{a_k}{r}\right)^{\beta+1}\right] & \dfrac{a_k}{(2\beta+1)c_{11}}\left[\left(\dfrac{r}{a_k}\right)^{\beta} - \left(\dfrac{a_k}{r}\right)^{\beta+1}\right] & F_u \\ \dfrac{\beta_1\beta_2}{(2\beta+1)c_{11}a_k}\left[\left(\dfrac{r}{a_k}\right)^{\beta-1} - \left(\dfrac{a_k}{r}\right)^{\beta+2}\right] & \dfrac{1}{(2\beta+1)c_{11}}\left[\beta_2^2\left(\dfrac{r}{a_k}\right)^{\beta-1} + \beta_1\left(\dfrac{a_k}{r}\right)^{\beta+2}\right] & F_\sigma \\ 0 & 0 & 1 \end{bmatrix}$$

Definition:

(a) $\beta = -\dfrac{1}{2} + \dfrac{1}{2}\sqrt{1 + 8(c_{22} + c_{23} - c_{12})}$

$\beta_1 = c_{11}\beta + c_{11} - 2c_{12}$, $\beta_2 = c_{11}\beta + 2c_{12}$

(b) ΔT = arbitrary temperature variation.

(c) a_k = the radial coordinate of the inner surface of the ith section. That is, the ith section begins at a_k and continues to r.

$$R(r) = \frac{\lambda_1}{c_{11}}\frac{d\,\Delta T}{dr} + \frac{2(\lambda_1 - \lambda_2)}{c_{11}}\frac{\Delta T}{r}$$

$$F_u = \frac{1}{2\beta+1}\left[r^{\beta}\int_{a_k}^{r}\xi^{-\beta+1}R(\xi)\,d\xi - r^{-\beta-1}\int_{a_k}^{r}\xi^{\beta+2}R(\xi)\,d\xi\right] + \frac{\lambda_1 a_k}{(2\beta+1)c_{11}}\left[\left(\frac{r}{a_k}\right)^{\beta} - \left(\frac{a_k}{r}\right)^{\beta+1}\right]\Delta T\Bigg|_{r=a_k}$$

$$F_\sigma = \frac{1}{2\beta+1}\left[\beta_2 r^{\beta-1}\int_{a_k}^{r}\xi^{-\beta+1}R(\xi)\,d\xi + \beta_1 r^{-\beta-2}\int_{a_k}^{r}\xi^{\beta+2}R(\xi)\,d\xi\right] - \lambda_1\,\Delta T + \frac{\lambda_1}{(2\beta+1)c_{11}}\left[\beta_2\left(\frac{r}{a_k}\right)^{\beta-1} + \beta_1\left(\frac{a_k}{r}\right)^{\beta+2}\right]\Delta T\Bigg|_{r=a_k}$$

10.6 Transfer Matrices

The transfer matrices for spherical elements are provided in Tables 10-3 to 10-9. If desired, the transfer matrices for segments extending from radial coordinate a_k to r can be employed for a center (holeless) segment by considering a small hole to be at the center. The boundary condition for this imaginary hole should be fixed.

10.7 Point Matrices

The transfer matrices that take into account concentrated (point) occurrences are listed in Table 10-10.

10.8 Loading Functions

The loading functions are provided in the transfer matrix tables for most common loadings. These functions can be calculated for other loadings from the formulas:

$$
\begin{aligned}
F_u &= -\int_{a_i}^{r} q_r(\xi, a_i)\, U_{u\sigma}(r, \xi)\, d\xi = -\int_{a_i}^{r} q_r(r, \xi)\, U_{u\sigma}(\xi, a_i)\, d\xi \\
F_\sigma &= -\int_{a_i}^{r} q_r(\xi, a_i)\, U_{\sigma\sigma}(r, \xi)\, d\xi = -\int_{a_i}^{r} q_r(r, \xi)\, U_{\sigma\sigma}(\xi, a_i)\, d\xi
\end{aligned}
\tag{10.10}
$$

TABLE 10-6 Massless Anisotropic Spherical Segment without Center Hole (Center Segment)

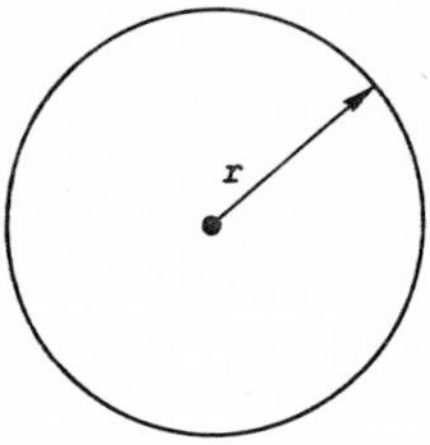

$$
\mathbf{U}_i = \begin{bmatrix} 0 & r^{\beta} & F_u \\ 0 & \beta_2 r^{\beta-1} & F_\sigma \\ 0 & 0 & 1 \end{bmatrix}
$$

Definitions:

$\beta \geq 1$, β_1, β_2, R, ΔT are the same as given in Table 10-5.

$$
F_u = \frac{1}{2\beta + 1}\left[r^{\beta}\int^{r} r^{-\beta+1} R(r)\, dr - r^{-\beta-1}\int^{r} r^{\beta+2} R(r)\, dr\right]
$$

$$
F_\sigma = \frac{1}{2\beta + 1}\left[\beta_2 r^{\beta-1}\int^{r} r^{-\beta+1} R(r)\, dr + \beta_1 r^{-\beta-2}\int^{r} r^{\beta+2} R(r)\, dr\right] - \lambda_1 \Delta T
$$

This $\mathbf{U}_i$ is not truly a transfer matrix, as the vector at $r = 0$ contains arbitrary constants rather than state variables. However, the vector at other radii is the state vector. All operations remain the same as those for the usual transfer matrices.

TABLE 10-7 Spherical Segment with Mass

$$\mathbf{U}_i = \begin{bmatrix} \frac{1}{e_1\sqrt{\beta r}}\left[e_3(a_k)J_\gamma(\beta r) - e_2(a_k)Y_\gamma(\beta r)\right] & \frac{1}{e_1\beta\sqrt{a_k r}}\left[J_\gamma(\beta a_k)Y_\gamma(\beta r) - Y_\gamma(\beta a_k)J_\gamma(\beta r)\right] & F_u \\ \frac{1}{e_1}\left[e_3(a_k)e_2(r) - e_2(a_k)e_3(r)\right] & \frac{1}{e_1\sqrt{\beta a_k}}\left[e_3(r)J_\gamma(\beta a_k) - e_2(r)Y_\gamma(\beta a_k)\right] & F_\sigma \\ 0 & 0 & 1 \end{bmatrix}$$

Definitions:

(a) Isotropic

$$\gamma = \frac{3}{2},\quad \beta^2 = \frac{\rho\omega^2}{\lambda + 2G}$$

$$e_1 = e_3(a_k)\frac{J_{3/2}(\beta a_k)}{\sqrt{\beta a_k}} - e_2(a_k)\frac{Y_{3/2}(\beta a_k)}{\sqrt{\beta a_k}}$$

$$e_2(r) = \frac{\beta(\lambda + 2G)}{(\beta r)^{3/2}}\left[\frac{3\lambda + 2G}{\lambda + 2G}J_{3/2}(\beta r) - \beta r J_{5/2}(\beta r)\right]$$

$$e_3(r) = \frac{\beta(\lambda + 2G)}{(\beta r)^{3/2}}\left[\frac{3\lambda + 2G}{\lambda + 2G}Y_{3/2}(\beta r) - \beta r Y_{5/2}(\beta r)\right]$$

(b) Anisotropic

$$\gamma = \frac{1}{2}\left[\frac{8(c_{22} + c_{23} - c_{12})}{c_{11}} + 1\right]^{1/2}$$

$$\beta^2 = \omega^2\rho/c_{11}$$

$$\gamma_1 = (\gamma - \tfrac{1}{2}) + 2c_{12}/c_{11}$$

$$e_1 = e_3(a_k)\frac{J_\gamma(\beta a_k)}{\sqrt{\beta a_k}} - e_2(a_k)\frac{Y_\gamma(\beta a_k)}{\sqrt{\beta a_k}}$$

$$e_2(r) = \frac{\beta c_{11}}{(r\beta)^{3/2}}\left[\gamma_1 J_\gamma(\beta r) - \beta r J_{\gamma+1}(\beta r)\right]$$

$$e_3(r) = \frac{\beta c_{11}}{(r\beta)^{3/2}}\left[\gamma_1 Y_\gamma(\beta r) - \beta r Y_{\gamma+1}(\beta r)\right]$$

$J_\gamma(\beta r)$, $Y_\gamma(\beta r)$ are Bessel functions of order γ of the first and second kinds, respectively.

a_k is the radial coordinate of the inner surface of the ith section. That is, the ith section begins at a_k and continues to r.

TABLE 10-8 Spherical Segment without Center Hole (Center Segment), including Mass

Definitions: $\beta = \sqrt{\dfrac{\omega^2 \rho}{\lambda + 2G}}$

$J_{3/2}(\beta r)$, $J_{5/2}(\beta r)$ are Bessel functions of the first kind.

$$\mathbf{U}_i = \left[\begin{array}{c|c|c} 0 & \dfrac{3\sqrt{\pi}}{\sqrt{2}(3\lambda + 2G)} \cdot \dfrac{J_{3/2}(\beta r)}{\beta^{3/2} r^{1/2}} & F_u \\ \hline 0 & \dfrac{3(\lambda + 2G)\sqrt{\pi}}{\sqrt{2}(3\lambda + 2G)} \dfrac{1}{(\beta r)^{3/2}} \left[\dfrac{3\lambda + 2G}{\lambda + 2G} J_{3/2}(\beta r) - \beta r J_{5/2}(\beta r)\right] & F_\sigma \\ \hline 0 & 0 & 1 \end{array}\right]$$

Here the notation $q_r(r, a_i)$ is employed to indicate a loading intensity (force/length^3) that begins at radius a_i and extends to radius r. The notation $U_{kj}(r, a_i)$ has a similar meaning. $U_{kj}(r, a_i)$ denotes a transfer matrix for a spherical segment that begins at radius a_i and extends to radius r. Do not use U_{kj} from transfer matrices for spherical segments with no center hole.

TABLE 10-9 Anisotropic Spherical Segment without Center Hole (Center Segment), including Mass

Definitions:

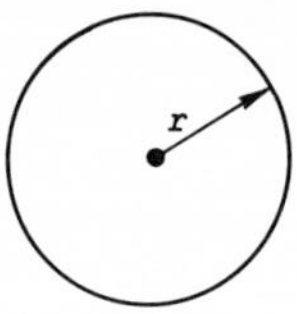

$$\gamma = \frac{1}{2}\left[\frac{8(c_{22} + c_{23} - c_{12})}{c_{11}} + 1\right]^{1/2} > \frac{3}{2}$$

$\beta^2 = \omega^2 \rho / c_{11}$

$J_\gamma(\beta r)$ are Bessel functions of order γ of the first kind.

This $\mathbf{U}_i$ is not truly a transfer matrix. See the note in Table 10-6.

$$\mathbf{U}_i = \left[\begin{array}{c|c|c} 0 & \dfrac{J_\gamma(\beta r)}{(\beta r)^{1/2}} & F_u \\ \hline 0 & \dfrac{c_{11}}{\beta^{1/2} r^{3/2}} \left[\left(\gamma - \dfrac{1}{2} + 2\dfrac{c_{12}}{c_{11}}\right) J_\gamma(\beta r) - \beta r J_{\gamma+1}(\beta r)\right] & F_\sigma \\ \hline 0 & 0 & 1 \end{array}\right]$$

TABLE 10-10 Point Matrices at $r = a_i$

Definitions:

(a) Radial Pressure (force/length^2)

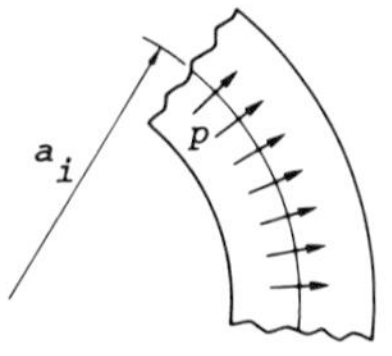

$$\overline{\mathbf{U}}_i = \begin{bmatrix} 1 & 0 & 0 \\ -M_i\omega^2 & 1 & -p \\ 0 & 0 & 1 \end{bmatrix}$$

(b) Concentrated Mass

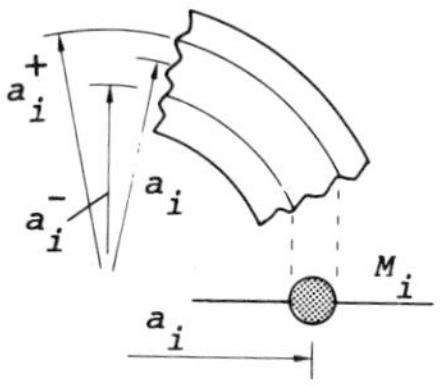

M_i is the mass per unit spherical surface area. M_i can be calculated as

$$M_i = \frac{(a_i^+)^3 - (a_i^-)^3}{3a_i^2}\rho \approx \Delta a\,\rho$$

where ρ is the mass per unit volume

$$\Delta a = a_i^+ - a_i^-$$

10.9 Other Information

Sphere problems can be solved by using procedures outlined in Chap. 8 if σ (Chap. 10) is substituted for P (Chap. 8). This applies to static problems as outlined in Section 8.11, free dynamics problems in Section 8.12, forced harmonic motion in Section 8.13, and the transient response of Section 8.14 if N_n of Eq. (8.18), $\eta_n(0)$ of Eq. (8.19), and $f_n(t)$ of Eq. (8.20) are replaced by

$$N_n = \int_{a_0}^{a_L} \rho r^2 u_n^2 \, dr \tag{10.11}$$

$$\eta_n(0) = \int_{a_0}^{a_L} \rho r^2 u(r, 0) u_n \, dr \tag{10.12}$$

$$\begin{aligned} f_n(t) = \int_{a_0}^{a_L} \left[q_r(r, t) r^2 u_n - \alpha \, \Delta T(r, t) \, d_n \right] dr + a_0^2 u(a_0, t)\sigma_n(a_0) \\ - a_L^2 u(a_L, t)\sigma_n(a_L) - a_0^2 p(a_0, t) u_n(a_0) + a_{L}^2 p(a_L, t) u_n(a_L) \end{aligned} \tag{10.13}$$

$$d_n = \begin{cases} \dfrac{3\lambda + 2G}{\lambda + 2G}\left(-4Gru_n - r^2\sigma_n\right) & \text{for isotropic material} \\[2ex] 2r\left(\dfrac{\lambda_1}{c_{11}} c_{12} - \lambda_{12}\right) u_n - \dfrac{\lambda_1}{c_{11}} \sigma_n & \text{for anisotropic material} \end{cases} \tag{10.14}$$

For proportional damping $c' = c_\rho \rho$, so that a term $c_\rho \rho \, \partial u / \partial t$ should be added to the right-hand sides of Eqs. (10.8*b*) and (10.9*b*). Set $c_\rho = 0$ for no damping. In the case of nonproportional damping for both isotropic and anisotropic materials, add the term $c' \, \partial u / \partial t$ to the right-hand sides of Eqs. (10.8*b*) and (10.9*b*) and replace Eq. (8.23) by

$$N_n = 2s_n \int_{a_0}^{a_L} \rho r^2 u_n^2 \, dr + \int_{a_0}^{a_L} c' r^2 u_n^2 \, dr \tag{10.15}$$

and Eq. (8.26) by

$$\eta_n(0) = \int_{a_0}^{a_L} \left[\rho s_n u(r, 0) u_n(r) + c' u(r, 0) u_n(r) + \rho \frac{\partial u}{\partial t}(r, 0) u_n(r) \right] r^2 \, dr \tag{10.16}$$

For discrete damping with dashpot constants c_i use

$$\int_{a_0}^{a_L} c' r^2 u_n^2 \, dr = \sum_i c_i a_i^2 u_n^2(a_i) \tag{10.17}$$

in N_n and

$$\int_{a_0}^{a_L} c' u(r, 0) u_n(r) r^2 \, dr = \sum_i c_i a_i^2 u(a_i, 0) u_n(a_i) \tag{10.18}$$

in $\eta_n(0)$.

For isotropic Voigt-Kelvin material continue to apply the formulas of Chap. 8.

10.10 Applications

Since static and dynamic problems for the radial motion of spheres are solved in much the same fashion as similar problems for other members, the example of previous chapters should suffice as references in setting up sphere problems. A few additional examples follow:

EXAMPLE 10.2 Thermally Loaded Sphere with No Center Hole A solid (holeless) sphere is subject to a change in temperature of $\Delta T(r)$. The surface of the sphere is free. Find the variation of stresses and displacements in the radial direction.

The responses are formed from the transfer matrix of Table 10-4.

$$\begin{aligned} u &= \sigma_0 \frac{r}{3\lambda + 2G} + \frac{3\lambda + 2G}{\lambda + 2G} \frac{\alpha}{r^2} \int_0^r \xi^2 \, \Delta T(\xi) \, d\xi \\ \sigma &= \sigma_0 - \frac{4G(3\lambda + 2G)}{\lambda + 2G} \frac{\alpha}{r^3} \int_0^r \xi^2 \, \Delta T(\xi) \, d\xi \end{aligned} \tag{1}$$

The initial parameter is given in Table 8-9, with $P = \sigma$, as

$$\sigma_0 = -F_\sigma / \nabla = -F_\sigma \big|_{r=a_L} \Big/ U_{\sigma\sigma} \big|_{r=a_L} \tag{2}$$

From (1)

$$\sigma_0 = \frac{4G(3\lambda + 2G)}{\lambda + 2G} \frac{a}{a_L^3} \int_0^{a_L} \xi^2 \, \Delta T(\xi) \, d\xi \tag{3}$$

The expressions of (1) for u and σ are now fully determined for a prescribed variation $\Delta T(r)$.

EXAMPLE 10.3 Natural Frequencies of Anisotropic Sphere Find the natural frequencies for a solid anisotropic sphere of radius a_L. The outer surface remains unconstrained.

The frequencies are those values of ω that make ∇ of Table 8-9 equal to zero. With $P = \sigma$, Table 8-9 gives $\nabla = U_{\sigma\sigma}|_{r=a_L}$. From Table 10-9,

$$\nabla = U_{\sigma\sigma}|_{r=a_L} = \frac{c_{11}}{\beta^{1/2}a_L^{3/2}}\left[\left(\gamma - \frac{1}{2} + 2\frac{c_{12}}{c_{11}}\right)J_\gamma(\beta a_L) - \beta a_L J_{\gamma+1}(\beta a_L)\right] = 0$$

or

$$\eta J_\gamma(\beta a_L) - \beta a_L J_{\gamma+1}(\beta a_L) = 0 \tag{1}$$

where

$$\gamma = \frac{1}{2}\left[\frac{8(c_{22}+c_{23}-c_{12})}{c_{11}} + 1\right]^{1/2} \qquad \beta^2 = \omega^2\rho/c_{11} \qquad \eta = \gamma - \frac{1}{2} + 2c_{12}/c_{11}$$

Figure 9-2 provides a plot of the first natural frequency for given values of γ and η.

C. COMPUTER PROGRAMS AND EXAMPLES

10.11 Benchmark Examples

Complicated sphere problems should be solved with the assistance of a computer program. The following example is provided as a benchmark problem against which a reader's own computer program can be checked. The computer program SPHERE was used for this example. SPHERE can be used to compute the radial displacement, stresses, and natural frequencies in a sphere of any complexity. The sphere can be formed of layers of different materials with arbitrary mechanical or thermal loading and boundary conditions.

EXAMPLE 10.4 Sphere with Internal and External Pressure Find the displacements and stresses in a sphere with an inner radius of 1 in and outer radius of 3 in. The pressure on the inner surface is 30,000 lb/in^2 and on the outer circumference is 10,000 lb/in^2. Let $E = 30(10^6)$ lb/in^2, $\nu = 0.3$.

The computer results are shown in Fig. 10-1.

RADIAL LOCATION	DISPLACEMENT	RADIAL STRESS	TANGENTIAL STRESS
1.00000E+00	3.26923E-04	-3.00000E+04	1.15385E+03
1.50000E+00	1.53846E-05	-1.53846E+04	-6.15385E+03
2.00000E+00	-1.33654E-04	-1.18269E+04	-7.93269E+03
2.50000E+00	-2.35692E-04	-1.05600E+04	-8.56615E+03
3.00000E+00	-3.19231E-04	-1.00000E+04	-8.84615E+03

Fig. 10-1 Results for Example 10.4.

REFERENCES

10.1 Eason, G.: "On the Vibration of Anisotropic Cylinders and Spheres," *Appl. Sci. Res., Part A, Mech. Heat*, vol. 12, 1962, pp. 81–85.

10.2 Lekhnitskii, S.: *Theory of Elasticity of an Anisotropic Elastic Body*, Holden-Day, San Francisco, 1963.

ELEVEN

Circular Plates

This chapter treats the static, stability, and dynamic analyses of circular plates. The loading, which can be mechanical or thermal, is transverse to the plate. Formulas for plates of constant thickness with radially symmetric loading are handled in Part A. The formulas for more complex plates with symmetric or asymmetric loading are provided in Part B. Included are plates of variable cross section and composite material. Part C contains some examples of problems solved by a computer program.

The formulas of this chapter apply to plates represented by the classical (Kirchhoff) theory of bending. This theory is based on the assumptions that the plate thickness is small compared with other dimensions and the deflections are small compared with the thickness of the plate.

A. SIMPLE PLATES

11.1 Notation and Conventions

v	Transverse deflection (length)
θ	Rotation or slope of the deflection curve (radians)
M	Bending moment (per unit length) along a circular arc (force)
V	Effective shear force (per unit length) along a circular arc (force/length)
h	Thickness of plate (length)
E	Modulus of elasticity of the material (force/length^2)

ν	Poisson's ratio
r	Radial coordinate (length)
a_0	Inner radius of plate (length)
a_L	Outer radius of plate (length)
ϕ	Circumferential coordinate (radians)

$D = Eh^3/[12(1-\nu^2)]$

w	Distributed transverse force, loading intensity (force/length^2)
W	Applied line transverse force at specified radius (force/length)
W_T	Concentrated transverse force
C	Applied line moment at specified radius (force · length/length)
w_1	Magnitude of distributed force, uniform in r direction (force/length^2)
P	Normal compressive (in-plane, axial) force per unit length of circular arc (force/length). Replace P by $-P$ for a tensile force.
z	The vertical distance from the middle surface of the plate, positive downwards
M_ϕ	Bending moment (per unit length) on a ϕ face (force · length/length)
$M_{r\phi}$	Twisting moment (per unit length) along a circular arc (force · length/length)
V_ϕ	Effective shear force (per unit length) on a ϕ face (force/length).
σ_r, σ_ϕ	Normal stresses (force/length^2)
ρ	Mass per unit area (mass/length^2, force · time^2/length^3)
ω	Natural frequency (radians/second)

Positive deflection v, slope θ, internal bending moment M, and internal shear forces V are displayed in Fig. 11-1.

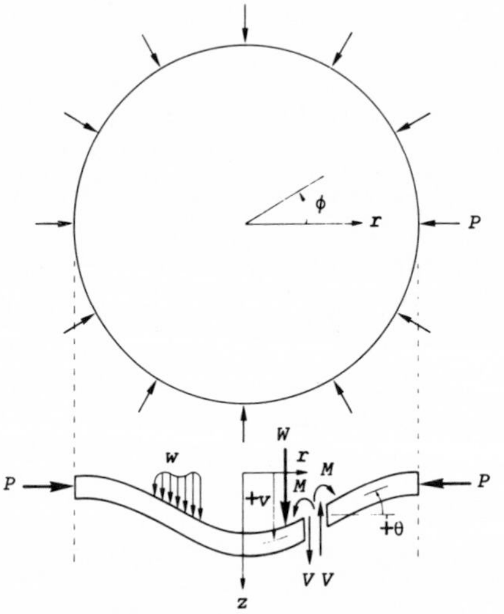

Fig. 11-1 Positive displacement v, slope θ, moment M, shear force V, and applied loading.

11.2 Stresses on a Cross Section

The tables of this chapter provide the deflection, slope, bending moment, and shear force. Also, the computer program is available for computing these variables. Once the internal moments and forces are known the stresses can be derived with the formulas given in this section. The radial stress is

$$\sigma_r = \frac{Mz}{h^3/12} \tag{11.1}$$

Once v is known, the moment M_ϕ can be calculated using the appropriate formula of Eqs. (11.3). Then

$$\sigma_\phi = \frac{M_\phi z}{h^3/12} \tag{11.2}$$

11.3 Differential Equations

The fundamental equations of motion for the bending of a symmetrically loaded, isotropic circular plate of constant thickness are

$$\left(\frac{d^2}{dr^2} + \frac{1}{r}\frac{d}{dr}\right)\left(\frac{d^2v}{dr^2} + \frac{1}{r}\frac{dv}{dr}\right) = \frac{w}{D}$$

$$M_r = M = -D\left(\frac{d^2v}{dr^2} + \frac{\nu}{r}\frac{dv}{dr}\right) \qquad M_\phi = -D\left(\nu\frac{d^2v}{dr^2} + \frac{1}{r}\frac{dv}{dr}\right)$$

$$M_{r\phi} = 0 \qquad V_\phi = 0 \tag{11.3}$$

$$V_r = V = -D\left(\frac{d^3v}{dr^3} + \frac{1}{r}\frac{d^2v}{dr^2} - \frac{1}{r^2}\frac{dv}{dr}\right)$$

11.4 Plate with no Center Hole

For a plate with rotationally symmetric loading and with no center hole (for example, Fig. 11-2), Eqs. (11.3) are integrated to give the deflection, slope, bending moment, and shear force:

Deflection:	$v = v_0 - M_0 \dfrac{r^2}{2D(1+\nu)} + F_v$	(11.4*a*)
Slope:	$\theta = M_0 \dfrac{r}{D(1+\nu)} + F_\theta$	(11.4*b*)
Bending moment:	$M = M_0 + F_M$	(11.4*c*)
Shear force:	$V = F_V$	(11.4*d*)

Note that both θ and V are zero at the center of the plate, $r = 0$. The F_v, F_θ, F_M, F_V are loading functions given in Table 11-1. If there is more than one loading on the plate, the F_v, F_θ, F_M, F_V functions are formed by adding the terms of Table 11-1 for each load. The initial parameters v_0, M_0 (values of v, M at the center of the plate, $r = 0$) are provided in Table 11-2.

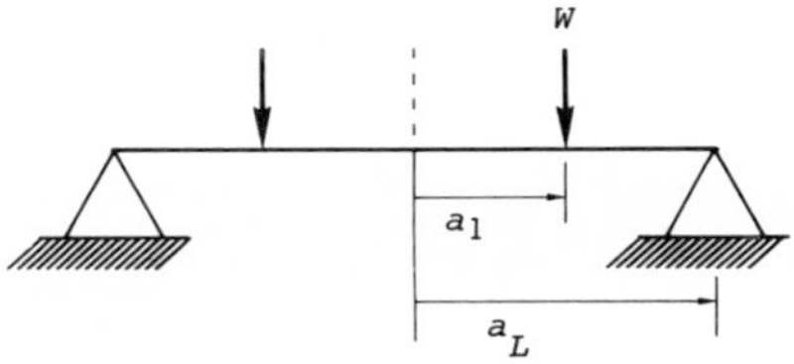

Fig. 11-2 Example 11.1.

TABLE 11-1 Loading Functions F_v, F_θ, F_M, F_V for Simple Plates of Constant Thickness (Eqs. 11.4, 11.5, and 11.6)

	Concentrated Force (Applies for plates with no center hole) W_T (force) 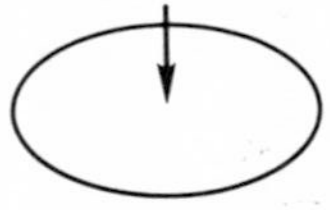	Concentrated Line Force W (force/length) 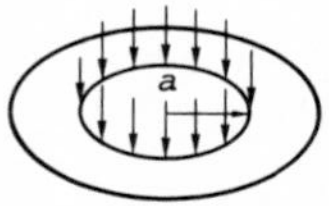	Uniform Loading 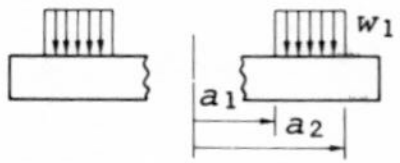	Ramp Loading 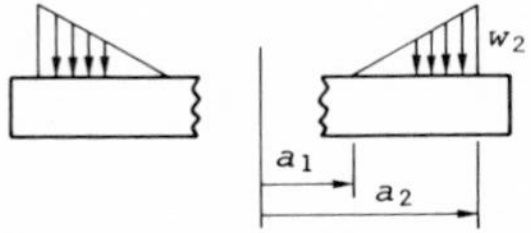 $\Delta\ell = a_2 - a_1$	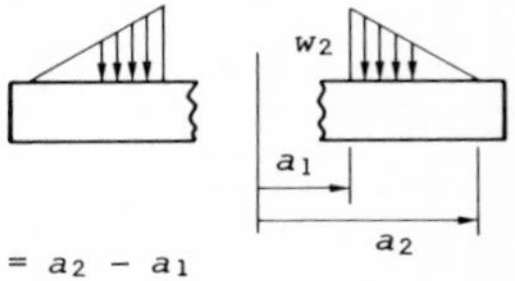 $\Delta\ell = a_2 - a_1$
$F_v(r)$	$\frac{W_T}{8\pi D} r^2(\ln r - 1)$	$\langle r-a\rangle^0 \frac{Wa}{4D} [(r^2 + a^2) \ln \frac{r}{a} - (r^2 - a^2)]$	$\frac{w_1}{8D} [F_{vw_1}(r,a_1) - F_{vw_1}(r,a_2)]$	$\frac{w_2}{D\,\Delta\ell} [F_{vw_2}(r,a_1) - F_{vw_2}(r,a_2)] - w_2F_{vw_1}(r,a_2)$	$-\frac{w_2}{D\,\Delta\ell} [F_{vw_2}(r,a_1) - F_{vw_2}(r,a_2)] + w_2F_{vw_1}(r,a_1)$
$F_\theta(r)$	$-\frac{W_T}{4\pi D} r(\ln r - \frac{1}{2})$	$-\langle r-a\rangle^0 \frac{Wa}{2D} \left[r \ln\frac{r}{a} - \frac{1}{2r}(r^2 - a^2)\right]$	$-\frac{w_1}{4D} [F_{\theta w_1}(r,a_1) - F_{\theta w_1}(r,a_2)]$	$-\frac{w_2}{D\,\Delta\ell} [F_{\theta w_2}(r,a_1) - F_{\theta w_2}(r,a_2)] - w_2F_{\theta w_1}(r,a_2)$	$\frac{w_2}{D\,\Delta\ell} [F_{\theta w_2}(r,a_1) - F_{\theta w_2}(r,a_2)] + w_2F_{\theta w_1}(r,a_1)$
$F_M(r)$	$-\frac{W_T}{4\pi} \left[(1+\nu)\ln r + \frac{1-\nu}{2}\right]$	$-\langle r-a\rangle^0 \frac{Wa}{2} \left[(1+\nu)\ln\frac{r}{a} + \frac{1-\nu}{2}\left(1 - \frac{a^2}{r^2}\right)\right]$	$-\frac{w_1}{4} [F_{Mw_1}(r,a_1) - F_{Mw_1}(r,a_2)]$	$-\frac{w_2}{\Delta\ell} [F_{Mw_2}(r,a_1) - F_{Mw_2}(r,a_2)] - w_2F_{Mw_1}(r,a_2)$	$\frac{w_2}{\Delta\ell} [F_{Mw_2}(r,a_1) - F_{Mw_2}(r,a_2)] + w_2F_{Mw_1}(r,a_1)$
$F_V(r)$	$-\frac{W_T}{2\pi r}$	$-\langle r-a\rangle^0 \frac{Wa}{r}$	$-\frac{w_1}{2} [F_{Vw_1}(r,a_1) - F_{Vw_1}(r,a_2)]$	$-\frac{w_2}{\Delta\ell} [F_{Vw_2}(r,a_1) - F_{Vw_2}(r,a_2)] - w_2F_{Vw_1}(r,a_2)$	$\frac{w_2}{\Delta\ell} [F_{Vw_2}(r,a_1) - F_{Vw_2}(r,a_2)] + w_2F_{Vw_1}(r,a_1)$

	Concentrated Ring Moment (force-length/length)	Arbitrary Loading
	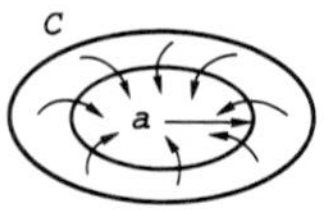	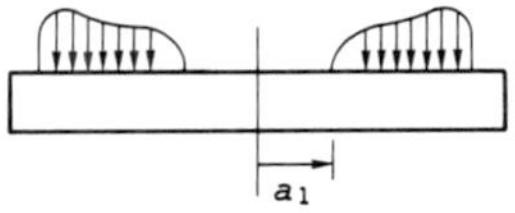
$F_V(r)$	$-\langle r - a\rangle^0 \frac{C}{2D}\left(a^2 \ln \frac{r}{a} - \frac{r^2 - a^2}{2}\right)$	$\int_{a_1}^{r} \frac{1}{r}\int r \int \frac{1}{r}\int \frac{wr}{D}\, dr$
$F_\theta(r)$	$-\langle r - a\rangle^0 \frac{C}{2D} \frac{r^2 - a^2}{r}$	$-\frac{1}{r}\int_{a_1}^{r} r\int \frac{1}{r}\int \frac{wr}{D}\, dr$
$F_M(r)$	$-\langle r - a\rangle^0 \frac{C}{2}\left[(1-\nu)\left(\frac{a}{r}\right)^2 + 1 + \nu\right]$	$-\int_{a_1}^{r} \frac{1}{r}\int wr\, dr + \frac{1-\nu}{r^2}\int_{a_1}^{r} r\int \frac{1}{r}\int wr\, dr$
$F_V(r)$	0	$-\frac{1}{r}\int_{a_1}^{r} wr\, dr$

$$F_{vw_1}(r,a_i) = \langle r - a_i\rangle^0 \left[\frac{r^4}{8} - \frac{5a_i^4}{8} + \frac{a_i^2 r^2}{2} - a_i^2\left(r^2 + \frac{a_i^2}{2}\right)\ln \frac{r}{a_i}\right]$$

$$F_{\theta w_1}(r,a_i) = \langle r - a_i\rangle^0 \left(\frac{r^3}{4} - \frac{a_i^4}{4r} - a_i^2 r \ln \frac{r}{a_i}\right)$$

$$F_{Mw_1}(r,a_i) = \langle r - a_i\rangle^0 \left[\frac{3+\nu}{4} r^2 - a_i^2 + \frac{(1-\nu)a_i^4}{4r^2} - (1+\nu)a_i^2 \ln \frac{r}{a_i}\right]$$

$$F_{Vw_1}(r,a_i) = \langle r - a_i\rangle^0 \frac{r^2 - a_i^2}{r}$$

$$F_{vw_2}(r,a_i) = \langle r - a_i\rangle^0 \left[\frac{r^5}{225} - \frac{a_i r^4}{64} - \frac{a_i^3 r^2}{144} + \frac{29a_i^5}{1600} + \frac{a_i^3}{8}\left(\frac{r^2}{3} + \frac{a_i^2}{10}\right)\ln \frac{r}{a_i}\right]$$

$$F_{\theta w_2}(r,a_i) = \langle r - a_i\rangle^0 \left(\frac{r^4}{45} - \frac{a_i r^3}{16} + \frac{a_i^3 r}{36} + \frac{a_i^5}{80r} + \frac{a_i^3}{12} r \ln \frac{r}{a_i}\right)$$

$$F_{Mw_2}(r,a_i) = \langle r - a_i\rangle^0 \left(\frac{4+\nu}{45} r^3 - \frac{3+\nu}{16} a_i r^2 + \frac{4+\nu}{36} a_i^3 - \frac{1-\nu}{80} \frac{a_i^5}{r^2} + \frac{1+\nu}{12} a_i^3 \ln \frac{r}{a_i}\right)$$

$$F_{Vw_2}(r,a_i) = \langle r - a_i\rangle^0 \left(\frac{r^2}{3} - \frac{a_i r}{2} + \frac{a_i^3}{6r}\right)$$

TABLE 11-2 Integration Constants for Plate with No Center Hole (Eqs. 11.4 and 11.5)

Outer Edge → / Center Condition ↓	1. Simply Supported 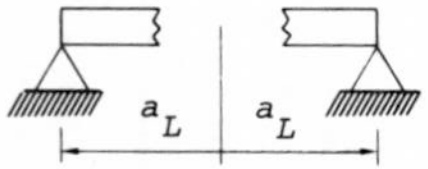	2. Fixed 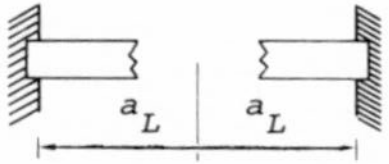	3. Free 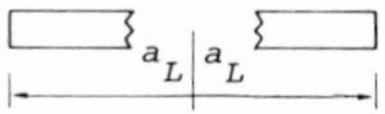	4. Guided 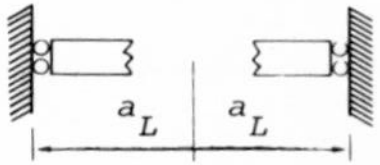
1. No Center Support 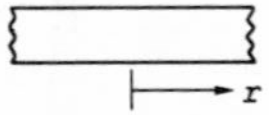	$V_0 = -\frac{a_L^2}{2D(1+\nu)} F_M\Big\rvert_{r=a_L} - F_V\Big\rvert_{r=a_L}$ $M_0 = -F_M\Big\rvert_{r=a_L}$	$V_0 = -\frac{a_L}{2} F_\theta\Big\rvert_{r=a_L} - F_V\Big\rvert_{r=a_L}$ $M_0 = -\frac{D(1+\nu)}{a_L} F_\theta\Big\rvert_{r=a_L}$	Kinematically Unstable	Kinematically Unstable
2. Center Support 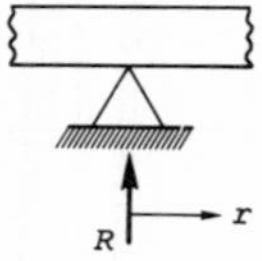	$R = -\frac{16(1+\nu)\pi D}{(3+\nu)a_L^2} F_V\Big\rvert_{r=a_L} - \frac{8\pi}{3+\nu} F_M\Big\rvert_{r=a_L}$ $C_1 = \frac{8(1+\nu)\ln a_L + 4(1-\nu)}{(3+\nu)a_L^2} F_V\Big\rvert_{r=a_L} + \frac{4(\ln a_L - 1)}{D(3+\nu)} F_M\Big\rvert_{r=a_L}$	$R = -\frac{16\pi D}{a_L^2} F_V\Big\rvert_{r=a_L} - \frac{8\pi D}{a_L} F_\theta\Big\rvert_{r=a_L}$ $C_1 = -\frac{8(\ln a_L - 1/2)}{a_L^2} F_V\Big\rvert_{r=a_L} - \frac{4(\ln a_L - 1)}{a_L} F_\theta\Big\rvert_{r=a_L}$	$R = -2\pi a_L F_V\Big\rvert_{r=a_L}$ $C_1 = \frac{2}{D(1+\nu)} F_M\Big\rvert_{r=a_L} - \frac{a_L}{D(1+\nu)}[(1+\nu)\ln a_L + \frac{1-\nu}{2}] F_V\Big\rvert_{r=a_L}$	$R = -2\pi a_L F_V\Big\rvert_{r=a_L}$ $C_1 = -\frac{a_L}{D}(\ln a_L - 1/2) F_V\Big\rvert_{r=a_L} + \frac{2}{a_L} F_\theta\Big\rvert_{r=a_L}$

EXAMPLE 11.1 Plate with Concentrated Load Find the deflection of the plate of Fig. 11-2 with a simply supported outer rim. W is a concentrated ring load; that is, it is a line load extending symmetrically around the plate at radius $r = a_1$.

The deflection is given by Eq. (11.4a) as

$$v = v_0 - M_0 \frac{r^2}{2D(1+\nu)} + F_v \tag{1}$$

The initial parameters v_0, M_0 are given in row 1, column 1 of Table 11-2 as

$$v_0 = -\frac{a_L^2}{2D(1+\nu)} F_M\big|_{r=a_L} - F_v\big|_{r=a_L} \qquad M_0 = -F_M\big|_{r=a_L} \tag{2}$$

The loading functions required to complete (1), (2) are found in Table 11-1 to be

$$\begin{aligned} F_v &= \langle r - a_1\rangle^0 \frac{Wa_1}{4D}\left[(r^2 + a_1^2)\ln\frac{r}{a_1} - (r^2 - a_1^2)\right] \\ F_M &= -\langle r - a_1\rangle^0 \frac{Wa_1}{2}\left[(1+\nu)\ln\frac{r}{a_1} + \frac{1-\nu}{2}\left(1 - \frac{a_1^2}{r^2}\right)\right] \end{aligned} \tag{3}$$

At $r = a_L$,

$$\begin{aligned} F_v\big|_{r=a_L} &= \frac{Wa_1}{4D}\left[(a_L^2 + a_1^2)\ln\frac{a_L}{a_1} - (a_L^2 - a_1^2)\right] \\ F_M\big|_{r=a_L} &= -\frac{Wa_1}{2}\left[(1+\nu)\ln\frac{a_L}{a_1} + \frac{1-\nu}{2}\left(1 - \frac{a_1^2}{a_L^2}\right)\right] \end{aligned} \tag{4}$$

Then, from (2), $M_0 = -F_M\big|_{r=a_L}$ and

$$\begin{aligned} v_0 &= -\frac{a_L^2}{2D(1+\nu)} F_M\big|_{r=a_L} - F_v\big|_{r=a_L} \\ &= -\frac{Wa_1}{4D}\left\{a_1^2 \ln\frac{a_L}{a_1} - \left[\frac{1-\nu}{2(1+\nu)} + 1\right](a_L^2 - a_1^2)\right\} \end{aligned} \tag{5}$$

Substitution of v_0 and M_0 in (1) gives the desired expression for v.

Note that the solution can be reduced to the case of a plate with a total center load W_T. Set $W = W_T/2\pi a_1$ and take the limit of v as $a_1 \to 0$. This leads to

$$v = \frac{W_T}{16\pi D}\left[(a_L^2 - r^2)\frac{3+\nu}{1+\nu} - 2r^2 \ln\frac{a_L}{r}\right]$$

The same result is obtained by employing at the outset the forcing functions for W_T of Table 11-1.

EXAMPLE 11.2 Plate with Concentrated and Distributed Loads Find the deflection and bending moment in the plate of Fig. 11-3. The forces W and w_1 are concentrated and distributed ring forces, respectively.

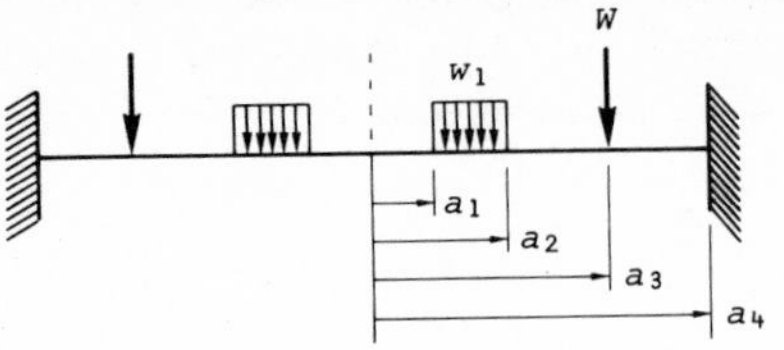

Fig. 11-3 Example 11.2.

From Eqs. (11.4*a* and *c*) the deflection and moment curves are given by

$$v = v_0 - M_0 \frac{r^2}{2D(1+\nu)} + F_v \tag{1}$$

$$M = M_0 + F_M \tag{2}$$

The initial parameters v_0, M_0 in (1) and (2) are obtained from row 1 of Table 11-2 for a fixed outer rim:

$$v_0 = -\frac{a_L}{2} F_\theta\big|_{r=a_L} - F_v\big|_{r=a_L} \qquad M_0 = -\frac{D(1+\nu)}{a_L} F_\theta\big|_{r=a_L} \tag{3}$$

These expressions require evaluation of the loading functions F_v, F_θ at $r = a_L$. These loading functions are taken from Table 11-1. For example,

$$F_\theta = -\frac{Wa_3}{2D}\langle r - a_3\rangle^0\left(r \ln \frac{r}{a_3} - \frac{r^2 - a_3^2}{2r}\right)$$

$$-\frac{w_1}{4D}\left[\langle r - a_1\rangle^0\left(\frac{r^3}{4} - \frac{a_1^4}{4r} - a_1^2 r \ln \frac{r}{a_1}\right) - \langle r - a_2\rangle^0\left(\frac{r^3}{4} - \frac{a_2^4}{4r} - a_2^2 r \ln \frac{r}{a_2}\right)\right]$$

and at $r = a_L$

$$F_\theta\big|_{r=a_L} = -W\frac{a_3 a_L}{2D}\left[\ln \frac{a_L}{a_3} - \frac{1}{2}\left(1 - \frac{a_3^2}{a_L^2}\right)\right]$$

$$-\frac{w_1}{4D}\left(\frac{a_2^4 - a_1^4}{4a_L} + a_2^2 a_L \ln \frac{a_L}{a_2} - a_1^2 a_L \ln \frac{a_L}{a_1}\right) \tag{4}$$

In the same fashion, the expression for $F_v\big|_{r=a_L}$ is calculated. Then v_0, M_0 of (3) can be evaluated. For example, the moment M_0 becomes

$$M_0 = (1+\nu)\left\{W\frac{a_3}{2}\left[\ln \frac{a_L}{a_3} - \frac{1}{2}\left(1 - \frac{a_3^2}{a_L^2}\right)\right]\right.$$

$$\left. + \frac{w_1}{4a_L}\left(\frac{a_2^4 - a_1^4}{4a_L} + a_2^2 a_L \ln \frac{a_L}{a_2} - a_1^2 a_L \ln \frac{a_L}{a_1}\right)\right\} \tag{5}$$

The formulas for deflection and moment are now given by (1) and (2) with the loading functions F_v, F_M taken from Table 11-1. In the case of M, with F_M from Table 11-1, (2) becomes

$$M = M_0 - W\frac{a_3}{2}\langle r - a_3\rangle^0\left[(1+\nu)\ln\frac{r}{a_3} + \frac{1-\nu}{2}\,\frac{r^2 - a_3^2}{r^2}\right]$$

$$-\frac{w_1}{4}\left\{\langle r - a_1\rangle^0\left[\frac{3+\nu}{4}r^2 - a_1^2 + \frac{(1-\nu)a_1^4}{4r^2} - (1+\nu)a_1^2\ln\frac{r}{a_1}\right]\right.$$

$$\left.-\langle r - a_2\rangle^0\left[\frac{3+\nu}{4}r^2 - a_2^2 + \frac{(1-\nu)a_2^4}{4r^2} - (1+\nu)a_2^2\ln\frac{r}{a_2}\right]\right\}$$

where M_0 is given by (5).

11.5 Plate with Rigid Support at the Center

For a plate with rotationally symmetric loading and with a center rigid support (for example, Fig. 11-4), the deflection, slope, bending moment, and shear force are given by

Deflection:	$v = -\frac{R}{8\pi D}r^2(\ln r - 1) + C_1\frac{r^2}{4} + F_v$	(11.5*a*)
Slope:	$\theta = \frac{R}{4\pi D}r\left(\ln r - \frac{1}{2}\right) - C_1\frac{r}{2} + F_\theta$	(11.5*b*)
Bending moment:	$M = \frac{R}{4\pi}\left[(1+\nu)\ln r + \frac{1-\nu}{2}\right] - \frac{C_1 D}{2}(1+\nu) + F_M$	(11.5*c*)
Shear force:	$V = \frac{R}{2\pi r} + F_V$	(11.5*d*)

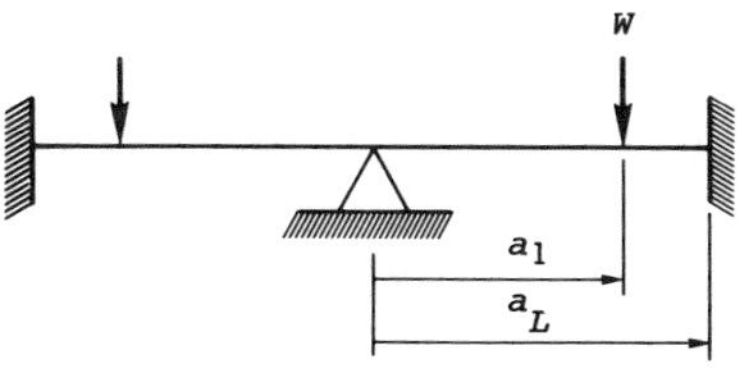

Fig. 11-4 Example 11.3.

The F_v, F_θ, F_M, F_V are loading functions given in Table 11-1. If there is more than one load on the plate, the F_v, F_θ, F_M, F_V functions are formed by adding the terms of Table 11-1 for each load. The constant R is the concentrated support reaction force (positive upwards) which is provided in Table 11-2 for various outer edge boundary conditions. The constant C_1 is also given in Table 11-2.

EXAMPLE 11.3 Plate with Concentrated Ring Load Determine the deflection caused in a plate, which is fixed on the outer rim and rigidly supported at the center, by a concentrated ring force (Fig. 11-4).

From Eq. (11.5*a*) the deflection is expressed by

$$v = -\frac{R}{8\pi D} r^2(\ln r - 1) + C_1 \frac{r^2}{4} + F_v \tag{1}$$

Table 11-1 gives F_v for the concentrated ring force W:

$$F_v = \langle r - a_1 \rangle^0 \frac{W a_1}{4D} \left[(r^2 + a_1^2) \ln \frac{r}{a_1} - (r^2 - a_1^2) \right] \tag{2}$$

According to Table 11-2, the reaction R and constant C_1 are given by

$$\begin{aligned} R &= -\frac{16\pi D}{a_L^2} F_v|_{r=a_L} - \frac{8\pi D}{a_L} F_\theta|_{r=a_L} \\ C_1 &= -\frac{8(\ln a_L - 1/2)}{a_L^2} F_v|_{r=a_L} - \frac{4(\ln a_L - 1)}{a_L} F_\theta|_{r=a_L} \end{aligned} \tag{3}$$

Insertion of $F_v|_{r=a_L}$ from (2) and $F_\theta|_{r=a_L}$ from Table 11-1 into (1) gives

$$\begin{aligned} C_1 &= \frac{W a_1}{D} (1 - \beta^2 + 2\beta^2 \ln \beta)\left[\ln a_0 + (1 - \beta^2) \ln \beta\right] \\ R &= 2\pi W a_1 (1 - \beta^2 + 2\beta^2 \ln \beta) \end{aligned} \tag{4}$$

where $\beta = a_1/a_L$. Substitution of (2) and (4) into (1) provides the expression for deflection at any radius r.

11.6 Plate with Center Hole

For a plate with rotationally symmetric loading and with a center hole (for example, Fig. 11-5) the deflection, slope, bending moment, and shear force are given by

Deflection:

$$\begin{aligned} v = v_0 &+ \theta_0 \left[-\frac{1}{2}(1 + \nu) a_0 \ln \frac{r}{a_0} - (1 - \nu) \frac{(r^2 - a_0^2)}{4a_0} \right] \\ &+ M_0 \left(\frac{a_0^2}{2D} \ln \frac{r}{a_0} - \frac{r^2 - a_0^2}{4D} \right) \\ &+ V_0 \frac{a_0}{4D} \left[-(a_0^2 + r^2) \ln \frac{r}{a_0} + (r^2 - a_0^2) \right] + F_v \\ &= v_0 U_{vv} + \theta_0 U_{v\theta} + M_0 U_{vM} + V_0 U_{vV} + F_v \end{aligned} \tag{11.6a}$$

[Equation (11.6a) is introduced because the same notation is employed in Table 11-3.]

Slope:
$$\theta = \theta_0\left[(1+\nu)\frac{a_0}{2r} + (1-\nu)\frac{r}{2a_0}\right] + M_0\frac{1}{2Dr}(r^2 - a_0^2)$$
$$+ V_0\left[\frac{a_0 r}{2D}\ln\frac{r}{a_0} - \frac{a_0}{4Dr}(r^2 - a_0^2)\right] + F_\theta$$
$$= \theta_0 U_{\theta\theta} + M_0 U_{\theta M} + V_0 U_{\theta V} + F_\theta \qquad (11.6b)$$

Note that $U_{\theta v} = 0$.

Bending moment:
$$M = \theta_0(1-\nu^2)\frac{Da_0}{2}\left(\frac{1}{a_0^2} - \frac{1}{r^2}\right) + M_0\left(\frac{1-\nu}{2}\frac{a_0^2}{r^2} + \frac{1+\nu}{2}\right)$$
$$+ V_0\frac{a_0}{2}\left[(1+\nu)\ln\frac{r}{a_0} + (1-\nu)\frac{r^2 - a_0^2}{2r^2}\right] + F_M$$
$$= \theta_0 U_{M\theta} + M_0 U_{MM} + V_0 U_{MV} + F_M \qquad (11.6c)$$

Note that $U_{Mv} = 0$.

Shear force:
$$V = V_0\frac{a_0}{r} + F_V = V_0 U_{VV} + F_V \qquad (11.6d)$$

Note that $U_{Vv} = U_{V\theta} = U_{VM} = 0$.

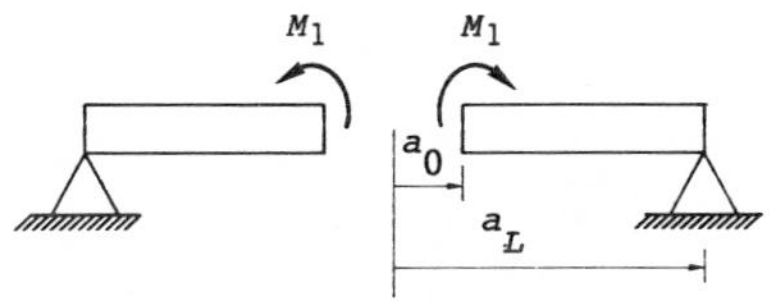

Fig. 11-5 Example 11.4.

The F_v, F_θ, F_M, F_V are loading functions given in Table 11-1. If there is more than one load on the plate, the F_v, F_θ, F_M, F_V functions are formed by adding the terms of Table 11-1 for each load. The initial parameters v_0, θ_0, M_0, V_0 (values of v, θ, M, V at the center edge of the plate, $r = a_0$) are provided in Table 11-3.

EXAMPLE 11.4 Ring with Moment on Inner Rim Find the deflection of the plate shown in Fig. 11-5.

The deflection is given by Eq. (11.6a) with the initial parameters taken from Table 11-3 and loading functions from Table 11-1. The inner edge is considered to be free because the moment applied there is taken into account as a loading function and not as a boundary condition. The outer edge is simply supported.

TABLE 11-3 Initial Parameters for Plate with Center Hole (Eqs. 11.6)

Inner Edge ↓ / Outer Edge →	1. Simply Supported 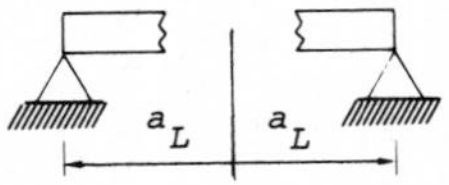	2. Fixed 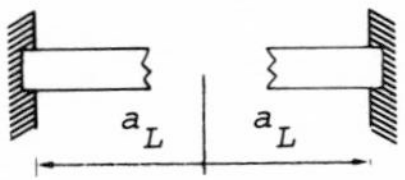	3. Free 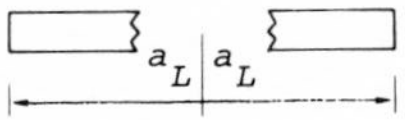	4. Guided 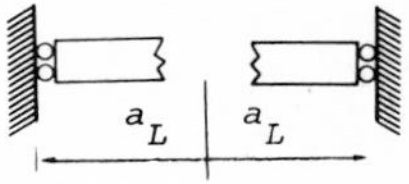
1. Simply Supported 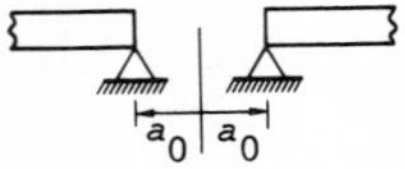 $v_0 = 0,\ M_0 = 0$	$\theta_0 = (F_M U_{vV} - F_V U_{MV})/\nabla$ $V_0 = (F_V U_{M\theta} - F_M U_{v\theta})/\nabla$ $\nabla = U_{v\theta} U_{MV} - U_{M\theta} U_{vV}$	$\theta_0 = (F_\theta U_{vV} - F_V U_{\theta V})/\nabla$ $V_0 = (F_V U_{\theta\theta} - F_\theta U_{v\theta})/\nabla$ $\nabla = U_{v\theta} U_{\theta V} - U_{\theta\theta} U_{vV}$	$\theta_0 = (F_V U_{MV} - F_M U_{VV})/\nabla$ $V_0 = (F_M U_{V\theta} - F_V U_{M\theta})/\nabla$ $\nabla = U_{M\theta} U_{VV} - U_{V\theta} U_{MV}$	$\theta_0 = (F_V U_{\theta V} - F_\theta U_{VV})/\nabla$ $V_0 = (F_\theta U_{V\theta} - F_V U_{\theta\theta})/\nabla$ $\nabla = U_{\theta\theta} U_{VV} - U_{\theta V} U_{V\theta}$
2. Fixed 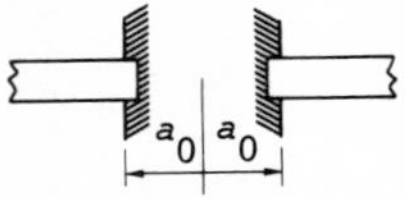 $v_0 = 0,\ \theta_0 = 0$	$M_0 = (F_M U_{vV} - F_V U_{MV})/\nabla$ $V_0 = (F_V U_{MM} - F_M U_{vM})/\nabla$ $\nabla = U_{vM} U_{MV} - U_{MM} U_{vV}$	$M_0 = (F_\theta U_{vV} - F_V U_{\theta V})/\nabla$ $V_0 = (F_V U_{\theta M} - F_\theta U_{vM})/\nabla$ $\nabla = U_{vM} U_{\theta V} - U_{\theta M} U_{vV}$	$M_0 = (F_V U_{MV} - F_M U_{VV})/\nabla$ $V_0 = (F_M U_{VM} - F_V U_{MM})/\nabla$ $\nabla = U_{MM} U_{VV} - U_{VM} U_{MV}$	$M_0 = (F_V U_{\theta V} - F_\theta U_{VV})/\nabla$ $V_0 = (F_\theta U_{VM} - F_V U_{\theta M})/\nabla$ $\nabla = U_{\theta M} U_{VV} - U_{\theta V} U_{VM}$
3. Free 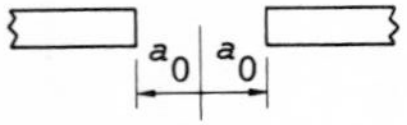 $M_0 = 0,\ V_0 = 0$	$v_0 = (F_M U_{v\theta} - F_V U_{M\theta})/\nabla$ $\theta_0 = (F_V U_{Mv} - F_M U_{vv})/\nabla$ $\nabla = U_{vv} U_{M\theta} - U_{Mv} U_{v\theta}$	$v_0 = (F_\theta U_{v\theta} - F_V U_{\theta\theta})/\nabla$ $\theta_0 = (F_V U_{\theta v} - F_\theta U_{vv})/\nabla$ $\nabla = U_{vv} U_{\theta\theta} - U_{\theta v} U_{v\theta}$	$v_0 = (F_V U_{M\theta} - F_M U_{v\theta})/\nabla$ $\theta_0 = (F_M U_{Vv} - F_V U_{Mv})/\nabla$ $\nabla = U_{Mv} U_{v\theta} - U_{M\theta} U_{Vv}$	$v_0 = (F_V U_{\theta\theta} - F_\theta U_{v\theta})/\nabla$ $\theta_0 = (F_\theta U_{Vv} - F_V U_{\theta v})/\nabla$ $\nabla = U_{\theta v} U_{v\theta} - U_{\theta\theta} U_{Vv}$

4. Guided $\theta_0 = 0,\ V_0 = 0$	$v_0 = (F_M U_{vM} - F_v U_{MM})/\nabla$ $M_0 = (F_v U_{Mv} - F_M U_{vv})/\nabla$ $\nabla = U_{vv}U_{MM} - U_{Mv}U_{vM}$	$v_0 = (F_\theta U_{vM} - F_v U_{\theta M})/\nabla$ $M_0 = (F_v U_{\theta v} - F_\theta U_{vv})/\nabla$ $\nabla = U_{vv}U_{\theta M} - U_{vM}U_{\theta v}$	$v_0 = (F_V U_{MM} - F_M U_{VM})/\nabla$ $M_0 = (F_M U_{Vv} - F_V U_{Mv})/\nabla$ $\nabla = U_{Mv}U_{VM} - U_{MM}U_{Vv}$	$v_0 = (F_V U_{\theta M} - F_\theta U_{VM})/\nabla$ $M_0 = (F_\theta U_{Vv} - F_V U_{\theta v})/\nabla$ $\nabla = U_{\theta v}U_{VM} - U_{\theta M}U_{Vv}$
5. Rigid Insert with Total Load W_T 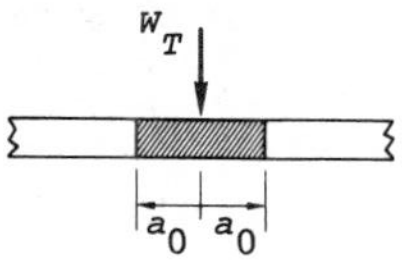	$v_0 = \left[\frac{W_T}{2\pi a_0}(U_{MM}U_{vV} - U_{vM}U_{MV}) + U_{vM}F_M - U_{MM}F_v\right]/\nabla$ $M_0 = \left[\frac{W_T}{2\pi a_0}(U_{vv}U_{MV} - U_{Mv}U_{vV}) + U_{Mv}F_v - U_{vv}F_M\right]/\nabla$ $V_0 = -W_T/(2\pi a_0),\quad \theta_0 = 0$ $\nabla = U_{vv}U_{MM} - U_{Mv}U_{vM}$	$v_0 = \left[\frac{W_T}{2\pi a_0}(U_{\theta M}U_{vV} - U_{vM}U_{\theta V}) + U_{vM}F_\theta - U_{\theta M}F_v\right]/\nabla$ $M_0 = \left[\frac{W_T}{2\pi a_0}(U_{vv}U_{\theta V} - U_{vV}U_{\theta v}) + U_{\theta v}F_v - U_{vv}F_\theta\right]/\nabla$ $V_0 = -W_T/(2\pi a_0),\quad \theta_0 = 0$ $\nabla = U_{vv}U_{\theta M} - U_{\theta v}U_{vM}$	$v_0 = \left[\frac{W_T}{2\pi a_0}(U_{MV}U_{VM} - U_{MM}U_{VV}) + U_{MM}F_V - U_{VM}F_M\right]/\nabla$ $M_0 = \left[\frac{W_T}{2\pi a_0}(U_{Mv}U_{VV} - U_{MV}U_{Vv}) + U_{Vv}F_M - U_{Mv}F_V\right]/\nabla$ $v_0 = -W_T/(2\pi a_0),\quad \theta_0 = 0$ $\nabla = U_{Mv}U_{VM} - U_{MM}U_{Vv}$	$v_0 = \left[\frac{W_T}{2\pi a_0}(U_{\theta V}U_{VM} - U_{\theta M}U_{VV}) + U_{\theta M}F_V - U_{VM}F_\theta\right]/\nabla$ $M_0 = \left[\frac{W_T}{2\pi a_0}(U_{\theta v}U_{VV} - U_{\theta V}U_{Vv}) + U_{Vv}F_\theta - U_{\theta v}F_V\right]/\nabla$ $v_0 = -W_T/(2\pi a_0),\quad \theta_0 = 0$ $\nabla = U_{\theta v}U_{VM} - U_{\theta M}U_{Vv}$

Definitions:

$$U_{kj} = U_{kj}\Big|_{r=a_L} \qquad k,j = v,\ \theta,\ M,\ V$$

$$F_k = F_k\Big|_{r=a_L} \qquad k = v,\ \theta,\ M,\ V$$

Then, $M_0 = V_0 = 0$ and

$$v_0 = (U_{v\theta}F_M - U_{M\theta}F_v)/\nabla|_{r=a_L} \qquad \theta_0 = (U_{Mv}F_v - U_{vv}F_M)/\nabla|_{r=a_L}$$
$$\nabla = (U_{vv}U_{M\theta} - U_{Mv}U_{v\theta}) \tag{1}$$

The relevant loading functions are

$$F_v = \frac{M_1}{2D}\left(a_0^2 \ln \frac{r}{a_0} - \frac{r^2 - a_0^2}{2}\right) \qquad F_M = \frac{M_1}{2}\left[(1-\nu)\frac{a_0^2}{r^2} + 1 + \nu\right] \tag{2}$$

where the functions $\langle r - a_0 \rangle$ have been ignored as the plate begins at $r = a_0$. From (2) and Eqs. (11.6), (1) becomes

$$v_0 = -\frac{M_1}{D}\left[\frac{a_0^2}{2(1+\nu)} + \frac{a_L^2 a_0^2}{(1-\nu)(a_L^2 - a_0^2)} \ln \frac{a_L}{a_0}\right]$$
$$\theta_0 = -\frac{M_1}{D}\frac{a_L^2 a_0^2}{a_L^2 - a_0^2}\left[\frac{1}{(1-\nu)a_0} + \frac{a_0}{(1+\nu)a_L^2}\right] \tag{3}$$

TABLE 11-4 Value of C for Eq. (11.7) (Ref. 11.1)

Type of Plate	C
1. Outer Edge Simply Supported, No Center Hole	4.191
2. Outer Edge Fixed, No Center Hole	14.68
3. Outer Edge ($r = a_L$) Fixed, Inner Edge ($r = a_0$) Free	a_0/a_L: 0.0, 0.1, 0.2, 0.3, 0.4; C: 14.68, 14.00, 13.26, 14.4*, 17.5*
4. Outer Edge ($r = a_L$) Simply Supported, Inner Edge ($r = a_0$) Free	a_0/a_L: 0.0, 0.1, 0.2, 0.3, 0.4; C: 4.19, 4.00, 3.70, 3.1*, 2.9*

*These values are approximate. For values of a_0/a_L greater than 0.3 use a circular plate computer program that can compute modes higher than the first symmetric mode.

TABLE 11-5 Values of C_{mn} for Eq. (11.8), $\nu = 0.3$ (Ref. 11.2)

n	Simply Supported Outer Edge			Clamped Outer Edge				
	$m = 0$	$m = 1$	$m = 2$	$m = 0$	$m = 1$	$m = 2$	$m = 3$	$m = 4$
0	4.977	13.94	25.65	10.216	21.26	34.88	51.04	69.67
1	29.76	48.51	70.14	39.77	60.82	84.58	111.01	140.11
2	74.20	102.80	134.33	89.10	120.08	153.81	190.30	229.52
3	138.34	176.84	218.24	158.18	199.06	242.71	289.17	338.41

Substitution of F_v from (2) and v_0, θ_0 from (3) in Eq. (11.6*a*) gives the deflection

$$v_0 = \frac{M_1}{D}\left[\frac{a_L^2 a_0^2}{(1-\nu)(a_L^2 - a_0^2)} \ln \frac{r}{a_L} + \frac{a_0^2(r^2 - a_L^2)}{2(1+\nu)(a_L^2 - a_0^2)}\right]$$

11.7 Stability

The critical compressive in-plane (radial) force $P = P_{cr}$ applied at the outer edge of a circular plate in the r direction is given by

$$P_{cr} = C\frac{D}{a_L^2} \tag{11.7}$$

where C is as listed in Table 11-4 for various boundary conditions.

11.8 Free Dynamic Response—Natural Frequencies

The lowest natural frequencies for the transverse vibration of uniform-thickness circular plates with no holes are given by

$$\omega_{mn} = \frac{C_{mn}}{a_L^2}\sqrt{\frac{D}{\rho}} \tag{11.8}$$

where C_{mn} is provided in Table 11-5.

B. COMPLEX PLATES

11.9 Notation for Complex Plates

k Modulus of elastic foundation (force/length^3)

k_1 Line ring extension spring constant (force/length^2)

k_1^*	Line ring rotary spring constant (force · length/length · radian)
Q_r	Shear force per unit length of circular arc (force/length)
P, P_ϕ	In-plane compression forces in the r, ϕ directions (force/length). Replace P by $-P$ for a tensile force. Both P and P_ϕ can be calculated for a particular r from the disk formulas of Chap. 8.
$D_r, D_\phi, D_{r\phi}$	Flexural rigidities; see Table 11-6.
ν_r, ν_ϕ	Poisson's ratio in the r, ϕ directions
E_r, E_ϕ	Moduli of elasticity of the material in the r, ϕ directions (force/length^2)
K_r, K_ϕ	Extensional rigidities; see Table 11-6.
G	Shear modulus of elasticity (force/length^2)
α_r, α_ϕ	Coefficients of thermal expansion in the r, ϕ directions (length/length · degree); $\alpha_r = \alpha_\phi = \alpha$ for isotropic material.
ΔT	Temperature change (degrees), that is, the temperature rise with respect to the reference temperature

$$M_{Tr} = \int_{-h/2}^{h/2} \frac{E_r(\alpha_r + \nu_\phi a_\phi)}{1 - \nu_r \nu_\phi} \Delta T \, z \, dz$$

$$M_{T\phi} = \int_{-h/2}^{h/2} \frac{E_\phi(\alpha_\phi + \nu_r \alpha_r)}{1 - \nu_r \nu_\phi} \Delta T \, z \, dz$$

$$M_T = \int_{-h/2}^{h/2} E\alpha(1 - \nu) \, \Delta T \, z \, dz$$

M_i	Ring lumped mass (mass/length)
P_{Li}	Ring lumped in-plane radial force (force · length/length)
$P_{\phi i}$	Ring lumped in-plane circumferential force (force · length/length)
c_1	Magnitude of distributed moment, uniform in radial direction (force · length/length^2)
M_{T1}	Magnitude of distributed thermal moment, uniform in radial direction
$\dfrac{\Delta w}{\Delta \ell}$	Gradient of distributed force, linearly varying in radial direction (force/length^3)
$\dfrac{\Delta c}{\Delta \ell}$	Gradient of distributed moment, linearly varying in radial direction (force · length/length^3)
$\dfrac{\Delta M_T}{\Delta \ell}$	Gradient of distributed thermal moment, linearly varying in radial direction
t	Time
ϵ_m^j	Constant (Table 11-13) that accounts for variations of distributed loads in the ϕ direction.

$B = 2D_{r\phi} + D_r \nu_\phi = 2D_{r\phi} + D_\phi \nu_r$

c'	External or viscous damping coefficient (mass/(length^2 · time), force · time/length^3)
c^*	Rotary viscous damping coefficient (mass/time, force · time/length)
c_ρ	Proportional viscous damping coefficient (1/time). If c' is chosen to be proportional to the mass, that is, $$c' = c_\rho\left(\frac{m^2 \rho i_r^2}{r^2} + \rho\right) \quad \text{and} \quad c^* = c_\rho \rho i_\phi^2$$ then c_ρ is the constant of proportionality. See Eqs. (11.15) for the definition of m.
c_i	Discrete linear ring damping (dashpot) constant (force · time/length^2)

TABLE 11-6 Material Constants

Plate	Constants
1. Homogeneous isotropic	$\nu_r = \nu_\phi = \nu$, $D_r = D_\phi = D = Eh^3/[12(1 - \nu^2)]$, $G = E/[2(1+\nu)]$, $D_{r\phi} = D(1-\nu)/2$ $K_r = K_\phi = K = Eh/(1-\nu^2)$, $\alpha_r = \alpha_\phi = \alpha$
2. Homogeneous orthotropic	$D_j = E_j h^3/[12(1 - \nu_r\nu_\phi)]$, $K_j = E_j h/(1 - \nu_r\nu_\phi)$, $j = r,\phi$ $D_{r\phi} = Gh^3/12$
3. Continuously composite, isotropic	$D_r = D_\phi = D = \int_{-h/2}^{h/2} \frac{Ez^2}{1-\nu^2}\,dz$, $D_{r\phi} = \int_{-h/2}^{h/2} \frac{Ez^2}{2(1+\nu)}\,dz$, $K_r = K_\phi = K = \int_{-h/2}^{h/2} \frac{E}{1-\nu^2}\,dz$ $\alpha_r = \alpha_\phi = \alpha$
4. Continuously composite, orthotropic	$D_j = \frac{1}{1-\nu_r\nu_\phi} \int_{-h/2}^{h/2} \frac{E_j z^2}{1-\nu_r\nu_\phi}\,dz$, $K_j = \int_{-h/2}^{h/2} \frac{E_j}{1-\nu_r\nu_\phi}\,dz$, $j = r,\phi$ $D_{r\phi} = \int_{-h/2}^{h/2} Gz^2\,dz$
5. Layered	$D_j = 2\sum_i \left(\frac{E_j}{1-\nu_r\nu_\phi}\right)_i \int_{\Delta h_i} z^2\,dz$, $K_j = 2\sum_i \left(\frac{E_j}{1-\nu_r\nu_\phi}\right)_i \Delta h_i$, $j = r,\phi$ $D_{r\phi} = 2\sum_i (G)_i \int_{\Delta h_i} z^2\,dz$, Δh_i = thickness of ith layer The summation extends over half of the plate thickness.

c_i^* Discrete rotary ring damping constant (force · time)

i_r, i_ϕ The radii of gyration of the mass about the radial and tangential axes. For isotropic, homogeneous material set $i_r^2 = i_\phi^2 = h^2/12$.

$\mathbf{U}_i$ Field matrix of the ith segment

$\overline{\mathbf{U}}_i$ Point matrix at $x = a_i$

The notation for transfer matrices is

$$\mathbf{U}_i = \begin{bmatrix} U_{vv} & U_{v\theta} & U_{vM} & U_{vV} & F_v \\ U_{\theta v} & U_{\theta\theta} & U_{\theta M} & U_{\theta V} & F_\theta \\ U_{Mv} & U_{M\theta} & U_{MM} & U_{MV} & F_M \\ U_{Vv} & U_{V\theta} & U_{VM} & U_{VV} & F_V \\ 0 & 0 & 0 & 0 & 1 \end{bmatrix}$$

11.10 Differential Equations

The fundamental equations of motion of a circular plate that may be subjected to a symmetric loading are given in first-order form as

$$\frac{\partial v}{\partial r} = -\theta$$

$$\frac{\partial \theta}{\partial r} = \frac{M}{D_r} + \nu_\phi\left(\frac{1}{r^2}\frac{\partial^2 v}{\partial \phi^2} - \frac{\theta}{r}\right) + \frac{M_{Tr}}{D_r}$$

$$\frac{\partial M}{\partial r} = -\left(1 - \frac{\nu_r D_\phi}{D_r}\right)\frac{M}{r} + V - \left[D_\phi(1 - \nu_r\nu_\phi) + 4D_{r\phi}\right]\frac{1}{r^3}\frac{\partial^2 v}{\partial \phi^2} - P\theta$$
$$+ \frac{D_\phi(1 - \nu_r\nu_\phi)}{r^2}\theta - \frac{4D_{r\phi}}{r^2}\frac{\partial^2\theta}{\partial\phi^2} - \frac{1}{r}\left(M_{T\phi} - \frac{\nu_r D_\phi}{D_r}M_{Tr}\right) + \rho i_\phi^2\frac{\partial^2\theta}{\partial t^2}$$

$$\frac{\partial V}{\partial r} = -\frac{V}{r} - \frac{\nu_r D_\phi}{r^2 D_r}\frac{\partial^2 M}{\partial \phi^2} + \frac{D_\phi}{r^4}(1 - \nu_r\nu_\phi)\frac{\partial^4 v}{\partial\phi^4} - \frac{4D_{r\phi}}{r^4}\frac{\partial^2 v}{\partial\phi^2}$$
$$-\left[D_\phi(1 - \nu_r\nu_\phi) + 4D_{r\phi}\right]\frac{\partial^2\theta}{r^3\,\partial\phi^2}$$
$$+\frac{1}{r^2}\frac{\partial}{\partial\phi}\left(P_\phi\frac{\partial v}{\partial\phi}\right) + \frac{1}{r^2}\left(\frac{\partial^2 M_{T\phi}}{\partial\phi^2} - \frac{\nu_r D_\phi}{D_r}\frac{\partial^2 M_{Tr}}{\partial\phi^2}\right)$$
$$+ kv - \frac{\rho i_r^2}{r^2}\frac{\partial^4 v}{\partial t^2\partial\phi^2} + \rho\frac{\partial^2 v}{\partial t^2} - w(r, \phi, t) \tag{11.9}$$

In the case of an isotropic plate with only transverse inertia taken into account, these equations are frequently combined to form the fourth-order relationships

$$D\,\nabla^4 v + \frac{\partial}{\partial r}\left(P\frac{\partial v}{\partial r}\right) - \frac{P}{r}\frac{\partial v}{\partial r} + \frac{1}{r^2}\frac{\partial^2}{\partial\phi^2}\left(P_\phi\frac{\partial v}{\partial\phi}\right) + kv + \rho\frac{\partial^2 v}{\partial t^2} = w - \nabla^2 M_T \tag{11.10}$$

where

$$\nabla^2 = \frac{\partial^2}{\partial r^2} + \frac{1}{r}\frac{\partial}{\partial r} + \frac{1}{r^2}\frac{\partial^2}{\partial \phi^2}$$

and $\nabla^4 = (\nabla^2)^2$.

The in-plane forces $P = P_r$, P_ϕ are in compression. These equations are appropriate for a plate with tensile in-plane forces if P and P_ϕ are replaced by $-P$ and $-P_\phi$.

Stresses and Other Variables After the responses v, θ, M, V are computed, other variables and stresses can be calculated from the following formulas. The circumferential and twisting moments are found from

$$M_\phi = -D_\phi\left(\frac{1}{r}\frac{\partial v}{\partial r} + \frac{1}{r^2}\frac{\partial^2 v}{\partial \phi^2} + \nu_r \frac{\partial^2 v}{\partial r^2}\right) - M_{T\phi} \tag{11.11}$$

$$M_{r\phi} = M_{\phi r} = 2D_{r\phi}\left(\frac{1}{r}\frac{\partial^2 v}{\partial r \partial \phi} - \frac{1}{r^2}\frac{\partial v}{\partial \phi}\right) \tag{11.12}$$

The shear force Q_r is given by

$$Q_r = V + \frac{1}{r}\frac{\partial M_{r\phi}}{\partial \phi} \tag{11.13}$$

The radial, circumferential, and shear stresses are

$$\sigma_r = \overline{E}_r\left[-\frac{\overline{P}}{K_r} + \frac{\overline{M}_z}{D_r} - (\alpha_r + \nu_\phi \alpha_\phi)\,\Delta T\right] \tag{11.14a}$$

$$\sigma_\phi = \overline{E}_\phi\left[-\frac{\overline{P}_\phi}{K_\phi} + \frac{\overline{M}_\phi z}{D_\phi} - (\alpha_\phi + \nu_r \alpha_r)\,\Delta T\right] \tag{11.14b}$$

$$\tau_{r\phi} = -G_{r\phi}\frac{M_{r\phi} z}{D_{r\phi}} \tag{11.14c}$$

where

$$\overline{P} = P + P_T \qquad \overline{P}_\phi = P_\phi + P_{T\phi}$$

$$\overline{M} = M + M_{Tr} \qquad \overline{M}_\phi = M_\phi + M_{T\phi}$$

$$P_T = -\int_{-h/2}^{h/2} \frac{E_r(\alpha_r + \nu_\phi \alpha_\phi)}{1 - \nu_r \nu_\phi}\,\Delta T\,dz \qquad P_{T\phi} = -\int_{-h/2}^{h/2} \frac{E_\phi(\alpha_\phi + \nu_r \alpha_r)}{1 - \nu_r \nu_\phi}\,\Delta T\,dz$$

$P_{r\phi}$ is the in-plane shear force. See Table 11-6 for the definitions of K_r, K_ϕ.

$$\overline{E}_r = \frac{E_r}{1 - \nu_r \nu_\phi}, \qquad \overline{E}_\phi = \frac{E_\phi}{1 - \nu_r \nu_\phi}$$

Elimination of the ϕ Derivatives The ϕ derivatives in these equations are eliminated by expanding the variables in a Fourier series of the form

$$
\begin{aligned}
v(r,\phi,t) &= \sum_{m=0}^{\infty} \left[v_m^c(r,t)\cos m\phi + v_m^s(r,t)\sin m\phi \right] \\
\theta(r,\phi,t) &= \sum_{m=0}^{\infty} \left[\theta_m^c(r,t)\cos m\phi + \theta_m^s(r,t)\sin m\phi \right] \\
M(r,\phi,t) &= \sum_{m=0}^{\infty} \left[M_m^c(r,t)\cos m\phi + M_m^s(r,t)\sin m\phi \right] \\
V(r,\phi,t) &= \sum_{m=0}^{\infty} \left[V_m^c(r,t)\cos m\phi + V_m^s(r,t)\sin m\phi \right]
\end{aligned}
\tag{11.15}
$$

which for symmetrical motion ($m = 0$) reduce to

$$
v(r,t) = v_0^c(r,t) \qquad \theta(r,t) = \theta_0^c(r,t)
$$
$$
M(r,t) = M_0^c(r,t) \qquad V(r,t) = V_0^c(r,t)
$$

The mechanical loading $w(r, \phi, t)$ and thermal loading $M_T(r, \phi, t)$ or M_{Tr}, $M_{T\phi}$ are also expanded in a Fourier series.

$$
\begin{aligned}
w(r,\phi,t) &= \sum_{m=0}^{\infty} \left[w_m^c(r,t)\cos m\phi + w_m^s(r,t)\sin m\phi \right] \\
M_T(r,\phi,t) &= \sum_{m=0}^{\infty} \left[M_{Tm}^c(r,t)\cos m\phi + M_{Tm}^s(r,t)\sin m\phi \right]
\end{aligned}
\tag{11.16}
$$

so that

$$
\begin{aligned}
w_0^c(r,t) &= \frac{1}{2\pi}\int_0^{2\pi} w(r,\phi,t)\,d\phi \\
w_m^c(r,t) &= \frac{1}{\pi}\int_0^{2\pi} w(r,\phi,t)\cos m\phi\,d\phi \qquad m>0 \\
w_m^s(r,t) &= \frac{1}{\pi}\int_0^{2\pi} w(r,\phi,t)\sin m\phi\,d\phi \qquad m>0
\end{aligned}
\tag{11.17}
$$

Similar relations apply for M_{T0}^c, M_{Tm}^c, M_{Tm}^s.

The expressions w_0^c, w_m^c, w_m^s and M_{T0}^c, M_{Tm}^c, M_{Tm}^s will be referred to as the *transformed loading functions*. These terms (Eqs. 11.17) are found by multiplying both sides of Eq. (11.16) by cos $m\phi$ and integrating from $\phi = 0$ to $\phi = 2\pi$ and by repeating the manipulation with sin $m\phi$.

Several rules are useful in evaluating Eqs. (11.17). If $w(r, \phi, t)$ is an odd function of ϕ, that is, if $w(r, \phi, t) = -w(r, -\phi, t)$, then

$$
w_m^c(r,t) = 0 \qquad \text{for} \qquad m = 0, 1, 2, \ldots
$$

and Eqs. (11.15) are sine series. For even functions of ϕ, that is, if $w(r, \phi, t)$

$= w(r, -\phi, t)$, then

$$w_m^s(r, t) = 0 \qquad \text{for} \qquad m = 0, 1, 2, \ldots$$

and Eqs. (11.15) contain only cosine terms.

Insertion of Eqs. (11.15) and (11.16) into the equations of motion leads to differential equations which are functions of r and t only. Thus, Eqs. (11.9) become

$$\frac{\partial v_m^j}{\partial r} = -\theta_m^j \qquad j = c \text{ or } s \tag{11.18a}$$

$$\frac{\partial \theta_m^j}{\partial r} = \frac{M_m^j}{D_r} - \frac{\nu_\phi}{r}\theta_m^j - \frac{\nu_\phi m^2}{r^2} v_m^j + \frac{M_{Trm}^j}{D_r} \tag{11.18b}$$

$$\frac{\partial M_m^j}{\partial r} = -\left(1 - \frac{\nu_r D_\phi}{D_r}\right)\frac{M_m^j}{r} + V_m^j + \left[D_\phi(1 - \nu_r\nu_\phi) + 4D_{r\phi}\right]\frac{m^2}{r^3} v_m^j$$
$$+ \left[D_\phi(1 - \nu_r\nu_\phi)\frac{1}{r^2} + 4D_{r\phi}\frac{m^2}{r^2} - P\right]\theta_m^j$$
$$- \frac{1}{r}\left(M_{T\phi m}^j - \frac{\nu_r D_\phi}{D_r} M_{Trm}\right) + \rho i_\phi^2 \frac{\partial^2 \theta_m^j}{\partial t^2} \tag{11.18c}$$

$$\frac{\partial V_m^j}{\partial r} = -\frac{V_m^j}{r} - \frac{\nu_r D_\phi m^2}{r^2 D_r} M_m^j + \frac{1}{r^4}\left[D_\phi(1 - \nu_r\nu_\phi)m^4 + 4D_{r\phi}m^2\right]v_m^j$$
$$+ \left[D_\phi(1 - \nu_r\nu_\phi) + 4D_{r\phi}\right]\frac{m^2}{r^3}\theta_m^j + \frac{P_\phi m^2}{r^2} v_m^j - \frac{m^2}{r^2}\left(M_{T\phi m} - \frac{\nu_r D_\phi}{D_r} M_{Trm}\right)$$
$$+ k v_m^j + \frac{m^2 \rho i_r^2}{r^2}\frac{\partial^2 v_m^j}{\partial t^2} + \rho\frac{\partial^2 v_m^j}{\partial t^2} - w_m^j(r, t) \tag{11.18d}$$

Transformed Loading Functions The information in this chapter can be used to develop the functions v_m^j, θ_m^j, M_m^j, V_m^j, $j = c$ or s. These functions placed in Eqs. (11.15) yield the complete expressions for v, θ, M, and V. Loadings, including concentrated forces and thermal loads, must be transformed in the sense of Eqs. (11.17) before the equations of motion are solved. This transformation has been implemented for the sizable array of loadings taken into account in the tables of this chapter.

11.11 Transfer Matrices for Uniform Segments

The transfer (field) matrices for plate segments are provided in Tables 11-7 to 11-12. Loads taken into account are uniformly (w_1, c_1, M_{T1}) and linearly ($\Delta w/\Delta \ell$, $\Delta c/\Delta \ell$, $\Delta M_T/\Delta \ell$) distributed in the radial direction. The circumferential (ϕ) variation of distributed loads is accounted for by inclusion of a constant ϵ_m^j (Table 11-13) in the loading functions; for example, see Table 11-10.

TABLE 11-7 Massless Isotropic Center Segment (No Center Hole)

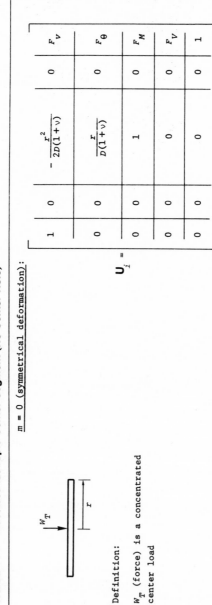

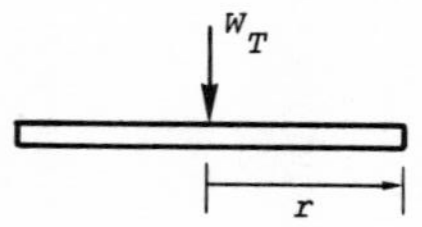

Definition:

W_T (force) is a concentrated center load

$m = 0$ (symmetrical deformation):

$$
\mathbf{U}_i = \begin{bmatrix} 1 & 0 & -\dfrac{r^2}{2D(1+\nu)} & 0 & F_v \\ 0 & 0 & \dfrac{r}{D(1+\nu)} & 0 & F_\theta \\ 0 & 0 & 1 & 0 & F_M \\ 0 & 0 & 0 & 0 & F_V \\ 0 & 0 & 0 & 0 & 1 \end{bmatrix}
$$

$$F_v = \varepsilon_0^c \left[w_1 \frac{r^4}{64D} + \frac{\Delta w}{\Delta \ell} \frac{r^5}{225D} + c_1 \frac{r^3}{9D} + \frac{\Delta c}{\Delta \ell} \frac{r^4}{32D} - \frac{M_{T1} r^2}{2D(1+\nu)} - \frac{r^3}{9D} \frac{\Delta M_T}{\Delta \ell} \right] + \frac{W_T}{8\pi D} r^2 (\ln r - 1)$$

$$F_\theta = -\varepsilon_0^c \left[-w_1 \frac{r^3}{16D} - \frac{\Delta w}{\Delta \ell} \frac{r^4}{45D} - c_1 \frac{r^2}{3D} - \frac{\Delta c}{\Delta \ell} \frac{r^3}{8D} + \frac{M_{T1} r}{D(1+\nu)} + \frac{r^2}{3D} \frac{\Delta M_T}{\Delta \ell} \right] - \frac{W_T}{4\pi D} r \left(\ln r - \frac{1}{2} \right)$$

$$F_M = -\varepsilon_0^c \left[-w_1 \frac{r^2}{16}(3+\nu) - \frac{\Delta w}{\Delta \ell} \frac{r^3}{45}(4+\nu) - c_1 r(2+\nu) - \frac{\Delta c}{\Delta \ell} \frac{r^2}{8}(3+\nu) + \frac{\nu - 1}{3} \frac{\Delta M_T}{\Delta \ell} r \right] - \frac{W_T}{4\pi} \left[(1+\nu) \ln r + \frac{1-\nu}{2} \right]$$

$$F_V = \varepsilon_0^c \left[-w_1 \frac{r}{2} - \frac{\Delta w}{\Delta \ell} \frac{r^2}{3} \right] - \frac{W_T}{2\pi r}$$

$m = 1$:

F_v, F_θ, F_M, F_V should be taken from Table 11-10 for $m = 1$ and $a_k = 0$.

$$\mathbf{U}_i = \begin{bmatrix} 0 & -r & 0 & -\dfrac{r^3}{2D(3+\nu)} & F_v \\ 0 & 1 & 0 & \dfrac{3r^2}{2D(3+\nu)} & F_\theta \\ 0 & 0 & 0 & r & F_M \\ 0 & 0 & 0 & 1 & F_V \\ 0 & 0 & 0 & 0 & 1 \end{bmatrix}$$

$m \geq 2$: The $\mathbf{U}_i$ in this table are not truly transfer matrices as the vector at radius $r = 0$ contains arbitrary constants rather than state variables. However, the vector at other radii is the state vector. All operations remain the same as those for the usual transfer matrices.

$$\mathbf{U}_i = \begin{bmatrix} r^m & r^{m+2} & 0 & 0 & F_v \\ -mr^{m-1} & -(m+2)r^{m+1} & 0 & 0 & F_\theta \\ -Dm(m-1)(1-\nu)r^{m-2} & -D(1+m)(m+2-\nu m+2\nu)r^m & 0 & 0 & F_M \\ Dm^2(m-1)(1-\nu)r^{m-3} & Dm(1+m)[m(1-\nu)-4]r^{m-1} & 0 & 0 & F_V \\ 0 & 0 & 0 & 0 & 1 \end{bmatrix}$$

F_v, F_θ, F_M, F_V should be taken from Table 11-10 for $m \geq 2$ and $a_k = 0$.

TABLE 11-8 Massless Rigid Center Segment (No Center Hole)

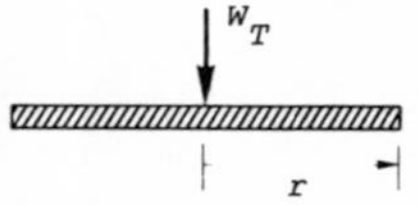

<u>$m = 0$ (symmetrical configuration):</u>

Definitions:

w_T (force) is a concentrated center load.

$$\mathbf{U}_i = \begin{bmatrix} 1 & 0 & 0 & 0 & F_v \\ 0 & 0 & 0 & 0 & F_\theta \\ 0 & 0 & 1 & 0 & F_M \\ 0 & 0 & 0 & 0 & F_V \\ 0 & 0 & 0 & 0 & 1 \end{bmatrix}$$

$F_v = 0$

$F_\theta = 0$

$$F_M = \varepsilon_0^c\left[w_1 \frac{r^2}{16}(3 + \nu) - \frac{\Delta w}{\Delta \ell}\frac{r^3}{45}(4 + \nu) - c_1 r(2 + \nu) - \frac{\Delta c}{\Delta \ell}\frac{r^2}{8}(3 + \nu)\right]$$

$$- \frac{w_T}{4\pi}\left[(1 + \nu)\ln r + \frac{1 - \nu}{2}\right]$$

$$F_V = \varepsilon_0^c\left(-w_1\frac{r}{2} - \frac{\Delta w}{\Delta \ell}\frac{r^2}{3}\right) - \frac{w_T}{2\pi r}$$

<u>$m = 1$:</u>

$$\mathbf{U}_i = \begin{bmatrix} 0 & -r & 0 & 0 & F_v \\ 0 & 1 & 0 & 0 & F_\theta \\ 0 & 0 & 0 & r & F_M \\ 0 & 0 & 0 & 1 & F_V \\ 0 & 0 & 0 & 0 & 1 \end{bmatrix}$$

F_v, F_θ, F_M, F_V should be taken from Table 11-10 for $m = 1$ with $a_k = 0$.

<u>$m \geq 1$:</u> $\mathbf{U}_i \equiv 0$

TABLE 11-9 Massless Center Segment (No Center Hole) with Rigid Support at the Center

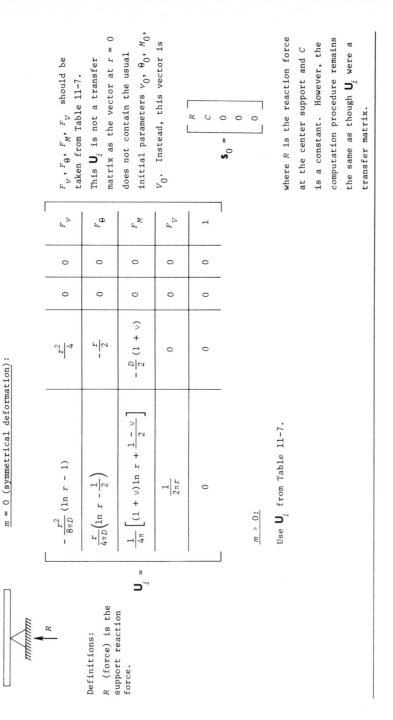

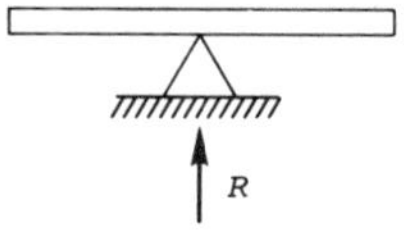

Definitions:

R (force) is the support reaction force.

$m = 0$ (symmetrical deformation):

$$\mathbf{U}_i = \begin{bmatrix} -\frac{r^2}{8\pi D}(\ln r - 1) & \frac{r^2}{4} & 0 & 0 & F_v \\ \frac{r}{4\pi D}\left(\ln r - \frac{1}{2}\right) & -\frac{r}{2} & 0 & 0 & F_\theta \\ \frac{1}{4\pi}\left[(1+\nu)\ln r + \frac{1-\nu}{2}\right] & -\frac{D}{2}(1+\nu) & 0 & 0 & F_M \\ \frac{1}{2\pi r} & 0 & 0 & 0 & F_V \\ 0 & 0 & 0 & 0 & 1 \end{bmatrix}$$

F_v, F_θ, F_M, F_V should be taken from Table 11-7.

This $\mathbf{U}_i$ is not a transfer matrix as the vector at $r = 0$ does not contain the usual initial parameters v_0, θ_0, M_0, V_0. Instead, this vector is

$$\mathbf{s}_0 = \begin{bmatrix} R \\ C \\ 0 \\ 0 \\ 0 \end{bmatrix}$$

where R is the reaction force at the center support and C is a constant. However, the computation procedure remains the same as though $\mathbf{U}_i$ were a transfer matrix.

$m > 0$:

Use $\mathbf{U}_i$ from Table 11-7.

TABLE 11-10 Massless Segment with Center Hole

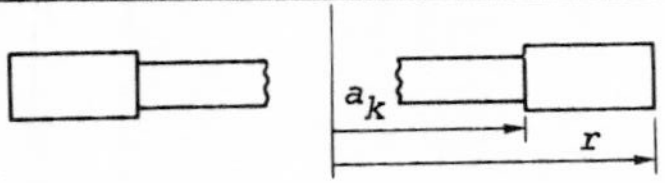

Definition: a_k is the radial coordinate of the inner surface of the ith section. That is, the ith section begins at a_k and continues to r.

$\underline{m = 0 \text{ (symmetrical deformation)}:}$

$$U_i = \begin{bmatrix}
1 & -\dfrac{1+\nu}{2}a_k\ln\dfrac{r}{a_k} - \dfrac{1-\nu}{4a_k}(r^2 - a_k^2) & \dfrac{a_k^2}{2D}\ln\dfrac{r}{a_k} - \dfrac{1}{4D}(r^2 - a_k^2) & -\dfrac{a_k(r^2+a_k^2)}{4D}\ln\dfrac{r}{a_k} + \dfrac{a_k}{4D}(r^2 - a_k^2) & F_v \\[2ex]
0 & \dfrac{1+\nu}{2}\dfrac{a_k}{r} + \dfrac{1-\nu}{2}\dfrac{r}{a_k} & \dfrac{r^2 - a_k^2}{2Dr} & \dfrac{a_k r}{2D}\ln\dfrac{r}{a_k} - \dfrac{a_k}{4Dr}(r^2 - a_k^2) & F_\theta \\[2ex]
0 & \dfrac{D(1-\nu^2)}{2}\dfrac{r^2 - a_k^2}{a_k r^2} & \dfrac{1-\nu}{2}\left(\dfrac{a_k}{r}\right)^2 + \dfrac{1+\nu}{2} & \dfrac{1+\nu}{2}a_k\ln\dfrac{r}{a_k} + \dfrac{(1-\nu)a_k}{4r^2}(r^2 - a_k^2) & F_M \\[2ex]
0 & 0 & 0 & \dfrac{a_k}{r} & F_V \\[2ex]
0 & 0 & 0 & 0 & 1
\end{bmatrix}$$

$F_v = \epsilon_0^c\left[\dfrac{w_1}{8D}F_{vw_1}(r,a_k) + \dfrac{\Delta w}{\Delta \ell}\dfrac{1}{D}F_{vw_2}(r,a_k) + \dfrac{c_1}{4D}F_{vc_1}(r,a_k) + \dfrac{\Delta c}{\Delta \ell}\dfrac{1}{4D}F_{vc_2}(r,a_k) + M_{T1}U_{vM} + \dfrac{\Delta M_T}{\Delta \ell}\left(\dfrac{a_k^3 - r^3}{9D} - \dfrac{a_k^3}{3D}\ln\dfrac{a_k}{r}\right)\right]$

$F_\theta = \epsilon_0^c\left[-\dfrac{w_1}{4D}F_{\theta w_1}(r,a_k) - \dfrac{\Delta w}{\Delta \ell}\dfrac{1}{D}F_{\theta w_2}(r,a_k) + \dfrac{c_1}{4D}F_{\theta c_1}(r,a_k) + \dfrac{\Delta c}{\Delta \ell}\dfrac{1}{4D}F_{\theta c_2}(r,a_k) + M_{T1}U_{\theta M} + \dfrac{\Delta M_T}{\Delta \ell}\dfrac{(r^2 - a_k^3/r)}{3D}\right]$

$F_M = \epsilon_0^c\left\{-\dfrac{w_1}{4}F_{Mw_1}(r,a_k) - \dfrac{\Delta w}{\Delta \ell}F_{Mw_2}(r,a_k) + c_1 F_{Mc_1}(r,a_k) + \dfrac{\Delta c}{\Delta \ell}F_{Mc_2}(r,a_k) - M_{T1}(1-\nu)\dfrac{r^2 - a_k^2}{2a_k^2} + \dfrac{\Delta M_T}{\Delta \ell}\dfrac{(\nu - 1)}{3}(r + a_k^3/r^2)\right\}$

$F_V = \epsilon_0^c\left[-\dfrac{w_1}{2}F_{Vw_1}(r,a_k) - \dfrac{\Delta w}{\Delta \ell}F_{Vw_2}(r,a_k)\right]$

where F_{vw_1}, F_{vw_2}, $F_{\theta w_1}$, $F_{\theta w_2}$, F_{Mw_1}, F_{Mw_2}, F_{Vw_1}, F_{Vw_2} are given in Table 11-1. The $\langle r - a_i \rangle^0$ terms can be ignored.

$$F_{vc_1}(r,a_k) = -\frac{2a_k^3}{3}\ln\frac{a_k}{r} + \frac{5a_k^3 + 4r^3 - 9r^2 a_k}{9} \qquad F_{vc_2}(r,a_k) = \frac{r^4}{8} - \frac{4a_k r^3}{9} + \frac{a_k^2 r^2}{2} - \frac{13a_k^4}{72} - \frac{a_k^4}{6}\ln\frac{r}{a_k}$$

$$F_{\theta c_1}(r,a_k) = -\left(\frac{4r^2}{3} - 2a_k r + \frac{2a_k^3}{3r}\right) \qquad F_{\theta c_2}(r,a_k) = -\left(\frac{r^3}{2} - \frac{4a_k r^2}{3} + a_k^2 r - \frac{a_k^4}{6r}\right)$$

$$F_{Mc_1}(r,a_k) = -\left(\frac{2+\nu}{3}r - \frac{1+\nu}{2}a_k - \frac{1-\nu}{6}\frac{a_k^3}{r^2}\right) \qquad F_{Mc_2}(r,a_k) = -\left(\frac{3+\nu}{8}r^2 - \frac{2+\nu}{3}a_k r + \frac{1+\nu}{4}a_k + \frac{1-\nu}{24}\frac{a_k^4}{r^2}\right)$$

$m = 1$:

$$\mathbf{U}_i = \left[\begin{array}{cccc|c}
\frac{(3-\nu)r}{4a_k} - \frac{(1-\nu)r^3}{8a_k^3} + \frac{(3+\nu)a_k}{8r} & -\frac{(1+\nu)r}{4} - \frac{(1-\nu)r^3}{8a_k^2} + \frac{(3+\nu)a_k^2}{8r} & -\frac{a_k r}{4D}\ln\frac{r}{a_k} - \frac{(r^2-a_k^2)(r^2-3a_k^2)}{16Da_k r} & \frac{a_k^2 r}{4D}\ln\frac{r}{a_k} - \frac{1}{16Dr}(r^4 - a_k^4) & F_v \\
-\frac{3-\nu}{4a_k} + \frac{3(1-\nu)r^2}{8a_k^3} + \frac{(3+\nu)a_k}{8r^2} & \frac{1+\nu}{4} + \frac{3(1-\nu)r^2}{8a_k^2} + \frac{(3+\nu)a_k^2}{8r^2} & \frac{a_k}{4D}\ln\frac{r}{a_k} + \frac{3(r^4 - a_k^4)}{16Da_k r^2} & -\frac{a_k^2}{4D}\ln\frac{r}{a_k} + \frac{(r^2-a_k^2)(3r^2-a_k^2)}{16Dr^2} & F_\theta \\
\frac{D(3+\nu)(1-\nu)}{4}\frac{r^4 - a_k^4}{a_k^3 r^3} & \frac{D(3+\nu)(1-\nu)}{4}\frac{r^4 - a_k^4}{a_k^2 r^3} & \frac{(3+\nu)r}{8a_k} + \frac{3(1-\nu)a_k^3}{8r^3} + \frac{(1+\nu)a_k}{4r} & \frac{(3+\nu)r}{8} - \frac{(1-\nu)a_k^4}{8r^3} - \frac{(1+\nu)a_k^2}{4r} & F_M \\
\frac{D(3+\nu)(1-\nu)}{4}\frac{r^4 - a_k^4}{a_k^3 r^4} & \frac{D(3+\nu)(1-\nu)}{4}\frac{r^4 - a_k^4}{a_k^2 r^4} & \frac{3+\nu}{8a_k} + \frac{3(1-\nu)a_k^3}{8r^4} - \frac{(3-\nu)a_k}{4r^2} & \frac{3+\nu}{8} - \frac{(1-\nu)a_k^4}{8r^4} + \frac{(3-\nu)a_k^2}{4r^2} & F_V \\
\hline
0 & 0 & 0 & 0 & 1
\end{array}\right]$$

TABLE 11-10 Massless Segment with Center Hole (*Continued*)

$$F_v = \epsilon_1^j\left\{\frac{w_1}{4D}\left(\frac{4r^4}{45} - \frac{a_k r^3}{4} + \frac{a_k^3 r}{9} + \frac{a_k^5}{20r} + \frac{a_k^3 r}{3}\ln\frac{r}{a_k}\right) + \frac{\Delta w/\Delta\ell}{4D}\left(\frac{r^5}{48} - \frac{4a_k r^4}{45} + \frac{a_k^2 r^3}{8} - \frac{7a_k^4 r}{144} - \frac{a_k^6}{120r} - \frac{a_k^4 r}{12}\ln\frac{r}{a_k}\right)\right.$$

$$\left. + \frac{c_1}{4D}\left[-\frac{r^3}{16} + \frac{a_k^2 r}{4} - \frac{3a_k^4}{16r} + \frac{r}{2}\left(\frac{r^2}{2} - a_k^2\right)\ln\frac{r}{a_k}\right] + \frac{\Delta c/\Delta\ell}{4D}\left[\frac{8r^4}{45} - \frac{3a_k r^3}{16} - \frac{a_k^3 r}{36} + \frac{3a_k^5}{80r} - \frac{a_k r}{2}\left(\frac{r^2}{2} - \frac{a_k^2}{3}\right)\ln\frac{r}{a_k}\right]\right\}$$

$$F_\theta = \epsilon_1^j\left\{-\frac{w_1}{4D}\left(\frac{16r^3}{45} - \frac{3a_k r^2}{4} + \frac{4a_k^3}{9} - \frac{a_k^5}{20r^2} + \frac{a_k^3}{3}\ln\frac{r}{a_k}\right) - \frac{\Delta w/\Delta\ell}{4D}\left(\frac{5r^4}{48} - \frac{16a_k r^3}{45} + \frac{3a_k^2 r^2}{8} - \frac{19a_k^4}{144} + \frac{a_k^6}{120r^2} - \frac{a_k^4}{12}\ln\frac{r}{a_k}\right)\right.$$

$$\left. - \frac{c}{4D}\left[\frac{r^2}{16} - \frac{a_k^2}{4} + \frac{3a_k^4}{16r^2} + \frac{1}{2}\left(\frac{3r^2}{2} - a_k^2\right)\ln\frac{r}{a_k}\right] - \frac{\Delta c/\Delta\ell}{4D}\left[\frac{32r^3}{45} - \frac{13a_k r^2}{16} + \frac{5a_k^3}{36} - \frac{3a_k^5}{80r^2} - \frac{a_k}{2}\left(\frac{3r^2}{2} - \frac{a_k^2}{3}\right)\ln\frac{r}{a_k}\right]\right\}$$

$$F_M = \epsilon_1^j\left\{-\frac{w_1}{4}\left[\frac{4(4+\nu)r^2}{15} - \frac{3+\nu}{2}a_k r + \frac{(1+\nu)a_k^3}{3r} + \frac{(1-\nu)a_k^5}{10r^3}\right] - \frac{\Delta w/\Delta\ell}{4}\left[\frac{5+\nu}{12}r^3 - \frac{4(4+\nu)}{15}a_k r^2 + \frac{3+\nu}{4}a_k^2 r - \frac{(1+\nu)a_k^4}{12r} - \frac{(1-\nu)a_k^6}{60r^3}\right]\right.$$

$$\left. - \frac{c_1}{4}\left[\frac{7+\nu}{8}r - \frac{(1+\nu)a_k^2}{2r} - \frac{3(1-\nu)a_k^4}{8r^3}\right] - \frac{\Delta c/\Delta\ell}{4}\left[\frac{8(4+\nu)r^2}{15} - \frac{19+5\nu}{8}a_k r + \frac{(1+\nu)a_k^3}{6r} + \frac{3(1-\nu)a_k^5}{40r^3} - \frac{3+\nu}{2}a_k r\ln\frac{r}{a_k}\right]\right\}$$

$$F_V = \epsilon_1^j\left\{-\frac{w_1}{4}\left[\frac{4(9+\nu)r}{15} - \frac{3+\nu}{2}a_k - \frac{(3-\nu)a_k^3}{3r^2} + \frac{(1-\nu)a_k^5}{10r^4}\right] - \frac{\Delta w/\Delta\ell}{4}\left[\frac{17+\nu}{12}r^2 - \frac{4(9+\nu)}{15}a_k r + \frac{3+\nu}{4}a_k^2 + \frac{(3-\nu)a_k^4}{12r^2} - \frac{(1-\nu)a_k^6}{60r^4}\right]\right.$$

$$\left. - \frac{c_1}{4}\left[-\frac{9-\nu}{8} + \frac{(3-\nu)a_k^2}{2r^2} - \frac{3(1-\nu)a_k^4}{8r^4} + \frac{3+\nu}{2}\ln\frac{r}{a_k}\right] - \frac{\Delta c/\Delta\ell}{4}\left[\frac{4(3+2\nu)r}{15} - \frac{3+5\nu}{8}a_k - \frac{(3-\nu)a_k^3}{6r^2} + \frac{3(1-\nu)a_k^5}{40r^4} - \frac{3+\nu}{2}\ln\frac{r}{a_k}\right]\right\}$$

$\underline{m > 2:}$

$$\mathbf{U}_i = \begin{bmatrix} \mathbf{H}(r)\mathbf{H}^{-1}(a_k) & \varepsilon_m^j\left[\mathbf{G}(r) - \mathbf{H}(r)\mathbf{H}^{-1}(a_k)\mathbf{G}(a_k)\right] \\ 0 & 1 \end{bmatrix}$$

$$\mathbf{H}(r) = \begin{bmatrix} r^m & r^{-m} & r^{m+2} & r^{-m+2} \\ -mr^{m-1} & mr^{-m-1} & -(m+2)r^{m+1} & (m-2)r^{-m+1} \\ -Dm(m-1)(1-\nu)r^{m-2} & -Dm(m+1)(1-\nu)r^{-m-2} & -D(m+1)(m+2-\nu m+2\nu)r^{m} & -D(m-1)(m-2-\nu m-2\nu)r^{-m} \\ Dm^2(m-1)(1-\nu)r^{m-3} & -Dm^2(m+1)(1-\nu)r^{-m-3} & Dm(m+1)(m-\nu m-4)r^{m-1} & -Dm(m-1)(m-\nu m+4)r^{-m-1} \end{bmatrix}$$

$$\mathbf{H}^{-1}(a_k) = \begin{bmatrix} \dfrac{m(1-\nu)+4}{8}a_k^{-m} & \dfrac{(m+2)(1-\nu)-4}{8m}a_k^{-m+1} & \dfrac{m-2}{8m(m-1)D}a_k^{-m+2} & \dfrac{a_k^{-m+3}}{8m(m-1)D} \\ -\dfrac{m(1-\nu)-4}{8}a_k^{m} & \dfrac{(m-2)(1-\nu)+4}{8m}a_k^{m+1} & -\dfrac{m+2}{8m(m+1)D}a_k^{m+2} & \dfrac{a_k^{m+3}}{8m(m+1)D} \\ -\dfrac{m(1-\nu)}{8}a_k^{-m-2} & -\dfrac{1-\nu}{8}a_k^{-m-1} & -\dfrac{a_k^{-m}}{8(m+1)D} & -\dfrac{a_k^{-m+1}}{8m(m+1)D} \\ \dfrac{m(1-\nu)}{8}a_k^{m-2} & -\dfrac{1-\nu}{8}a_k^{m-1} & \dfrac{a_k^{m}}{8(m-1)D} & -\dfrac{a_k^{m+1}}{8m(m-1)D} \end{bmatrix}$$

TABLE 11-10 Massless Segment with Center Hole (*Continued*)

$m = 2$:

$$
\mathbf{G}(r) = \begin{bmatrix}
\dfrac{w_1}{48D} r^4 \ln r + \dfrac{\Delta w}{\Delta \ell}\dfrac{1}{D}\left(\dfrac{r^5}{105} - \dfrac{a_k r^4 \ln r}{48}\right) \\[2ex]
-\dfrac{w_1}{48D} r^3 (4 \ln r + 1) + \dfrac{\Delta w}{\Delta \ell}\dfrac{1}{D}\left[-\dfrac{5r^4}{105} + \dfrac{a_k r^3 (4 \ln r + i)}{48}\right] \\[2ex]
-\dfrac{w_1}{48} r^2 \; (12 \ln r + 7 + \nu) + \dfrac{\Delta w}{\Delta \ell}\left[-\dfrac{(20+\nu) r^3}{105} + \dfrac{a_k r^2 (12 \ln r + 7 + \nu)}{48}\right] \\[2ex]
-\dfrac{w_1}{12} r[3(1+\nu)\ln r + 3 + \nu] + \dfrac{\Delta w}{\Delta \ell}\left\{-\dfrac{(47+16\nu) r^2}{105} + \dfrac{a_k r[3(1+\nu)\ln r + 3 + \nu]}{12}\right\}
\end{bmatrix}
$$

$m = 3$:

$$
\mathbf{G}(r) = \begin{bmatrix}
-\dfrac{w_1 r^4}{35D} + \dfrac{\Delta w}{\Delta \ell}\dfrac{1}{D}\left(\dfrac{r^5 \ln r}{96} + \dfrac{a_k r^4}{35}\right) \\[2ex]
\dfrac{w_1 4r^3}{35D} + \dfrac{\Delta w}{\Delta \ell}\dfrac{1}{D}\left[-\dfrac{(5 \ln r + 1) r^4}{96} - \dfrac{a_k 4r^3}{35}\right] \\[2ex]
\dfrac{w_1}{35} r^2 \; (12 - 5\nu) + \dfrac{\Delta w}{\Delta \ell}\left\{-\dfrac{r^3[4(5-\nu)\ln r + 9 + \nu]}{96} - \dfrac{a_k r^2 (12 - 5\nu)}{35}\right\} \\[2ex]
\dfrac{w_1}{35} r(13 - 27\nu) + \dfrac{\Delta w}{\Delta \ell}\left\{-\dfrac{r^2[12(1+3\nu)\ln r + 37 + 9\nu]}{96} + \dfrac{a_k r (13 - 27\nu)}{35}\right\}
\end{bmatrix}
$$

$m = 4$:

$$
\mathbf{G}(r) = \begin{bmatrix}
-\dfrac{w_1 r^4 \ln r}{96D} + \dfrac{\Delta w}{\Delta \ell}\dfrac{1}{D}\left(-\dfrac{r^5}{63} + \dfrac{a_k r^4 \ln r}{96}\right) \\[2ex]
\dfrac{w_1}{96D} r^3 \; (4 \ln r + 1) + \dfrac{\Delta w}{\Delta \ell}\dfrac{1}{D}\left[\dfrac{5r^4}{63} - \dfrac{a_k r^3}{96}(4 \ln r + 1)\right] \\[2ex]
\dfrac{w_1 r^2}{96}\left[12(1-\nu)\ln r + 7 + \nu\right] + \dfrac{\Delta w}{\Delta \ell}\left\{\dfrac{r^3}{63}(20 - 11\nu) - \dfrac{a_k r^2}{96}\left[12(1-\nu)\ln r + 7 + \nu\right]\right\} \\[2ex]
\dfrac{w_1}{6}\left[3(\nu - 1)\ln r + \nu\right] r + \dfrac{\Delta w}{\Delta \ell}\left\{-\dfrac{r^2}{63}(37 - 64\nu) - \dfrac{[3(\nu - 1)\ln r + \nu] r a_k}{6}\right\}
\end{bmatrix}
$$

$m = 5$:

$$
\mathbf{G}(r) = \begin{bmatrix}
\dfrac{w_1 r^4}{189D} + \dfrac{\Delta w}{\Delta \ell}\dfrac{1}{D}\left(-\dfrac{r^5 \ln r}{160} - \dfrac{a_k r^4}{189}\right) \\[2ex]
-\dfrac{w_1 4r^3}{189D} + \dfrac{\Delta w}{\Delta \ell}\dfrac{1}{D}\left[\dfrac{r^4 (5 \ln r + 1)}{160} + \dfrac{4a_k r^3}{189}\right] \\[2ex]
-\dfrac{w_1}{189} r^2 \; (12 - 21\nu) + \dfrac{\Delta w}{\Delta \ell}\left\{\dfrac{r^3}{160}[20(1-\nu)\ln r + 9 + \nu] + \dfrac{a_k r^2}{189}(12 - 21\nu)\right\} \\[2ex]
\dfrac{w_1 r}{189}(93 - 75\nu) + \dfrac{\Delta w}{\Delta \ell}\left\{-\dfrac{r^2}{32}[(1-\nu)\ln r - 1 - 5\nu] - \dfrac{r a_k}{189}(93 - 75\nu)\right\}
\end{bmatrix}
$$

TABLE 11-10 Massless Segment with Center Hole *(Continued)*

$m \geq 6$:

$$
G(r) = \begin{bmatrix}
\dfrac{w_1 r^4}{D}\left[\dfrac{1}{(m^2-4)(m^2-16)}\right] + \dfrac{\Delta w}{\Delta \ell}\dfrac{1}{D}\left[\dfrac{r^5}{(m^2-9)(m^2-25)} - \dfrac{r^4 a_k}{(m^2-4)(m^2-16)}\right] \\
-\dfrac{w_1 4r^3}{D}\left[\dfrac{1}{(m^2-4)(m^2-16)}\right] + \dfrac{\Delta w}{\Delta \ell}\dfrac{1}{D}\left[-\dfrac{5r^4}{(m^2-9)(m^2-25)} + \dfrac{4r^3 a_k}{(m^2-4)(m^2-16)}\right] \\
w_1 r^2 \dfrac{(m^2\nu - 12 - 4\nu)}{(m^2-4)(m^2-16)} + \dfrac{\Delta w}{\Delta \ell}\left[\dfrac{r^3(m^2\nu - 5\nu - 20)}{(m^2-9)(m^2-25)} - \dfrac{r^2 a_k (m^2\nu - 12 - 4\nu)}{(m^2-4)(m^2-16)}\right] \\
w_1 r \dfrac{(5m^2 - 3m^2\nu - 32)}{(m^2-4)(m^2-16)} + \dfrac{\Delta w}{\Delta \ell}\left[\dfrac{r^2(7m^2 - 4m^2\nu - 75)}{(m^2-9)(m^2-25)} - \dfrac{r a_k (5m^2 - 3m^2\nu - 32)}{(m^2-4)(m^2-16)}\right]
\end{bmatrix}
$$

11.12 Point Matrices

The transfer matrices that take into account concentrated (point) occurrences are listed in Tables 11-14 and 11-15. Point matrices for in-span indeterminates (rigid supports, moment releases, angle guides, and shear releases) are provided in Table 2-15. For plates, these in-span indeterminates are line conditions in that they occur at $r = a_i$, extending completely around the circumference of the plate.

11.13 Loading Functions

The loading functions are provided in the previous transfer matrix tables for most common loadings. These functions can be calculated for other loadings from the formulas

$$F_v = -\int_{a_k}^{r} w_m^j(\xi, a_k) U_{vV}(r, \xi)\, d\xi = -\int_{a_k}^{r} w_m^j(r, \xi) U_{vV}(\xi, a_k)\, d\xi$$

$$F_\theta = -\int_{a_k}^{r} w_m^j(\xi, a_k) U_{\theta V}(r, \xi)\, d\xi = -\int_{a_k}^{r} w_m^j(r, \xi) U_{\theta V}(\xi, a_k)\, d\xi$$

$$F_M = -\int_{a_k}^{r} w_m^j(\xi, a_k) U_{MV}(r, \xi)\, d\xi = -\int_{a_k}^{r} w_m^j(r, \xi) U_{MV}(\xi, a_k)\, d\xi$$

$$F_V = -\int_{a_k}^{r} w_m^j(\xi, a_k) U_{VV}(r, \xi)\, d\xi = -\int_{a_k}^{r} w_m^j(r, \xi) U_{VV}(\xi, a_k)\, d\xi$$

The notation $w_m^j(r, a_k)$ is employed to indicate a transformed loading function that begins at radius a_k and extends to radius r. For example, for a loading that varies linearly in the radial direction, $w_m^j(r, a_k) = (\Delta w/\Delta \ell)(r - a_k)\epsilon_m^j$. The notation $U_{ij}(r, a_k)$ has a similar meaning. $U_{ij}(r, a_k)$ indicates a transfer matrix element for a ring section that begins at radius a_k and extends to radius r. For example, from Table 11-10 for $m = 0$,

$$U_{vV}(r, a_k) = \frac{a_k}{4D}\left[-(r^2 + a_k^2)\ln\frac{r}{a_k} + (r^2 - a_k^2)\right]$$

Do not use U_{ij} from transfer matrices for a solid plate segment (no center hole). Instead, use U_{ij} from transfer matrices for ring segments, for example, from Table 11-10.

TABLE 11-11 Massless Center Segment (No Center Hole) and Compressive In-plane (Axial) Force P, Symmetric Case ($m = 0$)

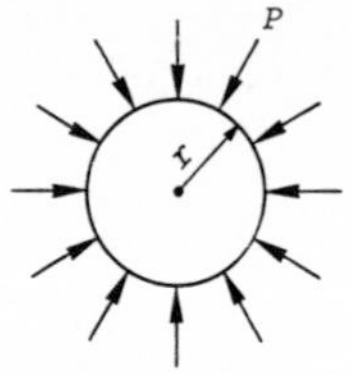

Definition:

$\alpha^2 = P/D$

$J_m(\alpha r)$, $m = 0, 1$ are Bessel functions.

$\mathbf{U}_i =$

1	0	$[J_0(\alpha r) - 1]/(D\alpha^2)$	0	F_V
0	0	$J_1(\alpha r)/(D\alpha)$	0	F_θ
0	0	$J_0(\alpha r) - (1 - \nu)J_1(\alpha r)/(\alpha r)$	0	F_M
0	0	0	0	F_V
0	0	0	0	1

11.14 Static Response

The procedure described here permits the transformed variables v^j_m, θ^j_m, M^j_m, V^j_m, $j = c, s$, to be calculated for simple or complex plates. These variables are then placed in the series of Eq. (11.15) to find the deflection v, slope θ, radial moment M, and shear force V. Finally, other variables of interest such as M_ϕ, $M_{r\phi}$, Q_r are computed from Eqs. (11.11), (11.12), and (11.13).

The technique for calculating v^j_m, θ^j_m, M^j_m, V^j_m resulting from static loading is the same as the procedure given in Section 2.16 for beams. In brief, the state vector

$$\mathbf{s} = \begin{bmatrix} v^j_m \\ \theta^j_m \\ M^j_m \\ V^j_m \\ 1 \end{bmatrix} \tag{11.19}$$

at any point along the plate is found by progressive multiplication of the transfer matrices for all occurrences from the inner edge to the outer edge up to that point. That is, the state at any point j is given by

$$\mathbf{s}_j = \mathbf{U}_j\mathbf{U}_{j-1} \cdots \overline{\mathbf{U}}_k \cdots \mathbf{U}_2\mathbf{U}_1\mathbf{s}_0 \tag{11.20}$$

where $\mathbf{s}_0 = \mathbf{s}_{r=a_0}$. If the outer edge of the plate occurs at station n, the state variables at the outer edge of the plate become

$$\mathbf{s}_{r=a_L} = \mathbf{s}_n = \mathbf{U}_n\mathbf{U}_{n-1} \cdots \overline{\mathbf{U}}_k \cdots \mathbf{U}_2\mathbf{U}_1\mathbf{s}_0 = \mathbf{U}\mathbf{s}_0 \tag{11.21}$$

The matrix $\mathbf{U}$ is the overall or global transfer matrix. The initial parameters v_0, θ_0, M_0, V_0 composing $\mathbf{s}_0$ are found from the boundary conditions that occur at the inner and outer edges of the plate.

TABLE 11-12 Center Segment (No Center Hole) with Mass, Symmetric Case ($m = 0$)

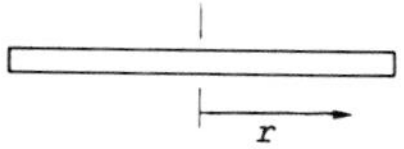

Definition:

$\alpha^2 = \sqrt{\omega^2 \rho / D}$

$\mathbf{U}_i =$

$[J_0(\alpha r) + I_0(\alpha r)]/2$	0	$[J_0(\alpha r) - I_0(\alpha r)]/(2D\alpha^2)$	0	F_V
$\alpha[J_1(\alpha r) - I_1(\alpha r)]/2$	0	$[J_1(\alpha r) + I_1(\alpha r)]/(2D\alpha)$	0	F_θ
$D\alpha\{\alpha[J_0(\alpha r) - I_0(\alpha r)]$ $-(1 - \nu)[J_1(\alpha r) - I_1(\alpha r)]/r\}/2$	0	$\{J_0(\alpha r) + I_0(\alpha r)$ $-(1 - \nu)[J_1(\alpha r) + I_1(\alpha r)]/[\alpha r]\}/2$	0	F_M
$-D\alpha^3[J_1(\alpha r) + I_1(\alpha r)]/2$	0	$-\alpha[J_1(\alpha r) - I_1(\alpha r)]/2$	0	F_V
0	0	0	0	1

TABLE 11-13 Transformed Loading Function Coefficients for Distributed Loads

Loading (Although forces are shown in the diagrams below, the formulas for ε_m^j apply as well to moments with the same distribution in the ϕ direction.)	$\varepsilon_m^j \qquad j = c,s$
1. Distributed Load Constant in the ϕ Direction. 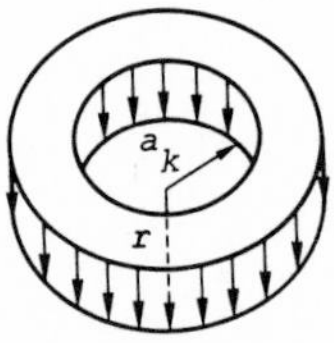	$\varepsilon_0^c = 1$ $\varepsilon_m^j = 0 \qquad m > 0$
2. Distributed Load Constant in the ϕ Direction, Covering $\phi = \phi_1$ to $\phi = \phi_2$ 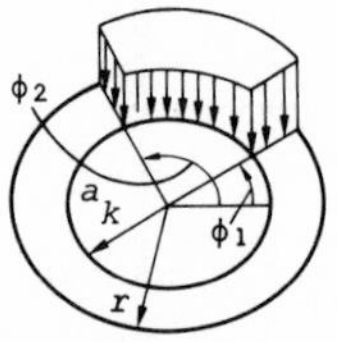	$\varepsilon_0^c = \dfrac{\phi_2 - \phi_1}{2\pi}$ $\varepsilon_m^c = \dfrac{1}{m\pi}\,(\sin m\phi_2 - \sin m\phi_1) \qquad m > 0$ $\varepsilon_m^s = -\dfrac{1}{m\pi}\,(\cos m\phi_2 - \cos m\phi_1) \qquad m > 0$
3. Distributed Load Ramp in the ϕ Direction. 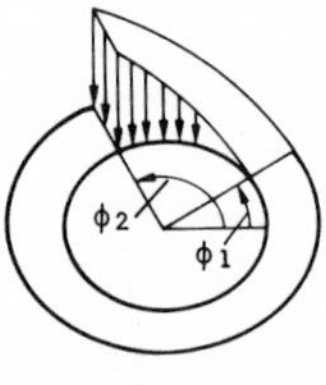	$\varepsilon_0^c = \dfrac{1}{4\pi}\,(\phi_2 - \phi_1)^2$ $\varepsilon_m^c = \dfrac{1}{m\pi}[(\phi_2 - \phi_1)\sin m\phi_2$ $\quad + \dfrac{1}{m}(\cos m\phi_2 - \cos m\phi_1)] \qquad m > 0$ $\varepsilon_m^s = \dfrac{1}{m\pi}[(\phi_1 - \phi_2)\cos m\phi_2$ $\quad + \dfrac{1}{m}(\sin m\phi_2 - \sin m\phi_1)] \qquad m > 0$
4. Harmonic Load, $\cos\phi$ 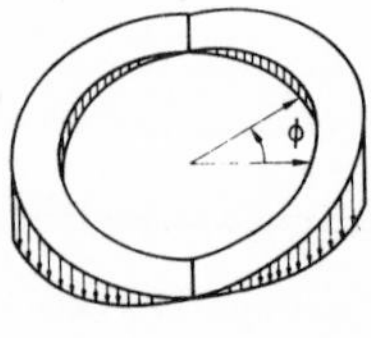	$\varepsilon_0^c = 0$ $\varepsilon_1^c = 1$ $\varepsilon_m^c = 0 \qquad m > 1$ $\varepsilon_m^s = 0 \qquad m > 0$

TABLE 11-14 Point Matrices at $r = a_i$

Definitions: $j = c$ or s

(a) Concentrated Force W_T (force)

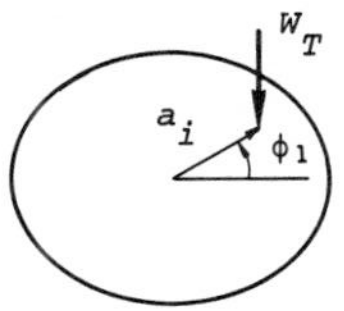

$W_0^c = W_T / 2\pi a_i$

$W_m^c = (W_T \cos m\phi_1)/\pi a_i \qquad m > 0$

$W_m^s = (W_T \sin m\phi_1)/\pi a_i \qquad m > 0$

$$
\bar{\mathbf{U}}_i = \begin{bmatrix}
1 & 0 & 0 & \frac{1}{k_2} & 0 \\
0 & 1 & \frac{1}{k_2^*} & 0 & 0 \\
0 & -M_i \omega^2 i_\phi^2 + k_1^* + P_{Li} & 1 & 0 & c_m^j \\
P_{\phi i} + k_1 - M_i \omega^2 - M_i \omega^2 m^2 i_r^2 & -M_i \Omega^2 a_i & 0 & 1 & -W_m^j \\
0 & 0 & 0 & 0 & 1
\end{bmatrix}
$$

(b) Multiple Concentrated Forces

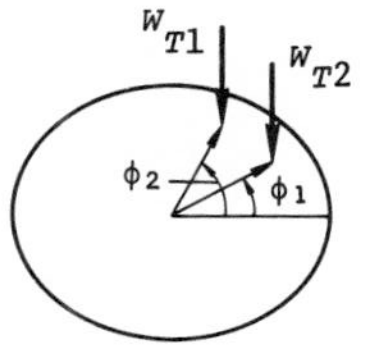

$W_0^c = (W_{T1} + W_{T2})/2\pi a_i$

$W_m^c = (W_{T1} \cos m\phi_1 + W_{T2} \cos m\phi_2)/\pi a_i \qquad m > 0$

$W_m^s = (W_{T1} \sin m\phi_1 + W_{T2} \sin m\phi_2)/\pi a_i \qquad m > 0$

(c) Uniform Line Ring Load W (force/length in ϕ direction)

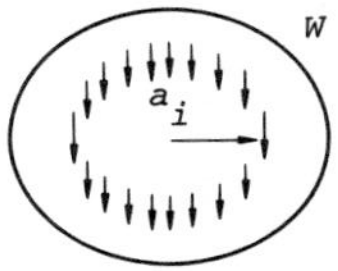

$W_0^c = W, \quad W_m^c = W_m^s = 0 \qquad m > 0$

(d) Uniform Line Load W (force/length in ϕ direction)

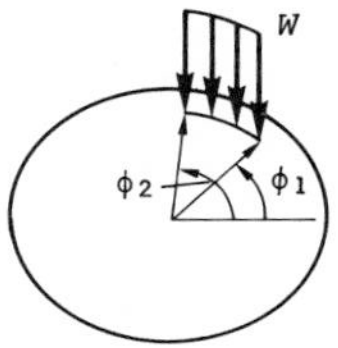

$W_0^c = W(\phi_2 - \phi_1)/2\pi$

$W_m^c = W(\sin m\phi_2 - \sin m\phi_1)/m\pi \qquad m > 0$

$W_m^s = -W(\cos m\phi_2 - \cos m\phi_1)/m\pi \qquad m > 0$

TABLE 11-14 Point Matrices at $r = a_i$ (*Continued*)

(e) Linearly Varying Line Load

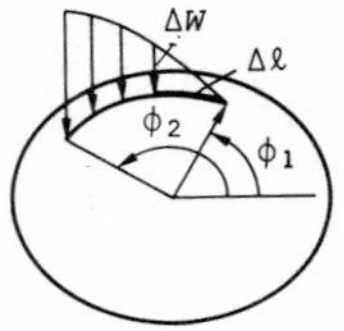

$$W_0^c = \frac{\Delta W}{\Delta \ell}\frac{a_i}{4\pi}(\phi_2 - \phi_1)^2$$

$$W_m^c = \frac{\Delta W}{\Delta \ell}\frac{a_i}{m\pi}\left[(\phi_2 - \phi_1)\sin m\phi_2 + \frac{1}{m}(\cos m\phi_2 - \cos m\phi_1)\right] \quad m > 0$$

$$W_m^s = \frac{\Delta W}{\Delta \ell}\frac{a_i}{m\pi}\left[(\phi_1 - \phi_2)\cos m\phi_2 + \frac{1}{m}(\sin m\phi_2 - \sin m\phi_1)\right] \quad m > 0$$

(f) Multiple Concentrated Moment C_T(force-length)

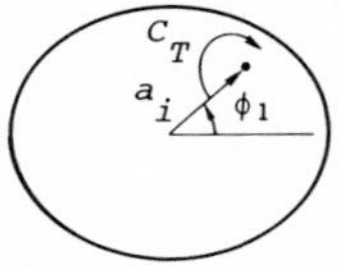

$$C_0^c = C_T/2\pi a_i$$

$$C_m^c = (C_T \cos m\phi_1)/\pi a_i \quad m > 0$$

$$C_m^s = (C_T \sin m\phi_1)/\pi a_i \quad m > 0$$

(g) Line Moment C (force-length/length in ϕ direction)

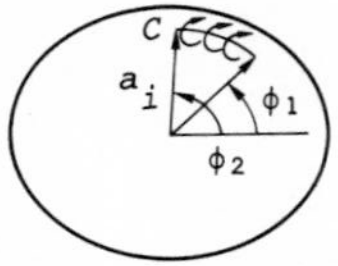

$$C_0^c = C(\phi_2 - \phi_1)/2\pi$$

$$C_m^c = C(\sin m\phi_2 - \sin m\phi_1)/m\pi \quad m > 0$$

$$C_m^s = -C(\cos m\phi_2 - \cos m\phi_1)m\pi \quad m > 0$$

(h) Line Ring Spring k_1 (force/length^2)

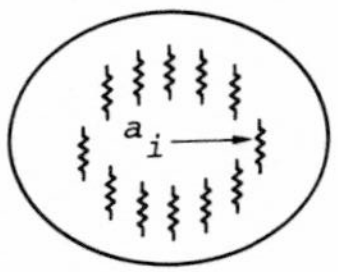

(i) Rotary Line Ring Spring (force-length/length)

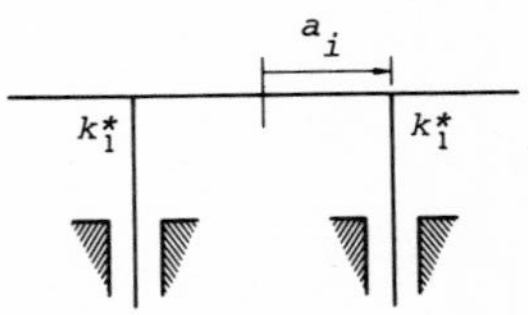

(j) Linear Line Ring Hinge
(shear release with spring)

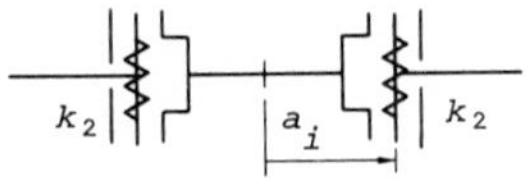

(moment release with rotary spring)

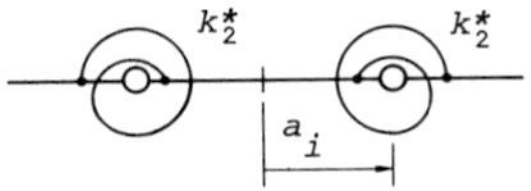

(l) Ring Lumped Mass M_i (mass/length in ϕ direction)

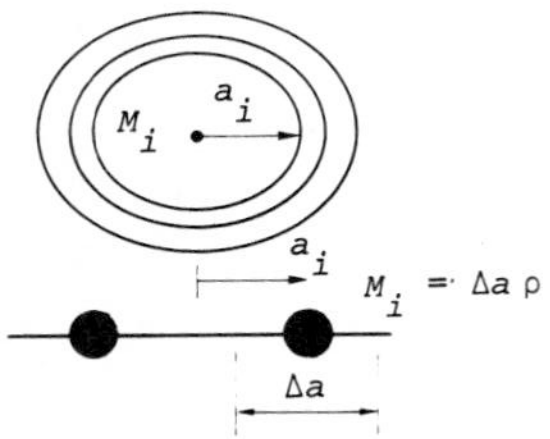

Rotary inertia terms contain i_r, i_ϕ.

(m) Lumped Line In-Plane Compressive Force

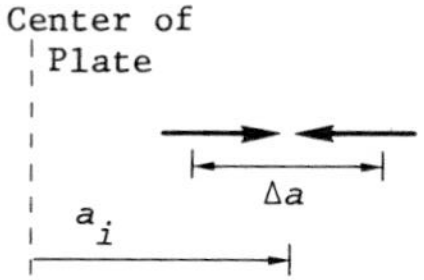

$$P_{Li} = P\,\Delta a$$

$$P_{\phi i} = P_\phi m^2\,\Delta a/r^2$$

Substitute $-P_{Li}$, $-P_{\phi i}$ for P_{Li}, $P_{\phi i}$ **if the force is tensile. Values of** P, P_ϕ **are obtained using the disk analysis of Chapter** 8.

(n) **Ω is the speed of rotation of a rotating circular plate.**

See Table 2-14 for values of k_1, k_1^*, etc. for more complex spring systems, including those with masses.

TABLE 11-15 Point Matrix for Circular Reinforcing Ring at $r = a_i$

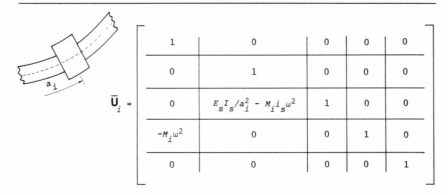

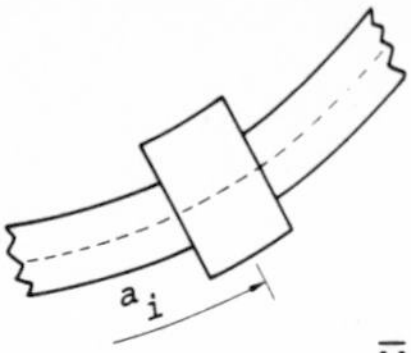

$\bar{\mathbf{U}}_i$ =

1	0	0	0	0
0	1	0	0	0
0	$E_s I_s / a_i^2 - M_i i_s \omega^2$	1	0	0
$-M_i \omega^2$	0	0	1	0
0	0	0	0	1

Definitions:

The reinforcing ring must be symmetric in the z direction about the middle plane of the plate. This transfer matrix applies for $m = 0$ (symmetrical deformation).

M_i is the mass per unit circumferential length of the ring.

I_s is the area moment of inertia of the ring about the middle plane of the plate.

i_s is the radius of gyration of the ring about the circumferential axis.

E_s is the modulus of elasticity of the ring.

Formulas for the Initial Parameters The application of the boundary conditions to Eq. (11.21) for calculation of v_0, θ_0, M_0, V_0 suffices to determine the initial parameters for all plates. Note that v_0, θ_0, M_0, V_0 must be computed for $j = c$ and $j = s$ as well as for each m. Formulas for calculating the initial parameters are listed in rows 1, 2, 3, 4 of Table 11-3 for plates with a center hole. In this table values of v_0, θ_0, M_0, V_0 are listed according to boundary conditions. For plates with no center hole, use Table 11-16. The U_{kj} and F_k in these tables are elements of the overall or global transfer matrix extending from $r = a_0$ or 0 to $r = a_L$, that is, the elements of $\mathbf{U}$ of Eq. (11.21).

EXAMPLE 11.5 Plate with Line Load Find the deflection everywhere in the plate of Fig. 11-3 if only the ring load W is placed on the plate. This load is of uniform magnitude in the ϕ direction.

The deflection is obtained from Eq. (11.20). This requires the computation of $\mathbf{s}_0$, the initial parameter vector. The initial parameters are calculated in terms of the global transfer matrix of Eq. (11.21), that is,

$$\mathbf{U} = \mathbf{U}_2 \bar{\mathbf{U}}_1 \mathbf{U}_1 \tag{1}$$

where $\mathbf{U}_1$ is the transfer matrix of the center segment as given in Table 11-7, with zeros in the fifth column; $\bar{\mathbf{U}}_1$ is the point matrix for the ring load W as

TABLE 11-16 Initial Parameters for Plate with No Center Hole

Center Condition ↓ / Outer Edge →	m	1. Simply Supported 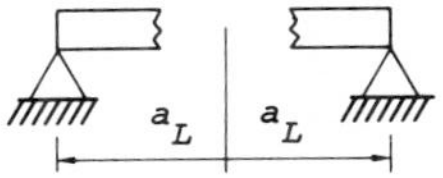	2. Fixed 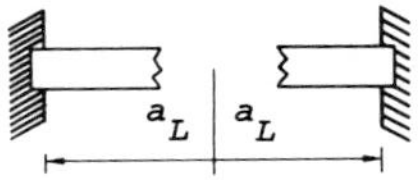	3. Free 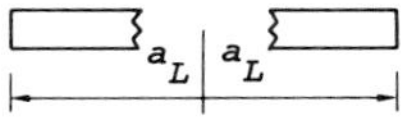	4. Guided 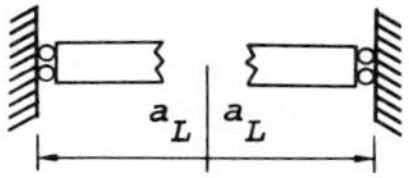
1. No Center Support* 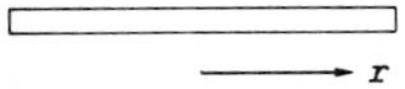	0	Use row 4 of Table 11-3			
	1	Use row 1 of Table 11-3			
	≥ 2	Use row 2 of Table 11-3			
2. Center Support** 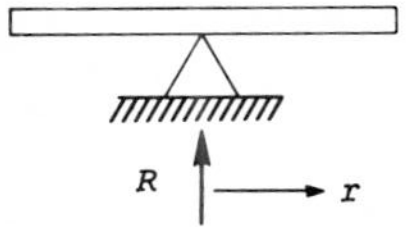	0	$R = (F_M U_{v\theta} - F_V U_{M\theta})/\nabla$ $C = (F_V U_{Mv} - F_M U_{vv})/\nabla$ $\nabla = U_{vv} U_{M\theta} - U_{v\theta} U_{Mv}$	$R = (F_\theta U_{v\theta} - F_V U_{\theta\theta})/\nabla$ $C = (F_V U_{\theta v} - F_\theta U_{vv})/\nabla$ $\nabla = U_{vv} U_{\theta\theta} - U_{v\theta} U_{\theta v}$	$R = (F_V U_{M\theta} - F_M U_{V\theta})/\nabla$ $C = (F_M U_{Vv} - F_V U_{Mv})/\nabla$ $\nabla = U_{Mv} U_{V\theta} - U_{M\theta} U_{Vv}$	$R = (F_V U_{\theta\theta} - F_\theta U_{V\theta})/\nabla$ $C = (F_\theta U_{Vv} - F_V U_{\theta v})/\nabla$ $\nabla = U_{\theta v} U_{V\theta} - U_{\theta\theta} U_{Vv}$
	1	Use row 1 of Table 11-3			
	≥ 2	Use row 2 of Table 11-3			

* The transfer matrix for the inner (center) segment should be taken from Tables 11-7, 8, 11, or 12.

** The transfer matrix for the inner (center) segment should be taken from Table 11-9. See this table for the definition of R and C.

Definitions: $U_{kj} = U_{kj}\big|_{r=a_L}$, $k,j = v,\theta,M,V$; $F_k = F_k\big|_{r=a_L}$, $k = v,\theta,M,V$.

given in Table 11-14, case (c); $\mathbf{U}_2$ is the transfer matrix of the ring segment from $r = a_3$ to $r = a_L$ of Table 11-10, with zeroes in the fifth column.

As noted in Table 11-14, case (c), all loading terms W_m^j are zero except W_0^c which is equal to W. This means that only $m = 0$ terms need be considered in the Eq. (11.15) expansion and only the $m = 0$ transfer matrices need be taken from Tables 11-7 and 11-10. This restriction to $m = 0$ terms is to be expected, since $m = 0$ corresponds to the symmetric deformation which will result from the symmetric loading W.

Before carrying out the multiplication of $\mathbf{U}_1$, $\overline{\mathbf{U}}_1$, $\mathbf{U}_2$ as indicated in (1), the matrices must be evaluated at the proper radius. That is, in (1)

$$\mathbf{U}_1 = \mathbf{U}_1\big|_{r=a_3} \qquad \mathbf{U}_2 = \mathbf{U}_2\Big|_{\substack{r=a_L \\ a_k=a_3}}$$

Finally, (1) gives

$$\mathbf{U} = \begin{bmatrix} 1 & 0 & U_{vM} & 0 & F_v \\ 0 & 0 & U_{\theta M} & 0 & F_\theta \\ 0 & 0 & U_{MM} & 0 & F_M \\ 0 & 0 & 0 & 0 & F_V \\ 0 & 0 & 0 & 0 & 1 \end{bmatrix}$$

$$\begin{aligned}
U_{vM} &= -\frac{a_3}{2D(1+\nu)} + \frac{a_3}{D(1+\nu)}\left[-\frac{1+\nu}{2}a_3 \ln\frac{a_L}{a_3} - \frac{1-\nu}{4a_3}\left(a_L^2 - a_3^2\right)\right] \\
&\quad + \frac{a_3^2}{2D}\ln\frac{a_2}{a_3} - \frac{1}{4D}\left(a_L^2 - a_3^2\right) = -\frac{a_L^2}{2D(1+\nu)} \\
U_{\theta M} &= \frac{a_3}{D(1+\nu)}\left(\frac{1+\nu}{2}\frac{a_3}{a_L} + \frac{1-\nu}{2}\frac{a_L}{a_3}\right) + \frac{a_L^2 - a_3^2}{2Da_L} = \frac{a_L}{D(1+\nu)} \\
U_{MM} &= \frac{a_3}{D(1+\nu)}\left(D\frac{1-\nu^2}{2}\frac{a_L^2 - a_3^2}{a_L a_3^2}\right) + \frac{1-\nu}{2}\left(\frac{a_3}{a_L}\right)^2 + \frac{1+\nu}{2} = 1 \\
F_v &= W\left[\frac{a_3\left(a_L^2 + a_3^2\right)}{4D}\ln\frac{a_L}{a_3} - \frac{a_3}{4D}\left(a_L^2 - a_3^2\right)\right] \\
F_\theta &= -W\left[\frac{a_3 a_L}{2D}\ln\frac{a_L}{a_3} - \frac{a_3}{4Da_L}\left(a_L^2 - a_3^2\right)\right] \\
F_M &= -W\left[\frac{1+\nu}{2}a_3\ln\frac{a_L}{a_3} + \frac{1-\nu}{4a_L^2}a_3\left(a_L^2 - a_3^2\right)\right] \\
F_V &= -W\frac{a_3}{a_L}
\end{aligned} \tag{2}$$

From column 2 of Table 11-16, which refers to row 4 of Table 11-3, the initial parameters are given by

$$\begin{aligned}\theta_0 &= V_0 = 0\\ v_0 &= (F_\theta U_{vM} - F_v U_{\theta M})/\nabla\,|_{r=a_L}\\ M_0 &= (F_v U_{\theta v} - F_\theta U_{vv})/\nabla\,|_{r=a_L}\\ \nabla &= (U_{vv}U_{\theta M} - U_{\theta v}U_{vM})_{r=a_L}\end{aligned} \tag{3}$$

For this problem with symmetric responses ($m = 0$), v_0, θ_0, M_0, V_0 need only be computed once (for $j = c$, $m = 0$). For nonsymmetric responses it would be necessary to calculate v_0, θ_0, M_0, V_0 for $j = c$ and $j = s$ for each m. Substitution of the matrix elements of (2) into (3) gives

$$\nabla = \frac{a_L}{D(1+\nu)}$$

$$v_0 = \frac{Wa_3}{8D}\left(a_L^2 - a_3^2 - 2a_3^2 \ln \frac{a_L}{a_3}\right)$$

$$M_0 = \frac{Wa_3(1+\nu)}{4}\left(2 \ln \frac{a_L}{a_3} - \frac{a_L^2 - a_3^2}{a_L^2}\right)$$

The deflection is now obtained using Eq. (11.20). For $r < a_3$, $\mathbf{s}_j = \mathbf{U}_1\mathbf{s}_0$ or, with $\mathbf{U}_1$ taken from Table 11-7,

$$v = v_0^c = v_0 \cdot 1 - M_0 \frac{r^2}{2D(1+\nu)} = \frac{Wa_3}{8D}\left[(a_L^2 - a_3^2)\left(1 + \frac{r^2}{a_L^2}\right) - 2(r^2 + a_3^2) \ln \frac{a_L}{a_3}\right]$$

and for $r > a_3$, $\mathbf{s}_j = \mathbf{U}_2\overline{\mathbf{U}}_1\mathbf{U}_1\mathbf{s}_0$ and

$$v = v_0^c = \frac{Wa_3}{8D}\left[(a_L^2 + a_3^2)\left(1 - \frac{r^2}{a_L^2}\right) + 2(r^2 + a_3^2) \ln \frac{r}{a_L}\right]$$

11.15 Stability

The critical in-plane radial force $P = P_{cr}$ is calculated in the same way that the buckling axial loads are found for beams. This procedure is described in Section 2.17. In finding the critical load, the variables v_m^j, θ_m^j, M_m^j, V_m^j are used instead of v, θ, M, V of Chap. 2. Applied transverse loadings do not influence the critical in-plane load analysis and hence, as can be seen from Eqs. (11.18), if $w_m^c = w_m^s = 0$, there is no difference between sine and cosine responses, that is, $v_m^c = v_m^s$, $\theta_m^c = \theta_m^s$, $M_m^c = M_m^s$, $V_m^c = V_m^s$. The critical load does not necessarily occur for the symmetric response case with $m = 0$. Therefore, it may be necessary to carry out the stability analysis for several values of m to find the lowest P_{cr}. As in Chap. 2, the transfer matrix formulation permits P to vary in the r direction.

For complicated plates the in-plane force should be discretized according to the lumped-force entry of Table 11-14. For any in-plane loading, including in-plane loads that vary continuously or in a piecewise-constant fashion in the radial direction, the disk analysis of Chap. 8 provides the forces P and P_θ at any r. These forces are calculated at a series of closely spaced radial coordinates r, and a point matrix of Table 11-14, case m, is formed for each set of values of P and P_θ. The plate is thus modeled by lumped forces (Table 11-14) connected by forceless segments (for example, according to the transfer matrix of Table 11-10). Many other effects can be incorporated by employing the proper transfer matrix. For example, use of the appropriate entries in Table 11-14 permits such effects as mass, springs, and in-span supports to be taken into account.

A frequently occurring situation is that of a plate subjected to in-plane forces at several radii r with the magnitudes of the forces remaining in a constant ratio with a nominal value. Then the nominal value that leads to buckling is found with a stability analysis.

EXAMPLE 11.6 Instability of a Plate with No Center Hole Find the buckling loads of a uniform circular plate with no center hole. The compressive in-plane force P is applied at the outer rim, which can be clamped or simply supported.

The in-plane force leading to instability is found by determining the lowest value of the force P which makes the characteristic equation equal to zero. The characteristic equation arises upon application of the boundary conditions to the response relations. The characteristic equation is also equal to ∇ of Table 11-16. The appropriate transfer matrix is given in Table 11-11. Use of this table requires the assumption that the critical in-plane force corresponds to an axially symmetric response ($m = 0$). For a clamped outer rim, Table 11-16 gives

$$\nabla = (U_{vv}U_{\theta M} - U_{vM}U_{\theta v})|_{r=a_L} \tag{1}$$

Inserting the transfer matrix elements of Table 11-11, we find that (1) becomes

$$\nabla = 1 \cdot J_1(\alpha a_L)/D\alpha - 0 = 0$$

or $J_1(\alpha a_L) = 0$, where $\alpha^2 = P/D$. The lowest value of αa_L for which $J_1(\alpha a_L) = 0$ leads to

$$P_{cr} = 14.68\,\frac{D}{a_L^2}$$

If the plate is simply supported on the outer rim, then from Table 11-16

$$\nabla = (U_{vv}U_{MM} - U_{Mv}U_{vM})|_{r=a_L}$$

Insertion of the transfer matrix gives

$$\nabla = 1 \cdot \left[J_0(\alpha a_L) - (1-\nu)J_1(\alpha a_L)/(\alpha a_L)\right] - 0$$

Setting $\nabla = 0$ yields

$$J_0(\alpha a_L) = \frac{1-\nu}{\alpha a_L}J_1(\alpha a_L)$$

The lowest value of P that satisfies this relationship is

$$P_{cr} = 4.191 \frac{D}{a_L^2}$$

if $\nu = 0.3$.

11.16 Free Dynamic Response—Natural Frequencies

The natural frequencies ω_{mn} for circular plates are found by applying the procedure for beams of Section 2.18 to the v_m^j, θ_m^j, M_m^j, V_m^j that are formed from the transfer matrices of this chapter. The primary difference between solutions for the two lies in the number of natural frequencies. For each value of m, there are n, $n = 1, 2, 3, \ldots,$ values of ω. Note that since there are no applied loadings involved in a free dynamic analysis—that is, $w_m^c = w_m^s = 0$—there is no difference between sine and cosine responses—that is, $v_m^c = v_m^s$, $\theta_m^c = \theta_m^s$, $M_m^c = M_m^s$, $V_m^c = V_m^s$.

The mode shapes are found by placing the natural frequency ω_{mn} in v_m^j, θ_m^j, M_m^j, V_m^j of Eq. (11.20) after the initial parameters have been evaluated. The variations in the ϕ direction can be found by using $\cos m\phi$ or $\sin m\phi$, for example,

$$v_m^c \cos m\phi, \quad \theta_m^c \cos m\phi, \quad M_m^c \cos m\phi, \quad V_m^c \cos m\phi \tag{11.22}$$

Most plate vibration problems are best solved with a lumped-mass model. This is accomplished by connecting lumped masses (from Table 11-14) with massless segments (from, for example, Table 11-10).

EXAMPLE 11.7 Vibration of a Plate with No Center Hole Find the natural frequencies for the symmetrical free motion of a circular plate with no central hole.

For symmetric motion $m = 0$. The frequencies are those values of $\omega_{0n} = \omega_n$ $(n = 1, 2, \ldots, \infty)$ that make ∇ of Table 11-16 equal to zero. For a plate clamped at $r = a_L$, Table 11-16 gives

$$\nabla = (U_{vv} U_{\theta M} - U_{vM} U_{\theta v})|_{r=a_L}$$

According to the transfer matrix of Table 11-12,

$$\begin{aligned}\nabla &= \frac{J_0(\alpha a_L) + I_0(\alpha a_L)}{2} \frac{J_1(\alpha a_L) + I_1(\alpha a_L)}{2D\alpha} \\ &\quad - \frac{J_0(\alpha a_L) - I_0(\alpha a_L)}{2D\alpha^2} \frac{J_1(\alpha a_L) - I_1(\alpha a_L)}{2/\alpha} \\ &= \left[I_0(\alpha a_L) J_1(\alpha a_L) + J_0(\alpha a_L) I_1(\alpha a_L) \right] / (2D\alpha)\end{aligned}$$

For $\nabla = 0$ with $\alpha_n^4 = \omega_n^2 \rho / D$, it is found that $\alpha_n a_L = 3.190$, 6.306, 9.425, 12.560,

15.710 and $\alpha_n - \alpha_{n-1} = \pi / a_L$ for higher frequencies. The frequencies are given by

$$\omega_n = \alpha_n^2 \sqrt{D/\rho}$$

11.17 Forced Harmonic Motion

If the forcing functions vary as sin Ωt or cos Ωt, where Ω is the frequency of the loading, the state variables v_m^j, θ_m^j, M_m^j, and V_m^j also vary as sin Ωt or cos Ωt. The spatial r distribution of v_m^j, θ_m^j, M_m^j, and V_m^j is found by setting up a static solution using those transfer matrices containing ω. In these transfer matrices ω should be replaced by Ω. The solution procedure follows that described in Chap. 2 for beams, and then utilizes Eq. (11.15).

11.18 Dynamic Response Due to Arbitrary Loading

In terms of a modal solution, the time-dependent state variables v_m^j, θ_m^j, M_m^j, V_m^j ($j = c$ or s) resulting from arbitrary dynamic loading are expressed by

Displacement method:

$$\begin{aligned} v_m^j(r, t) &= \sum_n A_n(t) v_{mn}^j(r) \\ \theta_m^j(r, t) &= \sum_n A_n(t) \theta_{mn}^j(r) \\ M_m^j(r, t) &= \sum_n A_n(t) M_{mn}^j(r) \\ V_m^j(r, t) &= \sum_n A_n(t) V_{mn}^j(r) \end{aligned} \tag{11.23a}$$

Acceleration Method:

$$\begin{aligned} v_m^j(r, t) &= v_{sm}^j(r, t) + \sum_n B_n(t) v_{mn}^j(r) \\ \theta_m^j(r, t) &= \theta_{sm}^j(r, t) + \sum_n B_n(t) \theta_{mn}^j(r) \\ M_m^j(r, t) &= M_{sm}^j(r, t) + \sum_n B_n(t) M_{mn}^j(r) \\ V_m^j(r, t) &= V_{sm}^j(r, t) + \sum_n B_n(t) V_{mn}^j(r) \end{aligned} \tag{11.23b}$$

where

$$A_n(t) = \frac{\eta_n(t)}{N_n} \qquad B_n(t) = \frac{\xi_n(t)}{N_n} \tag{11.24}$$

The complete dynamic response is obtained by placing these values of $v_m^j(r, t)$, $\theta_m^j(r, t)$, $M_m^j(r, t)$, $V_m^j(r, t)$ from Eqs. (11.23) in the summation of Eqs. (11.15). The resulting equations then represent the total responses $v(r, \phi, t)$, $\theta(r, \phi, t)$, $M(r, \phi, t)$, and $V(r, \phi, t)$.

No Damping or Proportional Damping For viscous damping, the terms

$$c' \frac{\partial v_m^j}{\partial t} \qquad \text{and} \qquad c^* \frac{\partial \theta_m^j}{\partial t}$$

should be added to the right-hand sides of Eqs. (11.18d) and (11.18c), respectively. If the damping is "proportional," then $c' = c_\rho(\rho + m^2\rho i_r^2/r^2)$ and $c^* = c_\rho \rho i_\phi^2$. Set $c' = 0$ and $c^* = 0$, for no damping.

In Eqs. (11.23b) the terms with subscript s are static solutions, as given in Section 11.14, that are determined for the applied loading at each point in time. The terms v_{mn}^j, θ_{mn}^j, M_{mn}^j, and V_{mn}^j are the mode shapes, that is, v_m^j, θ_m^j, M_m^j, V_m^j with $\omega = \omega_{mn}$, $n = 1, 2, \ldots$.

The quantities in Eqs. (11.24) are defined as

$$N_n = \int_{a_0}^{a_L}\left[\left(\rho + \frac{m^2\rho i_r^2}{r^2}\right)(v_{mn}^j)^2 + \rho i_\phi^2(\theta_{mn}^j)^2\right] r\,dr \tag{11.25a}$$

$$\eta_n(t) = e^{-\zeta_n\omega_{mn}t}\left(\cos\alpha_n t + \frac{\zeta_n\omega_{mn}}{\alpha_n}\sin\alpha_n t\right)\eta_n(0) + e^{-\zeta_n\omega_{mn}t}\frac{\sin\alpha_n t}{\alpha_n}\frac{\partial\eta_n}{\partial t}(0)$$

$$+ \int_0^t f_n(\tau)e^{-\zeta_n\omega_{mn}(t-\tau)}\frac{\sin\alpha_n(t-\tau)}{\alpha_n}\,d\tau \tag{11.25b}$$

$\xi_n(t)$ is taken from $\eta_n(t)$ by replacing

$$\eta_n(0) \quad \text{by} \quad \eta_n(0) - \frac{f_n(0)}{\omega_{mn}^2}$$

$$\frac{\partial\eta_n}{\partial t}(0) \quad \text{by} \quad \frac{\partial\eta_n}{\partial t}(0) - \frac{1}{\omega_{mn}^2}\frac{\partial f_n}{\partial t}(0)$$

$$f_n(\tau) \quad \text{by} \quad -\frac{1}{\omega_{mn}^2}\left[\frac{\partial^2}{\partial\tau^2} + c_\rho\frac{\partial}{\partial\tau} + (\omega_{mn}\zeta_n)^2\right]f_n(\tau)$$

and adding the term $\zeta_n^2 f_n(t)/[(1-\zeta_n^2)\omega_{mn}^2]$. In Eqs. (11.25), $\alpha_n = \omega_{mn}\sqrt{1-\zeta_n^2}$, $\zeta_n = c_\rho/2\omega_{mn}$.

If $\zeta_n > 1$, replace sin by sinh, cos by cosh, and $\sqrt{1-\zeta_n^2}$ by $\sqrt{\zeta_n^2 - 1}$; $\zeta_n = 0$ for zero viscous damping. Then

$$\eta_n(0) = \int_{a_0}^{a_L}\left[\left(\rho\frac{m^2 i_r^2}{r^2} + \rho\right)v_m^j(r,0)v_{mn}^j + \rho i_\phi^2\theta_m^j(r,0)\theta_{mn}^j\right] r\,dr \tag{11.26}$$

$$f_n(t) = \int_{a_0}^{a_L}\left\{\left[w_m^j + \frac{m^2}{r^2}\left(M_{T\phi m} - \frac{\nu_r D_\phi}{D_r}M_{Trm}\right)\right]v_{mn}^j\right.$$

$$+ \left[\frac{1}{r}\left(M_{T\phi m}^j - \frac{\nu_r D_\phi}{D_r}M_{Trm}\right)\right]\theta_{mn}^j \tag{11.27a}$$

$$\left. + \frac{M_{Trm}^j}{D_r}M_{mn}^j\right\} r\,dr + h_n(a_k, t)$$

It should be noted that although they are not explicitly so indicated, such functions as N_n, f_n, and η_n are functions of m.

In the case of the acceleration method, $f_n(t)$ can alternatively be expressed as

$$f_n(t) = \omega_{mn}^2 \int_{a_0}^{a_L} \left[\left(\rho \frac{m^2 i_r^2}{r^2} + \rho \right) v_{sm}^j v_{mn}^j + \rho i_\phi^2 \theta_{sm}^j \theta_{mn}^j \right] r\, dr \qquad (11.27b)$$

The function $h_n(a_k, t)$ accounts for nonhomogeneous displacement $v_m^j(a_k, t)$ or rotation $\theta_m^j(a_k, t)$ conditions at $r = a_k$, such as those produced by supports or prescribed time-dependent displacements located in-span or on the boundary:

$$h_n(a_k, t) = -v_m^j(a_k, t)\Delta V_{mn}^j(a_k) - \theta_m^j(a_k, t)\Delta M_{mn}^j(a_k) \qquad (11.28a)$$

$$\Delta V_{mn}^j(a_k) = V_{mn}^j(a_k^-) - V_{mn}^j(a_k^+)$$

$+\,(-)$ means just to the right (left) of $x = a_k$

$$\Delta M_{mn}^j(a_k) = M_{mn}^j(a_k^-) - M_{mn}^j(a_k^+)$$

If the outer rim $r = a_L$ has a nonzero (for example, time-dependent) displacement, then Eq. (11.28a) reduces for $a_k = a_L$ to

$$h_n(a_k, t) = -v_m^j(a_L, t)V_{mn}^j(a_L) - \theta_n^j(a_L, t)M_{mn}^j(a_L) \qquad (11.28b)$$

If the inner rim ($r = 0$ or a_0) has a nonhomogeneous displacement, then for $a_k = a_0$, Eq. (11.28a) reduces to

$$h_n(a_k, t) = +v_m^j(a_0, t)V_{mn}^j(a_0) + \theta_m^j(a_0, t)M_{mn}^j(a_0) \qquad (11.28c)$$

Nonproportional Damping The formulas of this chapter need considerable adjustment in order to make them applicable to materials with nonproportional viscous damping. As noted above, Eqs. (11.18c) and (11.18d) should be supplemented by the terms $c^* \partial\theta_m^j/\partial t$ and $c' \partial v_m^j/\partial t$, respectively, on the right-hand sides. Replace the frequency ω_{mn} by s_{mn}, where s_{mn} is a complex number. Also the mode shapes and some of the transfer matrices are complex functions. The real part of s_{mn} is commonly referred to as the damping exponent, whereas the imaginary part of s_{mn} is called the frequency of the damped free vibration.

Equations (11.23) and (11.24) remain valid but now the static solutions of Eqs. (11.23b) refer to

$$v_{sm}^j(r, t) = -\sum_n \frac{v_{mn}^j(r) f_n(t)}{s_{mn} N_n}$$

where $v_{mn}^j(r)$ and N_n are the mode shape and norm associated with the complex eigenvalue, s_{mn}. Similar definitions of the static solutions apply for the other state variables.

The definitions of this section must be adjusted so that

$$N_n = 2s_{mn}\int_{a_0}^{a_L}\left[\left(\rho\frac{m^2i_r^2}{r^2}+\rho\right)(v_{mn}^j)^2+\rho i_\phi^2(\theta_{mn}^j)\right]r\,dr$$

$$+\int_{a_0}^{a_L}\left[c'(v_{mn}^j)^2+c^*(\theta_{mn}^j)^2\right]r\,dr \tag{11.29}$$

$$\eta_n(t) = e^{s_{mn}t}\eta_n(0)+\int_0^t e^{s_{mn}(t-\tau)}f_n(\tau)\,d\tau \tag{11.30}$$

$$\xi_n(t) = e^{s_{mn}t}\left[\frac{1}{s_{mn}^2}\frac{\partial f_n}{\partial t}(0)+\frac{1}{s_{mn}}f_n(0)+\eta_n(0)\right]$$

$$-\frac{1}{s_{mn}^2}\frac{\partial f_n}{\partial t}(t)+\frac{1}{s_{mn}^2}\int_0^t e^{s_{mn}(t-\tau)}\frac{\partial^2 f_n}{\partial\tau^2}(\tau)\,d\tau \tag{11.31}$$

$$\eta_n(0) = \int_{a_0}^{a_L}\left[s_{mn}\left(\rho\frac{m^2i_r^2}{r^2}+\rho\right)v_m^j(r,0)v_{mn}^j(r)+s_{mn}\rho i_\phi^2\theta_m^j(r,0)\theta_{mn}^j(r)\right.$$

$$+c'v_m^j(r,0)v_{mn}^j(r)+c^*\theta_m^j(r,0)\theta_{mn}^j(r)$$

$$\left.+\left(\rho\frac{m^2i_r^2}{r^2}+\rho\right)\frac{\partial v_m^j(r,0)}{\partial t}v_{mn}^j(r)+\rho i_\phi^2\frac{\partial\theta_m^j(r,0)}{\partial t}\theta_{mn}^j(r)\right]r\,dr \tag{11.32}$$

In the case of the acceleration method, $f_n(t)$ can alternatively be expressed as

$$f_n(t) = -s_{mn}\int_{a_0}^{a_L}\left[c'v_{sm}^j(r,t)v_{mn}^j(r)+c^*\theta_{sm}^j(r,t)\theta_{mn}^j(r)\right]r\,dr$$

$$-s_{mn}^2\int_{a_0}^{a_L}\left[\left(\rho\frac{m^2i_r^2}{r^2}+\rho\right)v_{sm}^j(r,t)v_{mn}^j(r)+\rho i_\phi^2\theta_{sm}^j(r,t)\theta_{mn}^j(r)\right]r\,dr \tag{11.33}$$

For discrete linear and rotary ring damping with dashpot constants c_i (force · time/length^2) and c_i^* (force · time) at $r = a_i$, use the point matrix of Table 11-14 for $r = a_i$ with k_1 replaced by $k_1 + sc_i$ and ω^2 by $-s^2$. Similar replacements apply for c_i^*. Also, in N_n set

$$\int_{a_0}^{a_L}\left[c'(v_{mn}^j)^2+c^*(\theta_{mn}^j)^2\right]r\,dr = \sum_i c_ia_i\left[v_{mn}^j(a_i)\right]^2+\sum_i c_i^*a_i\left[\theta_{mn}^j(a_i)\right]^2 \tag{11.34}$$

and in $\eta_n(0)$ set

$$\int_{a_0}^{a_L} \left[c' v_m^j(r,0) v_{mn}^j(r) + c^* \theta_m^j(r,0) \theta_{mn}^j(r) \right] r\, dr$$

$$= \sum_i c_i a_i v_m^j(a_i, 0) v_{mn}^j(a_i) + \sum_i c_i^* a_i \theta_m^j(a_i, 0) \theta_{mn}^j(a_i) \qquad (11.35)$$

Voigt-Kelvin Material The formulas of this chapter are applicable to a plate of Voigt-Kelvin material. In the equations of motion, Eqs. (11.18), replace E by $E(1 + \epsilon\, \partial/\partial t)$ where ϵ is the Voigt-Kelvin damping coefficient, and where similar behavior in dilation and shear is assumed. If the plate is resting on a proportional-damped foundation, retain the

$$c_\rho \left(\rho \frac{m^2 i_r^2}{r^2} + \rho \right) \frac{\partial v_m^j}{\partial t} \qquad \text{and} \qquad c_\rho \rho i_\phi^2 \frac{\partial \theta_m^j}{\partial t}$$

terms on the right-hand sides of Eqs. (11.18d) and (11.18c), respectively. The formulas of this section for proportional damping apply if c_ρ is replaced by $c_\rho + \epsilon(\omega_{mn})^2$ and ζ_n is redefined as $\zeta_n = c_\rho/(2\omega_{mn}) + \epsilon\omega_{mn}/2$. Continue to use the same undamped mode shapes employed previously, where ω_{mn} is the corresponding nth undamped frequency.

C. COMPUTER PROGRAMS AND EXAMPLES

11.19 Benchmark Examples

Because a series summation is involved, circular plates with unsymmetrical loads should be solved with the assistance of a computer program. It is also recommended that a computer program be used for plates of variable thickness. The following examples are provided as benchmark examples against which the reader's own program can be checked. The computer program CIRCULARPLATE was used for these examples. CIRCULARPLATE calculates the deflection, slope, bending moments, shear force, and twisting moment for static and steady-state conditions, the critical in-plane radial load and mode shape for stability, and the natural frequencies and mode shapes for transverse vibrations. The plate can be formed of ring segments of different thickness or material properties with any mechanical or thermal loading, including the in-span ring supports, foundations, and boundary conditions.

EXAMPLE 11.8 Static Response of Plate with Uniform Loading Find the static deflection, slope, bending moments, shear force, and twisting moment for a plate with a center hole. The plate is 0.0125 m thick and has a 0.025 m inner radius, 0.125 m outer radius. Both inner and outer rims are clamped. A uniform load of 13,790 N/m^2 is applied over the whole plate. Let $\nu = 0.3$, $E = 207$ GN/m^2.

The results for this symmetrically loaded plate are shown in Fig. 11-6.

RADIAL LOCATION	DISPL	SLOPE	MOMENT	SHEAR	PHI MOMENT	TWIST MOMENT
.025	0.	0.	1.8597E+01	1.3395E+03	5.5791E+00	0.
.050	6.549E-08	2.936E-06	-2.6571E+00	4.1100E+02	-7.9712E-01	0.
.075	9.825E-08	-5.373E-07	-5.4952E+00	-1.3497E+01	-1.6486E+00	0.
.100	4.922E-08	-2.819E-06	-7.3093E-01	-3.1200E+02	-2.1928E-01	0.
.125	-5.083E-21	-1.234E-19	9.5098E+00	-5.6010E+02	2.8529E+00	0.

Fig. 11-6 Partial output for Example 11.8.

EXAMPLE 11.9 Static Response of a Plate Subject to Unsymmetric Loading Find the static deflections, slope, moments, and shear force of a circular plate with a center hole. The plate is clamped on its inner edge and is free on its outer edge. It is 0.5 in thick and has a 20-in inner radius, 30-in outer radius. A concentrated load of 10,000 lb acts at the position $\phi = 0°$ of the outer edge of the plate. Let $\nu = 0.3$, $E = 3(10^7)$ lb/in^2. Print results at ϕ locations of 0° and 45°.

Some results are given in Fig. 11-7.

FOR PHI EQUAL TO 0.00000

RADIAL LOCATION	DISPL	SLOPE	MOMENT	SHEAR	PHI MOMENT	TWIST MOMENT
20.000	0.	0.	6.8413E+03	1.3753E+03	2.0524E+03	0.
25.000	1.774E-01	6.094E-02	2.3933E+03	9.6937E+02	-3.8082E+02	0.
30.000	5.490E-01	8.308E-02	-8.1464E-12	1.0080E+03	-2.5538E+03	0.

FOR PHI EQUAL TO 45.00000

RADIAL LOCATION	DISPL	SLOPE	MOMENT	SHEAR	PHI MOMENT	TWIST MOMENT
20.000	0.	0.	9.9846E+01	-7.8713E+01	2.9954E+01	0.
25.000	6.011E-03	2.530E-03	1.9834E+02	-3.2607E+00	1.2909E+02	1.3181E+02
30.000	2.292E-02	3.810E-03	1.8418E-11	1.2808E+02	1.6298E+00	7.1937E+01

Fig. 11-7 Some results for Example 11.9.

EXAMPLE 11.10 Plate with Variable Cross Section Calculate the response of the plate of Fig. 11-8. Some results are shown in Fig. 11-9.

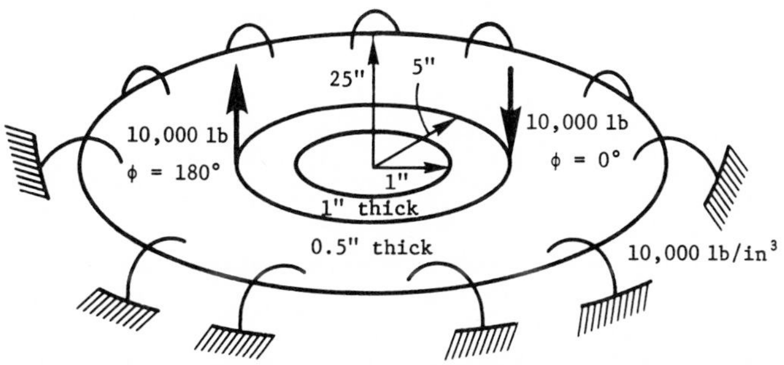

$E = 1(10^7)$ lb/in^2, $\nu = 0.3$

Fig. 11-8 Example 11.10.

FOR PHI EQUAL TO 0.00000

RADIAL LOCATION	DISPL	SLOPE	MOMENT	SHEAR	PHI MOMENT	TWIST MOMENT
1.000	0.	0.	1.6657E+04	2.0414E+04	4.9972E+03	0.
5.000	4.717E-02	1.610E-02	-9.3444E+02	-9.9025E+02	-5.2043E+02	8.0474E-05
25.000	2.678E-01	1.101E-02	1.1373E-03	1.0051E-04	1.3016E+01	-4.5671E-05

FOR PHI EQUAL TO 45.00000

RADIAL LOCATION	DISPL	SLOPE	MOMENT	SHEAR	PHI MOMENT	TWIST MOMENT
1.000	0.	0.	9.9813E+03	8.8388E+03	2.9944E+03	0.
5.000	2.493E-02	7.442E-03	1.5945E+02	7.4378E+00	2.0160E+02	2.9286E+02
25.000	1.795E-01	7.294E-03	8.0508E-04	7.1055E-05	7.7523E+00	5.1998E+01

Fig. 11-9 Partial output for Example 11.10.

EXAMPLE 11.11 Natural Frequencies of a Plate with No Center Hole Find the first three frequencies of a solid circular plate with a thickness of 0.5 in, a radius of 100 in, and a density of 0.28 lb/in^3. The plate is clamped on the outside edge. The material constants are $\nu = 0.3$, $E = 3(10^7)$ lb/in^2.

The first three frequencies of symmetrical modes are 5.10 Hz, 19.41 Hz, and 42.39 Hz. For the first three unsymmetrical modes, the frequencies are 11.24 Hz, 30.14 Hz, and 49.21 Hz.

EXAMPLE 11.12 Buckling of a Constant Stress Plate Find the force P that should be applied uniformly in compression throughout a plate in order to cause instability.

The critical force is found to be:

a_0/a_L	P_{cr}	*Mode number m corresponding to instability*
0	$(5.2)^2D/a_L^2$	0
0.02	$(5.9)^2D/a_L^2$	0
0.05	$(6.2)^2D/a_L^2$	0
0.1	$(6.7)^2D/a_L^2$	1
0.5	$(15.2)^2D/a_L^2$	4
0.7	$(20.3)^2D/a_L^2$	6

EXAMPLE 11.13 Buckling of a Plate with Inner Pressure Consider a plate with a center hole at which a pressure P is applied. Find the force P that will buckle the plate. Let $a_0/a_L = 0.4$.

The critical load is given below where C is clamped, S is simply supported, and F is free.

Support condition, inner-outer	P_{cr}
C-C	$(7.9)^2 D/a_L^2$
C-S	$(6.8)^2 D/a_L^2$
C-F	$(3.0)^2 D/a_L^2$
S-C	$(4.8)^2 D/a_L^2$
S-S	$(3.8)^2 D/a_L^2$
S-F	$(1.1)^2 D/a_L^2$
F-C	$(2.0)^2 D/a_L^2$
F-S	$(1.1)^2 D/a_L^2$

REFERENCES

11.1 Bares, R.: *Tables pour le calcul des dalles et des parois*, Dunod, Paris, 1969.
11.2 Leissa, A. W.: *Vibration of Plates*, NASA SP-160, 1969.

TWELVE

Rectangular Plates

This chapter treats the static, stability, and dynamic analyses of rectangular plates. The loading, which can be mechanical or thermal, is transverse to the plate. Response formulas for simply supported plates and fundamental stress formulas are handled in Part A. The formulas for more complex plates are provided in Part B. Included are stiffened orthotropic plates of variable cross section and composite materials. Because of the complexity of the solutions, the example problems provided in Part B are for quite simple loadings. However, the versatility of the formulas and methodology is demonstrated by the example problems in Part C where plates of variable thickness and with in-span supports are treated with a computer program.

The formulas of this chapter apply to plates represented by the classical (Kirchhoff) theory of plate bending. This theory is based on the assumptions that the plate thickness is small compared with the other dimensions, and the deflections are small compared with the thickness of the plate. The plate must be hinged on two opposite boundaries. In most cases the remaining two opposite edges can have any conditions.

A. SIMPLE PLATES

12.1 Notation and Conventions

v	Transverse deflection (length)
θ	Angle or slope of the deflection curve (radians)

M	Bending moment (per unit length) along an x edge (force)
V	"Equivalent" transverse force (per unit length) along an x edge (force/length)
h	Thickness of plate (length)
E	Modulus of elasticity of the material (force/length^2)
ν	Poisson's ratio
L	Length of plate in the x direction
L_y	Width of plate in the y direction
D	$Eh^3/12(1-\nu^2)$
D_x, D_y, D_{xy}	Material constants, flexural rigidities
w	Distributed transverse force, loading intensity (force/length^2)
$P = P_x$	Normal (in-plane, axial) force (per unit length) along an x edge (force/length)
P_y	Normal (in-plane, axial) force (per unit length) along a y edge (force/length)
W_T	Concentrated transverse force
W	Transverse line force (force/length)
C_T	Applied concentrated moment (force · length)
C	Applied line moment (force · length/length)
w_1	Magnitude of distributed force, uniform in x direction (force/length^2)
$\dfrac{\Delta w}{\Delta \ell}$	Gradient of distributed force, linearly varying in x direction (force/length^3)
M_y	Bending moment (per unit length) along a y edge (force)
M_{xy}	Twisting moment (per unit length) along an x or y edge (force)
V_y	"Equivalent" transverse force (per unit length) along a y edge (force/length)
Q_x	Transverse shear force (per unit length) along an x edge (force/length)
Q_y	Transverse shear force (per unit length) along a y edge (force/length)
σ_x, σ_y	Normal stresses (force/length^2)
τ_{xy}, τ_{xz}, τ_{yz}	Shear stresses (force/length^2)
x, y, z	Right-handed coordinate system
ρ	Mass per unit area (mass/length^2, force · time^2/length^3)
ω	Natural frequency (radians/sec)

Positive deflection v, internal bending moment M, and shear force V are displayed in Fig. 12-1.

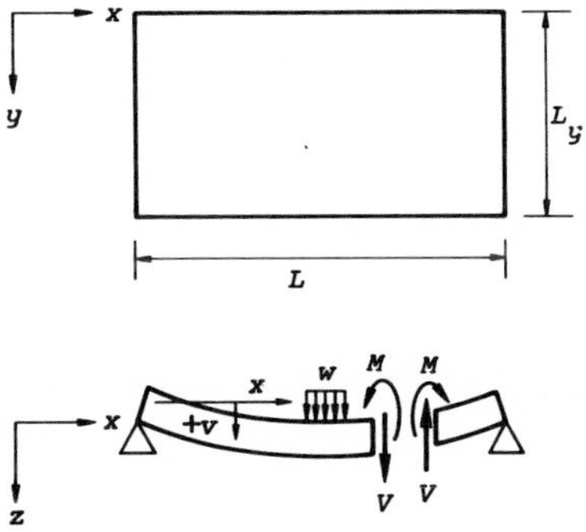

Fig. 12-1 Positive deflection v, internal bending moment M, and shear force V.

12.2 Stresses on a Cross Section

The tables of this chapter provide the deflections, slopes, bending moments, twisting moments, and transverse forces. Once the internal moments and forces are known the stresses can be derived using the formulas given in this section. The normal stresses in the x and y directions are

$$\sigma_x = \frac{Mz}{h^3/12} \qquad \sigma_y = \frac{m_y z}{h^3/12} \tag{12.1}$$

The shear stresses are

$$\begin{aligned} \tau_{xy} &= \frac{M_{xy} z}{h^3/12} \\ \tau_{xz} &= \frac{3Q_x}{2h}\left[1-\left(\frac{z}{h/2}\right)^2\right] \qquad \tau_{yz} = \frac{3Q_y}{2h}\left[1-\left(\frac{z}{h/2}\right)^2\right] \end{aligned} \tag{12.2}$$

12.3 Simply Supported Plate

The fundamental equations of motion for the bending of an isotropic, uniform rectangular plate are

$$\begin{aligned} &\frac{\partial^4 v}{\partial x^4} + 2\frac{\partial^4 v}{\partial x^2 \partial y^2} + \frac{\partial^4 v}{\partial y^4} = \frac{w}{D} \\ &M = -D\left(\frac{\partial^2 v}{\partial x^2} + \nu\frac{\partial^2 v}{\partial y^2}\right) \qquad M_y = -D\left(\frac{\partial^2 v}{\partial y^2} + \nu\frac{\partial^2 v}{\partial x^2}\right) \\ &M_{xy} = (1-\nu)D\frac{\partial^2 v}{\partial x \partial y} \\ &V = -D\left[\frac{\partial^3 v}{\partial x^3} + (2-\nu)\frac{\partial^3 v}{\partial x \partial y^2}\right] \qquad V_y = -D\left[\frac{\partial^3 v}{\partial y^3} + (2-\nu)\frac{\partial^3 v}{\partial x^2 \partial y}\right] \\ &Q_x = V + \frac{\partial M_{xy}}{\partial y} \qquad Q_y = V_y - \frac{\partial M_{yx}}{\partial x} \end{aligned} \tag{12.3}$$

If all edges of the plate are simply supported, the deflection, slopes, moments, and shear forces are given by

Deflection:

$$v = \sum_{n=1}^{\infty}\sum_{m=1}^{\infty} K_{nm} \sin\frac{n\pi x}{L}\sin\frac{m\pi y}{L_y} \tag{12.4a}$$

Slopes:

$$\begin{aligned} \theta &= -\sum_{n=1}^{\infty}\sum_{m=1}^{\infty} K_{nm}\frac{n\pi}{L}\cos\frac{n\pi x}{L}\sin\frac{m\pi y}{L_y} \\ \theta_y &= \sum_{n=1}^{\infty}\sum_{m=1}^{\infty} K_{nn}\frac{m\pi}{L_y}\sin\frac{n\pi x}{L}\cos\frac{m\pi y}{L_y} \end{aligned} \tag{12.4b}$$

Bending moments: $M = \pi^2 D \sum_{n=1}^{\infty} \sum_{m=1}^{\infty} K_{nm}\left[\left(\frac{n}{L}\right)^2 + \nu\left(\frac{m}{L_y}\right)^2\right] \sin\frac{n\pi x}{L} \sin\frac{m\pi y}{L_y}$

$$M_y = \pi^2 D \sum_{n=1}^{\infty} \sum_{m=1}^{\infty} K_{nm}\left[\left(\frac{m}{L_y}\right)^2 + \nu\left(\frac{n}{L}\right)^2\right] \sin\frac{n\pi x}{L} \sin\frac{m\pi y}{L_y} \qquad (12.4c)$$

Twisting moment: $M_{xy} = \pi^2 D(1-\nu) \sum_{n=1}^{\infty} \sum_{m=1}^{\infty} K_{nm} \frac{mn}{LL_y} \cos\frac{n\pi x}{L} \cos\frac{n\pi y}{L_y}$

(12.4d)

Shear forces:

$$V = \pi^3 D \sum_{n=1}^{\infty} \sum_{m=1}^{\infty} K_{nm}\left[\left(\frac{n}{L}\right)^3 + (2-\nu)\frac{n}{L}\left(\frac{m}{L_y}\right)^2\right] \cos\frac{n\pi x}{L} \sin\frac{m\pi y}{L_y}$$

$$V_y = \pi^3 D \sum_{n=1}^{\infty} \sum_{m=1}^{\infty} K_{nm}\left[\left(\frac{m}{L_y}\right)^3 + (2-\nu)\frac{m}{L_y}\left(\frac{n}{L}\right)^2\right] \sin\frac{n\pi x}{L} \cos\frac{m\pi y}{L_y} \qquad (12.4e)$$

In these formulas $K_{nm} = \dfrac{a_{nm}}{D\pi^4\left[(n^2/L^2) + (m^2/L_y^2)\right]^2}$ (12.5)

The parameters a_{nm} are given in Table 12-1 for various loadings. It is usually sufficiently accurate to include only a few terms in these expansions.

For concentrated loads such as cases 8, 9, 11, and 12 of Table 12-1, the series for the shear forces, Eqs. (12.4e), do not converge at the position of loading, for example, $x = a, y = b$. This indicates that the shear force becomes infinite when the area supporting the load approaches zero. In practice, however, the load is always supported by a small area which is not zero. It may be better to model the load by a rectangular uniform load such as case 3 of Table 12-1.

EXAMPLE 12.1 Plate with Uniformly Distributed Load Find the displacements, forces, and moments in a rectangular plate subject to a uniformly distributed load w_1. The sides are simply supported with lengths L and $L_y = 2L$.

The desired responses are given by Eqs. (12.4) with K_{nm} taken from Eq. (12.5) and Table (12-1). We find $a_{nm} = 16w_1/(\pi^2 nm)$, $n, m = 1, 3, 5, \ldots,$ and

$$K_{nm} = \frac{16w_1/(\pi^2 nm)}{D\pi^4\left[(n^2/L^2) + (m^2/L_y^2)\right]^2} = \frac{16w_1 L^4}{D\pi^6 nm\left[n^2 + m^2/4\right]^2} \qquad (1)$$

$n, m = 1, 3, 5, \ldots$

Thus, from Eq. (12.4a), the deflection is given by

$$v = \frac{16w_1 L^4}{D\pi^6} \sum_{n=1}^{\infty} \sum_{m=1}^{\infty} \frac{\sin(n\pi x/L)\sin(m\pi y/2L)}{nm\left[n^2 + m^2/4\right]^2} \qquad n, m = 1, 3, 5, \ldots \qquad (2)$$

TABLE 12-1 a_{nm} **for Eq. (12.5)**

Loading	a_{nm}
1. Uniform load w_1 over the whole plate	$\frac{16w_1}{\pi^2 nm} \quad n,m = 1,3,5,\ldots$
2. Linearly varying load 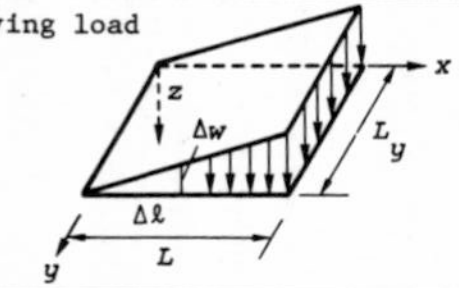	$(-1)^n \frac{8L}{mn\pi} \frac{\Delta w}{\Delta \ell} \quad m = 1,3,5,\ldots \quad n = 1,2,3,4,5,\ldots$
3. Uniform rectangular load 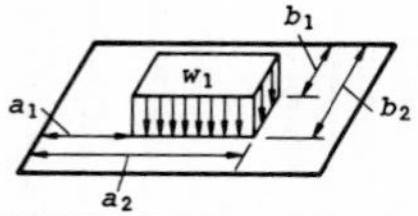	$\frac{4w_1}{mn\pi^2}\left(\cos\frac{n\pi a_1}{L} - \cos\frac{n\pi a_2}{L}\right)\left(\cos\frac{m\pi b_1}{L_y} - \cos\frac{m\pi b_2}{L_y}\right)$
4. 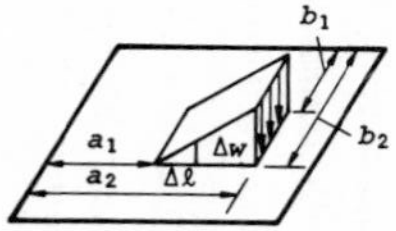	$\frac{4}{nm\pi}\frac{\Delta w}{\Delta \ell}\left(\cos\frac{m\pi b_1}{L_y} - \cos\frac{m\pi b_2}{L_y}\right)\left[(a_1 - a_2)\cos\frac{n\pi a_2}{L} + \frac{L}{m\pi}\left(\sin\frac{n\pi a_2}{L} - \sin\frac{n\pi a_1}{L}\right)\right]$
5. 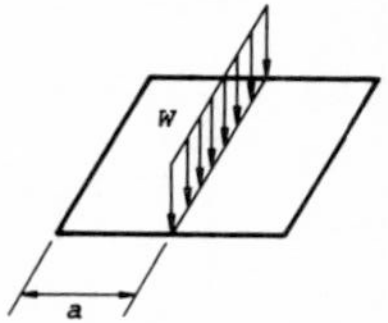	$\frac{8W}{\pi Lm}\sin\frac{n\pi a}{L} \quad m = 1,3,5,\ldots \quad n = 1,2,3,\ldots$ If this line load begins at $y = b_1$ and ends at $y = b_2$, then for a_{nm} use $\frac{4W}{m\pi L}\sin\frac{n\pi a}{L}\left(\cos\frac{m\pi b_1}{L_y} - \cos\frac{m\pi b_2}{L_y}\right)$
6. Linc load 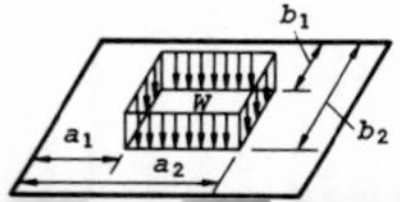	$\frac{4W}{LL_y}\left[\frac{L_y}{m\pi}\left(\sin\frac{n\pi a_1}{L} + \sin\frac{n\pi a_2}{L}\right)\left(\cos\frac{m\pi b_1}{L_y} - \cos\frac{m\pi b_2}{L_y}\right)\right.$ $\left. + \frac{L}{n\pi}\left(\cos\frac{n\pi a_1}{L} - \cos\frac{n\pi a_2}{L}\right)\left(\sin\frac{m\pi b_1}{L_y} + \sin\frac{m\pi b_2}{L_y}\right)\right]$

	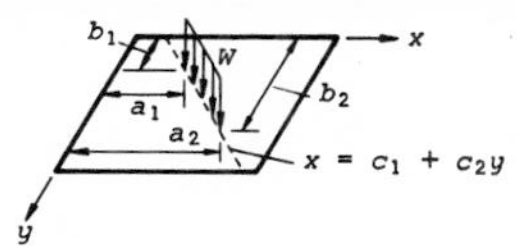	$\frac{2W}{LL_y c_3}\left\{\left[\sin\left(c_3 b_2 - \frac{n\pi c_1}{L}\right) - \sin\left(c_3 b_1 - \frac{n\pi c_1}{L}\right)\right] - \left[\sin\left(c_4 b_2 + \frac{n\pi c_1}{L}\right) - \sin\left(c_4 b_1 + \frac{n\pi c_1}{L}\right)\right]\right\}$ $c_3 = \frac{m\pi L - n\pi c_2 L_y}{LL_y}$ $\quad c_4 = \frac{m\pi L + n\pi c_2 L_y}{LL_y}$
8.	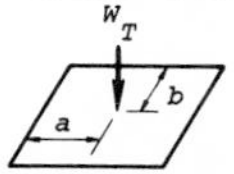	$\frac{4W_T}{LL_y}\sin\frac{n\pi a}{L}\sin\frac{m\pi b}{L_y}$
9.	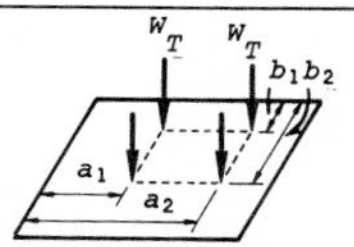	$\frac{4W_T}{LL_y}\left(\sin\frac{n\pi a_{\dot{1}}}{L} + \sin\frac{n\pi a_2}{L}\right)\left(\sin\frac{m\pi b_1}{L_y} + \sin\frac{m\pi b_2}{L_y}\right)$
10.	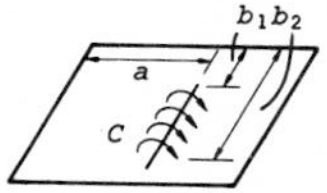	$-\frac{4nC}{mL^2}\cos\frac{n\pi a}{L}\left(\cos\frac{m\pi b_1}{L_y} - \cos\frac{m\pi b_2}{L_y}\right)$
11.	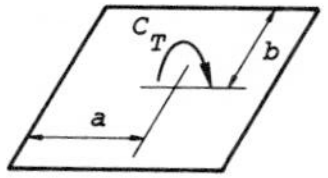	$-\frac{4nC_T}{L^2 L_y}\cos\frac{n\pi a}{L}\sin\frac{m\pi b}{L_y}$
12.	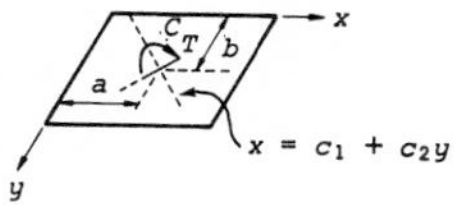	$\frac{4\pi C_T}{LL_y}\frac{\left(\frac{c_2 m}{L_y}\cos\frac{m\pi b}{L_y}\sin\frac{m\pi a}{L} + \frac{n}{L}\sin\frac{m\pi b}{L_y}\cos\frac{n\pi a}{L}\right)}{(1 + c_2^2)^{1/2}}$

The maximum deflection occurs at the center, $x = L/2, y = L$. From (2),

$$v_{\max} = \frac{16w_1L^4}{D\pi^6}(0.640 - 0.032 - 0.004 + 0.004 + \cdots) \approx 0.0101 w_1L^4/D$$

Note that the deflection series converges rapidly so that summation of two terms provides accuracy sufficient for practical purposes.

The maximum moments are found in a similar fashion. They too occur at $x = L/2, y = L$. Examination of the moment expressions shows that they converge more slowly than the deflection expansion. At the center, four terms provide sufficient accuracy. More than four terms are required as the moments are computed closer to the edges.

The shear forces are determined by placing (1) in Eq. (12.4*e*). The reactions along an edge can be found from the resulting expression. For example, from Eq. (12.4*e*) at $x = 0$, the reaction along the $x = 0$ edge is

$$V|_{x=0} = \frac{16w_1L}{\pi^3}\sum_{n=1}^{\infty}\sum_{m=1}^{\infty}\frac{n^2 + (2-\nu)(m^2/4)}{m[n^2 + m^2/4]^2}\sin\frac{m\pi y}{2L} \quad n, m = 1, 3, 5, \ldots$$

12.4 Stability

The critical in-plane (axial) force $(P_y)_{cr}$ in the y direction of a plate is given by (Ref. 12.3)

Isotropic:	$(P_y)_{cr} = K\dfrac{\pi^2 D}{L^2}$ or $(\sigma_y)_{cr} = \dfrac{(P_y)_{cr}}{h}$	(12.6*a*)
Orthotropic:	$(P_y)_{cr} = K\dfrac{\pi^2\sqrt{D_x D_y}}{L^2}$	(12.6*b*)

where K is listed in Table 12-2 for various boundary and in-plane force conditions. The number of half waves in the y and x directions are m and n, respectively. Select values of m and n of 1, 2, or 3 so that $(P_y)_{cr}$ assumes the least possible value. In these tables the aspect ratio is given by $\alpha = L_y/L$. Also, in case 1 the ratio P_x/P_y is assumed to be a prescribed constant for a particular problem.

EXAMPLE 12.2 Buckling Load for Uniform Plate Find the critical in-plane load $(P_y)_{cr}$ in a simply supported isotropic plate for which L_y is less than L. Also, $P_x = 0$.

The critical load is given by Eq. (12.6*a*) with (from case 1, Table 12-2)

$$K = \left(\frac{m}{\alpha} + \frac{\alpha n^2}{m}\right)^2 \tag{1}$$

TABLE 12-2 Constants for Stability Formulas (Eqs. 12.6)

Definition: $\alpha = L_y/L$

Plate	K	
	Isotropic	Orthotropic, take B, D_x, D_y from Table 12-4
1. All sides simply supported 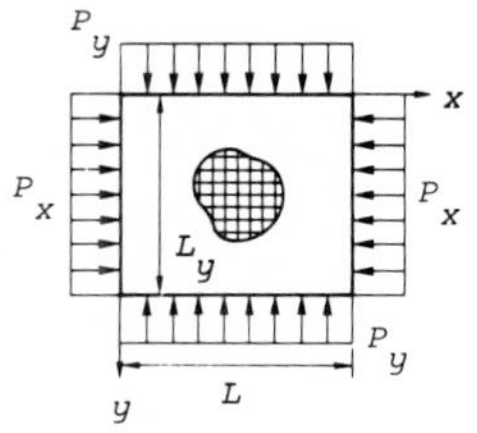	$\dfrac{\left(\dfrac{m}{\alpha} + \dfrac{\alpha n^2}{m}\right)^2}{1 + \dfrac{P_x}{P_y}\left(\dfrac{\alpha n}{m}\right)^2}$	$\dfrac{\sqrt{\dfrac{D_y}{D_x}}\left(\dfrac{m}{\alpha}\right)^2 + \dfrac{2Bn^2}{\sqrt{D_x D_y}} + \sqrt{\dfrac{D_x}{D_y}}\left(\dfrac{\alpha n^2}{m}\right)^2}{1 + \dfrac{P_x}{P_y}\left(\dfrac{\alpha n}{m}\right)^2}$
2. Two sides simply supported ($y = 0, L_y$) and two sides fixed ($x = 0, L$) 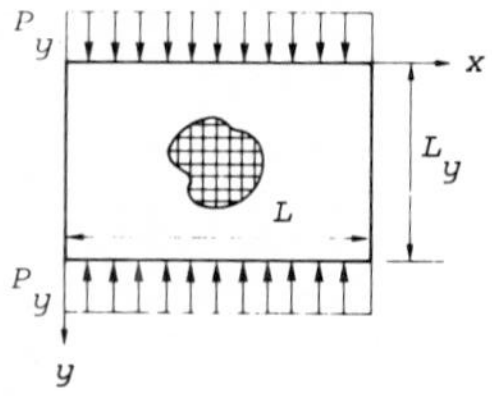	$\left(\dfrac{m}{\alpha}\right)^2 + \dfrac{8}{3}n^2 + \dfrac{16}{3}\left(\dfrac{\alpha n^2}{m}\right)^2$	$\sqrt{\dfrac{D_y}{D_x}}\left(\dfrac{m}{\alpha}\right)^2 + \dfrac{8Bn^2}{3\sqrt{D_x D_y}} + \dfrac{16}{3}\sqrt{\dfrac{D_x}{D_y}}\left(\dfrac{\alpha n^2}{m}\right)^2$

For $n = 1$ and $\alpha < 1$, the second term of (1) is always less than the first term. Then the minimum value of K[and $(P_y)_{cr}$] is obtained by letting $m = 1$. Thus,

$$(P_y)_{cr} = \frac{D\pi^2}{L^2}\left(\frac{1}{\alpha} + \alpha\right)^2$$

12.5 Free Dynamic Response—Natural Frequencies

Isotropic:

The natural frequencies ω_{mn}, $m, n = 1, 2, 3, \ldots,$ for a uniform plate which is simply supported on the $y = 0, L_y$ edges are given by

$$\omega_{mn} = \frac{k_{mn}}{L_y^2}\sqrt{\frac{D}{\rho}} \tag{12.7a}$$

where k_{mn} are given in Table 12-3*a*.

Orthotropic:

The natural frequencies of an orthotropic plate having either simply supported or clamped sides are given by (from Ref. 12.5)

$$\omega_{mn} = \frac{1}{L_y^2}\sqrt{\frac{1}{\rho}\left[D_x(\alpha k_1)^4 + 2D_{xy}\alpha^2 k_3 + D_y(k_2)^4\right]} \tag{12.7b}$$

where $\alpha = L_y/L$ and k_1, k_2, k_3 are given in Table 12-3*b*. See Table 12-4 for expressions for D_x, D_y, D_{xy} for a variety of rectangular plate types.

B. COMPLEX PLATES

12.6 Notation for Complex Plates

k	Modulus of elastic foundation (force/length^3)
k_i	Line extension spring constant (force/length^2)
k_i^*	Line rotary spring constant (force · length/length)
D_x, D_y, D_{xy}	Flexural rigidities; see Table 12-4.
ν_x, ν_y	Poisson's ratios in the x, y directions
E_x, E_y	Moduli of elasticity of the material in x, y directions (force/length^2)
α_x, α_y	Coefficients of thermal expansion in the x, y directions (length/length · degree). For isotropic material $\alpha_x = \alpha_y = \alpha$.
ΔT	Temperature change (degrees), that is, the temperature rise with respect to the reference temperature.

$$M_{Tx} = \int_{-h/2}^{+h/2} \frac{E_x(\alpha_x + \nu_y\alpha_y)}{(1 - \nu_x\nu_y)}\,\Delta T\, z\, dz$$

$$M_{Ty} = \int_{-h/2}^{+h/2} \frac{E_y(\alpha_y + \nu_x\alpha_x)}{(1 - \nu_x\nu_y)}\,\Delta T\, z\, dz$$

TABLE 12-3a Constants for Natural Frequency Formula for Isotropic Plates, Eq. (12.7a)

Definitions:
(a) C: clamped   S: simply supported
    F: free
(b) $\alpha = L_y/L$

Edges at $x = 0,L$	Value of k_{mn}	Source (Ref.)
S,S	$\pi^2(m^2 + \alpha^2 n^2)$	
C,C	$k_{11} = \pi^2\sqrt{1 + 2.5\alpha^2 + 5.14\alpha^4}$, $\quad k_{21} = 4\pi^2\sqrt{1 + 0.625\alpha^2 + 0.321\alpha^4}$, $\quad k_{12} = \pi^2\sqrt{1 + 9.32\alpha^2 + 39.06\alpha^4}$	12.4
C,S	$k_{11} = \pi^2\sqrt{1 + 2.33\alpha^2 + 2.44\alpha^4}$, $\quad k_{12} = 4\pi^2\sqrt{1 + 0.582\alpha^2 + 0.152\alpha^4}$, $\quad k_{21} = \pi^2\sqrt{1 + 8.69\alpha^2 + 25.63\alpha^4}$	12.4
S,F	$\alpha = 1.0 \quad 1.5 \quad 2.0 \quad 2.2$ $k_{11} = 12.9 \quad 17.2 \quad 23.2 \quad 26.2$	12.2
F,F	<table><tr><td>α \ k_{mn}</td><td>k_{11}</td><td>k_{12}</td><td>k_{21}</td><td>k_{22}</td></tr><tr><td>0.5</td><td>9.87</td><td>11.6</td><td>39.48</td><td>41.18</td></tr><tr><td>0.75</td><td>9.87</td><td>13.71</td><td>39.48</td><td>43.56</td></tr><tr><td>1.0</td><td>9.87</td><td>16.13</td><td>39.48</td><td>46.73</td></tr><tr><td>1.5</td><td>9.87</td><td>21.26</td><td>39.48</td><td>54.84</td></tr><tr><td>2.0</td><td>9.87</td><td>27.52</td><td>39.48</td><td>64.54</td></tr></table> $(\nu = 0.3)$	12.2

TABLE 12-3*b* Constants for Natural Frequency Formula for Orthotropic Plates, Eq. (12.7*b*)

Definitions: Except for the case of *S*, *S*, *S*, *S* edges, the values of k_1, k_2, k_3 are approximate.

(a) *C*: clamped *S*: simply supported (b) $\gamma_0 = n\pi$, $\gamma_1 = (n + 1/4)\pi$, $\gamma_2 = (n + 1/2)\pi$, $\gamma_3 = m\pi$, $\gamma_4 = (m + 1/4)\pi$, $\gamma_5 = (m + 1/2)\pi$

Edges at $x = 0, L$	Edges at $y = 0, L_y$	n	m	k_1	k_2	k_3
C,*C*	*C*,*C*	1 1 2,3,4,.. 2,3,4,..	1 2,3,4,.. 1 2,3,4,..	4.730 4.730 γ_2 γ_2	4.730 γ_5 4.730 γ_5	151.3 $12.30\gamma_5(\gamma_5 - 2)$ $12.30\gamma_2(\gamma_2 - 2)$ $\gamma_2\gamma_5(\gamma_2 - 2)(\gamma_5 - 2)$
C,*C*	*C*,*S*	1,2,3,.. 1,2,3,..	1 2,3,4,..	4.730 γ_2	γ_4 γ_4	$12.30\gamma_4(\gamma_4 - 1)$ $\gamma_2\gamma_4(\gamma_2 - 2)(\gamma_4 - 1)$
C,*C*	*S*,*S*	1 2,3,4,..	1,2,3,.. 1,2,3,..	4.730 γ_2	γ_3 γ_3	$12.30\gamma_3^2$ $\gamma_2\gamma_3^2(\gamma_2 - 2)$
C,*S*	*S*,*C*	1,2,3,..	1,2,3,..	γ_1	γ_4	$\gamma_1\gamma_4(\gamma_1 - 1)(\gamma_4 - 1)$
C,*S*	*S*,*S*	1,2,3,..	1,2,3,..	γ_1	γ_3	$\gamma_1\gamma_3^2(\gamma_1 - 1)$
S,*S*	*S*,*S*	1,2,3,..	1,2,3,..	γ_0	γ_3	$\gamma_0^2\gamma_3^2$

TABLE 12-4 Material Constants

General Definitions: $B = (D_x \nu_y + D_y \nu_x + 4D_{xy})/2, \quad D_{xy} = (2B - D_x \nu_y - D_y \nu_x)/4$

Plate	Constants
1. Homogeneous isotropic	$\nu_x = \nu_y = \nu, \quad D_x = D_y = D, \quad D_{xy} = (1 - \nu)D/2, \quad \alpha_x = \alpha_y = \alpha$ $D = Eh^3/[12(1 - \nu^2)], \; B = D$
2. Homogeneous orthotropic	$D_x = \frac{E_x h^3}{12(1 - \nu_x \nu_y)}, \quad D_y = \frac{E_y h^3}{12(1 - \nu_x \nu_y)}, \quad D_{xy} = \frac{Gh^3}{12}$
3. Isotropic plate with equidistant stiffeners in one direction (Ref. 12.1) 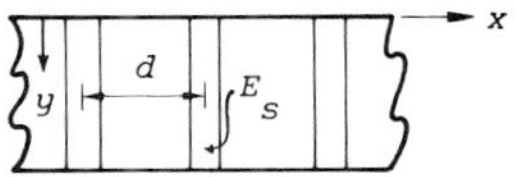 	$\nu_x = \nu_y = \nu, \quad B = \frac{Eh^3}{12(1 - \nu^2)}$
Stiffeners on two sides 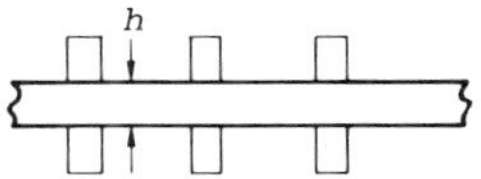 	$D_x = \frac{Eh^3}{12(1 - \nu^2)}, \quad D_y = D_x + \frac{E_s I_s}{d}$ I_s is the moment of inertia of a stiffener taken about the middle axis of the cross section of the plate.
Stiffeners on one side 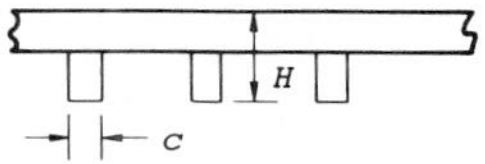 	$D_x = \frac{Eh^3 d}{12[d - c + c(h/H)^3]}, \quad D_y = \frac{EI_s}{d}, \quad D_{xy} = D'_{xy} + \frac{GJ}{2d}$ GJ is the torsional rigidity of a rib. D'_{xy} is the torsional rigidity of the slab without the ribs. I_s is the moment of inertia of a T section of width d.

TABLE 12-4 Material Constants (*Continued*)

4. Isotropic plate with equidistant stiffeners in two directions (Ref. 12.1) The axes of the ribs are parallel to the principal directions.	$\nu_x = \nu_y = \nu, \quad D_x = \dfrac{Eh^3}{12(1-\nu^2)} + \dfrac{E_1 I_1}{d_1}, \quad D_y = \dfrac{Eh^3}{12(1-\nu^2)} + \dfrac{E_2 I_2}{d_2}, \quad D_{xy} = \dfrac{Eh^3}{12(1-\nu^2)}$ I_1 is the moment of inertia about the plate's middle surface of stiffener in the x direction, E_1 is the modulus of elasticity of this stiffener, d_1 is the spacing of these stiffeners. I_2, E_2, d_2 are the corresponding constants for stiffeners lying in the y direction.
5. Corrugated plate (Ref. 12.1) 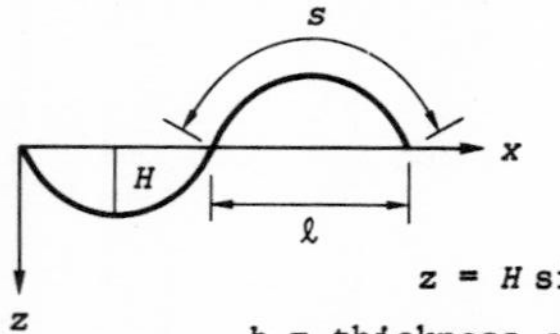 $z = H \sin \pi x/\ell$ h = thickness of sheet $s = \ell[1 + \pi^2 H^2/(4\ell^2)]$	$\nu_x = \nu_y = \nu, \quad D_x = \dfrac{\ell}{s}\dfrac{Eh^3}{12(1-\nu^2)}, \quad D_y = EI, \quad B = \dfrac{s}{\ell}\dfrac{Eh^3}{12(1+\nu)}$ I is the mean moment of inertia in the x,y plane per unit length. $I = 0.5hH^2(1 - 0.81/C_1), \quad C_1 = 1 + 2.5(H/2\ell)^2$
6. Open gridworks (Ref. 12.2) 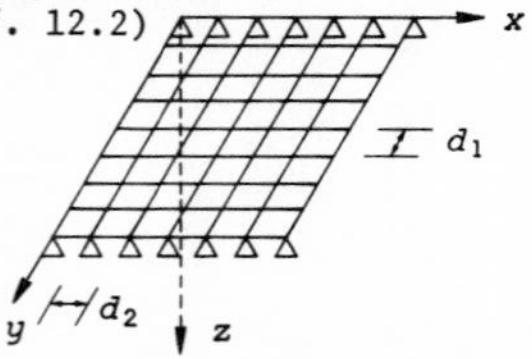 	$\nu_x = \nu_y = \nu, \quad D_x = \dfrac{EI_1}{d_1}, \quad D_y = \dfrac{EI_2}{d_2}, \quad B = \dfrac{GJ_1}{2d_1} + \dfrac{GJ_2}{2d_2}, \quad D_{xy} = \sqrt{D_x D_y}$ GJ_1, GJ_2 are torsional rigidities of the beams parallel to the x and y axes.
7. Concrete slab with steel reinforcement bars in both x and y directions (Ref. 12.2)	$\nu_x = \nu_y = \nu, \quad D_x = \dfrac{E_c}{1-\nu_c^2}\left[I_{cx} + \left(\dfrac{E_s}{E_c} - 1\right) I_{sx}\right], \quad D_y = \dfrac{E_c}{1-\nu_c^2}\left[I_{cy} + \left(\dfrac{E_s}{E_c} - 1\right) I_{sy}\right]$ $D_{xy} = (1-\nu_c)\sqrt{D_x D_y}/2$

ν_c is Poisson's ratio for concrete, E_c, E_s are the

I_{cx}, I_{cy} are moments of inertia of the slab material about the neutral axis in a section where x = constant, y = constant.

I_{sx}, I_{sy} are moments of inertia of the reinforcement bars about the neutral axis in a section where x = constant, y = constant.

8. Concrete slab stiffened by concrete ribs (Ref. 12.2)

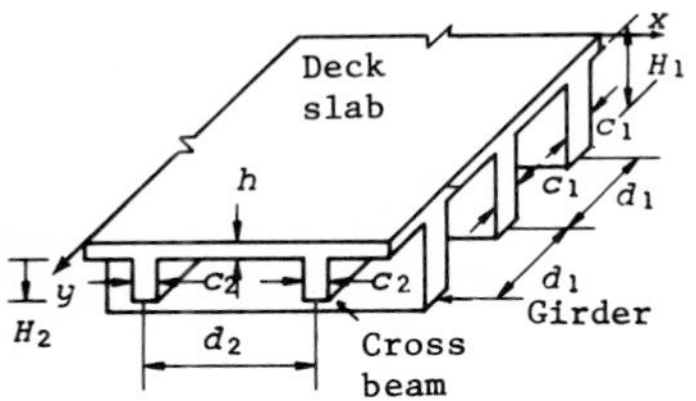

$$D_x = \frac{EI_x}{d_1}, \quad D_y = \frac{EI_y}{d_2}, \quad B = \frac{Eh^3}{12(1-\nu_x\nu_y)} + \frac{G}{2}\left(\frac{H_1c_1^3\alpha_1k_1}{d_1} + \frac{H_2c_2^3\alpha_2k_2}{d_2}\right)$$

I_x, I_y are moments of inertia of the section for x,y = constant, with respect to the neutral axis

k_i are reduction factors, depending on c_i/H_i, which are inserted to reduce the torsional rigidities of the concrete ribs after cracks have developed. Values of α_i are provided for some values of c_i/H_i.

c_i/H_i	1.0	1.2	1.5	2.0	2.5	3.0	4.0	6.0	8.0	10.0	∞
α_i	0.140	0.166	0.196	0.229	0.249	0.263	0.281	0.299	0.307	0.313	0.333

9. Steel deck plate (Ref. 12.2)

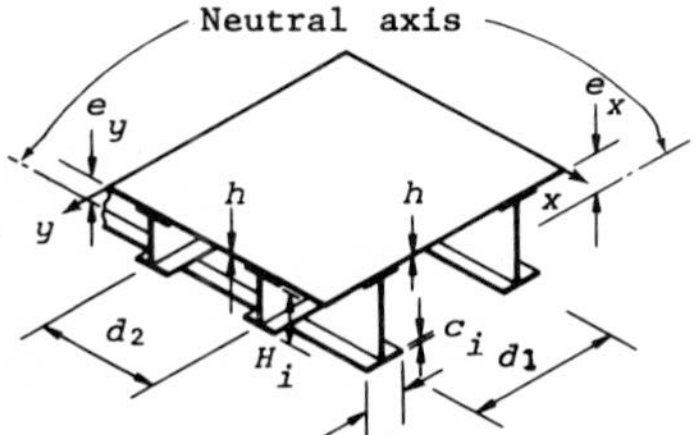

web thickness = c_i H_i

$$D_x = \frac{Eh^3}{12(1-\nu_x\nu_y)} + \frac{Ehe_x^2}{(1-\nu_x\nu_y)} + \frac{EI_x}{d_1}, \quad D_y = \frac{Eh^3}{12(1-\nu_x\nu_y)} + \frac{Ehe_y^2}{(1-\nu_x\nu_y)} + \frac{EI_y}{d_2}$$

$$B = \frac{Eh^3}{12(1-\nu_x\nu_y)} + \frac{G(\Sigma H_i c_i^3)}{6}\left(\frac{1}{d_1} + \frac{1}{d_2}\right)$$

I_x, I_y are the moments of inertia of the stiffeners with respect to their neutral axes in the x,y directions.

e_x, e_y are the distances from the middle plane of the plate to the neutral axes.

TABLE 12-4 Material Constants (*Continued*)

10. Composite (concrete-steel) slab (Ref. 12.2) 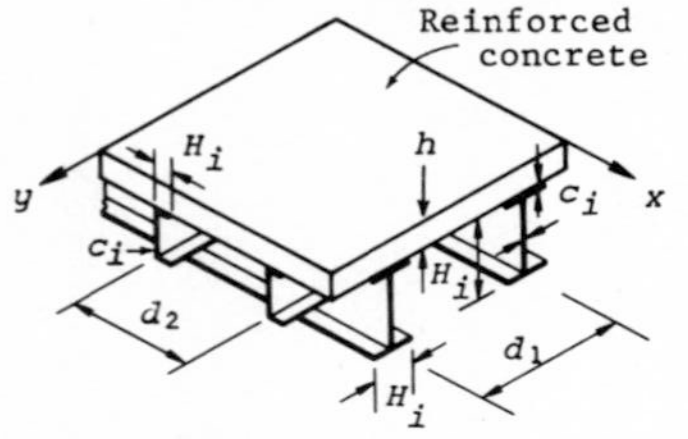	Use the formulas for a steel deck plate. First transform the concrete part of the slab into an equivalent steel plate.
11. Continuously composite, isotropic	$\nu_x = \nu_y = \nu, \quad D_x = D_y = D, \quad D_{xy} = (1 - \nu)D/2$ $D = \frac{1}{1 - \nu^2} \int_{-h/2}^{h/2} Ez^2\, dz$
12. Continuously composite, orthotropic	$D_x = \frac{1}{(1 - \nu_x \nu_y)} \int_{-h/2}^{h/2} E_x z^2\, dz, \quad D_y = \frac{1}{1 - \nu_x \nu_y} \int_{-h/2}^{h/2} E_y z^2\, dz, \quad D_{xy} = \int_{-h/2}^{h/2} G z^2\, dz$
13. Multiple isotropic layers (Ref. 12.2) 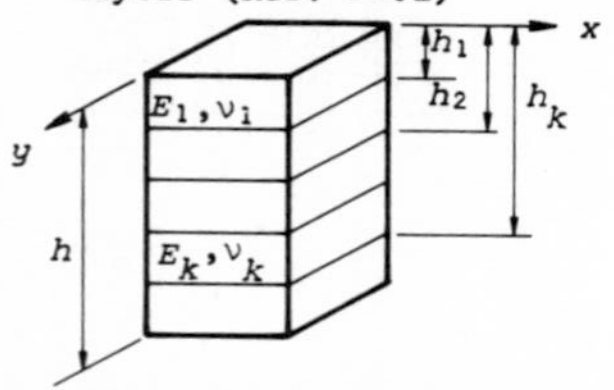	$\nu_x = \nu_y = \nu, \quad D_x = D_y = D$ $D = (Q_1 C - Q_2^2)/Q_1$ $Q_1 = \sum_{k=1} \frac{E_k}{1 - \nu_k^2}(h_k - h_{k-1}), \quad Q_2 = \frac{1}{2}\sum_{k=1} \frac{E_k}{1 - \nu_k^2}(h_k^2 - h_{k-1}^2)$ $C = \frac{1}{3}\sum_{k=1} \frac{E_k}{1 - \nu_k^2}(h_k^3 - h_{k-1}^3)$

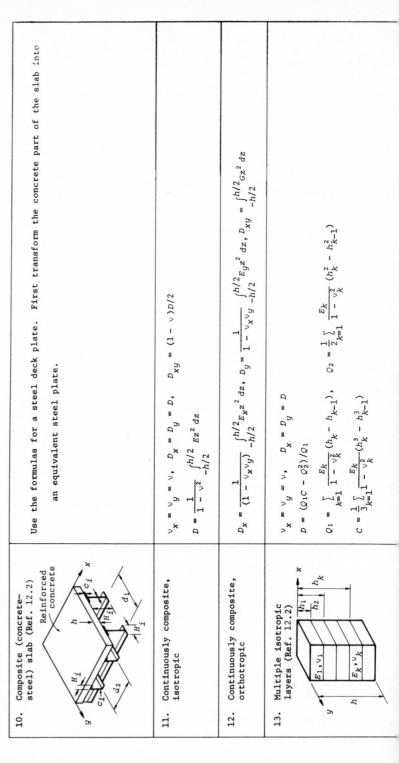

14. Symmetrically constructed with isotropic layers (Ref. 12.3) 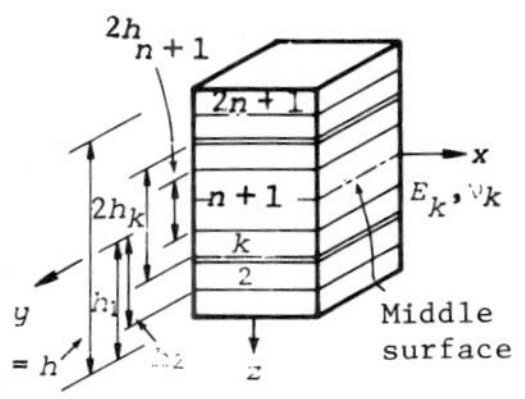 	$\nu_x = \nu_y = \nu, \quad D_x = D_y = D, \quad D_{xy} = (1 - \nu)D/2,$ $D = \frac{2}{3}\sum_{m=1}^{n+1} C_m, \quad \nu = \frac{2}{3D}\sum_{m=1}^{n+1} \nu_m C_m$ $C_m = \frac{E_m(h_m^3 - h_{m+1}^3)}{1 - \nu_m^2}$ with h_{n+2} set equal to zero E_m, ν_m are the Young's modulus and Poisson's ratio for the mth layer. This plate is constructed of an odd number of homogeneous layers symmetrically located about the middle layer.
15. Symmetrically constructed with orthotropic layers (Ref. 12.3). The figure and description for the plate with isotropic layers applies.	$\nu_x = \frac{2}{3D_y}\sum_{m=1}^{n+1} \nu_{ym}, \quad \nu_y = \nu_x D_y/D_x, \quad D_x = \frac{2}{3}\sum_{m=1}^{n+1} C_{xm}, \quad D_y = \frac{2}{3}\sum_{m=1}^{n+1} C_{ym},$ $D_{xy} = \frac{2}{3}\left[\sum_{m=1}^{n+1} G_m(h_m^3 - h_{m+1}^3) + G_{n+1}h_{n+1}^3\right], \quad C_{xm} = \frac{E_{xm}(h_m^3 - h_{m+1}^3)}{1 - \nu_{xm}\nu_{ym}},$ $C_{ym} = \frac{E_{ym}(h_m^3 - h_{m+1}^3)}{1 - \nu_{xm}\nu_{ym}}$ with h_{n+2} set equal to zero
16. Symmetrically constructed with identical orthotropic layers (Ref. 12.3). The figure and description for the plate with isotropic layers applies.	$\nu_x = \frac{2(2n+1)^3 \nu_{yn}}{Q_2}, \quad \nu_y = \frac{2(2n+1)^3 \nu_{yn}}{Q_1}, \quad D_x = \frac{E_x' h^3}{12(1 - \nu_x\nu_y)}, \quad D_y = \frac{E_y' h^3}{12(1 - \nu_x\nu_y)}, \quad D_{xy} = \frac{Gh^3}{12}$ $E_x' = \frac{E_x}{2(2n+1)^3}\,\frac{1 - \nu_x\nu_y}{1 - \nu_{xn}\nu_{yn}}\,Q_1, \quad E_y' = \frac{E_y}{2(2n+1)^3}\,\frac{1 - \nu_x\nu_y}{(1 - \nu_{xn}\nu_{yn})E_y/E_x}\,Q_2$ $Q_1 = (2n+1)^3(1 + E_y/E_x) + [3(2n+1)^2 - 2](1 - E_y/E_x),$ $Q_2 = (2n+1)^3(1 + E_y/E_x) - [3(2n+1)^2 - 2](1 - E_y/E_x)$
17. Plywood (Ref. 12.1) The x axis is parallel the face grain.	$\nu_x = E''/E_y', \quad \nu_y = E''/E_x',$ $D_x = E_x' h^3/12, \quad D_y = E_y' h^3/12,$ $D_{xy} = Gh^3/12$

	E_x'	E_y'	E''	G
Maple, 5-ply	1.87	0.60	0.073	0.159
Afara, 3-ply	1.96	0.165	0.043	0.110
Gaboon (Okoume), 3-ply	1.28	0.11	0.014	0.085
Birch, 3- and 5-ply	2.00	0.167	0.077	0.17
Birch with bakelite membranes	1.70	0.85	0.061	0.10

t Time

ϵ_m Constant (Table 12-8) that accounts for the variation of distributed loads in the y direction.

$B = (D_x \nu_y + D_y \nu_x + 4D_{xy})/2$

$G \approx \dfrac{\sqrt{E_x E_y}}{2\left(1+\sqrt{\nu_x \nu_y}\right)}$ Shear modulus of elasticity (force/length²)

c' External or viscous damping coefficient (mass/time · length², force · time/length³)

c_ρ Proportional viscous damping coefficient (1/time). If c' is chosen to be proportional to the mass, that is, $c' = c_\rho \rho$, then c_ρ is the constant of proportionality.

c_i Discrete line damping (dashpot) constant (force · time/length²)

M_i Line lumped mass (mass/length)

P_{Li} Line lumped in-plane force (force · length/length). This can be calculated as $P_{Li} = \Delta a P$, where Δa is a small x segment of plate with in-plane force P.

ℓ Length of segment, span of transfer matrix

c_1 Magnitude of distributed moment, uniform in x direction (force · length/length²)

M_{T1} Magnitude of distributed thermal moment, uniform in x direction

$\dfrac{\Delta c}{\Delta \ell}$ Gradient of distributed moment, linearly varying in x direction (force · length/length³)

$\dfrac{\Delta M_T}{\Delta \ell}$ Gradient of distributed thermal moment, linearly varying in x direction

Ω Frequency of steady-state forces and responses (radians/time)

$\mathbf{U}_i$ Field matrix of the ith segment

$\overline{\mathbf{U}}_i$ Point matrix at $x = a_i$

The notation for transfer matrices is

$$\mathbf{U}_i = \begin{bmatrix} U_{vv} & U_{v\theta} & U_{vM} & U_{vV} & F_v \\ U_{\theta v} & U_{\theta\theta} & U_{\theta M} & U_{\theta V} & F_\theta \\ U_{Mv} & U_{M\theta} & U_{MM} & U_{MV} & F_M \\ U_{Vv} & U_{V\theta} & U_{VM} & U_{VV} & F_V \\ 0 & 0 & 0 & 0 & 1 \end{bmatrix}$$

12.7 Differential Equations

The fundamental equations of motion for the bending of a composite orthotropic rectangular plate with compressive in-plane forces P, P_y are

$$M = M_x = -D_x\left(\frac{\partial^2 v}{\partial x^2} + \nu_y \frac{\partial^2 v}{\partial y^2}\right) - M_{Tx} \qquad M_y = -D_y\left(\frac{\partial^2 v}{\partial y^2} + \nu_x \frac{\partial^2 v}{\partial x^2}\right) - M_{Ty}$$

$$M_{xy} = 2D_{xy}\frac{\partial^2 v}{\partial x \partial y} = -M_{yx} \tag{12.8}$$

$$Q_x = \frac{\partial M_x}{\partial x} + \frac{\partial M_{yx}}{\partial y} \qquad Q_y = \frac{\partial M_y}{\partial y} - \frac{\partial M_{xy}}{\partial x}$$

$$\frac{\partial Q_x}{\partial x}+\frac{\partial Q_y}{\partial y}-P_y\frac{\partial^2 v}{\partial y^2}-P\frac{\partial^2 v}{\partial x^2}=-w+kv+\rho\frac{\partial^2 v}{\partial t^2}$$

$$V=V_x=Q_x-\frac{\partial M_{xy}}{\partial y}\qquad V_y=Q_y+\frac{\partial M_{yx}}{\partial x}$$

Frequently these equations are combined to form a fourth-order relationship

$$D_x\frac{\partial^4 v}{\partial x^4}+2B\frac{\partial^4 v}{\partial x^2\partial y^2}+D_y\frac{\partial^4 v}{\partial y^4}+P\frac{\partial^2 v}{\partial x^2}+P_y\frac{\partial^2 v}{\partial y^2}+kv+\rho\frac{\partial^2 v}{\partial t^2}$$

$$=w-\left(\frac{\partial^2 M_{Tx}}{\partial x^2}+\frac{\partial^2 M_{Ty}}{\partial y^2}\right)\tag{12.9}$$

where $B=(D_x\nu_y+D_y\nu_x+4D_{xy})/2$.

In first-order form the equations for the variables v, θ, M, and V become

$$\frac{\partial v}{\partial x}=-\theta\tag{12.10a}$$

$$\frac{\partial \theta}{\partial x}=\nu_y\frac{\partial^2 v}{\partial y^2}+\frac{M}{D_x}+\frac{M_{Tx}}{D_x}\tag{12.10b}$$

$$\frac{\partial M}{\partial x}=-4D_{xy}\frac{\partial^2\theta}{\partial y^2}+V\tag{12.10c}$$

$$\frac{\partial V}{\partial x}=D_y(1-\nu_x\nu_y)\frac{\partial^4 v}{\partial y^4}+(P_y-\nu_y P)\frac{\partial^2 v}{\partial y^2}-\frac{1}{D_x}\left(P+\nu_x D_y\frac{\partial^2}{\partial y^2}\right)M$$

$$+kv+\rho\frac{\partial^2 v}{\partial t^2}-\frac{1}{D_x}\left(P+\nu_x D_y\frac{\partial^2}{\partial y^2}\right)M_{Tx}+\frac{\partial^2 M_{Ty}}{\partial y^2}-w(x,y,t)\tag{12.10d}$$

The in-plane forces $P=P_x$, P_y are in compression. These equations are appropriate for a plate with tensile in-plane forces if P and P_y are replaced by $-P$ and $-P_y$.

Stresses on a Cross Section The formulas of this chapter yield displacements, internal forces, and moments. These variables can be placed in the formulas of Table 12-5 to compute the stresses.

Elimination of y Derivatives The y derivatives in the response equations are eliminated by expanding the variables v, θ, M, V in a sine series.

$$v(x,y,t)=\sum_{m=1}^{\infty}v_m(x,t)\sin\frac{m\pi y}{L_y}\tag{12.11a}$$

$$\theta(x,y,t)=\sum_{m=1}^{\infty}\theta_m(x,t)\sin\frac{m\pi y}{L_y}\tag{12.11b}$$

$$M(x,y,t)=\sum_{m=1}^{\infty}M_m(x,t)\sin\frac{m\pi y}{L_y}\tag{12.11c}$$

$$V(x,y,t)=\sum_{m=1}^{\infty}V_m(x,t)\sin\frac{m\pi y}{L_y}\tag{12.11d}$$

TABLE 12-5 Stresses

Plate	Stresses
1. Isotropic or orthotropic with material properties that do not vary through the depth of the plate.	$\sigma_x = -\frac{P}{h} + \frac{Mz}{h^3/12}$, $\sigma_y = -\frac{P_y}{h} + \frac{M_y z}{h^3/12}$, $\tau_{xy} = \frac{P_{xy}}{h} - \frac{M_{xy} z}{h^3/12}$, P_{xy} is an in-plane shear force $\tau_{xz} = \frac{3Q_x}{2h}\left[1 - \left(\frac{z}{h/2}\right)^2\right]$, $\tau_{yz} = \frac{3Q_y}{2h}\left[1 - \left(\frac{z}{h/2}\right)^2\right]$
2. Orthotropic with material properties that vary through the depth of the plate (Ref. 12.3). This variation must be symmetric about the middle surface. The in-plane force P_{xy} is ignored.	$\sigma_x = -\frac{E_x P}{\int_{-h/2}^{h/2} E_x\,dz} + \frac{E_x Mz}{\int_{-h/2}^{h/2} E_x z^2\,dz}$, $\sigma_y = -\frac{E_y P_y}{\int_{-h/2}^{h/2} E_y\,dz} + \frac{E_y M_y z}{\int_{-h/2}^{h/2} E_y z^2\,dz}$, $\tau_{xy} = \frac{-GM_{xy} z}{\int_{-h/2}^{h/2} G z^2\,dz}$ $\tau_{xz} = \frac{\partial}{\partial x}\left[\frac{\partial^2 v}{\partial x^2}\int_{-h/2}^{z}\frac{E_x z}{1 - \nu_x \nu_y}dz + \frac{\partial^2 v}{\partial y^2}\int_{-h/2}^{z}\left(\frac{E_x \nu_y}{1 - \nu_x \nu_y} + 2G\right)z\,dz\right]$ $\tau_{yz} = \frac{\partial}{\partial y}\left[\frac{\partial^2 v}{\partial x^2}\int_{-h/2}^{z}\left(\frac{E_x \nu_y}{1 - \nu_x \nu_y} + 2G\right)z\,dz + \frac{\partial^2 v}{\partial y^2}\int_{-h/2}^{z}\frac{E_y z}{1 - \nu_x \nu_y}dz\right]$
3. Layered plate. The stresses in the mth layer are given (Ref. 12.3). 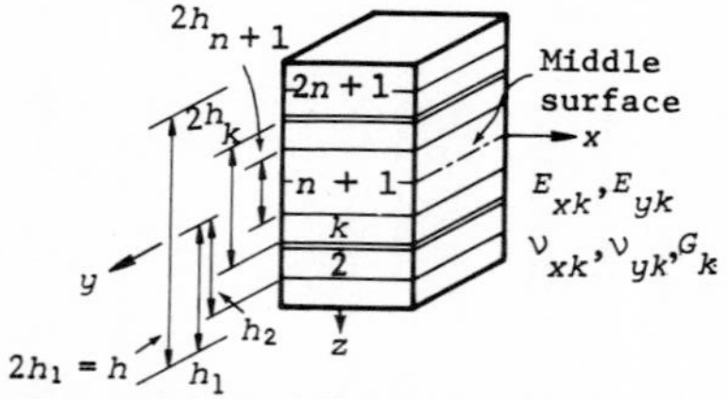	$\sigma_{xm} = -zC_{xm}\left(\frac{\partial^2 v}{\partial x^2} + \nu_{ym}\frac{\partial^2 v}{\partial y^2}\right)$ $\sigma_{ym} = -zC_{ym}\left(\frac{\partial^2 v}{\partial y^2} + \nu_{ym}\frac{\partial^2 v}{\partial x^2}\right)$ $\tau_{xym} = -2zG_m\frac{\partial^2 v}{\partial x \partial y}$ For a symmetrically constructed plate the other shear stresses for $m = 2, 3, \ldots, n+1$ (stresses are symmetrically distributed) are

In-plane forces are ignored.	$\tau_{xzm} = \frac{z^2}{2}\frac{\partial}{\partial x}\left[C_{xm}\frac{\partial^2 v}{\partial x^2} + (C_{xm}\nu_{ym} + 2G_m)\frac{\partial^2 v}{\partial y^2}\right] - \frac{h^2}{2}\frac{\partial}{\partial x}\left[C_{11m}\frac{\partial^2 v}{\partial x^2} + (C_{12m} + 2C_{66m})\frac{\partial^2 v}{\partial y^2}\right]$ $\tau_{yzm} = \frac{z^2}{2}\frac{\partial}{\partial y}\left[(C_{xm}\nu_{ym} + 2G_m)\frac{\partial^2 v}{\partial y} + C_{ym}\frac{\partial^2 v}{\partial y^2}\right] - \frac{h^2}{2}\frac{\partial}{\partial y}\left[\left(C_{12m} + 2C_{66m}\right)\frac{\partial^2 v}{\partial x^2} + C_{22m}\frac{\partial^2 v}{\partial y^2}\right]$ $C_{xm} = \frac{E_{xm}}{1 - \nu_{xm}\nu_{ym}}, \quad C_{ym} = \frac{E_{ym}}{1 - \nu_{xm}\nu_{ym}}, \quad C_{111} = C_{x1}, \; C_{221} = C_{y1}, \quad C_{121} = C_{x1}\nu_{y1}, \; C_{661} = G_1$ $C_{11m} = \frac{1}{h^2}\left[\sum_{k=1}^{m-1} C_{xk}(h_k^2 - h_{k+1}^2) + C_{xm}h_m^2\right], \quad C_{22m} = \frac{1}{h^2}\left[\sum_{k=1}^{m-1} C_{yk}(h_k^2 - h_{k+1}^2) + C_{ym}h_m^2\right]$ $C_{12m} = \frac{1}{h^2}\sum_{k=1}^{m-1} C_{xk}\nu_{ym}(h_k^2 - h_{k+1}^2) + C_{xm}\nu_{ym}h_m^2\Big], \quad C_{66m} = \frac{1}{h^2}\left[\sum_{k=1}^{m-1} G_k(h_k^2 - h_{k+1}^2) + G_m h_m^2\right]$
4. **Thermal loading**	$\sigma_x = \frac{Mz}{h^3/12} - \frac{E_x(\alpha_x + \nu_y\alpha_y)}{1 - \nu_x\nu_y}\Delta T, \quad \sigma_y = \frac{M_y z}{h^3/12} - \frac{E_y(\alpha_y + \nu_x\alpha_x)}{1 - \nu_x\nu_y}\Delta T$

Since these expansions lead to conditions of $v = M_y = 0$ along $y = 0$, $y = L_y$, the plate under consideration is simply supported along $y = 0$ and $y = L_y$.

The mechanical and thermal loadings are also expanded in the sine series

$$w(x, y, t) = \sum_{m=1}^{\infty} w_m(x, t) \sin \frac{m\pi y}{L_y}$$
$$M_T(x, y, t) = \sum_{m=1}^{\infty} M_{Tm}(x, t) \sin \frac{m\pi y}{L_y} \tag{12.12a}$$

so that

$$w_m(x, t) = \frac{2}{L_y} \int_0^{L_y} w(x, y, t) \sin \frac{m\pi y}{L_y}\, dy$$
$$M_{Tm}(x, t) = \frac{2}{L_y} \int_0^{L_y} M_T(x, y, t) \sin \frac{m\pi y}{L_y}\, dy \tag{12.12b}$$

The quantity M_T is intended to represent M_{Tx} or M_{Ty}. The terms w_m amd M_{Tm} will be referred to as *transformed loading functions*. The expressions for w_m and M_{Tm} of Eqs. (12.12*b*) are found by multiplying both sides of Eqs. (12.12*a*) by $\sin \frac{n\pi y}{L_y}$ and integrating from $y = 0$ to $y = L_y$.

Insertion of Eqs. (12.11) and (12.12*a*) in the equations of motion (Eqs. 12.10) leads to differential equations of motion which are functions of x and t only. Thus, Eqs. (12.10) become

$$\frac{\partial v_m}{\partial x} = -\theta_m \tag{12.13a}$$

$$\frac{\partial \theta_m}{\partial x} = -\nu_y \left(\frac{m\pi}{L_y}\right)^2 v_m + \frac{M_m}{D_x} + \frac{M_{Txm}}{D_x} \tag{12.13b}$$

$$\frac{\partial M_m}{\partial x} = 4D_{xy}\left(\frac{m\pi}{L_y}\right)^2 \theta_m + V_m \tag{12.13c}$$

$$\begin{aligned}\frac{\partial V_m}{\partial x} &= \left[D_y(1-\nu_x\nu_y)\left(\frac{m\pi}{L_y}\right)^4 + (P\nu_y - P_y)\left(\frac{m\pi}{L_y}\right)^2 + k\right] v_m \\ &\quad + \left[\frac{-P}{D_x} + \frac{D_y}{D_x}\nu_x\left(\frac{m\pi}{L_y}\right)^2\right] M_m + \rho\frac{\partial^2 v_m}{\partial t^2} - \frac{1}{D_x}\left[P - \nu_x D_y\left(\frac{m\pi}{L_y}\right)^2\right] M_{Txm} \\ &\quad - \left(\frac{m\pi}{L_y}\right)^2 M_{Tym} - w_m(x, t)\end{aligned} \tag{12.13d}$$

Transformed Loading Functions The information in this chapter can be used to develop the functions v_m, θ_m, M_m, V_m. These functions placed in Eqs. (12.11) yield the complete expressions for v, θ, M and V. Other variables of interest, for example, M_y or M_{xy}, can then be derived with Eqs. (12.8). Loadings, including

TABLE 12-6 Massless, Isotropic Plate

$$
\mathbf{U}_i = \begin{bmatrix}
-\frac{\beta(1-\nu)}{2}\ell s + c & -\frac{1}{2}\left[(1+\nu)\frac{s}{\beta} + (1-\nu)\ell c\right] & -\frac{\ell s}{2D\beta} & \frac{1}{2D\beta^2}\left(\frac{s}{\beta} - \ell c\right) \\
-\frac{\beta^2}{2}\left[(1+\nu)\frac{s}{\beta} - (1-\nu)\ell c\right] & \frac{\beta(1-\nu)}{2}\ell s + c & \frac{1}{2D}\left(\frac{s}{\beta} + \ell c\right) & \frac{\ell s}{2D\beta} \\
\frac{D\beta^3(1-\nu)^2}{2}\ell s & \frac{D\beta^2}{2}\left[(3-2\nu-\nu^2)\frac{s}{\beta} + (1-\nu)^2\ell c\right] & \frac{\beta(1-\nu)}{2}\ell s + c & \frac{1}{2}\left[(1+\nu)\frac{s}{\beta} + (1-\nu)\ell c\right] \\
\frac{D\beta^4}{2}\left[(3-2\nu-\nu^2)\frac{s}{\beta} - (1-\nu)^2\ell c\right] & -\frac{D\beta^3(1-\nu)^2}{2}\ell s & \frac{\beta^2}{2}\left[(1+\nu)\frac{s}{\beta} - (1-\nu)\ell c\right] & -\frac{\beta(1-\nu)}{2}\ell s + c
\end{bmatrix}
$$

$$
F_V = \frac{\varepsilon_m}{2D\beta^4}\left[w_1(\beta\ell s - 2c + 2) + \frac{\Delta w}{\Delta\ell}\left(-\frac{3}{\beta}s + \ell c + 2\ell\right) - c_1\beta^2\left(\frac{s}{\beta} - \ell c\right) + \frac{\Delta c}{\Delta\ell}(\beta\ell s - 2c + 2) - M_{T1}2\beta^2\left(-\frac{\beta\ell s}{2} - 2 + 2c\right) - \frac{\Delta M_T}{\Delta\ell}2\beta(2s - 2\beta\ell - \beta\ell c)\right]
$$

$$
F_\theta = \frac{\varepsilon_m}{2D\beta^4}\left[w_1\beta^2\left(\frac{s}{\beta} - \ell c\right) - \frac{\Delta w}{\Delta\ell}(\beta\ell s - 2c + 2) - c_1\beta^3\ell s + \frac{\Delta c}{\Delta\ell}\beta^2\left(\frac{s}{\beta} - \ell c\right) - M_{T1}2\beta^3\left(\frac{3s}{2} - \frac{\beta\ell c}{2}\right) - \frac{\Delta M_T}{\Delta\ell}2\beta^4\left(2c - 2 - \frac{\beta\ell s}{2}\right)\right]
$$

$$
F_M = \frac{\varepsilon_m}{\beta^2}\left[-w_1\left(\frac{1-\nu}{2}\beta\ell s + \nu c - \nu\right) + \frac{\Delta w}{\Delta\ell}\left(\frac{1-3\nu}{2}\frac{s}{\beta} - \frac{1-\nu}{2}\ell c + \nu\ell\right) - c_1\beta^2\left(\frac{1+\nu}{2}\frac{s}{\beta} + \frac{1-\nu}{2}\ell c\right) - \frac{\Delta c}{\Delta\ell}\left(\frac{1-\nu}{2}\beta\ell s + \nu c - \nu\right)\right.
$$

$$
\left. - M_{T1}\beta^2(1-\nu)\left(\frac{\beta\ell s}{2} + 2 - 2c\right) - \frac{\Delta M_T}{\Delta\ell}\beta^2(1-\nu)\left(2\ell - \frac{2s}{\beta}\right)\right]
$$

$$
F_V = \varepsilon_m\left[w_1\left(\frac{1-\nu}{2}\ell c - \frac{3-\nu}{2\beta}s\right) + \frac{\Delta w}{\Delta\ell}\left(\frac{1-\nu}{2\beta}\ell s - \frac{2-\nu}{\beta^2}c + \frac{2-\nu}{\beta^2}\right) + c_1\left(\frac{1-\nu}{2}\beta\ell s - c + 1\right) - \frac{\Delta c}{\Delta\ell}\left(\frac{3-\nu}{2}\frac{s}{\beta} - \frac{1-\nu}{2}\ell c - \ell\right)\right.
$$

$$
\left. - M_{T1}2(1-\nu)\beta s - \frac{\Delta M_T}{\Delta\ell}2(1-\nu)(c-1)\right]
$$

Definitions:

$s = \sinh\beta\ell, \quad c = \cosh\beta\ell$

$\beta = m\pi/L_y$

TABLE 12-7 Isotropic Plate with Mass or Axial Force

$$
\mathbf{U}_i = \begin{bmatrix}
\frac{1}{g}(\eta_1 c + \eta_2 C) & -\frac{1}{g}\left(\frac{\eta_2}{a}s + \frac{\eta_1}{b}S\right) & -\frac{1}{gD}(c - C) & -\frac{1}{gD}\left(\frac{s}{a} - \frac{S}{b}\right) \\
-\frac{1}{g}(\eta_1 as - \eta_2 bS) & \frac{1}{g}(\eta_2 c + \eta_1 C) & \frac{1}{gD}(as + bS) & \frac{1}{gD}(c - C) \\
-\frac{D}{g}\eta_1\eta_2(c - C) & \frac{D}{g}\frac{\eta_2^2}{a}s - \frac{\eta_1^2}{b}S & \frac{1}{g}(\eta_2 c + \eta_1 C) & \frac{1}{g}\left(\frac{\eta_2}{a}s + \frac{\eta_1}{b}S\right) \\
-\frac{D}{g}(\eta_1^2 as + \eta_2^2 bS) & \frac{D}{g}\eta_1\eta_2(c - C) & \frac{1}{g}(\eta_1 as - \eta_2 bS) & \frac{1}{g}(\eta_1 c + \eta_2 C)
\end{bmatrix}
$$

Definitions:

$s = \sinh a\ell$, $c = \cosh a\ell$

$S = \sin b\ell$, $C = \cos b\ell$

This table is not applicable for the case

$$\left(\frac{m\pi}{L_y}\right)^4 > \frac{P_y}{D}\left(\frac{m\pi}{L_y}\right)^2$$

For this case use the general

(a) Vibrating Plate

$$a^2 = \sqrt{\frac{\rho}{D}}\,\omega + \left(\frac{m\pi}{L_y}\right)^2, \quad b^2 = \sqrt{\frac{\rho}{D}}\,\omega - \left(\frac{m\pi}{L_y}\right)^2$$

$$g = 2\sqrt{\frac{\rho}{D}}\,\omega, \quad \eta_1 = \sqrt{\frac{\rho}{D}}\,\omega - (1-\nu)\left(\frac{m\pi}{L_y}\right)^2$$

$$\eta_2 = \sqrt{\frac{\rho}{D}}\,\omega + (1-\nu)\left(\frac{m\pi}{L_y}\right)^2, \quad h = \nu\left(\frac{m\pi}{L_y}\right)^2$$

(b) Plate with In-Plane Compressive Force P_y $(P = 0)$

$$a^2 = \frac{m\pi}{L_y}\sqrt{\frac{P_y}{D}} + \left(\frac{m\pi}{L_y}\right)^2, \quad b^2 = \frac{m\pi}{L_y}\sqrt{\frac{P_y}{D}} - \left(\frac{m\pi}{L_y}\right)^2$$

$$g = 2\,\frac{m\pi}{L_y}\sqrt{\frac{P_y}{D}}, \quad \eta_1 = \frac{m}{L_y}\sqrt{\frac{P_y}{D}} - (1-\nu)\left(\frac{m\pi}{L_y}\right)^2$$

$$\eta_2 = \frac{m\pi}{L_y}\sqrt{\frac{P_y}{D}} + (1-\nu)\left(\frac{m\pi}{L_y}\right)^2, \quad h = \nu\left(\frac{m\pi}{L_y}\right)^2$$

$$F_v = \frac{m}{gD}\left\{w_1\left(\frac{c}{a^2} + \frac{C}{b^2} - \frac{g}{a^2b^2}\right) + \frac{\Delta w}{\Delta \ell}\left(\frac{s}{a^3} + \frac{S}{b^3} - \frac{g\ell}{a^2b^2}\right) + c_1\left(\frac{s}{a} - \frac{S}{b}\right) + \frac{\Delta c}{\Delta \ell}\left(\frac{c}{a^2} + \frac{C}{b^2} - \frac{g}{a^2b^2}\right)\right.$$

$$\left. - M_{T1}\left[\eta_2\frac{c-1}{a^2} + \eta_1\frac{1-C}{b^2} + (1-\nu)\left(\frac{m\pi}{L_y}\right)^2\left(\frac{c-1}{a^2} - \frac{1-C}{b^2}\right)\right] - \frac{\Delta M_T}{\Delta \ell}\left[\eta_2\frac{s-a\ell}{a^3} + \eta_1\frac{b\ell+s}{b^3} + (1-\nu)\left(\frac{m\pi}{L_y}\right)^2\left(\frac{s-a\ell}{a^3} - \frac{b\ell+s}{b^3}\right)\right]\right\}$$

$$F_\theta = -\frac{\varepsilon_m}{gD}\left\{w_1\left(\frac{s}{a} - \frac{S}{b}\right) + \frac{\Delta w}{\Delta \ell}\left(\frac{c}{a^2} + \frac{C}{b^2} - \frac{g}{a^2b^2}\right) + c_1(c - C) + \frac{\Delta c}{\Delta \ell}\left(\frac{s}{a} - \frac{S}{b}\right)\right.$$

$$\left. - M_{T1}\left[\eta_2\frac{s}{a} - \eta_1\frac{S}{b} + (1-\nu)\left(\frac{m\pi}{L_y^2}\right)^2\left(\frac{s}{a} + \frac{S}{b}\right)\right] - \frac{\Delta M_T}{\Delta \ell}\left[\eta_2\frac{c-1}{a^2} - \eta_1\frac{C-1}{b^2} + (1-\nu)\left(\frac{m\pi}{L_y}\right)^2\left(\frac{c-1}{a^2} - \frac{C-1}{b^2}\right)\right]\right\}$$

$$F_M = -\frac{\varepsilon_m}{g}\left\{w_1\left(\frac{\eta_2}{a^2}c - \frac{\eta_1}{b^2}C + \frac{gh}{a^2b^2}\right) + \frac{\Delta w}{\Delta \ell}\left(\frac{\eta_2}{a^3}s - \frac{\eta_1}{b^3}S + \frac{gh\ell}{a^2b^2}\right) + c_1\left(\frac{\eta_2}{a}s + \frac{\eta_1}{b}S\right) + \frac{\Delta c}{\Delta \ell}\left(\frac{\eta_2}{a^2}c - \frac{\eta_1}{b^2}C + \frac{gh}{a^2b^2}\right)\right.$$

$$\left. - M_{T1}\left[\eta_2^2\frac{c-1}{a^2} + \eta_1^2\frac{C-1}{b^2} + (1-\nu)\left(\frac{m\pi}{L_y^2}\right)^2\left(\eta_2\frac{c-1}{a^2} - \eta_1\frac{C-1}{b^2}\right)\right] - \frac{\Delta M_T}{\Delta \ell}\left[\eta_2^2\frac{s-a\ell}{a^3} + \eta_1^2\frac{S-b\ell}{b^3} + (1-\nu)\left(\frac{m\pi}{L_y}\right)^2\left(\eta_2\frac{s-a\ell}{a^3} - \eta_1\frac{s-b\ell}{b^3}\right)\right]\right\}$$

$$F_v = -\frac{\varepsilon_m}{g}\left\{w_1\left(\frac{\eta_2}{a}s + \frac{\eta_1}{b}S\right) + \frac{\Delta w}{\Delta \ell}\left(\frac{\eta_1}{a^2}c - \frac{\eta_2}{b^2}C + \frac{a^4 - b^4 - gh}{a^2b^2}\right) + c_1\left(\eta_1 c + \eta_2 C - g\right) + \frac{\Delta c}{\Delta \ell}\left(\frac{\eta_1}{a}s + \frac{\eta_2}{b}S - g\ell\right)\right.$$

$$\left. - M_{T1}\left[\eta_1\eta_2\left(\frac{s}{a} + \frac{S}{b}\right) + (1-\nu)\left(\frac{m\pi}{L_y^2}\right)\left(\eta_1\frac{s}{a} - \eta_2\frac{S}{b}\right)\right] - \frac{\Delta M_T}{\Delta \ell}\left[\eta_1\eta_2\left(\frac{c-1}{a^2} + \frac{1-c}{b^2}\right) + (1-\nu)\left(\frac{m\pi}{L_y}\right)^2\left(\eta_1\frac{c-1}{a^2} + \eta_2\frac{1-C}{b^2}\right)\right]\right\}$$

TABLE 12-8 Transformed Loading Function Coefficients for Particular Loading Distributions in the y Direction

Loading (Force or Moment)	ε_m
1. Distributed Load Constant in y Direction 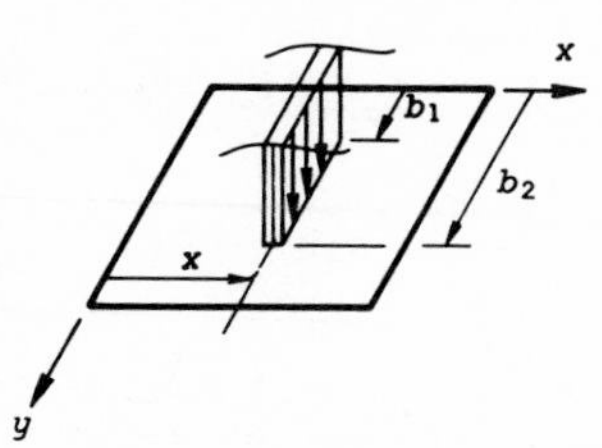	$\frac{2}{m\pi}\left(\cos\frac{m\pi b_1}{L_y} - \cos\frac{m\pi b_2}{L_y}\right)$ If $b_1 = 0$, $b_2 = L_y$: $\frac{2}{m\pi}(1 - \cos m\pi) = \begin{cases} 4/m\pi & \text{if } m = 1,3,5,7\ldots \\ 0 & \text{if } m = 2,4,6,8\ldots \end{cases}$
2. Distributed Load Ramp in y Direction 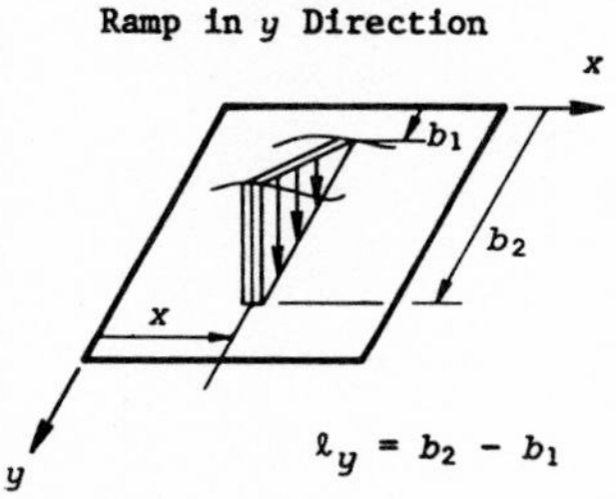 $\ell_y = b_2 - b_1$	$\frac{2}{m^2\pi^2}\left[-\ell_y m\pi \cos\frac{m\pi b_2}{L_y} + L_y\left(\sin\frac{m\pi b_2}{L_y} - \sin\frac{m\pi b_1}{L_y}\right)\right]$
3. Sinusoidal Load in y Direction 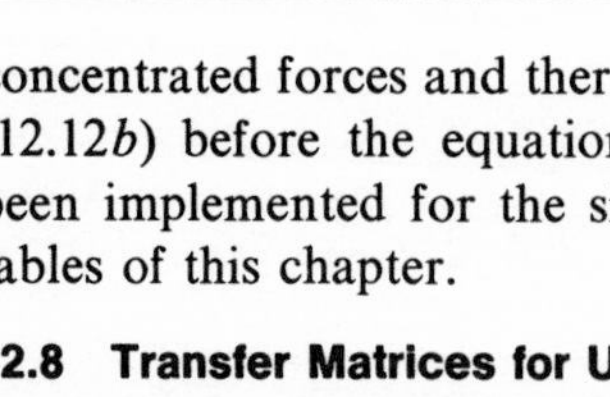	$\varepsilon_1 = 1$ $\varepsilon_m = 0, \quad m > 1$

concentrated forces and thermal loads, must be transformed in the sense of Eqs. (12.12*b*) before the equations of motion are solved. This transformation has been implemented for the sizable array of loadings taken into account in the tables of this chapter.

12.8 Transfer Matrices for Uniform Segments

The transfer (field) matrices for a plate segment of length ℓ are provided in Tables 12-6 and 12-7. In these tables, applied loadings, that is, forces w_1 and $\Delta w/\Delta\ell$ and moments c_1 and $\Delta c/\Delta\ell$, can vary in the x direction either uniformly or linearly. The coefficient ϵ_m of Table 12-8 is inserted in the loading functions of Tables 12-6 and 12-7 to account for the proper loading variation in the y direction.

12.9 General Transfer Matrix for Uniform Segments

The general solution of the equations of motion of Eq. (12.14) for a uniform panel is given in Table 12-9. This transfer matrix reduces for special cases to the matrices of Tables 12-6 and 12-7. To use Table 12-9, determine the relative magnitudes of various plate parameters and then select the proper table entry. Column 4 of the e_i definitions along with column $4a$ of the final page of Table 12-9 ($\zeta > 0$) can be used for all plates regardless of the magnitude of the plate parameters, but this will frequently lead to elements involving complex numbers in the transfer matrix. Accordingly, computations with column 4 of the e_i definitions will require computers with complex arithmetic capabilities. See Example 2.5 for further discussions of the use of column 4 entries.

12.10 Point Matrices

The transfer matrices that take into account concentrated point occurrences are listed in Tables 12-10 and 12-11. Point matrices for in-span indeterminates (rigid supports, moment releases, angle guides, and shear releases) are provided by Table 2-15. These in-span indeterminates are line conditions in that they occur at $x = a_i$, extending from $y = 0$ to $y = L_y$.

12.11 Loading Functions

The loading functions are provided in the previous transfer matrix tables for most common loadings. These functions can be calculated for other loadings from the formulas

$$F_v = -\int_0^\ell w_m(x) U_{vV}(\ell - x)\, dx + \int_0^\ell M_{Tm}(x)\left[\frac{U_{v\theta}(\ell - x)}{D} + \alpha_3 U_{vV}(\ell - x)\right] dx$$

$$F_\theta = -\int_0^\ell w_m(x) U_{\theta V}(\ell - x)\, dx + \int_0^\ell M_{Tm}(x)\left[\frac{U_{\theta\theta}(\ell - x)}{D} + \alpha_3 U_{\theta V}(\ell - x)\right] dx$$

$$F_M = -\int_0^\ell w_m(x) U_{MV}(\ell - x)\, dx + \int_0^\ell M_{Tm}(x)\left[\frac{U_{M\theta}(\ell - x)}{D} + \alpha_3 U_{MV}(\ell - x)\right] dx$$

$$F_V = -\int_0^\ell w_m(x) U_{VV}(\ell - x)\, dx + \int_0^\ell M_{Tm}(x)\left[\frac{U_{V\theta}(\ell - x)}{D} + \alpha_3 U_{VV}(\ell - x)\right] dx \qquad (12.14)$$

TABLE 12-9 General Plate Solution

$$\mathbf{U}_i = \begin{bmatrix} e_1 + (\zeta + \alpha_1)e_3 & -e_2 - (\zeta + \alpha_2)e_4 & -e_3/D_x & -e_4/D_x & F_v \\ -e_0 - (\zeta + \alpha_1)e_2 & e_1 + (\zeta + \alpha_2)e_3 & e_2/D_x & e_3/D_x & F_\theta \\ D_x[\lambda + \alpha_1(\zeta + \alpha_1)]e_3 & D_x[e_0 - (\alpha_1 - \zeta - \alpha_2)e_2 - \alpha_1(\zeta + \alpha_2)e_4] & e_1 - \alpha_1 e_3 & e_2 - \alpha_1 e_4 & F_M \\ D_x[(\lambda + \alpha_1\alpha_2 + \alpha_1\zeta)e_2 - \lambda(\alpha_2 - \alpha_1)e_4] & -D_x[\lambda + \alpha_2(\zeta + \alpha_2)]e_3 & e_0 - \alpha_2 e_2 & e_1 - \alpha_2 e_3 & F_V \\ 0 & 0 & 0 & 0 & 1 \end{bmatrix}$$

Definitions:

For orthotropic plates:

$$\lambda = \frac{1}{D_x}\left[k - \rho\omega^2 + D_y\left(\frac{m\pi}{L_y}\right)^4 - P_y\left(\frac{m\pi}{L_y}\right)^2\right]$$

$$\zeta = \frac{1}{D_x}\left[P - 2B\left(\frac{m\pi}{L_y}\right)^2\right] \qquad B = (D_x\nu_y + D_y\nu_x + 4D_{xy})/2$$

$$\alpha_1 = \nu_y\left(\frac{m\pi}{L_y}\right)^2$$

$$\alpha_2 = \left(\frac{4D_{yx}}{D_x} + \nu_y\right)\left(\frac{m\pi}{L_y}\right)^2$$

$$\alpha_3 = -\frac{P}{D_x} - \left(1 - \nu_x\frac{D_y}{D_x}\right)\left(\frac{m\pi}{L_y}\right)^2$$

For isotropic plates:

$$\lambda = \frac{1}{D}\left[k - \rho\omega^2 + D\left(\frac{m\pi}{L_y}\right)^4 - P_y\left(\frac{m\pi}{L_y}\right)^2\right]$$

$$\zeta = \frac{1}{D}\left[P - 2D\left(\frac{m\pi}{L_y}\right)^2\right]$$

$$\alpha_1 = \nu\left(\frac{m\pi}{L_y}\right)^2$$

$$\alpha_2 = (2 - \nu)\left(\frac{m\pi}{L_y}\right)^2$$

$$\alpha_3 = -\frac{P}{D} - (1 - \nu)\left(\frac{m\pi}{L_y}\right)^2$$

P, P_y are in compression. Replace ... by ... y y
if this force is in tension.

Column 4 of the third page and column 4a ($\zeta > 0$) of the fourth page of this table apply for any magnitude of λ, ζ. However, then the transfer matrix elements usually will be complex quantities and computations will require computers with complex arithmetic capabilities.

$$F_V = \varepsilon_m \left\{ w_1 e_5 + \frac{\Delta w}{\Delta \ell} e_6 + c_1 e_4 + \frac{\Delta c}{\Delta \ell} e_5 - M_{T1}\left[e_3 + (\zeta + \alpha_2) e_5 + \alpha_3 e_5 \right] - \frac{\Delta M_T}{\Delta \ell}\left[e_4 + (\zeta + \alpha_2) e_6 + \alpha_3 e_6 \right] \right\} / D_x$$

$$F_\theta = \varepsilon_m \left\{ -w_1 e_4 - \frac{\Delta w}{\Delta \ell} e_5 - c_1 e_3 - \frac{\Delta c}{\Delta \ell} e_4 + M_{T1}\left[e_2 + (\zeta + \alpha_2) e_4 + \alpha_3 e_4 \right] + \frac{\Delta M_T}{\Delta \ell}\left[e_3 + (\zeta + \alpha_2) e_5 + \alpha_3 e_5 \right] \right\} / D_x$$

$$F_M = \varepsilon_m \Bigg(w_1(-e_3 + \alpha_1 e_5) + \frac{\Delta w}{\Delta \ell}(-e_4 + \alpha_1 e_6) + c_1(-e_2 + \alpha_1 e_4) + \frac{\Delta c}{\Delta \ell}(-e_3 + \alpha_1 e_5) - M_{T1}\left\{ (\alpha_1 - \alpha_2) e_3 + [\lambda + \alpha_1(\zeta + \alpha_2)] e_5 - \alpha_3(e_3 - \alpha_1 e_5) \right\}$$
$$- \frac{\Delta M_T}{\Delta \ell}\left\{ (\alpha_1 - \alpha_2) e_4 + [\lambda + \alpha_1(\zeta + \alpha_2)] e_6 - \alpha_3(e_4 - \alpha_1 e_6) \right\} \Bigg)$$

$$F_V = \varepsilon_m \Bigg(w_1(-e_2 + \alpha_2 e_4) + \frac{\Delta w}{\Delta \ell}(-e_3 + \alpha_2 e_5) + c_1(-e_1 + 1 + \alpha_2 e_3) + \frac{\Delta c}{\Delta \ell}(-e_2 + \ell + \alpha_2 e_4) - M_{T1}\left\{ [\lambda + \alpha_2(\zeta + \alpha_2)] e_4 - \alpha_3(e_2 - \alpha_2 e_4) \right\}$$
$$- \frac{\Delta M_T}{\Delta \ell}\left\{ [\lambda + \alpha_2(\zeta + \alpha_2)] e_5 - \alpha_3(e_3 - \alpha_2 e_5) \right\} \Bigg)$$

TABLE 12-9 General Plate Solution (*Continued*)

	1. $\lambda < 0$	2. $\lambda = 0$	$\lambda > 0$ 3. $\lambda = \zeta^2/4$	4. $\lambda < \zeta^2/4$	5. $\lambda > \zeta^2/4$
e_0	$\frac{1}{g}(a^3C - b^3D)$	$-\zeta B$	$-\frac{\zeta}{4}(3C + A\ell)$	$-\frac{1}{g}(b^3D - a^3C)$	$-\lambda e_4 - \zeta e_2$
e_1	$\frac{1}{g}(a^2A + b^2B)$	A	$\frac{1}{2}(2A - B\ell)$	$\frac{1}{g}(b^2B - a^2A)$	$AB - \frac{b^2 - a^2}{2ab}CD$
e_2	$\frac{1}{g}(aC + bD)$	B	$\frac{1}{2}(C + A\ell)$	$\frac{1}{g}(bD - aC)$	$\frac{1}{2ab}(aAD + bBC)$
e_3	$\frac{1}{g}(A - B)$	$\frac{1}{\zeta}(1 - A)$	$\frac{C\ell}{2}$	$\pm\frac{1}{g}(A - B)$ + for $\zeta > 0$, − for $\zeta < 0$	$\frac{1}{2ab}CD$
e_4	$\frac{1}{g}\left(\frac{C}{a} - \frac{D}{b}\right)$	$\frac{1}{\zeta}(\ell - B)$	$\frac{1}{\zeta}(C - A\ell)$	$\pm\frac{1}{g}\left(\frac{C}{a} - \frac{D}{b}\right)$ + for $\zeta > 0$, − for $\zeta < 0$	$\frac{1}{2(a^2 + b^2)}\left(\frac{AD}{b} - \frac{BC}{a}\right)$
e_5	$\frac{1}{g}\left(\frac{A}{a^2} + \frac{B}{b^2}\right) - \frac{1}{a^2b^2}$	$\frac{1}{\zeta}\left(\frac{\ell^2}{2} - e_3\right)$	$\frac{2}{\zeta^2}(-2A - B\ell + 2)$	$\frac{1}{g}\left(\frac{B}{b^2} - \frac{A}{a^2}\right) + \frac{1}{a^2b^2}$	$\frac{1 - e_1}{\lambda} - \frac{\zeta}{\lambda}e_3$
e_6	$\frac{1}{g}\left(\frac{C}{a^3} + \frac{D}{b^3}\right) - \frac{\ell}{a^2b^2}$	$\frac{1}{\zeta}\left(\frac{\ell^3}{6} - e_4\right)$	$\frac{2}{\zeta^2}(-3C + A\ell + 2\ell)$	$\frac{1}{g}\left(\frac{D}{b^3} - \frac{C}{a^3}\right) + \frac{\ell}{a^2b^2}$	$\frac{\ell - e_2}{\lambda} - \frac{\zeta}{\lambda}e_4$

1. $A = \cosh a\ell$ $B = \cos b\ell$ $C = \sinh a\ell$ $D = \sin b\ell$ $g = a^2 + b^2$ $a^2 = \sqrt{\beta^4 + \zeta^2/4} - \tfrac{1}{2}\zeta$ $b^2 = \sqrt{\beta^4 + \zeta^2/4} + \tfrac{1}{2}\zeta$ $\beta^4 = -\lambda$	2a. $\zeta > 0$: $\alpha^2 = \zeta$ $A = \cos \alpha\ell$ $B = \dfrac{\sin \alpha\ell}{\alpha}$	3a. $\zeta > 0$: $\beta^2 = \zeta/2$ $A = \cos \beta\ell$ $B = \beta \sin \beta\ell$ $C = \dfrac{\sin \beta\ell}{\beta}$	4a. $\zeta > 0$: $g = b^2 - a^2$ $A = \cos a\ell$ $B = \cos b\ell$ $C = \sin a\ell$ $D = \sin b\ell$ $a^2 = \tfrac{1}{2}\zeta - \sqrt{\tfrac{1}{4}\zeta^2 - \lambda}$ $b^2 = \tfrac{1}{2}\zeta + \sqrt{\tfrac{1}{4}\zeta^2 - \lambda}$	5. $A = \cosh a\ell$ $B = \cos b\ell$ $C = \sinh a\ell$ $D = \sin b\ell$ $a^2 = \tfrac{1}{2}\sqrt{\lambda} - \tfrac{1}{4}\zeta$ $b^2 = \tfrac{1}{2}\sqrt{\lambda} + \tfrac{1}{4}\zeta$
	2b. $\zeta < 0$: $\alpha^2 = -\zeta$ $A = \cosh \alpha\ell$ $B = \dfrac{\sinh \alpha\ell}{\alpha}$	3b. $\zeta < 0$: $\beta^2 = -\tfrac{1}{2}\zeta$ $A = \cosh \beta\ell$ $B = -\beta \sinh \beta\ell$ $C = \dfrac{\sinh \beta\ell}{\beta}$	4b. $\zeta < 0$: $g = b^2 - a^2$ $A = \cosh a\ell$, $B = \cosh b\ell$ $C = \sinh a\ell$, $D = \sinh b\ell$ $a^2 = -\tfrac{1}{2}\zeta + \sqrt{\tfrac{1}{4}\zeta^2 - \lambda}$ $b^2 = -\tfrac{1}{2}\zeta - \sqrt{\tfrac{1}{4}\zeta^2 - \lambda}$	

Instructions:

To use this general transfer matrix, follow the steps:

1. Calculate the parameters λ, ζ.
2. Look up the appropriate e_i functions according to the magnitude of λ.
3. Substitute these e_i expressions into the transfer matrix of this table.
4. According to the distribution of the loading in the y direction, insert ε_m of Table 12-8 into the loading functions.

TABLE 12-10 Point Matrix for Loadings, Springs, and Masses at $x = a_i$

$$\bar{\mathbf{U}}_i = \begin{bmatrix} 1 & 0 & 0 & 1/k_2 & v_{1m} \\ 0 & 1 & 1/k_2^* & 0 & -\alpha_m \\ 0 & P_{Li} + k_1^* & 1 & 0 & -C_m \\ k_1 - M_i\omega^2 & 0 & 0 & 1 & -W_m \\ 0 & 0 & 0 & 0 & 1 \end{bmatrix}$$

Definitions:

(a) Concentrated Force W_T

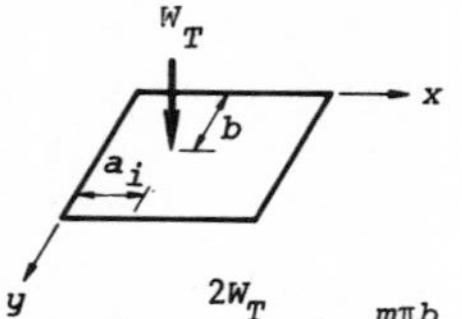

$$W_m = \frac{2W_T}{L_y} \sin \frac{m\pi b}{L_y}$$

(b) Line Force W (force/length in y direction)

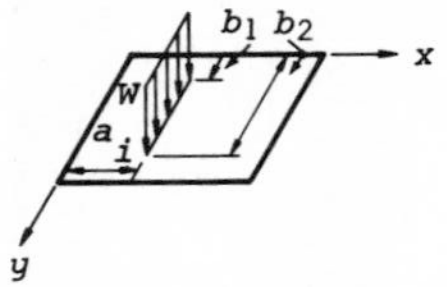

$$W_m = \frac{2W}{m\pi}\left(\cos\frac{m\pi b_1}{L_y} - \cos\frac{m\pi b_2}{L_y}\right)$$

(c) Linearly Varying Line Force

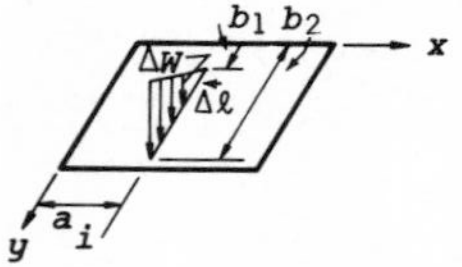

$$W_m = \frac{\Delta W}{\ldots}\,\frac{2}{\ldots}\left[(b_1 - b_2)m\pi \cos\frac{m\pi b_2}{\ldots} + L\left(\sin\frac{m\pi b_2}{\ldots} - \sin\frac{m\pi b_1}{\ldots}\right)\right]$$

(d) Concentrated Moment C_T (force-length)

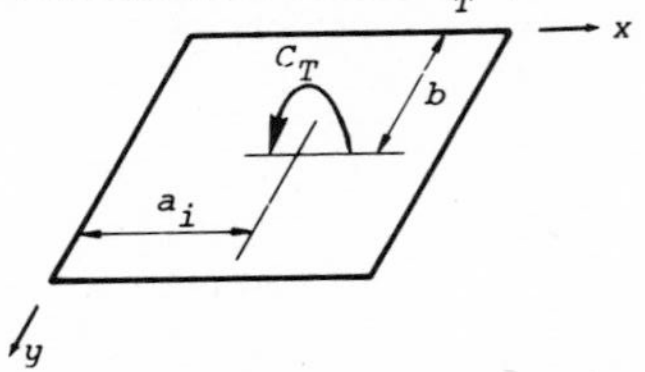

$$C_m = \frac{2C_T}{L_y} \sin \frac{m\pi b}{L_y}$$

(e) Line Moment C (force-length/length in y direction)

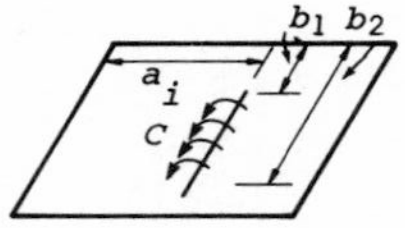

$$C_m = \frac{2C}{m\pi}\left(\cos\frac{m\pi b_1}{L_y} - \cos\frac{m\pi b_2}{L_y}\right)$$

(f) Jump in Level v_1 (length) and Change in Slope α (radians)

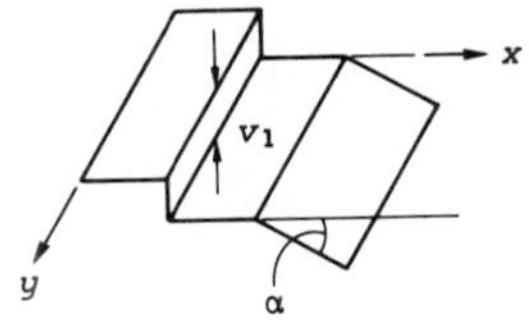

$$v_{1m} = \frac{2v_1}{m\pi}(1 - \cos m\pi)$$

$$\alpha_m = \frac{2\alpha}{m\pi}(1 - \cos m\pi)$$

(g) Springs: k_1 (force/length^2) and k_1^* (force-length/length^2). Values of k_1, k_1^* can be taken from Table 2-14 for various spring, flexible support combinations. For example: Line spring k_1

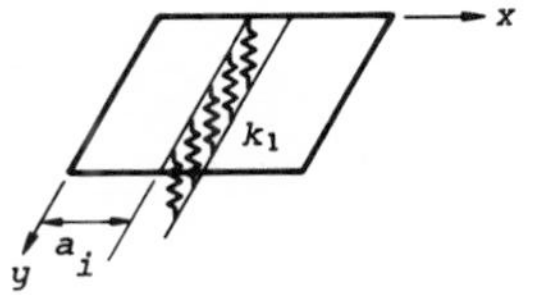

(h) Linear and Rotary Hinges k_2, k_2^*

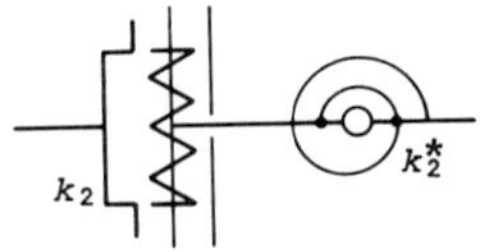

(i) Line Lumped Mass M_i (mass/length in y direction)

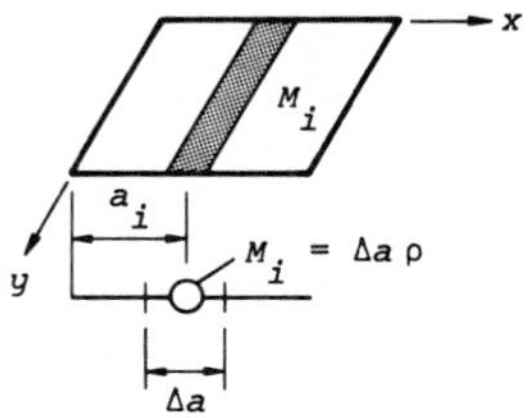

(j) Line Lumped Compressive In-plane Force P_{Li} (force-length/length)

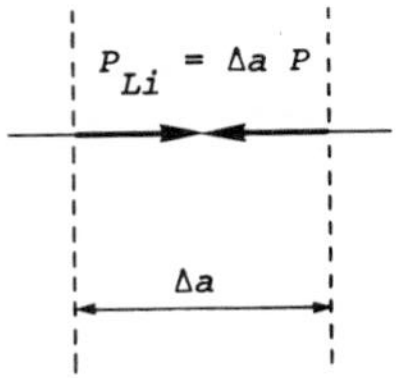

Substitute $-P$ for P if force is tensile. For a distributed in-plane force (force/length^2), P is found using the condition of equilibrium $\Sigma F_x = 0$.

TABLE 12-11 Point Matrix for Thin-Walled Stiffener at $x = a_i$

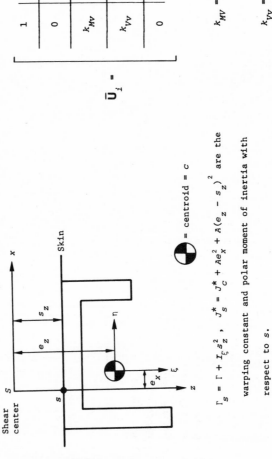

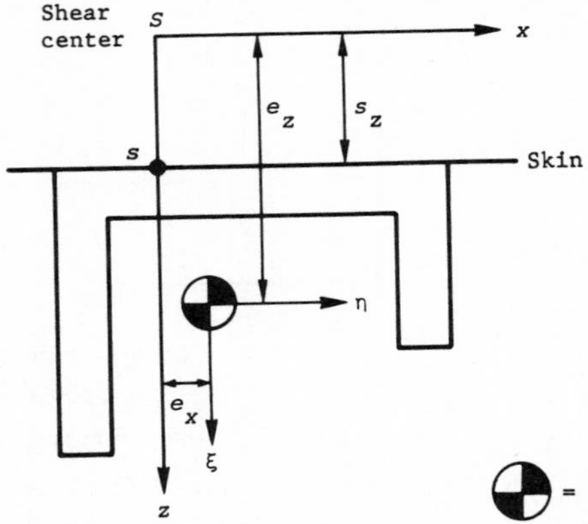

= centroid = c

$$\bar{\mathbf{U}}_i = \begin{bmatrix} 1 & 0 & 0 & 0 & 0 \\ 0 & 1 & 0 & 0 & 0 \\ k_{Mv} & k_{M\theta} & 1 & 0 & 0 \\ k_{Vv} & k_{V\theta} & 0 & 1 & 0 \\ 0 & 0 & 0 & 0 & 1 \end{bmatrix}$$

$\Gamma_s = \Gamma + I_\xi s_z^2$, $J_s^* = J_c^* + Ae_x^2 + A(e_z - s_z)^2$ are the warping constant and polar moment of inertia with respect to s.

J - torsional constant

M_i - mass per length along stiffener

ρ^* - mass density of stiffener

$$k_{Mv} = -k_{V\theta} = EI_{\eta\xi} s_z \left(\frac{m\pi}{L_y}\right)^4 - M_i e_x \omega^2$$

$$k_{Vv} = -EI_\eta \left(\frac{m\pi}{L_y}\right)^4 + M_i \omega^2$$

$$k_{M\theta} = E\Gamma_s \left(\frac{m\pi}{L_y}\right)^4 + GJ\left(\frac{m\pi}{L_y}\right)^2 - \rho^* J_s^* \omega^2$$

where $w_m(x)$ is the transformed applied distributed force, M_T is the thermal moment due to a temperature change ΔT across the plate depth, and

$$M_{Tx} = M_{Ty} = M_T = \frac{1}{1-\nu} \int_{-h/2}^{h/2} E\alpha \, \Delta T \, z \, dz$$

$$\alpha_3 = -\frac{P}{D} - (1-\nu)\left(\frac{m\pi}{L_y}\right)^2$$

The notation $U_{ij}(\ell - x)$ refers to the U_{ij} given in the transfer matrix tables with ℓ replaced by $\ell - x$. In each entry of Eqs. (12.14) it is permissible to switch the $(\ell - x)$ dependency from the transfer matrix element to the loading variable, for example,

$$F_v = -\int_0^{\ell} w_m(x) U_{vV}(\ell - x)\, dx = -\int_0^{\ell} w_m(\ell - x) U_{vV}(x)\, dx$$

where $U_{ij}(x)$ is U_{ij} from the tables with ℓ replaced by x. See Section 2.15 or 3.9 for examples of the use of the formulas of the type in Eqs. (12.14)

12.12 Static Response

The procedure for calculating the deflection, slope, moment, and shear force functions v_m, θ_m, M_m, V_m is identical to that described in Section 2.16 for beams. The transfer matrices are taken from the information of the present chapter. These are then employed in the progression multiplication described in Section 2.16. The boundary conditions at $x = 0$ and $x = L$ are applied as indicated in Section 2.16 to find the initial parameters. Table 2-19 and the formulas of Eqs. (2.31), (2.32) still apply for determining initial parameters. After forming the functions v_m, θ_m, M_m, V_m they are placed in Eqs. (12.11) to provide the final expressions for v, θ, M, V. The procedure applies for plates that are simply supported on the $y = 0$, $y = L_y$ edges, and with any boundary conditions on the $x = 0$, $x = L$ edges. Note that arbitrary variations in geometry and loading, as well as in-span supports, can be incorporated.

For plates with several changes in geometry and loading, the computations required for a solution are quite complicated even if only a few terms are employed in the series expansion. For these cases, it is recommended that a computer program be used.

EXAMPLE 12.3 Static Response of Simply Supported, Uniformly Loaded Plate Find the deflection of an isotropic rectangular plate subjected to a uniformly distributed load w_1. All sides are simply supported.

The deflection is given by Eq. (12.11*a*) with v_m taken from the appropriate transfer matrix. Thus,

$$v_m = v_0 U_{vv} + \theta_0 U_{v\theta} + M_0 U_{vM} + V_0 U_{vV} + F_v \tag{1}$$

in which the transfer matrix elements of Table 12-6 should be used. From Table

2-19 for simply supported ends, the initial parameters are given by

$$v_0 = M_0 = 0 \qquad \theta_0 = \frac{1}{\nabla}(F_M U_{vV} - F_v U_{MV})_{x=L}$$

$$V_0 = \frac{1}{\nabla}(F_v U_{M\theta} - F_M U_{v\theta})_{x=L} \tag{2}$$

$$\nabla = (U_{v\theta} U_{MV} - U_{M\theta} U_{vV})_{x=L}$$

With $\ell = x$, the transfer matrix elements in Table 12-6 that are needed in (1) and (2) are

$$U_{v\theta} = -\frac{1}{2}\left[(1+\nu)\frac{\sinh \beta x}{\beta} + (1-\nu)x \cosh \beta x\right]$$

$$U_{vV} = \frac{1}{2D\beta^2}\left(\frac{\sinh \beta x}{\beta} - x \cosh \beta x\right)$$

$$U_{M\theta} = \frac{D\beta^2}{2}\left[(3 - 2\nu - \nu^2)\frac{\sinh \beta x}{\beta} + (\nu^2 - 2\nu + 1)x \cosh \beta x\right] \tag{3}$$

$$U_{MV} = -U_{v\theta}$$

$$F_v = \frac{\epsilon w_1}{2D\beta^4}(\beta x \sinh \beta x - 2 \cosh \beta x + 2)$$

$$F_M = -\epsilon w_1\left(\frac{1-\nu}{2}x\frac{\sinh \beta x}{\beta} + \frac{\nu}{\beta^2}\cosh \beta x - \frac{\nu}{\beta^2}\right)$$

where $\beta = m\pi/L_y$. Since the loading is uniformly distributed in the y direction and extends the width of the plate, ϵ_m is given as case 1 of Table 12.8.

$$\epsilon_m = \begin{cases} 4/m\pi & \text{for odd } m \\ 0 & \text{for even } m \end{cases} \tag{4}$$

Substitution of (3) in (2) leads to

$$\nabla = -\frac{\sinh^2 \beta L}{\beta^2}$$

$$\theta_0 = \frac{1}{\nabla}\frac{\epsilon_m w_1}{2D\beta^4}(1 - \cosh \beta L)\left(L - \frac{\sinh \beta L}{\beta}\right) \tag{5}$$

$$V_0 = \frac{1}{\nabla}\frac{\epsilon_m w_1}{2\beta^2}(1 - \cosh \beta L)\left[L(\nu - 1) - (\nu - 3)\frac{\sinh \beta L}{\beta}\right]$$

Place $v_0 = M_0 = 0$ and (5) in (1)

$$v_m = \theta_0 U_{v\theta} + V_0 U_{vV} + F_v = \frac{\epsilon_m w_1}{D\beta^4}\left\{1 + \left[\frac{\beta x}{2} - \frac{1-\cosh\beta L}{\sinh\beta L} + \frac{\beta L(1-\cosh\beta L)}{2\sinh^2\beta L}\right]\sin\beta x - \left[1 - \frac{\beta x(1-\cosh\beta L)}{2\sinh\beta L}\right]\cos\beta x\right\}$$

$$= \frac{\epsilon_m w_1}{D\beta^4}\left[1 + \left(\frac{\beta x}{2} + \frac{\sinh\frac{\beta L}{2}}{\cosh\frac{\beta L}{2}} - \frac{\beta L}{4\cosh^2\frac{\beta L}{2}}\right)\sinh\beta x - \left(1 + \frac{\beta x\sinh\frac{\beta L}{2}}{2\cosh\frac{\beta L}{2}}\right)\cosh\beta x\right] \tag{6}$$

The final expression is found when (4) and (6) are substituted into Eq. (12.11a)

$$v = \sum_{m=1}^{\infty} v_m(x)\sin\frac{m\pi y}{L_y}$$

$$= \frac{4w_1 L_y^4}{\pi^5 D}\sum_{m=1,3,5,\ldots}\frac{1}{m^5}\left[1 + \left(\frac{\beta x}{2} + \tanh\frac{\beta L}{2} - \frac{\beta L}{4\cosh^2\frac{\beta L}{2}}\right)\sinh\beta x - \left(1 + \frac{\beta x}{2}\tanh\frac{\beta L}{2}\right)\cosh\beta x\right]\sin\frac{m\pi y}{L_y}$$

This is a rapidly converging series, so that often the first term suffices for most practical problems.

This same result is obtained if the transfer matrix elements are taken from the general transfer matrix of Table 12-9. To accomplish this, set $D_x = D_y = D$, $\nu_x = \nu_y = \nu$, $D_{xy} = (1-\nu)D/2$ (Table 12-4) and $k = \rho = P = P_y = 0$. Then $\lambda = (m\pi/L_y)^4$, $\zeta = -2(m\pi/L_y)^2$ so that $\lambda = \zeta^2/4$. The transfer matrix elements are taken from Table 12-9 for $\lambda > 0$, $\lambda = \zeta^2/4$, $\zeta < 0$, $\beta^2 = -\zeta/2$.

EXAMPLE 12.4 Sinusoidally Loaded Plate Determine the deflection of the square plate with sinusoidal loading of Fig. 12-2.

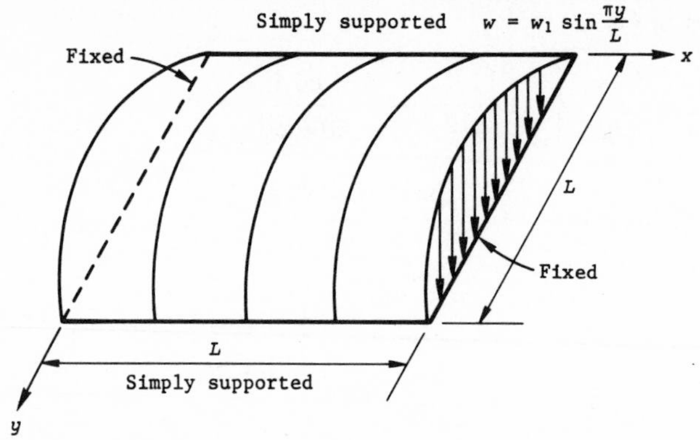

Fig. 12-2 Example 12.4.

The solution procedure is the same as for Example 12.3. The factor ϵ_m, which varies according to the distribution of loading in the y direction, must be selected as case 3 of Table 12-8. Also the initial parameters must be chosen to satisfy the fixed ends at $x = 0, L$.

Thus, from Table 12-8,

$$\epsilon_m = \begin{cases} 1 & \text{if } m = 1 \\ 0 & \text{otherwise} \end{cases} \tag{1}$$

From Table 2-19 for fixed ends,

$$\begin{aligned} v_0 = \theta_0 = 0 \qquad & M_0 = \frac{1}{\nabla}(F_\theta U_{vV} - F_v U_{\theta V})_{x=\ell=L} \\ V_0 &= \frac{1}{\nabla}(F_v U_{\theta M} - F_\theta U_{vM})_{x=\ell=L} \\ \nabla &= (U_{vM} U_{\theta V} - U_{\theta M} U_{vV})_{x=\ell=L} \end{aligned} \tag{2}$$

With $\ell = x$, the transfer matrix elements needed in (2) are

$$\begin{aligned} U_{vM} &= -\frac{x \sinh \beta x}{2D\beta} \qquad U_{vV} = \frac{1}{2D\beta^2}\left(\frac{\sinh \beta x}{\beta} - x \cosh \beta x\right) \\ U_{\theta M} &= \frac{1}{2D}\left(\frac{\sinh \beta x}{\beta} + x \cosh \beta x\right) \qquad U_{\theta V} = -U_{vM} \\ F_v &= \frac{\epsilon_m w_1}{2D\beta^4}(\beta x \sinh \beta x - 2 \cosh \beta x + 2) \\ F_\theta &= \frac{\epsilon_m w_1}{2D\beta^2}\left(\frac{\sinh \beta x}{\beta} - x \cosh \beta x\right) \end{aligned} \tag{3}$$

where $\beta^2 = (m\pi/L)^2$. Insertion of (3) in (2) leads to

$$\begin{aligned} \nabla &= \frac{1}{4D^2\beta^4}(\beta^2 L^2 - \sinh^2 \beta L) \\ M_0 &= \frac{\epsilon_m w_1}{\beta^2}\,\frac{\beta L - \sinh \beta L}{\beta L + \sinh \beta L}, \quad V_0 = \frac{\epsilon_m w_1}{\beta}\,\frac{\cosh \beta L - 1}{\beta L + \sinh \beta L} \end{aligned} \tag{4}$$

From Eq. (1) of Example 12.3,

$$v_m = M_0 U_{vM} + V_0 U_{vV} + F_v$$

$$= \frac{\epsilon_m w_1}{D\beta^4}\left[1 + \frac{\beta x \sinh \beta L + \cosh \beta L - 1}{\beta L + \sinh \beta L} \sinh \beta x \right.$$

$$\left. + \frac{\beta x(\cosh \beta L - 1) - \beta L - \sinh \beta L}{\beta L + \sinh \beta L} \cos \beta x \right] \tag{5}$$

Since ϵ_m vanishes for all but $m = 1$, where it is unity, the deflection according to Eq. (12.11a) is

$$v = \sum_{m=1}^{\infty} v_m \sin \frac{m\pi y}{L_y} = v_1 \sin \frac{\pi y}{L} \tag{6}$$

Note that $\beta L = \pi$ when $m = 1$, so we have the deflection

$$v = \frac{w_1}{D\beta^4}\left[1 + \frac{\beta x \sinh \pi + \cosh \pi - 1}{\pi + \sinh \pi} \sinh \beta x \right.$$

$$\left. + \frac{\beta x(\cosh \pi - 1) - \pi - \sinh \pi}{\pi + \sinh \pi} \cosh \pi \right] \sin \frac{\pi y}{L}$$

12.13 Stability

A variety of stability problems can be solved. The general transfer matrix of Table 12-9 is applicable for most stability problems. If P_y is zero, then the critical in-plane load $(P_x)_{cr} = P_{cr}$ is calculated in the same way that the buckling axial loads are found for beams. This procedure is described in Section 2.17. This critical load can be found from the variables v_m, θ_m, M_m, V_m, with $m = 1$, of this chapter rather than v, θ, M, V of Chap. 2. As in Chap. 2, the transfer matrix formulation permits P to vary in the x direction. There is no need to use Eqs. (12.11) except to calculate the deflected shape during buckling.

If P is zero, then $(P_y)_{cr}$ is determined in the same fashion as P_{cr}. The formulation requires that P_y, a compressive in-plane force, be applied at the simply supported edges $y = 0$, $y = L_y$. Table 12-7 contains the transfer matrix corresponding to this case.

If both P and P_y are nonzero, then combinations of P and P_y can be found that make the boundary condition determinant ∇ equal zero. Frequently, it is known that the forces P and P_y remain proportional to each other, as when $P_y = cP$ where c is a known constant of proportionality. Then the usual procedure for finding a buckling load is followed to compute P_{cr}.

EXAMPLE 12.5 Buckling of Plate with a Free Edge Find the in-plane force $(P_y)_{cr}$ of the plate of Fig. 12-3. The plate thickness is constant. The $y = 0$, $y = L_y$, $x = 0$ edges are simply supported. The $x = L$ edge is free.

The critical in-plane force $(P_y)_{cr}$ is found as the lowest value of P_y that makes

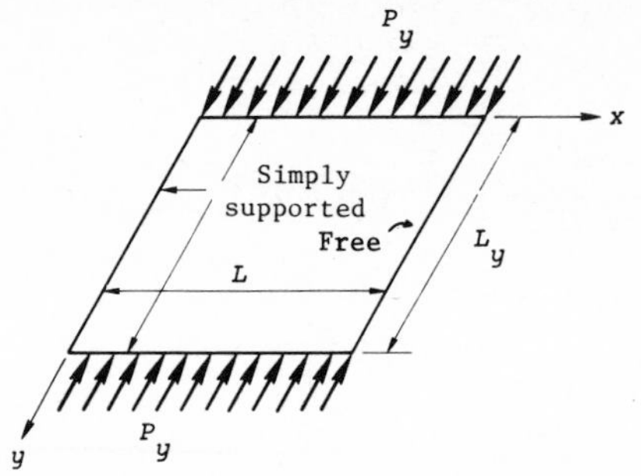

Fig. 12-3 Example 12.5.

the characteristic equation equal to zero. This characteristic equation is established from the boundary conditions, or it can be taken as ∇ from Table 2-19. Table 12-7 provides the required transfer matrix.

The transfer matrix of Table 12-7 applies to plates with simple supports at $y = 0$, $y = L_y$. From Table 2-19, the value of ∇ for a simply supported edge at $x = 0$ and a free edge at $x = L$ is

$$\nabla = (U_{M\theta} U_{VV} - U_{V\theta} U_{MV})_{x=\ell=L} \tag{1}$$

The transfer matrix elements are taken from Table 12-7 as

$$\begin{aligned}
U_{M\theta} &= \frac{D\alpha^2}{g}\left[\left(\sqrt{k} + \gamma\alpha\right)^2 \frac{\sinh a\ell}{a} - \left(\sqrt{k} - \gamma\alpha\right)^2 \frac{\sin b\ell}{b}\right] \\
U_{MV} &= \frac{\alpha}{g}\left[\left(\sqrt{k} + \gamma\alpha\right)^2 \frac{\sinh a\ell}{a} + \left(\sqrt{k} - \gamma\alpha\right) \frac{\sin b\ell}{b}\right] \\
U_{V\theta} &= \frac{D\alpha^2}{g}(k - \gamma^2\alpha^2)(\cosh a\ell - \cos b\ell) \\
U_{VV} &= \frac{\alpha}{g}\left[\left(\sqrt{k} - \gamma\alpha\right)\cosh a\ell + \left(\sqrt{k} + \gamma\alpha\right)\cos b\ell\right]
\end{aligned} \tag{2}$$

where $\alpha = m\pi / L_y$

$k = P_y / D$

$\gamma = 1 - \nu$

$$a = \alpha\sqrt{\frac{\sqrt{k}}{\alpha} + 1}$$

$$b = \alpha\sqrt{\frac{\sqrt{k}}{\alpha} - 1}$$

$g = a^2 + b^2$

In this case, the critical load corresponds to the lowest value of P_y that makes ∇ of (1) equal zero. We substitute (2) into (1) and find

$$\frac{b}{\alpha}\left(\frac{L_y\sqrt{k}}{m\pi} + \gamma\right)^2 \tanh\frac{m\pi aL}{\alpha L_y} + \frac{a}{\alpha}\left(\frac{L_y\sqrt{k}}{m\pi} - \gamma\right)^2 \tan\frac{m\pi bL}{\alpha L_y} = 0 \tag{3}$$

By letting $m = 1, 2, 3 \ldots$ in succession, we can solve this equation for $\sqrt{k}$ (and therefore P_y) for a specified aspect ratio L/L_y. The lowest value of P_y is obtained for $m = 1$. For $\nu = 0.25$, (3) leads to

$$(P_y)_{\text{cr}} = \frac{\pi^2 D}{L_y^2}\left(1 + 0.456\,\frac{L_y^2}{L^2}\right)$$

for long plates with $L_y/L > 10$.

12.14 Free Dynamic Response—Natural Frequencies

The natural frequencies ω_{mn} for rectangular plates are found by applying the procedure for beams of Section 2.18 to the v_m, θ_m, M_m, V_m that are formed with the transfer matrices of this chapter. The primary difference between solutions for the two lies in the number of natural frequencies. For each value of m, there are n, $n = 1, 2, 3, \ldots$, values of ω.

The mode shapes are found by placing the natural frequencies ω_{mn} in v_m, θ_m, M_m, V_m. The variations in the y direction can be included by introducing $\sin m\pi y/L_y$, that is,

$$v_m \sin\frac{m\pi y}{L_y} \qquad \theta_m \sin\frac{m\pi y}{L_y} \qquad M_m \sin\frac{m\pi y}{L_y} \qquad V_m \sin\frac{m\pi y}{L_y}$$

EXAMPLE 12.6 Natural Frequencies of Plate with Mixed Edge Conditions Find the natural frequencies of a plate simply supported at $y = 0, L_y$ and fixed at $x = 0, L$.

The boundary conditions lead to the frequency equation from which ω is determined. The formulas of the transfer matrices apply for simply supported edges on $y = 0, L_y$. For fixed edges at $x = 0$, $x = L$ the boundary conditions are $v|_{x=0} = \theta|_{x=0} = v|_{x=L} = \theta|_{x=L} = 0$. These lead to

$$\begin{aligned} M_0 U_{vM} + V_0 U_{vV} &= 0 \\ M_0 U_{\theta M} + V_0 U_{\theta V} &= 0 \end{aligned} \tag{1}$$

where the loading functions are ignored, since they do not enter the analysis for natural frequencies. The elements of U_{kj} are taken from the transfer matrix of Table 12-7 with $\ell = L$. Equations (1) yield

$$\nabla = \begin{vmatrix} U_{vM} & U_{vV} \\ U_{\theta M} & U_{\theta V} \end{vmatrix}_{x=\ell=L} = 0 \tag{2}$$

This same relation can be taken from Table 2-19 for fixed-fixed end conditions.

From Table 12-7,

$$\begin{aligned} U_{vM} &= -\frac{1}{gD}(\cosh aL - \cos bL) \qquad & U_{vV} &= -\frac{1}{gD}\left(\frac{1}{a}\sinh aL - \frac{1}{b}\sin bL\right) \\ U_{\theta M} &= \frac{1}{gD}(a \sinh aL + b \sin bL) \qquad & U_{\theta V} &= \frac{1}{gD}(\cosh aL - \cos bL) \end{aligned} \tag{3}$$

with $g = 2\omega\sqrt{\rho/D}$, $a^2 = \omega\sqrt{\rho/D} + (m\pi/L_y)^2$, $b^2 = \omega\sqrt{\rho/D} - (m\pi/L_y)^2$. Placing (3) in (2) gives the frequency equation

$$\nabla = \frac{1}{g^2D^2}\left(-2 + 2\cosh aL\cos bL + \frac{b^2 - a^2}{ab}\sinh aL\sin bL\right) = 0 \qquad (4)$$

Those values of ω that satisfy (4) are the natural frequencies. An array of frequencies ω are found for each value of m, $m = 1, 2, 3, \ldots$. These ω are labeled as ω_{mn}, $m = 1, 2, 3, \ldots$; $n = 1, 2, 3, \ldots$. For a square plate, that is, $L/L_y = 1$, (4) leads to a fundamental natural frequency of $\omega = \omega_{11} = 28.9510/(L^2\sqrt{\rho/D})$.

12.15 Forced Harmonic Motion

If the forcing functions vary as $\sin \Omega t$ or $\cos \Omega t$, where Ω is the frequency of the loading, the state variables v_m, θ_m, M_m, and V_m also vary as $\sin \Omega t$ or $\cos \Omega t$. The spatial x distribution of v_m, θ_m, M_m, and V_m is found by setting up a static solution using those transfer matrices containing ω. In these transfer matrices, ω should be replaced by Ω. The solution procedure follows that described in Chap. 2 for beams.

12.16 Dynamic Response Due to Arbitrary Loading

In terms of a modal solution, the time-dependent state variables v_m, θ_m, M_m, V_m resulting from arbitary dynamic loading are expressed by

Displacement method:

$$\begin{aligned} v_m(x,t) &= \sum_n A_n(t)v_{mn}(x) \\ \theta_m(x,t) &= \sum_n A_n(t)\theta_{mn}(x) \\ M_m(x,t) &= \sum_n A_n(t)M_{mn}(x) \\ V_m(x,t) &= \sum_n A_n(t)V_{mn}(x) \end{aligned} \qquad (12.15a)$$

Acceleration method:

$$\begin{aligned} v_m(x,t) &= v_{sm}(x,t) + \sum_n B_n(t)v_{mn}(x) \\ \theta_m(x,t) &= \theta_{sm}(x,t) + \sum_n B_n(t)\theta_{mn}(x) \\ M_m(x,t) &= M_{sm}(x,t) + \sum_n B_n(t)M_{mn}(x) \\ V_m(x,t) &= V_{sm}(x,t) + \sum_n B_n(t)V_{mn}(x) \end{aligned} \qquad (12.15b)$$

where

$$A_n(t) = \frac{\eta_n(t)}{N_n} \qquad B_n(t) = \frac{\xi_n(t)}{N_n} \qquad (12.16)$$

Equations (12.15) placed in Eqs. (12.11) provide the complete dynamic response. That is, these are the solutions to Eqs. (12.10).

No Damping or Proportional Damping For viscous damping a term $c'\partial v_m/\partial t$ should be added to the right-hand side of Eq. (12.13d). If the damping is "proportional," then $c' = c_\rho \rho$. Set $c' = 0$ for no damping.

In Eqs. (12.15b) the terms with subscript s are static solutions, as given in Section 12.12, that are determined for the applied loading at each point in time. The terms v_{mn}, θ_{mn}, M_{mn}, and V_{mn} of Eqs. (12.15a, b) are the mode shapes, that is, v_m, θ_m, M_m, V_m with $\omega = \omega_{mn}$, $n = 1, 2, \ldots$.

The quantities in Eq. (12.16) are defined as

$$N_n = \int_0^L \rho v_{mn}^2 \, dx \tag{12.17a}$$

$$\eta_n(t) = e^{-\zeta_n\omega_{mn}t}\left(\cos\alpha_n t + \frac{\zeta_n\omega_{mn}}{\alpha_n}\sin\alpha_n t\right)\eta_n(0) + e^{-\zeta_n\omega_{mn}t}\,\frac{\sin\alpha_n t}{\alpha_n}\,\frac{\partial\eta_n}{\partial t}(0)$$
$$+ \int_0^t f_n(\tau)e^{-\zeta_n\omega_{mn}(t-\tau)}\,\frac{\sin\alpha_n(t-\tau)}{\alpha_n}\,d\tau \tag{12.17b}$$

$\xi_n(t)$ is taken from $\eta_n(t)$ by replacing

$$\eta_n(0) \quad \text{by} \quad \eta_n(0) - \frac{f_n(0)}{\omega_{mn}^2}$$

$$\frac{\partial\eta_n}{\partial t}(0) \quad \text{by} \quad \frac{\partial\eta_n}{\partial t}(0) - \frac{1}{\omega_{mn}^2}\frac{\partial f_n}{\partial t}(0)$$

$$f_n(\tau) \quad \text{by} \quad -\frac{1}{\omega_{mn}^2}\left[\frac{\partial^2}{\partial\tau^2} + c_\rho\frac{\partial}{\partial\tau} + (\omega_{mn}\zeta_n)^2\right]f_n(\tau)$$

and adding the term $\zeta_n^2 f_n(t)/[(1-\zeta_n^2)\omega_{mn}^2]$. In these expressions, $\alpha_n = \omega_{mn}\sqrt{1-\zeta_n^2}$, $\zeta_n = c_\rho/2\omega_{mn}$.

If $\zeta_n > 1$, replace sin by sinh, cos by cosh, and $\sqrt{1-\zeta_n^2}$ by $\sqrt{\zeta_n^2 - 1}$. For zero viscous damping, $\zeta_n = 0$. Then

$$\eta_n(0) = \int_0^L \rho v(x, 0)v_n \, dx \tag{12.18}$$

$$f_n(t) = \int_0^L\left(\left\{w_m(x,t) + \frac{1}{D_x}\left[P - \nu_x D_y\left(\frac{m\pi}{L_y}\right)^2\right]M_{Txm} + \left(\frac{m\pi}{L_y}\right)^2 M_{Tym}\right\}v_{mn}\right.$$
$$\left. + \frac{M_{Txm}}{D_x}M_{mn}\right)dx + h_n(a_k, t) \tag{12.19a}$$

Note that although they are not explicitly so indicated, such functions as N_n, f_n, and η_n vary with m.

In the case of the acceleration method, $f_n(t)$ can alternatively be expressed as

$$f_n(t) = \omega_{mn}^2\int_0^L \rho v_{sm}(x,t)v_{mn}\,dx \tag{12.19b}$$

The function $h_n(a_k, t)$ accounts for nonhomogeneous displacement $v_m(a_k, t)$ or rotation $\theta_m(a_k, t)$ conditions at $x = a_k$, such as supports or prescribed time-dependent displacements located in-span or on the boundary:

$$h_n(a_k, t) = -v_m(a_k, t)\,\Delta V_{mn}(a_k) - \theta_m(a_k, t)\,\Delta M_{mn}(a_k)$$

$$\Delta V_{mn}(a_k) = V_{mn}(a_k^-) - V_{mn}(a_k^+) \qquad +(-) \text{ means just to the right (left) of } x = a_k \tag{12.20a}$$

$$\Delta M_{mn}(a_k) = M_{mn}(a_k^-) - M_{mn}(a_k^+)$$

If the right end ($x = L$) has a nonzero (for example, time-dependent) displacement, then Eq. (12.20a) reduces for $a_k = L$ to

$$h_n(a_k, t) = -v_m(L, t)V_{mn}(L) - \theta_n(L, t)M_{mn}(L) \tag{12.20b}$$

If the left end ($x = 0$) has a nonhomogeneous displacement, then Eq. (12.20a) reduces for $a_k = 0$ to

$$h_n(a_k, t) = +v_m(0, t)V_{mn}(0) + \theta_m(0, t)M_{mn}(0) \tag{12.20c}$$

Nonproportional Damping The formulas of this chapter need considerable adjustment in order to make them applicable to materials with nonproportional viscous damping. Equation (12.13d) should be supplemented by the term $c'\partial v_m/\partial t$ on the right-hand side. Replace the frequency ω_{mn} by s_{mn} where s_{mn} is a complex number. The real part of s_{mn} is commonly referred to as the damping exponent, whereas the imaginary part of s_{mn} is called the frequency of the damped free vibration. Also, the mode shapes and some of the transfer matrices are complex functions.

Equations (12.15) and (12.16) remain valid, but now the static solutions of Eq. (12.15b) are of the form

$$v_{sm}(x, t) = -\sum_n \frac{v_{mn}(x)f_n(t)}{s_{mn}N_n} \tag{12.21}$$

where $v_{mn}(x)$ and N_n are the mode shape and norm associated with the complex eigenvalue, s_{mn}. Expressions similar to Eq. (12.21) apply for the slope, moment, and shear force.

The definitions of the previous subsection must be adjusted so that

$$N_n = 2s_{mn}\int_0^L \rho v_{mn}^2\,dx + \int_0^L c' v_{mn}^2\,dx \tag{12.22}$$

$$\eta_n(t) = e^{s_{mn}t}\eta_n(0) + \int_0^t e^{s_{mn}(t-\tau)}f_n(\tau)\,d\tau \tag{12.23}$$

$$\xi_n(t) = e^{s_{mn}t}\left[\frac{1}{s_{mn}^2}\frac{\partial f_n}{\partial t}(0) + \frac{1}{s_{mn}}f_n(0) + \eta_n(0)\right] - \frac{1}{s_{mn}^2}\frac{\partial f_n}{\partial t}(t) + \frac{1}{s_{mn}^2}\int_0^t e^{s_{mn}(t-\tau)}\frac{\partial^2 f_n}{\partial \tau^2}(\tau)\,d\tau \tag{12.24}$$

$$\eta_n(0) = \int_0^L\left[\rho s_{mn}v_m(x, 0)v_{mn}(x) + c'v_m(x, 0)v_{mn}(x) + \rho\frac{\partial v_m}{\partial t}(x, 0)v_{mn}(x)\right]dx \tag{12.25}$$

In the case of the acceleration method, $f_n(t)$ can alternatively be expressed as

$$f_n(t) = -s_{mn}\int_0^L c' v_{sm}(x, t) v_{mn}(x)\, dx - s_{mn}^2 \int_0^L \rho v_{sm}(x, t) v_{mn}(x)\, dx \quad (12.26)$$

For discrete line damping with dashpot constants c_i (force · time/length^2) at $x = a_i$, use the point matrix of Table 12-10 for $x = a_i$ with k_1 replaced by $k_1 + sc_i$ and ω^2 by $-s^2$. Also, in N_n set

$$\int_0^L c' v_{mn}^2\, dx = \sum_i c_i v_{mn}^2(a_i) \quad (12.27)$$

an in $\eta_n(0)$ set

$$\int_0^L c' v_m(x, 0) v_{mn}(x)\, dx = \sum_i c_i v_m(a_i, 0) v_{mn}(a_i) \quad (12.28)$$

Voigt-Kelvin Material The formulas of this chapter are applicable to a plate of Voigt-Kelvin material. In the equations of motion, Eqs. (12.13), replace E by $E(1 + \epsilon\, \partial/\partial t)$ where ϵ is the Voigt-Kelvin damping coefficient, and where similar behavior in dilatation and shear is assumed. If the plate is externally damped (for example, if it is resting on a proportional-damped foundation), retain the $c_\rho \rho\, \partial v_m/\partial t$ term on the right-hand side of Eq. (12.13d). The response formulas of this section for proportional damping apply if c_ρ is replaced by $c_\rho + \epsilon(\omega_{mn})^2$ and ζ_n is redefined as $\zeta_n = c_\rho/(2\omega_{mn}) + \epsilon\omega_{mn}/2$. Continue to use the same undamped mode shapes employed previously, where ω_{mn} is the corresponding nth undamped frequency.

C. COMPUTER PROGRAMS AND EXAMPLES

12.17 Benchmark Examples

Complicated plate problems should be solved with the assistance of a computer program. The following examples are provided as benchmark examples against which a reader's own program can be checked. The computer program RECTANGULARPLATE was used for these examples. RECTANGULARPLATE calculates the deflection, slope, bending moments, twisting moment, and shear force of a plate for static and steady state conditions, the critical load and mode shapes for stability, and the natural frequencies and mode shapes for transverse vibrations. The plate can be formed of segments of different geometric and material (isotropic or orthotropic) properties with any mechanical or thermal loading, in-span supports, foundations, and boundary conditions.

EXAMPLE 12.7 Plate with Variable Thickness Find the deflection, slope, moments, and shear force in the plate of variable thickness shown in Fig. 12-4. Ignore the rigid in-span line support. The plate is 20 in wide. A 2-lb/in^2 load covers the plate. Let $E = 30(10^6)$ lb/in^2, $\nu = 0.3$.

Partial results are provided in Fig. 12-5.

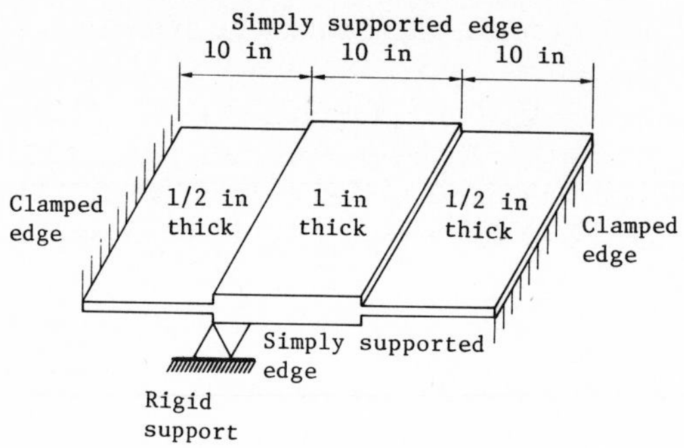

Fig. 12-4 Examples 12.7 and 12.8 (with no rigid support) and Example 12.9 (with rigid support).

FOR Y EQUAL TO 5.00000

X LOCATION	DISPL	SLOPE	MOMENT	SHEAR	Y MOMENT	TWIST MOMENT
0.00	0.	0.	-4.1757E+01	1.7557E+01	-1.2527E+01	0.
5.00	7.28E-04	-1.77E-04	8.8619E+00	7.0826E+00	8.9774E+00	-6.4370E+00
10.00	1.19E-03	1.10E-05	1.7719E+01	-1.3677E+00	8.2316E+01	3.2494E-01
15.00	1.16E-03	-3.29E-13	1.9150E+01	3.5722E-07	7.9779E+01	-5.2134E-07
20.00	1.19E-03	-1.10E-05	1.7719E+01	1.3677E+00	1.4941E+01	-4.0621E-02
25.00	7.28E-04	1.77E-04	8.8613E+00	-7.0823E+00	8.9780E+00	6.4364E+00
30.00	2.44E-07	-2.07E-07	-4.1799E+01	-1.7532E+01	-1.2493E+01	-3.9046E-02

FOR Y EQUAL TO 10.00000

X LOCATION	DISPL	SLOPE	MOMENT	SHEAR	Y MOMENT	TWIST MOMENT
0.00	0.	0.	-5.3743E+01	2.0601E+01	-1.6123E+01	0.
5.00	9.84E-04	-2.46E-04	1.0346E+01	9.5433E+00	9.6934E+00	-1.1611E-05
10.00	1.65E-03	1.04E-05	2.6201E+01	3.8393E-01	1.0470E+02	-1.2319E-06
15.00	1.63E-03	4.65E-13	2.6367E+01	-5.0518E-07	1.0511E+02	4.7550E-12
20.00	1.65E-03	-1.04E-05	2.6201E+01	-3.8396E-01	1.9965E+01	1.5402E-07
25.00	9.84E-04	2.46E-04	1.0347E+01	-9.5438E+00	9.6926E+00	1.1616E-05
30.00	-3.44E-07	2.92E-07	-5.3683E+01	-2.0636E+01	-1.6171E+01	3.6672E-07

Fig. 12-5 Partial output for Example 12.7.

EXAMPLE 12.8 Natural Frequencies of Plate with Variable Thickness Find the first three natural frequencies of the plate of Example 12.7. The frequencies should be the first three corresponding to a single half-wave mode shape in the y direction.

These frequencies are found to be 1472, 2940, and 5400 Hz.

EXAMPLE 12.9 Plate with In-Span Support Calculate the deflection, slope, moments, and shear force of the plate of Fig. 12-4. This plate, with a 2 lb/in^2 uniform load, rests on an in-span rigid support as shown. Let $E = 3(10^7)$ lb/in^2, $\nu = 0.3$.

Some output is given in Fig. 12-6.

FOR Y EQUAL TO 10.00000

X LOCATION	DISPL	SLOPE	MOMENT	SHEAR	Y MOMENT	TWIST MOMENT
0.00	0.	0.	-1.2469E+01	9.0750E+00	-3.7408E+00	0.
5.00	7.03E-05	1.72E-05	6.2587E+00	-1.5768E+00	1.7375E+00	8.4682E-07
10.00	-7.76E-15	-7.32E-05	-2.8751E+01	2.1029E+01	-8.6253E+00	-2.8443E-05
15.00	4.25E-04	-8.66E-05	1.3171E+01	1.1236E+01	2.8231E+01	-3.1827E-05
20.00	8.31E-04	-7.73E-05	1.5694E+01	3.5591E+00	1.0496E+01	-3.1349E-06
25.00	6.53E-04	1.41E-04	1.1650E+01	-6.2965E+00	7.5376E+00	6.3611E-06
30.00	-1.08E-09	-7.57E-09	-4.0094E+01	-1.7215E+01	-1.2029E+01	-9.4707E-09

Fig. 12-6 Partial output for Example 12.9.

REFERENCES

12.1 Timoshenko S., and Woinowsky-Krieger, S.: *Theory of Plates and Shells*, 2d ed., McGraw-Hill, New York, 1959.

12.2 Szilard, R.: *Theory and Analysis of Plates*, Prentice-Hall, New York, 1974.

12.3 Lekhnitskii, S. G.: *Anisotropic Plates*, 2d ed., Gordon and Breach, New York, 1968.

12.4 Bareš, R.: *Tables for the Analysis of Plates, Slabs and Diaphragms Based on the Elastic Theory*, Bauverlag, Wiesbaden, 1971.

12.5 Leissa, A.: *Vibration of Plates*, NASA SP-160, Washington, 1969.

THIRTEEN

Thin-Walled Cylinders

This chapter treats the static, stability, and dynamic radial analysis of circular cylindrical shells. The applied loading, which can be either mechanical or thermal, and all responses must be symmetrical about the cylinder's axis. The formulas of this chapter apply to cylinders for which the shell thickness is much smaller than the mean shell radius. Stresses in thick-walled cylinders can be found with the formulas of Chap. 9.

13.1 Notation and Conventions

v	Radial displacement, positive inwards toward shell center (length)
θ	Angle of rotation or slope of the displacement curve (radians)
M	Axial bending moment per unit of circumferential length (force · length/length)
V	Shear force per unit of circumferential length (force/length)
h	Thickness of shell (length)
R	Radius of the cylinder (length)
E	Modulus of elasticity of the material (force/length^2)
ν	Poisson's ratio
L	Length of shell in the x direction
ϕ	Circumferential coordinate (radians)

$D = Eh^3/12(1-\nu^2)$

w	Distributed transverse force, radial loading intensity (force/length^2)
$P = P_x$	Normal compressive (in-plane, axial) force per unit of circumferential length (force/length). Replace P by $-P$ for a tensile force.
k	Modulus of elastic foundation (force/length^3)
P_ϕ	Circumferential (hoop) membrane force per unit length (force/length)
M_ϕ	Circumferential bending moment per unit length (force · length/length)
D_V	Shear stiffness
D_x, D_ϕ	Flexural rigidities; see Table 13-1
K_x, K_ϕ, K	Extensional rigidities; see Table 13-1.
ν_x, ν_ϕ	Poisson's ratios in the x, ϕ directions
E_x, E_ϕ	Moduli of elasticity in the x, ϕ directions (force/length^2)
α_x, α_ϕ,	Coefficients of thermal expansion in the x, ϕ directions (length/length · degree). For isotropic material $\alpha_x = \alpha_\phi = \alpha$.
ΔT	Temperature change (degrees), that is, the temperature rise with respect to the reference temperature.
M_T	Thermal moment; see Table 13-1.
P_{Tx}, $P_{T\phi}$	Thermal forces; see Table 13-1.
σ_x, σ_ϕ	Normal stresses (force/length^2)
$x, y, z,$	Right-handed coordinate system
ρ	Mass per unit area (mass/length^2, force · time^2/length^3)
ω	Natural frequency (radians/second)
	Positive deflection v, internal moments, and internal forces are displayed in Fig. 13-1. Applied forces are positive if they are in the direction of positive coordinate axes.
$\mathbf{U}_i$	Field matrix of the ith segment
$\overline{\mathbf{U}}_i$	Point matrix at $x = a_i$

The notation for transfer matrices is

$$\mathbf{U}_i = \begin{bmatrix} U_{vv} & U_{v\theta} & U_{vM} & U_{vV} & F_v \\ U_{\theta v} & U_{\theta\theta} & U_{\theta M} & U_{\theta V} & F_\theta \\ U_{Mv} & U_{M\theta} & U_{MM} & U_{MV} & F_M \\ U_{Vv} & U_{V\theta} & U_{VM} & U_{VV} & F_V \\ 0 & 0 & 0 & 0 & 1 \end{bmatrix} \tag{13.1}$$

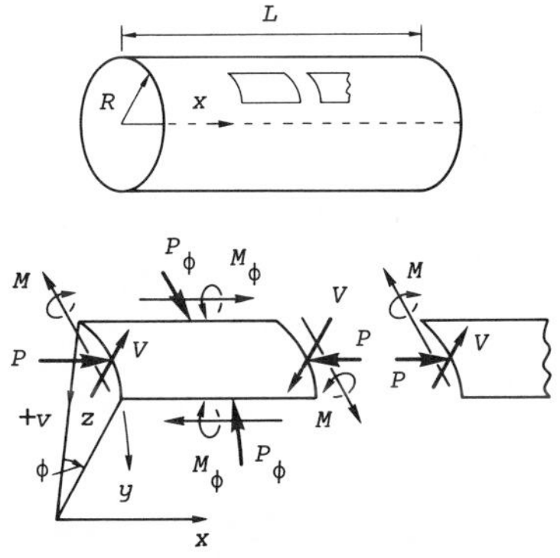

Fig. 13-1 Positive deflection, internal moments, and forces.

TABLE 13-1 Material Constants

Cylinder	Constants
1. Homogeneous Isotropic	$\nu_x = \nu_\phi = \nu,\ E_x = E_\phi = E,\ \alpha_x = \alpha_\phi = \alpha,\ D_x = D_\phi = D = \dfrac{Eh^3}{12(1 - \nu^2)}$ $K_x = K_\phi = K = \dfrac{Eh}{1 - \nu^2},\ P_{Tx} = P_{T\phi} = -\dfrac{1}{1 - \nu}\int_h E\alpha\ \Delta T\ dz,\ M_T = \dfrac{1}{1 - \nu}\int_h E\alpha\ \Delta T\ z\ dz$
2. Homogeneous Orthotropic	$D_x = \dfrac{E_x h^3}{12(1 - \nu_x\nu_\phi)},\ D_\phi = \dfrac{E_\phi h^3}{12(1 - \nu_x\nu_\phi)},\ K_x = \dfrac{E_x h}{1 - \nu_x\nu_\phi},\ K_\phi = \dfrac{E_\phi h}{1 - \nu_x\nu_\phi}$ $P_{Tx} = -\dfrac{E_x(\alpha_x + \nu_\phi\alpha_\phi)}{1 - \nu_x\nu_\phi}\int_h \Delta T\ dz,\ P_{T\phi} = -\dfrac{E_\phi(\alpha_\phi + \nu_x\alpha_x)}{1 - \nu_x\nu_\phi}\int_h \Delta T\ dz$ $M_T = \dfrac{E_x(\alpha_x + \nu_\phi\alpha_\phi)}{1 - \nu_x\nu_\phi}\int_h \Delta T\ z\ dz$
3. Continuously Composite	$D_x = \dfrac{1}{1 - \nu_x\nu_\phi}\int_h E_x z^2\ dz,\ D_\phi = \dfrac{1}{1 - \nu_x\nu_\phi}\int_h E_\phi z^2\ dz,\ K_x = \dfrac{1}{1 - \nu_x\nu_\phi}\int_h E_x\ dz,\ K_\phi = \dfrac{1}{1 - \nu_x\nu_\phi}\int_h E_\phi\ dz$ $P_{Tx} = -\dfrac{1}{1 - \nu_x\nu_\phi}\int_h E_x(\alpha_x + \nu_\phi\alpha_\phi)\ \Delta T\ dz,\ P_{T\phi} = -\dfrac{1}{1 - \nu_x\nu_\phi}\int_h E_\phi(\alpha_\phi + \nu_x\alpha_x)\ \Delta T\ dz$ $M_T = \dfrac{1}{1 - \nu_x\nu_\phi}\int_h E_x(\alpha_x + \nu_\phi\alpha_\phi) z\ \Delta T\ dz$

4. Symmetrically Constructed with Isotropic Layers (Ref. 13.1) 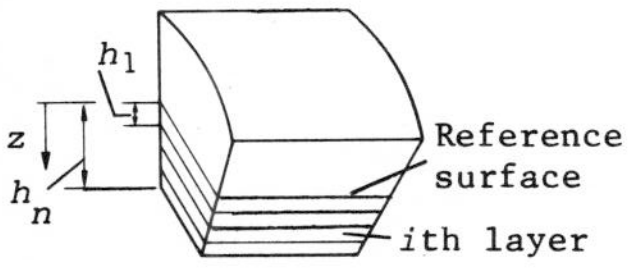 For *i*th layer, E_i, ν_i, α_i	The layers are arranged symmetrically with respect to the middle surface which is taken as the reference surface. $h_0 = 0$, $\bar{E}_i = E_i/(1 - \nu_i^2)$ $D_x = D_\phi = D = \frac{2}{3}\sum_{i=1}^{n} \bar{E}_i (h_i^3 - h_{i-1}^3)$, $K_x = K_\phi = K = 2\sum_{i=1}^{n} \bar{E}_i (h_i - h_{i-1})$ $M_T = 2\sum_{i=1}^{n} \bar{E}_i \alpha_i (1 + \nu_i) \int_{\Delta h_i} z\, \Delta T\, dz$, $P_{Tx} = P_{T\phi} = P_T = -2\sum_{i=1}^{n} \bar{E}_i \alpha_i (1 + \nu_i) \int_{\Delta h_i} \Delta T\, dz$
5. Symmetrically Constructed with Orthotropic Layers (Ref. 13.2) For the *i*th layer, E_{xi} $E_{\phi i}$, ν_{xi}, $\nu_{\phi i}$, α_{xi}, $\alpha_{\phi i}$	$D_x = \frac{2}{3}\sum_{i=1}^{n} \bar{E}_{xi}(h_i^3 - h_{i-1}^3)$, $D_\phi = \frac{2}{3}\sum_{i=1}^{n} \bar{E}_{\phi i}(h_i^3 - h_{i-1}^3)$, $K_x = 2\sum_{i=1}^{n} \bar{E}_{xi}(h_i - h_{i-1})$ $K_\phi = 2\sum_{i=1}^{n} \bar{E}_{\phi i}(h_i - h_{i-1})$, $M_T = 2\sum_{i=1}^{n} \bar{E}_{xi}(\alpha_{xi} + \nu_{\phi i}\alpha_{\phi i}) \int_{\Delta h_i} z\, \Delta T\, dz$ $P_{Tx} = -2\sum_{l=1}^{n} \bar{E}_{xi}(\alpha_{xi} + \nu_{\phi i}\alpha_{\phi i}) \int_{\Delta h_i} \Delta T\, dz$, $P_{T\phi} = -2\sum_{i=1}^{n} \bar{E}_{\phi i}(\alpha_{\phi i} + \nu_{xi}\alpha_{xi}) \int_{\Delta h_i} \Delta T\, dz$ $h_0 = 0$, $\bar{E}_{xi} = \frac{E_{xi}}{1 - \nu_{xi}\nu_{\phi i}}$, $\bar{E}_{\phi i} = \frac{E_{\phi i}}{1 - \nu_{xi}\nu_{\phi i}}$

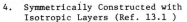

13.2 Stresses

The tables of this chapter provide the internal bending moment and shear force at any point along a cylinder. Once the bending moment and shear forces are known, the axial and circumferential stresses can be derived from the formulas given in this section.

Isotropic material:

$$\sigma_x = \frac{E}{1-\nu^2}\left[-\frac{P+P_{Tx}}{K} + \frac{M+M_T}{D}z - \alpha(1+\nu)\,\Delta T\right] \tag{13.2}$$

$$\sigma_\phi = \frac{E}{1-\nu^2}\left[-\frac{P_\phi+P_{T\phi}}{K} + \nu\frac{M+M_T}{D}z - \alpha(1+\nu)\,\Delta T\right] \tag{13.3}$$

with

$$P_\phi = \frac{Eh}{R}v + \nu(P+P_{Tx}) - P_{T\phi}$$

Orthotropic material:

$$\sigma_x = \frac{E_x}{1-\nu_x\nu_\phi}\left[-\frac{P+P_{Tx}}{K_x} + \frac{M+M_T}{D_x}z - (\alpha_x+\nu_\phi\alpha_\phi)\,\Delta T\right] \tag{13.4}$$

$$\sigma_\phi = \frac{E_\phi}{1-\nu_x\nu_\phi}\left[-\frac{P_\phi+P_{T\phi}}{K_\phi} + \nu_x\frac{M+M_T}{D_x}z - (\alpha_\phi+\nu_x\alpha_x)\,\Delta T\right] \tag{13.5}$$

with

$$P_\phi = K_\phi(1-\nu_x\nu_\phi)\frac{v}{R} + \frac{K_\phi\nu_x}{K_x}(P+P_{Tx}) - P_{T\phi}$$

The constants in these relations are defined in Table 13-1 for various shells.

13.3 Differential Equations

The fundamental equations of motion in first-order form for the radial, symmetric bending of a cylinder are

$$\frac{\partial v}{\partial x} = -\theta + \frac{V}{D_V} \tag{13.6a}$$

$$\frac{\partial \theta}{\partial x} = \frac{M+M_T}{D_x} \tag{13.6b}$$

$$\frac{\partial M}{\partial x} = V - P\theta + \rho r_y^2\frac{\partial^2\theta}{\partial t^2} - c(x,t) \tag{13.6c}$$

$$\frac{\partial V}{\partial x} = \frac{K_\phi(1-\nu_x\nu_\phi)}{R^2}v + \rho\frac{\partial^2 v}{\partial t^2} - w(x,t) + \frac{1}{R}\left[\frac{\nu_x K_\phi}{K_x}(P+P_{Tx}) - P_{T\phi}\right] \tag{13.6d}$$

TABLE 13-2 Equivalence of Cylinder and Beam

Beam (Eqs. (2.16))	Cylinder (Eqs. (13.6))
v	v
θ	θ
M	M
V	V
GA_s	D_V
EI	D_x
M_T	M_T
P	P
c'	c'
k	$\dfrac{K_\phi(1-\nu_x\nu_\phi)}{R^2}$ or $\dfrac{Eh}{R^2}$
$-w$	$-w + \dfrac{1}{R}\left[\dfrac{\nu_x K_\phi}{K_x}(P + P_{Tx}) - P_{T\phi}\right]$ or $-w + \dfrac{1}{R}(\nu P + \int_h E\alpha\, \Delta T\, dz)$

For an isotropic shell the final equation reduces to

$$\frac{\partial V}{\partial x} = \frac{Eh}{R^2} v + \rho \frac{\partial^2 v}{\partial t^2} - w(x, t) + \frac{1}{R}\left(\nu P + \int_h E\alpha\, \Delta T\, dz\right) \tag{13.7}$$

See Table 13-1 for the values of the constants in Eqs. (13.6) for several types of shells. Note that equations for a cylinder are basically the same as for a beam on an elastic foundation with a modulus $K_\phi(1 - \nu_x\nu_\phi)/R^2$ for an orthotropic shell and a modulus Eh/R^2 for an isotropic shell. Table 13-2 provides further details on this equivalence.

13.4 Transfer Matrices

The transfer matrices for a cylinder of length ℓ can be taken from the tables of Chap. 2 with the substitutions of Table 13-2. Some other transfer matrices are provided in Tables 13-3 and 13-4.

13.5 Displacement

Cylinder displacement problems can be solved with the information in Chap. 2 for beams if the substitutions of Table 13-2 are made. All formulas, tables, and examples for loadings, static solutions, free dynamics, and forced dynamics can

TABLE 13-3 Point Matrix for Elastic Ring Stiffener (Refs. 13.3, 13.4)

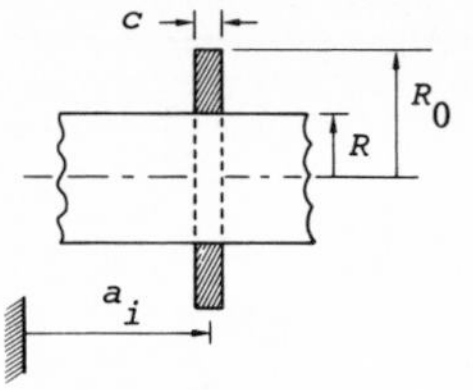

Definition:

$$\eta = \frac{R_0^2 + R^2}{R_0^2 - R^2}$$

$$\bar{\mathbf{U}}_i = \begin{bmatrix} 1 & 0 & 0 & 0 & 0 \\ 0 & 1 & 0 & 0 & 0 \\ 0 & \dfrac{Ec^3}{12R(\eta + \nu)} & 1 & 0 & 0 \\ \dfrac{Ec}{R(\eta + \nu)} & 0 & 0 & 1 & 0 \\ 0 & 0 & 0 & 0 & 1 \end{bmatrix}$$

be converted in this fashion. Even though no mention is made in Table 13-2, the formulas of Chap. 2 apply to the dynamic response of damped cylinders. For example, c' would be the external or viscous radial damping coefficient [mass/(time · length^2), force · time/length^3]. Because of the form of Eqs. (13.6), it is necessary in all cases to select beam formulas from Chap. 2 with an elastic foundation modulus k equal to $K_\phi(1 - \nu_x \nu_\phi)/R^2$ or Eh/R^2.

13.6 Applications

The examples of Chap. 2 serve to illustrate the application of the information in this chapter in solving cylinder problems. The following example demonstrates the use of the formulas of Chap. 2.

EXAMPLE 13.1 Cylinder with Internal Pressure Determine the deflection and moment along a uniform cylinder with fixed ends and with constant internal pressure p.

The desired variables can be calculated from the transfer matrix of Table 2-10 with $EI = D$ and $k = Eh/R^2$. In the notation of Table 2-10,

$$\lambda = \frac{k}{EI} = \frac{Eh}{DR^2} \qquad \beta^4 = \frac{\lambda}{4} = \frac{Eh}{4DR^2} \tag{1}$$

Since the internal pressure is directed in the negative z direction, we set w_1 of Table 2-10 equal to $-p$. The initial parameters for this fixed-fixed cylinder are given in Table 2-19 as

$$v_0 = \theta_0 = 0 \quad M_0 = \frac{1}{\nabla}(F_\theta U_{vV} - F_v U_{\theta V})_{x=L} \quad V_0 = \frac{1}{\nabla}(F_v U_{\theta M} - F_\theta U_{vM})_{x=L}$$

$$\nabla = (U_{vM} U_{\theta V} - U_{\theta M} U_{vV})_{x=L} \tag{2}$$

Insertion of the transfer matrix elements of Table 2-10 into (2) gives

$$\nabla = \frac{1}{8\beta^4 D^2}(\sin^2 \beta L - \sinh^2 \beta L)$$

$$M_0 = \frac{p}{2\beta^2}\,\frac{\sinh \beta L - \sin \beta L}{\sinh \beta L + \sin \beta L} \qquad V_0 = -\frac{p}{\beta}\,\frac{\cosh \beta L - \cos \beta L}{\sinh \beta L + \sin \beta L} \tag{3}$$

TABLE 13-4 Cylinder Segment with Linearly Varying Wall Thickness

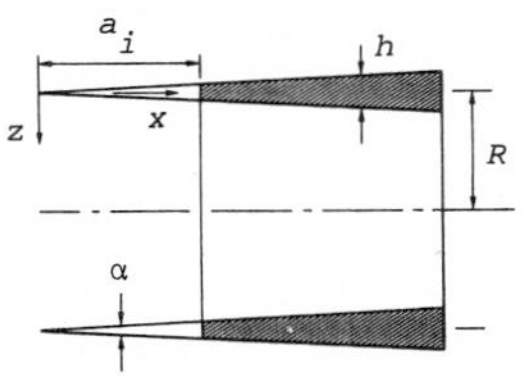

$$\mathbf{U}_i = \left[\begin{array}{c|c} \mathbf{H}(\eta)\,\mathbf{H}^{-1}(\eta_{a_i}) & \mathbf{F} \\ \hline \mathbf{O} & 1 \end{array}\right]$$

where **F** is the loading vector, $[F_V, F_\theta, F_M, F_V]$

O is a row vector of zeros, $[0\ 0\ 0\ 0]$

Definitions:

$h = \alpha x$

$\beta^4 = \dfrac{12(1 - \nu^2)}{R^2\alpha^2}, \quad \gamma = \dfrac{E\alpha^3}{48(1 - \nu^2)} = \dfrac{E\alpha}{4R^2\beta^4}$

$\eta = 2\beta\sqrt{x}, \quad \eta_{a_i} = 2\beta\sqrt{a_i}$

$$\mathbf{H}(\eta) = \left[\begin{array}{c|c|c|c}
\frac{2\beta}{\eta}\phi_1'(\eta) & \frac{2\beta}{\eta}\phi_2'(\eta) & \frac{2\beta}{\eta}\phi_3'(\eta) & \frac{2\beta}{\eta}\phi_4'(\eta) \\ \hline
-\frac{4\beta^3}{\eta^3}[\eta\phi_2(\eta) - 2\phi_1'(\eta)] & \frac{4\beta^3}{\eta^3}[\eta\phi_1(\eta) + 2\phi_2'(\eta)] & -\frac{4\beta^3}{\eta^3}[\eta\phi_4(\eta) - 2\phi_3'(\eta)] & \frac{4\beta^3}{\eta^3}[\eta\phi_3(\eta) + 2\phi_4'(\eta)] \\ \hline
-\frac{\gamma\eta}{2\beta}[\eta^2\phi_2'(\eta) - 4\eta\phi_2(\eta) + 8\phi_1'(\eta)] & \frac{\gamma\eta}{2\beta}[\eta^2\phi_1'(\eta) - 4\eta\phi_1(\eta) - 8\phi_2'(\eta)] & -\frac{\gamma\eta}{2\beta}[\eta^2\phi_4'(\eta) - 4\eta\phi_4(\eta) + 8\phi_3'(\eta)] & \frac{\gamma\eta}{2\beta}[\eta^2\phi_3'(\eta) - 4\eta\phi_3(\eta) - 8\phi_4'(\eta)] \\ \hline
\beta\gamma\eta[\eta\phi_1(\eta) + 2\phi_2'(\eta)] & \beta\gamma\eta[\eta\phi_2(\eta) - 2\phi_1'(\eta)] & \beta\gamma\eta[\eta\phi_3(\eta) + 2\phi_4'(\eta)] & \beta\gamma\eta[\eta\phi_4(\eta) - 2\phi_3'(\eta)]
\end{array}\right]$$

with $\phi_1(\eta) = \text{ber}\ \eta$, $\phi_2(\eta) = -\text{bei}\ \eta$, $\phi_3(\eta) = -\frac{2}{\pi}\text{kei}\ \eta$, $\phi_4(\eta) = -\frac{2}{\pi}\text{ker}\ \eta$, $' = \frac{d}{d\eta}$. The Kelvin functions ber η, bei η, kei η, ker η and their derivatives can be taken from mathematical tables for particular values of η.

The deflection and bending moment at any point x are given by

$$\begin{aligned} v &= M_0 U_{vM} + V_0 U_{vV} + F_v \\ M &= M_0 U_{MM} + V_0 U_{MV} + F_M \end{aligned} \tag{4}$$

where M_0, V_0 are taken from (3) and U_{vM}, U_{vV}, F_v, U_{MM}, U_{MV}, F_M are given in Table 2-10 with $\ell = x$. For example,

$$\begin{aligned} U_{vM} &= -\frac{1}{2\beta^2 D} \sinh \beta x \sin \beta x \\ U_{vV} &= -\frac{1}{4\beta^3 D} (\cosh \beta x \sin \beta x - \sinh \beta x \cos \beta x) \\ F_v &= -\frac{p}{4D\beta^4} (1 - \cosh \beta x \cos \beta x) \end{aligned} \tag{5}$$

13.7 Computer Programs and Benchmark Examples

Complicated cylinder problems should be solved with the assistance of a computer program. With the proper substitutions (Tables 13-2) a beam program can be used to find the deflection, slope, bending moment, and shear force for a cylinder. The following examples are provided as benchmark problems against which a reader's own computer program can be checked. The beam computer program BEAMRESPONSE was used for these examples.

EXAMPLE 13.2 Clamped Cylinder with Internal Pressure Determine the response of a thin-walled cylinder, clamped on both ends and subjected to a uniform internal pressure of $p = 100$ lb/in^2. For this cylinder $E = 30(10^6)$ lb/in^2, $\nu = 0.3$, $h = 0.25$ in, $R = 36$ in, $L = 20$ in.

This problem is equivalent to a clamped and uniformly loaded beam resting on an elastic foundation. The beam has a flexural rigidity of $Eh^3/12(1 - \nu^2) = 30(10^6)(0.00143)$, and the modulus of the elastic foundation is $Eh/R^2 = 5.787(10^3)$.

At $x = 10$ in, it is found that the deflection is -0.0179 in, the bending moment is 3.72 lb · in/in, and the shear force is $-5.82(10^{-11})$ lb/in.

EXAMPLE 13.3 Cylinder with Concentrated Loading Determine the static response of a cylinder with free ends subjected to a ring load of 500 lb/in at $x = L/2$. For this cylinder, $E = 30(10^6)$ psi, $\nu = 0.3$, $h = 0.1$ in, $R = 10$ in, $L = 20$ in.

This problem is equivalent to a free beam on an elastic foundation subjected to a concentrated load of 500 lb. For the equivalent beam $E = 30(10^6)$ lb/in^2, I (moment of inertia) $= h^3/12(1 - \nu^2) = 0.916(10^{-4})$in^4.

At the center ($x = L/2$), the deflection is 0.01 in, the slope is 0, the bending moment is 97.2 in · lb/in, and the shear force is -250 lb/in.

REFERENCES

13.1 Paul, B.: "Linear Bending Theory of Laminated Cylindrical Shells Under Axisymmetric Load," *J. Appl. Mech.*, vol. 30, pp. 98–102, 1963.

13.2 Boyd, D. E., and Kishore, B. R.: "Thermal Stresses in Axially Loaded Orthotropic Cylinders", *AIAA J.*, vol 6, pp. 980–983, 1968.

13.3 Olk, T. F.: "Beanspruchungen von Höhenflugzeug-Druckkammern im Bereich der Versteifungsringe," *Z. Flugwiss.*, vol 11, no. 2, pp. 45–58, 1963.

13.4 Olk, T. F.: "Beanspruchungen von Höhenflugzeug-Druckkammern im Bereich der Versteifungsringe," *Z. Flugwiss.*, vol. 11, no. 3, pp. 93–109, 1963.

FOURTEEN

Cross-Sectional Properties and Combined Stresses in Bars

The stress formulas for bars in the previous chapters apply to a single effect, for example, bending in the vertical plane or warping shear stresses. Formulas are given in standard references for the stresses resulting from combinations of bending and torsional loadings. Also readily available are formulas for unsymmetrical bending, that is, bending for other than symmetrical cross sections bent in the plane of symmetry.

Complicated stress analysis problems should be solved with the assistance of a computer program. A powerful method, the finite-element method, for computing sectional properties and stresses is detailed in Refs. 14.1, 14.2, and 14.3. The following examples are provided as benchmark examples against which a reader's own program can be checked. The computer program BEAMSTRESS was used for these examples. BEAMSTRESS calculates the cross-sectional properties, normal stresses, and shear stresses for a bar of any cross-sectional shape. The properties include area, centroid, moments of inertia about any axes, radii of gyration, shear center, shear deformation coefficients, torsional constant, and warping constant. Modulus weighted properties are calculated for composite sections. The stresses include normal stresses due to bending and warping and shear stresses due to torsion, warping, and transverse loads. For shear-related sectional properties and stresses, BEAMSTRESS employs the finite-element method of Refs. 14.1, 14.2, and 14.3.

EXAMPLE 14.1 Cross-Sectional Properties of a Square Section Find the torsional and warping constants for a square section, 1 in by 1 in.

The torsional constant is found to be 0.146 in^4 and the warping constant is $13(10^{-5})$ in^6.

EXAMPLE 14.2 Normal and Shear Stress Analysis Find normal and shear stresses and cross-sectional properties of the cross section in Fig. 14-1 subjected to the loadings:

- Bending moment about y axis = 15,000 in · lb
- Bending moment about z axis = 5,000 in · lb
- Shear force applied at $y = 4$ in, $z = 4$ in, with z component = 5000 lb, y component = 500 lb
- Twisting moment about x axis = 2000 in · lb

Do not include the effect of constrained warping. The modulus of elasticity is $29(10^6)$ lb/in^2 and Poisson's ratio is 0.27.

The quadrilateral mesh input to the program is shown. Some output is displayed in Fig. 14-2.

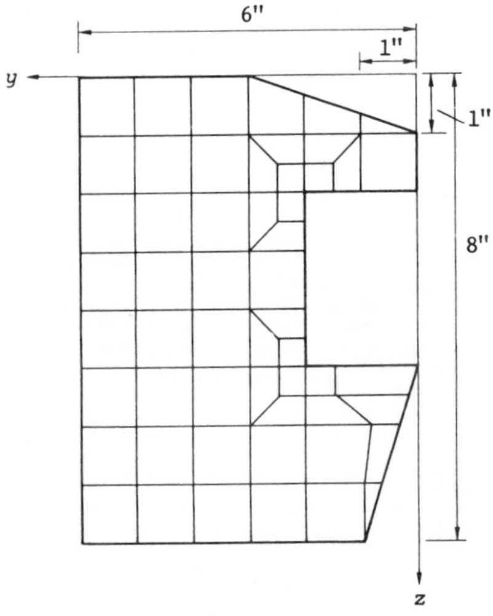

Fig. 14-1 Example 14.2.

EXAMPLE 14.3 Normal Stress Analysis of Z Section Find the normal stresses (axial and bending stresses) of a Z section with a height of 6 in and a thickness of 0.5 in. The flanges are 3 in wide. Suppose y, z coordinates are centroidal coordinates. The section is subject to a bending moment of 10,000 in · lb about the y axis, a bending moment of 2000 in · lb about the z axis, and a compressive axial force of 500 lb applied on the z axis 0.75 in below the centroid. The orientation of this Z section is the same as for the Z section of Fig. 14-4.

```
                    CROSS-SECTIONAL PROPERTIES

AREA OF SECTION...            39.00125E+00     WARPING CONSTANT...           28.49524E+01
Y-COORDINATE OF CENTROID...   34.87107E-01     TORSIONAL CONSTANT...         15.05642E+01
Z-COORDINATE OF CENTROID...   41.02493E-01     Y-RADIUS OF GYRATION          2.34847E+00
Y-MOMENT OF INERTIA...        21.51046E+01     Z-RADIUS OF GYRATION          1.52960E+00
Z-MOMENT OF INERTIA...        91.25051E+00     POLAR MOMENT OF INERTIA       3.06355E+02
PRODUCT OF INERTIA....       -69.43380E-01     Y-PRINCIPAL MOMENT OF INERTIA...  9.08625E+01
ANGLE TO PRINCIPAL AXES...   -86.80            Z-PRINCIPAL MOMENT OF INERTIA...  2.15493E+02
Y-COORDINATE OF SHEAR CENTER...  42.41476E-01  IYZ-PRINCIPAL                 0.
Z-COORDINATE OF SHEAR CENTER...  41.25179E-01  I MAX...                      2.15493E+02
SHEAR COEFFICIENT AYY...      11.76942E-01     I MIN...                      9.08625E+01
SHEAR COEFFICIENT AZZ...      14.66280E-01
SHEAR COEFFICIENT AYZ...     -57.74032E-04

                    STRESSES ON THE CROSS SECTION

 COORDINATES OF
ELEMENT CENTROID          NORMAL  STRESSES          SHEAR  STRESSES
   Y          Z                 SIGMA               TXY              TXZ
 4.500      4.500            -23.167E+00        44.732E-01       21.915E+01
 3.500      4.500             26.443E+00       -49.735E-01       21.942E+01
 2.750      4.625             72.167E+00       -25.672E+00       22.674E+01
 2.250      4.750             10.549E+01       -57.964E+00       27.484E+01
```

Fig. 14-2 Partial output for Example 14.2.

Some of the stresses are computed to be:

Coordinates, in		*Stress, lb/in²*
y	*z*	
1.625	3.000	14.9
0.000	3.000	2216.0
0.000	0.000	−83.3
0.000	−3.000	−2382.7
−1.625	−3.000	−181.6

EXAMPLE 14.4 Torsional Stress Analysis of Double Channel For the cross section of Fig. 14-3 find the shear stresses due to a 1000 in · lb twisting moment. Also,

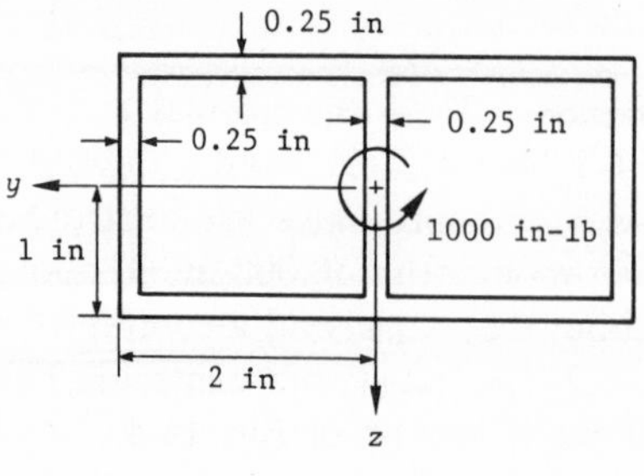

Fig. 14-3 Example 14.4.

find the torsional constant. Use a modulus of elasticity of $30(10^6)$ lb/in^2 and Poisson's ratio of 0.27. Do not take constrained warping effects into account.

The torsional constant is found to be 4.209 in^4. Some of the stresses are

Coordinates, in		*Stresses, lb/in^2*	
y	z	τ_{xy}	τ_{xz}
0.063	0.063	$-11.983(10^{-3})$	29.687
1.063	0.875	-301.64	$33.552(10^{-3})$
1.875	0.063	$-26.395(10^{-3})$	301.64

EXAMPLE 14.5 Restrained Warping Stresses in a Z Section Find the normal and shear stresses due to constrained warping and the cross-sectional properties of the Z section of Fig. 14-4 which is subject to a bimoment of 239,055 lb · in^2 and a warping torque of 10,000 lb · in. The beam's cross section is homogeneous, with modulus of elasticity and Poisson's ratio of $29(10^6)$ lb/in^2 and 0.27, respectively.

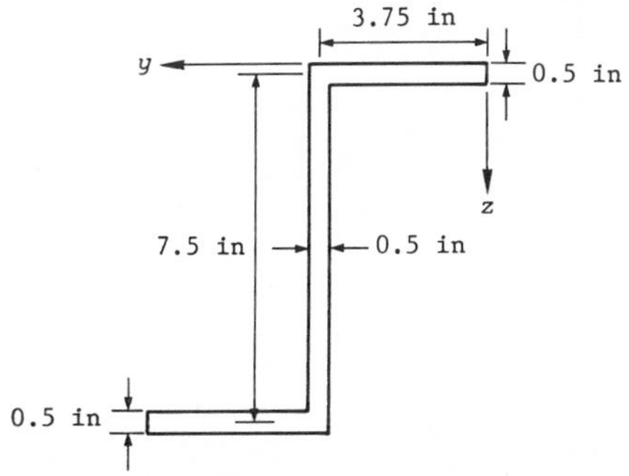

Fig. 14-4 Example 14.5.

The torsional constant is found to be 0.7865 in^4; the warping constant is 153.862 in^6. Some of the stresses are

Coordinates, in			*Warping stresses, lb/in^2*	
y	z	σ	τ_{xy}	τ_{xz}
0.500	0.250	$-1.343(10^4)$	$3.114(10^2)$	0.000
3.000	0.250	$1.133(10^3)$	$9.546(10^2)$	0.000
3.750	6.000	$5.373(10^3)$	0.000	$-4.496(10^2)$
5.000	7.750	$-1.780(10^3)$	$-9.478(10^2)$	0.000
6.500	7.750	$-1.052(10^4)$	$-3.114(10^2)$	0.000

EXAMPLE 14.6 Restrained Warping Stresses in a Wide Flange Section Find the normal and shear stresses due to constrained warping and the cross-sectional properties of the wide flange section of Fig. 14-5. This section is subject to a 10,000 lb · in^2 bimoment and a 20,000 lb · in warping torque. The modulus of elasticity is $29(10^6)$ lb/in^2 and Poisson's ratio is 0.27.

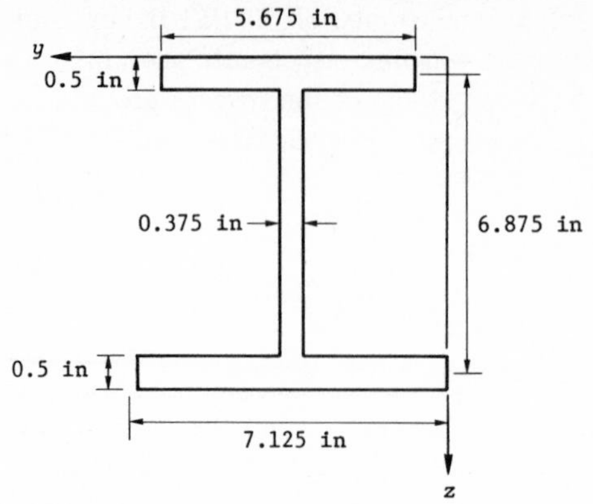

Fig. 14-5 Example 14.6.

Some calculated results are torsional constant = 0.7804 in^4, warping constant = 233.587 in^6.

Coordinates, in		*Warping stresses, lb/in²*		
y	z	σ	τ_{xy}	τ_{xz}
1.500	0.250	$-4.051(10^2)$	$7.186(10^2)$	0.000
3.000	0.250	$-1.096(10^2)$	$1.491(10^3)$	0.000
4.875	0.250	$2.573(10^2)$	$1.216(10^3)$	0.000
0.375	7.125	$3.087(10^2)$	$-2.452(10^2)$	0.000
2.625	7.125	$8.994(10^1)$	$-1.142(10^3)$	0.000
5.250	7.125	$-1.628(10^2)$	$-9.526(10^2)$	0.000

EXAMPLE 14.7 Shear Stress Analysis of a Composite Beam Find the torsional and warping constants for the square section shown in Fig. 14-6. As indicated, the cross section is composed of two materials.

The torsional constant is computed to be 812.3 in^4 and the warping constant is 258.8 in^6.

EXAMPLE 14.8 Stress Analysis of a Prestressed-Concrete Bridge Girder For the prestressed bridge girder illustrated in Fig. 14-7, compute the stresses, taking no constrained warping effects into account. Also find the cross-sectional properties. The modulus of elasticity is 27.79 GN/m^2 and Poisson's ratio is 0.15. On

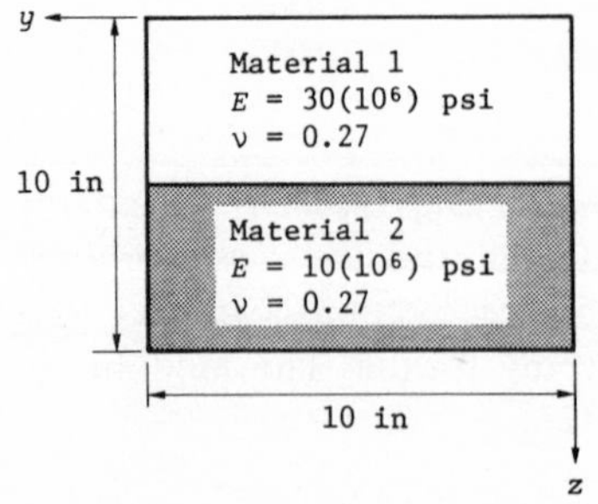

Fig. 14-6 Example 14.7.

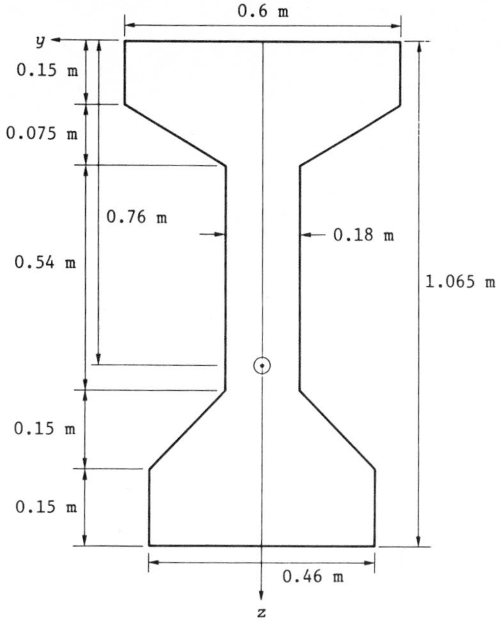

Fig. 14-7 Example 14.8.

the cross section there is a compressive axial force of 2224 kN because of prestressing at $y = 0$, $z = 0.76$ m. Also occurring on the cross section are a bending moment of 1.139 MN · m about the y axis and a shear force of 143.4 kN applied at $y = 0$, $z = 0.5325$ m in the z direction.

The torsional and warping constants are found to be 0.00513 m^4 and 0. 0006732 m^6, respectively. Typical stresses are:

Coordinates, m		*Stresses, $MN/m^2 = MPa$*		
y	z	σ	τ_{xy}	τ_{xz}
0.274	0.019	−13.03	−0.02139	0.01542
0.169	0.056	−12.54	−0.1014	0.07313
0.116	0.169	−11.08	−0.3102	0.2961
0.213	0.934	−1.131	0.04098	0.1182
0.022	1.046	0.3318	0.007399	0.03858

REFERENCES

14.1 Herrmann, L. R.: "Elastic Torsional Analysis of Irregular Shapes," *J. Eng. Mech. Div., Proc. Am. Soc. Civil Eng.*, vol. 91, pp. 11–19 1965.

14.2 Mason, W. E., and Herrmann, L. R.: "Elastic Shear Analysis of General Prismatic Beams," *J. Eng. Mech. Div., Proc. Am. Soc. Civ. Eng.*, vol. 94, pp. 965–983, 1968.

14.3 Chang, P. Y., Thasanatorn, C., and Pilkey, W. D.: "Restrained Warping Stresses in Thin-Walled Open Sections," *J. Struct. Div., Proc. Am. Soc. Civil Eng.*, vol. 101, pp. 2467–2472, 1975.

Appendix 1

Techniques for Deriving Transfer Matrices

This appendix introduces several basic techniques for deriving the transfer matrices from the differential equations of motion for structural members. It is not necessary to understand this material in order to apply the transfer matrices given in the various tables in this book. This section is intended for readers wishing to verify given transfer matrices, develop matrices for special cases, or learn more about the underlying theory of transfer matrix methods.

There are numerous ways to obtain the transfer matrices from the equations of motion. The most direct of available methods is simply to "rearrange" in the solution the arbitrary constants of integration in terms of the initial parameters (Ref. A1.1). Reference A1.1 also details techniques for deriving point matrices. Other more general methods are presented in the following subsections.

The partial differential equations of motion for beams are given by Eqs. (2.16). For the case of steady-state motion, the loadings and responses (deflection, slope, moment, and shear force) all vary harmonically, that is, all vary as sin ωt. Then Eqs. (2.16) reduce to the ordinary differential equations

$$\frac{dv}{dx} = -\theta + \frac{V}{GA_s} \qquad \frac{dM}{dx} = V + (k^* - P - \rho r^2\omega^2)\theta - c$$

$$\frac{d\theta}{dx} = \frac{M}{EI} + \frac{M_T}{EI} \qquad \frac{dV}{dx} = (k - \rho\omega^2)v - w \tag{A1.1}$$

These equations apply to the free vibration of a beam if the applied loadings c, w, M_T are set equal to zero. Also, they are appropriate for static responses if ω is set equal to zero.

A1.1 Laplace Transformation

In matrix notation, Eqs. (A1.1) can be written as

$$\frac{d\mathbf{s}}{dx} = \mathbf{A}\mathbf{s} + \mathbf{f} \tag{A1.2}$$

with

$$\mathbf{s} = \begin{bmatrix} v \\ \theta \\ M \\ V \end{bmatrix} \qquad \mathbf{f} = \begin{bmatrix} 0 \\ M_{T1}/EI \\ -\left(c_1 + \dfrac{\Delta c}{\Delta \ell}\right) \\ -\left(w_1 + \dfrac{\Delta w}{\Delta \ell}\right) \end{bmatrix} \tag{A1.3}$$

$$\mathbf{A} = \begin{bmatrix} 0 & -1 & 0 & 1/GA_s \\ 0 & 0 & 1/EI & 0 \\ 0 & k^* - P - \rho r^2\omega^2 & 0 & 1 \\ k - \rho\omega^2 & 0 & 0 & 0 \end{bmatrix}$$

where the uniform and linearly varying loading types, for example, c_1 and $\Delta c/\Delta \ell$, which appear frequently in this book are indicated instead of the more general quantities, for example, c of Eq. (A1.1).

The Laplace transform of Eq. (A1.2) is written

$$s\mathbf{s}(s) = \mathbf{s}(0) + \mathbf{A}\mathbf{s}(s) + \mathbf{f}(s) \tag{A1.4}$$

Then

$$\mathbf{s}(s) = (\mathbf{I}s - \mathbf{A})^{-1}\mathbf{s}(0) + (\mathbf{I}s - \mathbf{A})^{-1}\mathbf{f}(s) \tag{A1.5}$$

where $\mathbf{I}$ is a unit diagonal matrix. The inverse Laplace transform is

$$\mathbf{s}(x) = \mathcal{L}^{-1}\left[(\mathbf{I}s - \mathbf{A})^{-1}\right]\mathbf{s}(0) + \mathcal{L}^{-1}\left[(\mathbf{I}s - \mathbf{A})^{-1}\mathbf{f}(s)\right] \tag{A1.6}$$

The term $\mathcal{L}^{-1}[(\mathbf{I}s - \mathbf{A})^{-1}]$ is the 4×4 transfer matrix, and the second term in Eq. (A1.6) is the fifth column of the extended transfer matrix.

EXAMPLE A1.1 Euler-Bernoulli Beam with Shear Deformation In order to illustrate this method, choose the case of a beam considering shear deformation with ρ, k,

k^*, P, $\Delta w/\Delta \ell$, c_1, $\Delta c/\Delta \ell$, M_{T1} equal to zero. Then

$$\mathbf{A} = \begin{bmatrix} 0 & -1 & 0 & 1/GA_s \\ 0 & 0 & 1/EI & 0 \\ 0 & 0 & 0 & 1 \\ 0 & 0 & 0 & 0 \end{bmatrix} \qquad \mathbf{f} = \begin{bmatrix} 0 \\ 0 \\ 0 \\ -w_1 \end{bmatrix} \tag{1}$$

$$[\mathbf{I}s - \mathbf{A}]^{-1} = \begin{bmatrix} \frac{1}{s} & -\frac{1}{s^2} & -\frac{1}{s^3EI} & -\frac{1}{s^4EI} + \frac{1}{s^2GA_s} \\ 0 & \frac{1}{s} & \frac{1}{s^2EI} & \frac{1}{s^3EI} \\ 0 & 0 & \frac{1}{s} & \frac{1}{s^2} \\ 0 & 0 & 0 & \frac{1}{s} \end{bmatrix} \tag{2}$$

$$\mathfrak{L}^{-1}[(\mathbf{I}s - \mathbf{A})^{-1}] = \begin{bmatrix} 1 & -x & -\frac{x^2}{2EI} & -\frac{x^3}{6EI} + \frac{x}{GA_s} \\ 0 & 1 & \frac{x}{EI} & \frac{x^2}{2EI} \\ 0 & 0 & 1 & x \\ 0 & 0 & 0 & 1 \end{bmatrix} \tag{3}$$

$$(\mathbf{I}s - \mathbf{A})^{-1}\mathbf{f}(s) = \begin{bmatrix} \frac{w_1}{s^5EI} - \frac{w_1}{s^3GA_s} \\ -\frac{w_1}{s^4EI} \\ -\frac{w_1}{s^3} \\ -\frac{w_1}{s^2} \end{bmatrix} \tag{4}$$

$$\mathfrak{L}^{-1}[(\mathbf{I}s - \mathbf{A})^{-1}\mathbf{f}(s)] = \begin{bmatrix} \frac{w_1x^4}{24EI} - \frac{w_1x^2}{2GA_s} \\ -\frac{w_1x^3}{6EI} \\ -\frac{w_1x^2}{2} \\ -w_1x \end{bmatrix} \tag{5}$$

Combine Eqs. (A1.6), (3), and (5) in the extended transfer matrix form

$$\begin{bmatrix} v \\ \theta \\ M \\ V \\ 1 \end{bmatrix}_x = \begin{bmatrix} 1 & -x & -\dfrac{x^2}{EI} & -\dfrac{x^3}{EI} + \dfrac{x}{GA_s} & \dfrac{w_1 x^4}{24EI} - \dfrac{w_1 x^2}{2GA_s} \\ 0 & 1 & \dfrac{x}{EI} & \dfrac{x^2}{2EI} & -\dfrac{w_1 x^3}{6EI} \\ 0 & 0 & 1 & x & -\dfrac{w_1 x^2}{2} \\ 0 & 0 & 0 & 1 & -w_1 x \\ 0 & 0 & 0 & 0 & 1 \end{bmatrix} \begin{bmatrix} v \\ \theta \\ M \\ V \\ 1 \end{bmatrix}_0 \tag{6}$$

The Laplace transform method is straightforward. When some of the parameters ρ, k, k^*, P are not zero, the inversion of $(\mathbf{I}s - \mathbf{A})^{-1}$ can become complicated.

It is possible to avoid using the right-hand terms of Eq. (A1.6) to compute the loading function vector. Since only uniform and linearly varying w, c, M_T are considered, all of the elements in $\mathbf{f}(s)$ are a function of either $1/s$ or $1/s^2$. Then, from Eq. (A1.6), the elements of the fifth column of the extended transfer matrix are the inverse transform of the products of the Laplace transform load and $1/s$ or $1/s^2$

$$F_i = \mathfrak{L}^{-1}\left[U_{i\theta}(s)\,\frac{M_{T1}}{EI}\,\frac{1}{s} + U_{iM}(s)\left(c_1\,\frac{1}{s} + \frac{\Delta c}{\Delta \ell}\,\frac{1}{s^2}\right) + U_{iV}(s)\left(w_1\,\frac{1}{s} + \frac{\Delta w}{\Delta \ell}\,\frac{1}{s^2}\right)\right]$$

$$i = v, \theta, M, V \tag{A1.7}$$

or

$$F_i = \int_0^x \left[\frac{M_{T1}}{EI}\,U_{i\theta}(\xi) + c_1 U_{iM}(\xi) + w_1 U_{iV}(\xi) \right] d\xi + \int_0^x \left\{ \int_0^u \left[\frac{\Delta c}{\Delta \ell}\,U_{iM}(\xi) + \frac{\Delta w}{\Delta \ell}\,U_{iV}(\xi) \right] d\xi \right\} du \tag{A1.8}$$

where F_i is defined in Eq. (2.13) or (2.14). Thus it is possible to concentrate on calculating the 4×4 transfer matrix and then use Eq. (A1.8) to find the fifth column.

A1.2 Exponential Expansion

The solution of Eq. (A1.2) with $\mathbf{f} = 0$ and constant $\mathbf{A}$ is

$$\mathbf{s} = e^{\mathbf{A}x}\mathbf{s}_0 \tag{A1.9}$$

Thus, the transfer matrix **U** equals $e^{\mathbf{A}x}$. This transfer matrix can be expanded as

$$\mathbf{U} = e^{\mathbf{A}x} = \mathbf{I} + \mathbf{A}x + \mathbf{A}^2 \frac{x^2}{2} + \mathbf{A}^3 \frac{x^3}{3!} + \cdots \tag{A1.10}$$

Although this is a solution that is often reasonably easy to implement computationally, it is sometimes useful to simplify the expansion using the Cayley-Hamilton theorem ("a matrix satisfies its own characteristic equation," Ref. A1.2). A function of a square matrix of order n can be replaced by a polynomial in **A** of order $n - 1$. In the case of a beam,

$$e^{\mathbf{A}x} = c_0\mathbf{I} + c_1\mathbf{A}x + c_2(\mathbf{A}x)^2 + c_3(\mathbf{A}x)^3 \tag{A1.11}$$

The Cayley-Hamilton theorem permits **A** to be replaced by its characteristic values n_i, that is,

$$e^{n_i x} = c_0 + c_1 n_i x + c_2(n_i x)^2 + c_3(n_i x)^3, \qquad i = 1, 2, 3, 4 \tag{A1.12}$$

which can then be solved to find the functions c_0, c_1, c_2, c_3. Insertion of these expressions into Eq. (A1.11) provides the desired transfer matrix.

EXAMPLE A1.2 Vibrating Beam Consider a vibrating beam. Assume a solution to Eq. (A1.2) (with $\mathbf{f} = 0$) of the form e^{nx}. Substitution of this solution in Eq. (A1.2) gives the characteristic equation

$$|\mathbf{I}n - \mathbf{A}| = 0 \tag{1}$$

For the vibrating beam represented by the **A** matrix of Eq. (A1.3) (with $k = k^* = r = P = 1/GA_s = 0$),

$$|\mathbf{I}n - \mathbf{A}| = \begin{vmatrix} n & +1 & 0 & 0 \\ 0 & n & \dfrac{-1}{EI} & 0 \\ 0 & 0 & n & -1 \\ \rho\omega^2 & 0 & 0 & n \end{vmatrix} = n^4 - \lambda^4 = 0$$

where

$$\lambda^4 = \frac{\rho\omega^2}{EI}$$

The characteristic values are then

$$n_1 = \lambda \qquad n_2 = -\lambda \qquad n_3 = i\lambda \qquad n_4 = -i\lambda \tag{2}$$

Equations (A1.12) now become

$$\begin{aligned} e^{\lambda x} &= c_0 + c_1\lambda x + c_2(\lambda x)^2 + c_3(\lambda x)^3 \\ e^{-\lambda x} &= c_0 - c_1\lambda x + c_2(\lambda x)^2 - c_3(\lambda x)^3 \\ e^{i\lambda x} &= c_0 + ic_1\lambda x - c_2(\lambda x)^2 - ic_3(\lambda x)^3 \\ e^{-i\lambda x} &= c_0 - ic_1\lambda x - c_2(\lambda x)^2 + ic_3(\lambda x)^3 \end{aligned}$$

which can be solved for the c_j functions to give

$$\begin{aligned} c_0 &= \frac{1}{2}(\cosh \lambda x + \cos \lambda x) \\ c_1 &= \frac{1}{2\lambda}(\sinh \lambda x + \sin \lambda x) \\ c_2 &= \frac{1}{2\lambda^2}(\cosh \lambda x - \cos \lambda x) \\ c_3 &= \frac{1}{2\lambda^3}(\sinh \lambda x - \sin \lambda x) \end{aligned} \tag{3}$$

The desired transfer matrix follows, after considerable algebra, from Eq. (A1.11). Thus

$$\mathbf{U} = e^{\mathbf{A}x} = \begin{bmatrix} c_0 & -c_1 & -\frac{c_2}{EI} & -\frac{c_3}{EI} \\ -\lambda^4 c_3 & c_0 & \frac{c_1}{EI} & \frac{c_2}{EI} \\ -\lambda^4 EIc_2 & \lambda^4 EIc_3 & c_0 & c_1 \\ -\lambda^4 EIc_1 & \lambda^4 EIc_2 & \lambda^4 c_3 & c_0 \end{bmatrix} \tag{4}$$

A1.3 Matrizant

If the solution to the homogeneous form ($\mathbf{f} = 0$) of Eq. (A1.2) is written as the integral equation

$$\mathbf{s}(x) = \mathbf{s}_{a_0} + \int_{a_0}^{x} \mathbf{A}(\sigma_1)\mathbf{s}(\sigma_1)\, d\sigma_1 \tag{A1.13}$$

and the Picard integration

$$\mathbf{s}_{n+1} = \mathbf{s}_n + \int_{a_0}^{x} \mathbf{A}(\sigma_1)\mathbf{s}(\sigma_1)\, d\sigma_1$$

is employed, the transfer matrix can be developed in the *matrizant* series (Refs. A1.2, A1.3):

$$\mathbf{U} = \mathbf{I} + \int_{a_0}^{x} \mathbf{A}(\sigma_1)\, d\sigma_1 + \int_{a_0}^{x} \mathbf{A}(\sigma_1) \int_{a_0}^{\sigma_1} \mathbf{A}(\sigma_2)\, d\sigma_2\, d\sigma_1 + \cdots \tag{A1.14}$$

This series converges uniformly if the elements of $\mathbf{A}(x)$ are continuous functions of x. Rapid convergence is usually achieved for small intervals $(x - a_0)$. Note that this series again reduces to Eq. (A1.10) for constant $\mathbf{A}$ (and $a_0 = 0$).

EXAMPLE A1.3 Point Matrix for an In-Plane Force of a Circular Plate The matrizant method can be useful in establishing a point matrix. To demonstrate this consider the derivation of a point matrix for a lumped representation of the in-plane forces P, P_ϕ of a circular plate.

The **A** matrix for a circular plate is (Chap. 11)

$$\mathbf{A} = \begin{bmatrix} 0 & 1 & 0 & 0 \\ \nu \dfrac{m^2}{r^2} & -\dfrac{\nu}{r} & \dfrac{1}{D} & 0 \\ -\dfrac{Dm^2}{r^3}(3 - 2\nu - \nu^2) & P + \dfrac{D}{r^2}[1 - \nu^2 + 2m^2(1-\nu)] & -\dfrac{(1-\nu)}{r} & -1 \\ \dfrac{P_\phi m^2}{r^2} + \dfrac{D}{r^4}[2m^2(1-\nu) + m^4(1-\nu^2)] & -\dfrac{Dm^2}{r^3}(3 - 2\nu - \nu^2) & -\dfrac{\nu m^2}{r^2} & -\dfrac{1}{r} \end{bmatrix} \tag{1}$$

The variable r now replaces x.

To compute the point matrix at radius a_i it is necessary to find the integrals

$$\lim_{r \to a_i} \int_{a_i}^{r} \mathbf{A}(\sigma_1)\, d\sigma_1 \qquad \lim_{r \to a_i} \int_{a_i}^{r} \mathbf{A}(\sigma_1) \int_{a_i}^{\sigma_1} \mathbf{A}(\sigma_2)\, d\sigma_2\, d\sigma_1 \qquad \text{etc.} \tag{2}$$

It appears to follow from (1) that all elements in the matrix resulting from the first integral of (2) are zero. If, however, the elements containing P and P_ϕ are examined, it will be seen that they are not zero. This can be observed by reasoning what these integrals represent. The integral $\int_{a_i}^{r} P\, dr$ represents the total radial force exerted by the element of the plate. As $r \to a_i$ this total force does not vanish, but approaches a constant value

$$P_{Li} = P(r - a_i) \tag{3}$$

Similarly, for P_ϕ the total force cannot vanish and

$$P_{\phi i} = \frac{P_\phi m^2 (r - a_i)}{r a_i} \tag{4}$$

Thus

$$\lim_{r \to a_i} \int_{a_i}^{r} \mathbf{A}(\sigma_1)\, d\sigma_1 = \begin{bmatrix} 0 & 0 & 0 & 0 \\ 0 & 0 & 0 & 0 \\ 0 & P_{Li} & 0 & 0 \\ P_{\phi i} & 0 & 0 & 0 \end{bmatrix} \tag{5}$$

Consider the second integral of (2). Again the limiting process indicates that only those terms containing P and P_ϕ remain finite as $r \to a_i$. On taking the product

$$\mathbf{A}(\sigma_1) \int_{a_i}^{\sigma_i} \mathbf{A}(\sigma_2)\, d\sigma_2$$

we find that all terms vanish as $r \to a_i$, yielding

$$\lim_{r \to a_i} \int_{a_i}^{r} \mathbf{A}(\sigma_1) \int_{a_i}^{\sigma_i} \mathbf{A}(\sigma_2)\, d\sigma_2\, d\sigma_1 = \mathbf{0}$$

Similar reasoning shows that the remaining integrals are also zero. Finally, from

Eq. (A1.14) and (5), the desired point matrix is

$$\overline{\mathbf{U}}_i = \begin{bmatrix} 1 & 0 & 0 & 0 \\ 0 & 1 & 0 & 0 \\ 0 & P_{Li} & 1 & 0 \\ P_{\phi i} & 0 & 0 & 1 \end{bmatrix} \tag{6}$$

The magnitudes of P and P_ϕ required to compute P_{Li} and $P_{\phi i}$ of (3) and (4) are found with the disk theory of Chap. 8 for in-plane forces.

A1.4 Numerical Integration

In addition to the many analytical methods described here, there are numerous numerical methods for integrating Eqs. (A1.2) to provide the transfer matrix **U**. Typical methods suitable for integrating systems of first-order equations are Runge-Kutta and predictor-corrector. Numerical computation of transfer matrices is relatively inefficient. Hence, numerical methods are not recommended for cases that can be treated analytically. Only cases for which analytical solutions to Eqs. (A1.2) are difficult to derive should be treated numerically. Members with continuously varying cross-sectional properties and those with more than four equations of motion are examples of problems where numerical integration may be required. In the case of nonuniform members, often numerical integration can be avoided by employing a model with piecewise constant segments. For stability and free dynamics problems, the eigenvalues are calculated using an iterative search of the characteristic equation. This iteration combined with the inefficiency of numerical generation of the transfer matrices for each segment may lead to excessive computation time and, in addition, may result in numerical instabilities.

A1.5 A General Method

The technique described below is a general approach used to derive many of the transfer matrices contained in this book.

Combine Eqs. (A1.1), with the applied loads **f** set equal to zero, into a single fourth-order equation

$$\frac{d^4 v}{dx^4} + (\zeta - \eta)\frac{d^2 v}{dx^2} + (\lambda - \zeta\eta)v = 0 \tag{A1.15}$$

where $\zeta = (P + \rho r^2 \omega^2 - k^*)/EI$
$\eta = (k - \rho\omega^2)/GA_s$
$\lambda = (k - \rho\omega^2)/EI$

The Laplace transformation of Eq. (A1.15) gives

$$\begin{aligned} v(s)\left[s^4 + (\zeta - \eta)s^2 + (\lambda - \zeta\eta)\right] &= s^3 v(0) + s^2 v'(0) + s v''(0) + v'''(0) \\ &= (\zeta - \eta)\left[s v(0) + v'(0)\right] \end{aligned} \tag{A1.16}$$

The inverse transform of $v(s)$ leads to

$$v(x) = [e_1(x) + (\zeta - \eta)e_3(x)]v(0) + [e_2(x) + (\zeta - \eta)e_4(x)]v'(0) + e_3(x)v''(0) + e_4(x)v'''(0) \qquad \text{(A1.17)}$$

where

$$e_i(x) = \mathfrak{L}^{-1}\left[\frac{s^{4-i}}{s^4 + (\zeta - \eta)s^2 + \lambda - \zeta\eta}\right] \qquad \text{(A1.18)}$$

It follows from Eq. (A1.18) that

$$\begin{aligned} e_i(x) &= \frac{d}{dx}\, e_{i+1}(x) \qquad i = -2, -1, 0, 1, 2, 3 \\ e_{i+1}(x) &= \int_0^x e_i(\xi)\, d\xi \qquad i = 4, 5, 6 \end{aligned} \qquad \text{(A1.19)}$$

Equation (A1.17) and its derivatives can be written as

$$\begin{bmatrix} v(x) \\ v'(x) \\ v''(x) \\ v'''(x) \end{bmatrix} = \begin{bmatrix} e_1 + (\zeta - \eta)e_3 & e_2 + (\zeta - \eta)e_4 & e_3 & e_4 \\ e_0 + (\zeta - \eta)e_2 & e_1 + (\zeta - \eta)e_3 & e_2 & e_3 \\ e_{-1} + (\zeta - \eta)e_1 & e_0 + (\zeta - \eta)e_2 & e_1 & e_2 \\ e_{-2} + (\zeta - \eta)e_0 & e_{-1} + (\zeta - \eta)e_1 & e_0 & e_1 \end{bmatrix} \begin{bmatrix} v(0) \\ v'(0) \\ v''(0) \\ v'''(0) \end{bmatrix} \qquad \text{(A1.20)}$$

or

$$\mathbf{v}(x) = \mathbf{E}(x)\mathbf{v}(0) \qquad \text{(A1.21)}$$

Use Eq. (A1.1) to develop the notation $\mathbf{v}(x) = \mathbf{Ks}(x)$, where

$$\mathbf{K} = \begin{bmatrix} 1 & 0 & 0 & 0 \\ 0 & -1 & 0 & 1/GA_s \\ \eta & 0 & -1/EI & 0 \\ 0 & \zeta - \eta & 0 & -1/EI + \eta/GA_s \end{bmatrix} \qquad \text{(A1.22)}$$

Then Eq. (A1.21) appears as

$$\mathbf{s}(x) = \mathbf{K}^{-1}\mathbf{E}(x)\mathbf{Ks}(0) \qquad \text{(A1.23)}$$

By definition $\mathbf{K}^{-1}\mathbf{E}(x)\mathbf{K}$ is the desired transfer matrix. The fifth column of loading function terms in the extended transfer matrix can be obtained with Eq. (A1.7). This is a very useful, general approach in that it applies to all values of ζ, λ, and η. The required inversion, $\mathbf{K}^{-1}$, is not difficult to calculate.

EXAMPLE A1.4 Euler-Bernoulli Beam with Axial Force Find the transfer matrix for a simple beam with axial force P.

By definition (Eq. A1.15),

$$\lambda = 0 \qquad \zeta = P/EI \qquad 1/GA_s = 0 \qquad \eta = 0 \qquad (1)$$

From Eqs. (A1.18) and (A1.19)

$$e_1 = \mathfrak{L}^{-1}\left(\frac{s^3}{s^4+\zeta s^2}\right) = \mathfrak{L}^{-1}\left(\frac{s}{s^2+\zeta}\right) = \cos ax \qquad a^2 = \zeta$$

$$e_0 = -a \sin ax \qquad e_{-1} = -a^2 \cos ax \qquad e_{-2} = a^3 \sin ax$$

$$e_2 = \mathfrak{L}^{-1}\left(\frac{1}{s^2+\zeta}\right) = \frac{1}{a}\sin ax \tag{2}$$

$$e_3 = \mathfrak{L}^{-1}\left(\frac{1}{s^2+\zeta}\,\frac{1}{s}\right) = \int_0^x \frac{1}{a}\sin a\xi \, d\xi = \frac{1}{a^2}(1-\cos ax)$$

$$e_4 = \int_0^x \frac{1}{a^2}(1-\cos a\xi)\,d\xi = \frac{x}{a^2} - \frac{1}{a^3}\sin ax = \frac{ax - \sin ax}{a^3}$$

These can be combined to form $\mathbf{E}$. We note that $e_1 + \zeta e_3 = 1$, $e_2 + \zeta e_4 = x$, $e_0 + \zeta e_2 = 0$, $e_{-1} + \zeta e_1 = 0$, $e_{-2} + \zeta e_0 = 0$. From Eq. (A1.22),

$$\mathbf{K} = \begin{bmatrix} 1 & 0 & 0 & 0 \\ 0 & -1 & 0 & 0 \\ 0 & 0 & -1/EI & 0 \\ 0 & \zeta & 0 & -1/EI \end{bmatrix} \qquad \mathbf{K}^{-1} = \begin{bmatrix} 1 & 0 & 0 & 0 \\ 0 & -1 & 0 & 0 \\ 0 & 0 & -EI & 0 \\ 0 & -\zeta EI & 0 & -EI \end{bmatrix} \tag{3}$$

Then

$$\mathbf{K}^{-1}\mathbf{E}(x) = \begin{bmatrix} 1 & x & e_3 & e_4 \\ 0 & -1 & -e_2 & -e_3 \\ 0 & 0 & -EIe_1 & -EIe_2 \\ 0 & -\zeta EI & 0 & -EI \end{bmatrix} \tag{4}$$

$$\mathbf{K}^{-1}\mathbf{E}(x)\mathbf{K} = \begin{bmatrix} 1 & -x+\zeta e_4 & -\dfrac{e_3}{EI} & -\dfrac{e_4}{EI} \\ 0 & 1-\zeta e_3 & \dfrac{e_2}{EI} & \dfrac{e_3}{EI} \\ 0 & -\zeta EIe_2 & e_1 & e_2 \\ 0 & 0 & 0 & 1 \end{bmatrix}$$

Since $-x + \zeta e_4 = -x + x - a \sin ax = -a \sin ax = -e_2$, $1 - \zeta e_3 = \cos ax = e_1$, $\zeta \mathrm{EI} = P$,

$$\mathbf{U} = \mathbf{K}^{-1}\mathbf{E}(x)\mathbf{K} = \begin{bmatrix} 1 & -e_2 & -e_3/EI & -e_4/EI \\ 0 & e_1 & e_2/EI & e_3/EI \\ 0 & -Pe_2 & e_1 & e_2 \\ 0 & 0 & 0 & 1 \end{bmatrix} \tag{5}$$

It is useful to calculate the e_i functions from Eq. (A1.18) for various values of ζ, η, and λ. In doing so, note that the roots of the denominator in Eq. (A1.18) can be found. That is, the four roots of

$$s^4 + (\zeta - \eta)s^2 + \lambda - \zeta\eta = 0$$

are

$$s_{1,2,3,4} = \pm\sqrt{-\left(\frac{\zeta-\eta}{2}\right) \pm \sqrt{\left(\frac{\zeta-\eta}{2}\right)^2 - (\lambda - \eta\zeta)}} \tag{A1.24}$$

Taking advantage of these roots, we find $e_4(x)$ to be:

1. If $\zeta = \eta = \lambda = 0$, $s_{1,2,3,4} = 0$

$$e_4(x) = \mathcal{L}^{-1}\left(\frac{1}{s^4}\right) = \frac{x^3}{6}$$

2. $\eta = \lambda = 0$, $\zeta < 0$, $s_{1,2} = \pm a$, $a = \sqrt{-\zeta}$, $s_{3,4} = 0$

$$e_4(x) = \mathcal{L}^{-1}\left[\frac{1}{s^2(s^2 - a^2)}\right] = \frac{1}{a^3}(\sinh ax - ax)$$

3. If $\eta = \lambda = 0$, $\zeta > 0$, $s_{1,2} = \pm ia$, $a = \sqrt{\zeta}$, $s_{3,4} = 0$

$$e_4(x) = \mathcal{L}^{-1}\left[\frac{1}{s^2(s^2 + a^2)}\right] = \frac{1}{a^3}(ax - \sin ax) \tag{A1.25}$$

4. If $\eta = \zeta = 0$, $\lambda < 0$, $s_{1,2} = \pm a$, $s_{3,4} = \pm ia$, $a = \lambda^{1/4}$

$$e_4(x) = \mathcal{L}^{-1}\left(\frac{1}{s^4 - a^4}\right) = \frac{1}{2a^3}(\sinh ax - \sin ax)$$

5. If $\eta = \zeta = 0$, $\lambda > 0$, $s_{1,2,3,4} = \pm(a \pm ia)$, $a = (\lambda/4)^{1/4}$

$$e_4(x) = \mathcal{L}^{-1}\left(\frac{1}{s^4 + 4a^4}\right) = \frac{1}{4a^3}(\cosh ax \sin ax - \sinh ax \cos ax)$$

Other cases are given in Table 2-13. Given the above values of $e_4(x)$, the other $e_i(x)$ functions can be obtained from Eqs. (A1.19).

REFERENCES

A1.1 Pilkey, W. D., and Pilkey, O. H.: *Mechanics of Solids*, Quantum Publishers, New York, 1974.

A1.2 Pestel, E. C., and Leckie, F. A.: *Matrix Method in Elasto-Mechanics*, McGraw-Hill, New York, 1963.

A1.3 Wunderlich, W.: "Calculation of Shells of Revolution by Means of Transfer Matrices," *Ing. Arch.*, vol. 36, pp. 262–279, 1967.

Appendix 2

Procedure for Overcoming Numerical Difficulties

The determination of the response of structural members by the transfer matrix method is sometimes beset with numerical difficulties. For example, numerical instabilities may occur in the calculation of high natural frequencies, the static response of a flexible member on a stiff foundation, or the deformation of a very long member. This problem has been given considerable attention in the literature. See, for example, Ref. A2.1. An effective technique for overcoming these numerical problems is to combine the Riccati transformation with the transfer matrix. An interesting side effect of this combination is that less computer storage is required and improved computational time is achieved.

A2.1 Riccati Transformation

The transfer of state variables from station i to station $i + 1$, across segment i, for a member with n state variables can be written

$$\mathbf{s}_{i+1} = \mathbf{U}_i\mathbf{s}_i + \mathbf{F}_i \tag{A2.1}$$

where $\mathbf{s}$ is an $n \times 1$ matrix of state variables, $\mathbf{U}$ is an $n \times n$ transfer matrix, and $\mathbf{F}$ is an $n \times 1$ matrix of loading functions. Let Eq. (A2.1) be partitioned so that

$$\begin{bmatrix} \mathbf{f} \\ \hline \mathbf{e} \end{bmatrix}_{i+1} = \left[\begin{array}{c|c} \mathbf{U}_{11} & \mathbf{U}_{12} \\ \hline \mathbf{U}_{21} & \mathbf{U}_{22} \end{array}\right]_i \begin{bmatrix} \mathbf{f} \\ \hline \mathbf{e} \end{bmatrix}_i + \begin{bmatrix} \mathbf{F}_f \\ \hline \mathbf{F}_e \end{bmatrix}_i \tag{A2.2}$$

where **f** contains the $n/2$ state variables that are zero at the left-hand boundary and **e** contains the remaining $n/2$ complementary state variables. The $\mathbf{U}_{ij}$ are $(n/2) \times (n/2)$ submatrices, $\mathbf{F}_f$ is a $(n/2) \times 1$ submatrix of loading function terms corresponding to the **f** state variables and $\mathbf{F}_e$ contains the forcing terms for the **e** state variables.

A generalized Riccati transformation at station i can be defined as

$$\mathbf{f}_i = \mathbf{S}_i \mathbf{e}_i + \mathbf{P}_i \tag{A2.3}$$

which relates the **f** state variables to the **e** state variables. The $(n/2) \times (n/2)$ matrix **S** is the Riccati transfer matrix and the $(n/2) \times 1$ matrix **P** contains the loading function terms. Expansion of Eq. (A2.2) gives

$$\mathbf{f}_{i+1} = \mathbf{U}_{11i}\mathbf{f}_i + \mathbf{U}_{12i}\mathbf{e}_i + \mathbf{F}_{fi} \tag{A2.4}$$

and

$$\mathbf{e}_{i+1} = \mathbf{U}_{21i}\mathbf{f}_i + \mathbf{U}_{22i}\mathbf{e}_i + \mathbf{F}_{ei} \tag{A2.5}$$

Substitution of Eq. (A2.3) into Eq. (A2.5) and solving for $\mathbf{e}_i$ yields

$$\mathbf{e}_i = [\mathbf{U}_{21}\mathbf{S} + \mathbf{U}_{22}]_i^{-1}\mathbf{e}_{i+1} - [\mathbf{U}_{21}\mathbf{S} + \mathbf{U}_{22}]_i^{-1}[\mathbf{U}_{21}\mathbf{P} + \mathbf{F}_e]_i \tag{A2.6}$$

Use of Eqs. (A2.3 and (A2.6) to eliminate $\mathbf{f}_i$ and $\mathbf{e}_i$ from Eq. (A2.4) gives

$$\begin{aligned}\mathbf{f}_{i+1} = &[\mathbf{U}_{11}\mathbf{S} + \mathbf{U}_{12}]_i[\mathbf{U}_{21}\mathbf{S} + \mathbf{U}_{22}]_i^{-1}\mathbf{e}_{i+1} + [\mathbf{U}_{11}\mathbf{P} + \mathbf{F}_f]_i \\ &- [\mathbf{U}_{11}\mathbf{S} + \mathbf{U}_{12}]_i[\mathbf{U}_{21}\mathbf{S} + \mathbf{U}_{22}]_i^{-1}[\mathbf{U}_{21}\mathbf{P} + \mathbf{F}_e]_i\end{aligned} \tag{A2.7}$$

Equation (A2.7) may be written as

$$\mathbf{f}_{i+1} = \mathbf{S}_{i+1}\mathbf{e}_{i+1} + \mathbf{P}_{i+1} \tag{A2.8}$$

where

$$\mathbf{S}_{i+1} = [\mathbf{U}_{11}\mathbf{S} + \mathbf{U}_{12}]_i[\mathbf{U}_{21}\mathbf{S} + \mathbf{U}_{22}]_i^{-1} \tag{A2.9}$$

and

$$\mathbf{P}_{i+1} = [\mathbf{U}_{11}\mathbf{P} + \mathbf{F}_f]_i - \mathbf{S}_{i+1}[\mathbf{U}_{21}\mathbf{P} + \mathbf{F}_e]_i \tag{A2.10}$$

Equations (A2.9) and (A2.10) are the general recursive relationships for **S** and **P**. Since the left-hand boundary conditions are homogeneous, the initial conditions are

$$\mathbf{S}_0 = \mathbf{P}_0 = 0 \tag{A2.11}$$

As will be seen below, it is useful to be able to move in the other direction, from station $i + 1$ to station i. Then

$$\mathbf{e}_i = \mathbf{T}_i\mathbf{e}_{i+1} + \mathbf{Q}_i \tag{A2.12}$$

From Eq. (A2.6)

$$\mathbf{T}_i = [\mathbf{U}_{21}\mathbf{S} + \mathbf{U}_{22}]_i^{-1} \tag{A2.13}$$

and

$$\mathbf{Q}_i = -\mathbf{T}_i[\mathbf{U}_{21}\mathbf{P} + \mathbf{F}_e]_i \tag{A2.14}$$

T is a $(n/2) \times (n/2)$ matrix that transmits the **e** state variables and **Q** is a $(n/2) \times 1$ matrix containing loading function terms. Equations (A2.13) and (A2.14) are the general recursive relationships for **T** and **Q**.

With little more than a change in notation, point occurrences such as concentrated loads and in-span supports are incorporated in the Riccati transfer matrix methodology. Where the previous equations were derived for a segment, the point matrix is used for concentrated occurrences. Instead of the stations i and $i + 1$, the stations just to the left of i and just to the right of i are used in the previous equations for point occurrences.

Equations (A2.9), (A2.10), (A2.13), and (A2.14) are needed while moving from left to right through the structure and Eqs. (A2.12) and (A2.8) are solved while moving from right to left. This forward and backward movement in solving equations is equivalent to Gauss elimination.

A2.2 How to Use the Riccati Transfer Matrix

To start the method the transfer matrix is organized so that the submatrices in Eq. (A2.2) may be identified. Next, the structure is segmented according to two decisions. First, a reasonable but not essential decision is to choose the length of each segment so that all of the terms composing each element of the transfer matrix are approximately the same order of magnitude. Second, a station should be chosen where the state variables are to be printed out.

Now, the matrices **S** and **P** are determined at each station and the matrices **T** and **Q** are determined for each segment. The process starts at the left-hand boundary and proceeds from left to right through the member until the right-hand boundary is reached. All of the intermediate values of **S**, **P**, **T**, and **Q** are retained. At the right-hand boundary, the $n/2$ known state variables are applied to Eq. (A2.3), a procedure which determines the remaining $n/2$ state variables at that point. For eigenvalue problems $\mathbf{P} = \mathbf{Q} = 0$ and Eq. (A2.3) at the right boundary leads to a determinant, depending on the boundary condition, being set equal to zero for a nontrivial solution. This determinant is the characteristic equation that gives the frequencies or the buckling load.

Finally, the state variables at each station are determined by moving from right to left through the structure. Successive applications of Eq. (A2.12) give **e** at any station. Equation (A2.3) is now used to calculate **f** at any station. This completes the solution.

EXAMPLE A2.1 Vibrating Beam Use the combined Riccati transformation and the transfer matrix method to determine the frequencies of a vibrating beam with free-free boundaries.

Since the left end is free, $M_0 = V_0 = 0$ and the transfer matrix should be organized as in Eq. (A2.2)

$$\begin{bmatrix} M \\ V \\ \hline \theta \\ v \end{bmatrix}_{i+1} = \left[\begin{array}{cc|cc} U_{MM} & U_{MV} & U_{M\theta} & U_{Mv} \\ U_{VM} & U_{VV} & U_{V\theta} & U_{Vv} \\ \hline U_{\theta M} & U_{\theta V} & U_{\theta\theta} & U_{\theta v} \\ U_{vM} & U_{vV} & U_{v\theta} & U_{vv} \end{array}\right]_i \begin{bmatrix} M \\ V \\ \hline \theta \\ v \end{bmatrix}_i \tag{1}$$

where U_{jk} can be taken from Chap. 2 for the appropriate beam. In our notation

$$\mathbf{f} = \begin{bmatrix} M \\ V \end{bmatrix} \qquad \mathbf{e} = \begin{bmatrix} \theta \\ v \end{bmatrix} \tag{2}$$

Segment the beam and compute **S**, **P** at each station and **T**, **Q** for each segment from Eqs. (A2.9), (A2.10) and (A2.13), (A2.14). Retain these values. At the right-hand boundary, $i = m$, Eq. (A2.3) becomes

$$\mathbf{S}_m \mathbf{e}_m = \mathbf{0} \tag{3}$$

The frequencies ω_n are determined so that

$$\det |\mathbf{S}_m| = 0 \tag{4}$$

The mode shapes are computed using ω_n by moving from right to left through the structures with Eqs. (A2.3) and (A2.12).

A2.3 Indeterminate In-Span Conditions

An indeterminate in-span condition occurs at an intermediate point of a structure where one or more state variables are zero. Their complementary state variables are discontinuous, while all other state variables are continuous. The following scheme will be used to determine the general relationships that are necessary to transfer across an in-span condition.

Just to the left (superscript $-$) of the in-span occurrence, two new vectors of state variables, $\bar{\mathbf{f}}$ and $\bar{\mathbf{e}}$, will be derived from **f** and **e** so that $\bar{\mathbf{f}}$ contains all of the zero state variables at the in-span condition. Let

$$\hat{\mathbf{f}}_i^- = \mathbf{A}\mathbf{f}_i^- = \begin{bmatrix} \mathbf{a} \\ \hline \mathbf{b} \\ \hline \hat{\mathbf{c}} \end{bmatrix} \tag{A2.15}$$

where $\mathbf{A} = n/2 \times n/2$ transformation matrix

$\mathbf{a}$ = vector of the $n/2 - k - \ell$ continuous state variables in **f**.

$\mathbf{b}$ = vector of the k zero (at the left end) state variables originally in **f**

$\hat{\mathbf{c}}$ = vector of the complementary state variables of the ℓ zero state variables contained in **e**.

From Eqs. (A2.3) and (A2.15),

$$\hat{\mathbf{f}}_i^- = \hat{\mathbf{s}}_i^- \hat{\mathbf{e}}_i^- + \hat{\mathbf{P}}_i^- \tag{A2.16}$$

$$\text{with } \hat{\mathbf{s}}_i^- = \left[\mathbf{ASA}^{-1}\right]_i^- \qquad \hat{\mathbf{e}}_i^- = \mathbf{A}_i^- \hat{\mathbf{e}}_i^- \quad \hat{\mathbf{P}}_i^- = \mathbf{A}_i^- \mathbf{P}_i^-$$

Expansion of Eq. (A2.16) gives

$$\begin{matrix} & 1 & n/2-k-\ell & k & \ell \\ \begin{matrix} n/2-k-\ell \\ k \\ \ell \end{matrix} & \begin{bmatrix} \mathbf{a} \\ \mathbf{b} \\ \hat{\mathbf{c}} \end{bmatrix}_i^- = & \left[\begin{array}{c|c|c} \hat{\mathbf{S}}_{11} & \hat{\mathbf{S}}_{12} & \hat{\mathbf{S}}_{13} \\ \hline \hat{\mathbf{S}}_{21} & \hat{\mathbf{S}}_{22} & \hat{\mathbf{S}}_{23} \\ \hline \hat{\mathbf{S}}_{31} & \hat{\mathbf{S}}_{32} & \hat{\mathbf{S}}_{33} \end{array}\right]_i^- & \begin{bmatrix} \hat{\mathbf{a}} \\ \hat{\mathbf{b}} \\ \mathbf{c} \end{bmatrix}_i^- + & \begin{bmatrix} \mathbf{P}_a \\ \mathbf{P}_b \\ \mathbf{P}_{\hat{c}} \end{bmatrix}_i^- \end{matrix} \tag{A2.17}$$

where the stated variables with hats are the complementary variables to the state variables without hats. Rearrange Eq. (A2.17) so that all of the state variables without hats are grouped.

$$\begin{matrix} & 1 & n/2-k-\ell & k & \ell \\ \begin{matrix} n/2-k-\ell \\ k \\ \ell \end{matrix} & \begin{bmatrix} \mathbf{a} \\ \mathbf{b} \\ \mathbf{c} \end{bmatrix}_i^- = & \left[\begin{array}{c|c|c} \hat{\mathbf{S}}_{11} & \hat{\mathbf{S}}_{12} & \hat{\mathbf{S}}_{13} \\ \hline \hat{\mathbf{S}}_{21} & \hat{\mathbf{S}}_{22} & \hat{\mathbf{S}}_{23} \\ \hline -\hat{\mathbf{S}}_{33}^{-1}\hat{\mathbf{S}}_{31} & -\hat{\mathbf{S}}_{33}^{-1}\hat{\mathbf{S}}_{32} & \hat{\mathbf{S}}_{33}^{-1} \end{array}\right]_i^- & \begin{bmatrix} \hat{\mathbf{a}} \\ \hat{\mathbf{b}} \\ \hat{\mathbf{c}} \end{bmatrix}_i^- + & \begin{bmatrix} \mathbf{P}_a \\ \mathbf{P}_b \\ -\hat{\mathbf{S}}_{33}^{-1}\mathbf{P}_{\hat{c}} \end{bmatrix}^- \end{matrix} \tag{A2.18}$$

Equation (A2.18) is now written as

$$\bar{\mathbf{f}}_i^- = \bar{\mathbf{S}}_i^- \bar{\mathbf{e}}_i^- + \bar{\mathbf{P}}_i^- \tag{A2.19}$$

The in-span conditions are achieved by placing a kinematic constraint on the state variables that are to be zero. The transfer matrix across the in-span condition [to the right side (+)] is

$$\begin{bmatrix} \bar{\mathbf{f}} \\ \bar{\mathbf{e}} \end{bmatrix}_i^+ = \left[\begin{array}{c|c} \mathbf{I} & \mathbf{0} \\ \hline \mathbf{K} & \mathbf{I} \end{array}\right]_i \begin{bmatrix} \bar{\mathbf{f}} \\ \bar{\mathbf{e}} \end{bmatrix}_i^- \tag{A2.20}$$

where $\mathbf{K}$ is a $(n/2) \times (n/2)$ submatrix of kinematic constraints. For the condition of zero displacements and rotations, the kinematic constraint is a spring; and for forces and moments the kinematic constraint is the reciprocal of a spring. $\bar{\mathbf{S}}$, $\bar{\mathbf{P}}$, $\bar{\mathbf{T}}$, and $\bar{\mathbf{Q}}$ just to the right of the in-span condition are determined by taking the limit of these functions as $\mathbf{K}$ goes to infinity:

$$\bar{\mathbf{S}}_i^+ = \lim_{\mathbf{K}\to\infty} \bar{\mathbf{S}}_i^- \left([\mathbf{K}\bar{\mathbf{S}} + \mathbf{I}]_i^- \right)^{-1} \tag{A2.21}$$

$$\bar{\mathbf{P}}_i^+ = \lim_{\mathbf{K}\to\infty} \left\{ \bar{\mathbf{P}}_i^- - \bar{\mathbf{S}}_i^- \left([\mathbf{K}\bar{\mathbf{S}} + \mathbf{I}]_i^- \right)^{-1} [\mathbf{K}\bar{\mathbf{P}}]_i^- \right\} \tag{A2.22}$$

$$\bar{\mathbf{T}}_i^+ = \lim_{\mathbf{K}\to\infty} \left([\mathbf{K}\bar{\mathbf{S}} + \mathbf{I}]_i^- \right)^{-1} \tag{A2.23}$$

$$\bar{\mathbf{Q}}_i^+ = \lim_{\mathbf{K}\to\infty} -\left([\mathbf{K}\bar{\mathbf{S}} + \mathbf{I}]_i^- \right)^{-1} [\mathbf{K}\bar{\mathbf{P}}]_i^- \tag{A2.24}$$

This approach for treating indeterminate in-span conditions differs from that used in the usual transfer-matrix method where the discontinuity in the discontinuous state variable is introduced as a new unknown.

EXAMPLE A2.2 Beam with Rigid In-Span Support The process of moving across an indeterminate in-span condition will be illustrated by a beam resting on a rigid in-span support at station i.

To develop the necessary relationships to get from just to the left of station i to just to the right of station i, a spring of constant k is placed at station i. The in-span condition ($v_i = 0$) will be obtained by taking the limit as $k \to \infty$. Assume the left-hand boundary is clamped so that

$$\mathbf{f} = \begin{bmatrix} v \\ \theta \end{bmatrix} \quad \text{and} \quad \mathbf{e} = \begin{bmatrix} M \\ V \end{bmatrix} \tag{1}$$

In the notation of Eq. (A2.15), $\mathbf{a} = \theta$, $\mathbf{b} = v$, $\hat{\mathbf{c}} = 0$. Then Eq. (A2.15) becomes

$$\underbrace{\begin{bmatrix} \theta \\ v \end{bmatrix}_i^-}_{\hat{\mathbf{f}}_i^-} = \underbrace{\begin{bmatrix} 0 & 1 \\ 1 & 0 \end{bmatrix}}_{\mathbf{A}} \underbrace{\begin{bmatrix} v \\ \theta \end{bmatrix}_i^-}_{\mathbf{f}_i^-}$$

Note that, since $\underline{\hat{c}} = 0$,

$$\bar{\mathbf{f}}_i^- = \hat{\mathbf{f}}_i^- \tag{2}$$

A point just to the left of station i is governed by Eq. (A2.19). The point-transfer matrix in the form of (1) for a spring is

$$\bar{\mathbf{U}}_i = \left[\begin{array}{cc|cc} 1 & 0 & 0 & 0 \\ 0 & 1 & 0 & 0 \\ \hline 0 & 0 & 1 & 0 \\ 0 & k & 0 & 1 \end{array}\right] \tag{3}$$

From comparison of (3) with Eq. (A2.20), it follows that

$$\mathbf{K} = \begin{bmatrix} 0 & 0 \\ 0 & k \end{bmatrix} \tag{4}$$

Let

$$\mathbf{S} = \left[\begin{array}{c|c} S_{11} & S_{12} \\ \hline S_{21} & S_{22} \end{array}\right] \tag{5}$$

and substitute (4) in Eq. (A2.21)

$$\bar{\mathbf{S}}_i^+ = \lim_{k\to\infty} \left(\frac{1}{1 + kS_{11}} \begin{bmatrix} k \det \mathbf{S} + S_{22} & S_{21} \\ S_{12} & S_{11} \end{bmatrix} \right)_i^- = \begin{bmatrix} \det[\mathbf{S}]/S_{11} & 0 \\ 0 & 0 \end{bmatrix}_i^- \tag{6}$$

In a similar fashion, from Eqs. (A2.22), (A2.23), and (A2.24)

$$\overline{\mathbf{P}}_i^+ = \begin{bmatrix} P_2 - S_{21}P_1/S_{11} \\ 0 \end{bmatrix}^- \tag{7}$$

$$\overline{\mathbf{T}}_i^+ = \begin{bmatrix} 1 & 0 \\ -S_{12}/S_{11} & 0 \end{bmatrix}_i^- \tag{8}$$

$$\overline{\mathbf{Q}}_i^+ = \begin{bmatrix} 0 \\ -P_1/S_{11} \end{bmatrix}_i^- \tag{9}$$

The matrices necessary to move across the in-span condition are now known.

Table A2-1 gives the transformations for a beam with various left-hand boundary conditions.

A2.4 Computational Efficiency

Since the Riccati transfer-matrix method and the transfer-matrix method may be used to analyze the same structure with the same number of segments, the computational time and storage requirements of the two methods can be compared. Assume that the number of multiplications is proportional to the computational time. In terms of computational time for each segment:

	Riccati transfer-matrix method	Transfer-matrix method
Eigenvalue problems	$n^3/2$	n^3
Static or steady-state responses	$n^3/2 + n^2$	$n^3 + n^2$

and for storage requirements:

	Riccati transfer-matrix method	Transfer-matrix method
Eigenvalue problems	$n^2/2$	n^2
Static or steady-state responses	$n^2/2 + n$	$n^2 + n$

To understand how these entries are reached, consider the computational time for eigenvalue problems by the Riccati transfer-matrix method. In Eq. (A2.13), the matrix product and resulting matrix inverse require $2(n/2)^3$ multiplications.

TABLE A2-1 In-Span Conditions

In-Span Condition / Left-Hand Boundary Condition		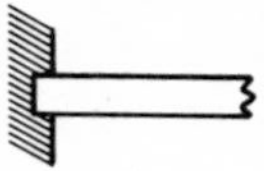 $f = \begin{bmatrix} V \\ \theta \end{bmatrix}$	 $f = \begin{bmatrix} V \\ M \end{bmatrix}$	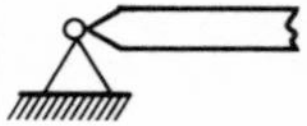 $f = \begin{bmatrix} V \\ M \end{bmatrix}$	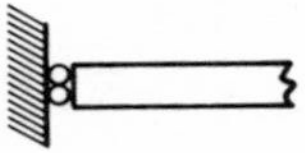 $f = \begin{bmatrix} \theta \\ V \end{bmatrix}$
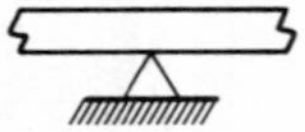	$\bar{f} =$ $\bar{S} =$ $\bar{P} =$ $\bar{T} =$ $\bar{Q} =$	$\begin{bmatrix} \theta \\ V \end{bmatrix}$ A_1 A_2 A_3 A_4	$\begin{bmatrix} M \\ V \end{bmatrix}$ A_{13} A_{14} A_{15} A_{16}	$\begin{bmatrix} M \\ V \end{bmatrix}$ A_1 A_2 A_3 A_4	$\begin{bmatrix} \theta \\ V \end{bmatrix}$ A_9 A_{10} A_{11} A_{12}
	$\bar{f} =$ $\bar{S} =$ $\bar{P} =$ $\bar{T} =$ $\bar{Q} =$	$\begin{bmatrix} V \\ \theta \end{bmatrix}$ A_5 A_6 A_7 A_8	$\begin{bmatrix} V \\ \theta \end{bmatrix}$ A_9 A_{10} A_{11} A_{12}	$\begin{bmatrix} V \\ \theta \end{bmatrix}$ A_9 A_{10} A_{11} A_{12}	$\begin{bmatrix} V \\ \theta \end{bmatrix}$ A_1 A_2 A_3 A_4
	$\bar{f} =$ $\bar{S} =$ $\bar{P} =$ $\bar{T} =$ $\bar{Q} =$	$\begin{bmatrix} V \\ M \end{bmatrix}$ A_9 A_{10} A_{11} A_{12}	$\begin{bmatrix} V \\ M \end{bmatrix}$ A_5 A_6 A_7 A_8	$\begin{bmatrix} V \\ M \end{bmatrix}$ A_5 A_6 A_7 A_8	$\begin{bmatrix} V \\ M \end{bmatrix}$ A_{13} A_{14} A_{15} A_{16}
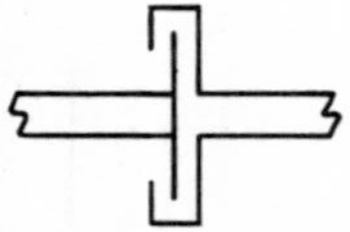	$\bar{f} =$ $\bar{S} =$ $\bar{P} =$ $\bar{T} =$ $\bar{Q} =$	$\begin{bmatrix} \theta \\ V \end{bmatrix}$ A_{13} A_{14} A_{15} A_{16}	$\begin{bmatrix} M \\ V \end{bmatrix}$ A_1 A_2 A_3 A_4	$\begin{bmatrix} M \\ V \end{bmatrix}$ A_{13} A_{14} A_{15} A_{16}	$\begin{bmatrix} \theta \\ V \end{bmatrix}$ A_5 A_6 A_7 A_8

Definitions

$$A_1 = \begin{bmatrix} \det S/s_{11} & 0 \\ 0 & 0 \end{bmatrix} \qquad A_2 = \begin{bmatrix} p_2 - s_{21}\, p_1/s_{11} \\ 0 \end{bmatrix}$$

$$A_3 = \begin{bmatrix} 1 & 0 \\ -s_{12}/s_{11} & 0 \end{bmatrix} \qquad A_4 = \begin{bmatrix} 0 \\ -p_1/s_{11} \end{bmatrix}$$

$$A_5 = \begin{bmatrix} \det S/s_{22} & 0 \\ 0 & 0 \end{bmatrix} \qquad A_6 = \begin{bmatrix} p_1 - s_{12}p_2/s_{22} \\ 0 \end{bmatrix}$$

$$A_7 = \begin{bmatrix} 1 & 0 \\ -s_{21}/s_{22} & 0 \end{bmatrix} \qquad A_8 = \begin{bmatrix} 0 \\ -p_2/s_{22} \end{bmatrix}$$

$$A_9 = \begin{bmatrix} s_{11} + s_{12}s_{21} & 0 \\ 0 & 0 \end{bmatrix} \qquad A_{10} = \begin{bmatrix} p_1 + s_{12}p_2 \\ 0 \end{bmatrix}$$

$$A_{11} = \begin{bmatrix} 1 & 0 \\ s_{21} & 0 \end{bmatrix} \qquad A_{12} = \begin{bmatrix} 0 \\ p_2 \end{bmatrix}$$

$$A_{13} = \begin{bmatrix} s_{12}s_{21} + s_{22} & 0 \\ 0 & 0 \end{bmatrix} \qquad A_{14} = \begin{bmatrix} s_{21}p_1 + p_2 \\ 0 \end{bmatrix}$$

$$A_{15} = \begin{bmatrix} 1 & 0 \\ s_{12} & 0 \end{bmatrix} \qquad A_{16} = \begin{bmatrix} 0 \\ p_1 \end{bmatrix}$$

Equation (A2.9) contains this matrix inverse and two additional matrix products for a total of $4(n/2)^3$ or $(n^3)/2$ multiplications.

These figures indicate that the Riccati transfer-matrix method requires about half the computational time and storage requirements of the usual transfer matrix method.

REFERENCES

A2.1 Uhrig, R.: "The Transfer Matrix Method Seen as One Method of Structural Analysis among Others," *J. Sound Vib.*, vol. 4, no. 2, pp. 136–148, 1966.

Index